NEKTON

NEKTON

by

Yu. G. Aleyev

DR. W. JUNK b.v. – PUBLISHERS – THE HAGUE 1977

ISBN-13: 978-94-010-1326-0 e-ISBN-13: 978-94-010-1324-6
DOI: 10.1007/978-94-010-1324-6

Softcover reprint of the hardcover 1st edition 1977
Translated from the Russian by B. M. Meerovich
Cover design: Max Velthuijs

CONTENTS

INTRODUCTION

1. Nekton as an ecomorphological type of biont

The term nekton was suggested and used for the first time in 1890 by E. Haeckel in his book Plankton-Studien. Etymologically the word nekton derives from the Greek νηκτηρ, i.e. swimming. As Haeckel defined it, nekton describes collectively all swimming animals that are 'free to choose their path', i.e. can resist a strong current of water and, distinct from planktonic animals, go where they wish.

While giving a general idea of the dividing line between plankton and nekton, Haeckel's definition, which has played an important role in shaping our ideas about nekton, today no longer provides a sufficient basis for ecological and functional morphological investigations, since it affords no possibility of quantitatively assessing either the boundary between plankton and nekton or that between nekton and other ecomorphological types of biont. Thus Parin (1968), proceeding from Haeckel's principle, believes that in the epipelagic zone of the ocean the minimum size of nektonic fishes with a well-developed capacity for active swimming may be between 15 and 30 cm, as fishes shorter than 15 cm are unable to counter oceanic currents. Meanwhile young *Leucaspius* (*Leucaspius delineatus*) only 1.5 cm long, observed by this writer in ponds near Moscow proved capable of active horizontal migrations across the entire body of water, which, if Haeckel's definition is accepted, brings the border between planktonic and nektonic fish in this case to between 1.5 and 2.0 cm.

Thus, in order to comply with Haeckel's concept, the border between plankton and nekton must depend appreciably on the velocity of the water: in one case fishes up to 15–30 cm long are regarded as plankton whereas in the other those no longer than 2–3 cm are defined as nekton, yet from the ecomorphological point of view the nektonic *Leucaspius* of the pond is not fundamentally different from the planktonic *Spratelloides*, *Stolephorus* or *Atherion* of the sea.

Many researchers (Zernov 1949; Zenkevich 1951; Barthelmes 1957; Kiselev 1969, and others) have pointed out the difficulties in distinguishing between plankton and nekton. Since this is based on evaluating the capacity of pelagic animals for active forward movement in water, it is apparent that such a distinction is basically a task for biohydromechanics and must be made in accordance with the general theory of hydromechanics. For the purpose of making specific the differences between plankton and nekton it is, therefore, advisable to examine their biohydrodynamic features, and first of all, the adaptations determining the characteristic magnitude of the hydrodynamic resistance of a biont's body (Aleyev 1972a). An analysis of this aspect of the matter (Aleyev 1972a) made it possible to create a new, biohydrodynamic conception of life forms of the pelagic zone which, apart from being helpful in a quantitative assessment of the dividing line between plankton and nekton, made possible the establishment of the basic factors, determining the development of all essential nektonic adaptations.

Any swimming animal encounters resistance on the part of the water; its magnitude depends on a number of variables, and in particular, is directly proportional to the square of the moving animal's speed relative to the surrounding water. As water is a viscous medium, around any body moving in it, that of a swimming animal included, there forms a boundary layer, i.e. a layer of water to which the moving body imparts some additional velocity. The boundary layer moves with the moving body, in this case the moving animal, forming what is called an added mass, and the speed of the water at every point in this layer is inversely proportional to the distance of that point from the moving body's surface. Since there takes place adhesion of water to the surface of the body, when making calculations within the coordinates system, related to the moving body, as is adopted in hydromechanics and is adopted further in the text, the flow velocity in the boundary layer increases from zero at the surface of the moving body to its full value in the outer stream, in which the fluid can be regarded (Schlichting 1956) as flowing without friction. Two states of the boundary layer are distinguished: the laminar, i.e. streamlined state, which is unstable and characterized by a streamlined motion of fluid particles along rectilinear trajectories, and the more stable turbulent one, in which the movement of the liquid is of a pulsating character, i.e. the fluid particles move along curved, tortuous trajectories. Under certain conditions the boundary layer changes from laminar to turbulent. However, the turbulent boundary layer in this case retains immediately adjacent to the surface of the body a very thin laminar sublayer, whose thickness (Martynov 1958) amounts to only about 1% of the boundary-layer total thickness. The water forming the boundary layer and flowing along the moving body to its rear tip creates behind the body the so-called hydrodynamic wake.

In the case of streamlined bodies, which all nektonic animals have, the boundary layer on the anterior part is very thin, owing to which it remains laminar over a certain length and only after thickening to a certain limit changes to a state of turbulence. The greater the distance from the frontal point of a streamlined body, the greater is the boundary-layer thickness, both in the laminar and the turbulent state, this thickening occurring much faster in a turbulent boundary layer. In the case of well-streamlined bodies the transition of the boundary layer from laminar to turbulent greatly depends on the smoothness of the body surface, since both isolated roughnesses and roughness formed by projections or depressions whose separation is of the same order as their height accelerate the transition to turbulence.

The magnitude of the hydrodynamic resistance encountered by the body (Prandtl 1951; Schlichting 1956; Martynov 1958, and others) is expressed in hydromechanics in the form of a dimensionless drag coefficient C_D, defined by the formula

$$C_D = 2\,F/\rho\,V^2 S, \tag{1}$$

where F is the drag force, V the speed of movement and ρ the density of the fluid in which the movement takes place (here water). As far as the value of S is concerned, in the case of completely immersed streamlined bodies whose frictional drag is greater than their form drag, which is true of all nektonic animals, it represents the area of the completely wetted body surface. Among the types of resistance to movement known to hydrodynamics, there are three that are characteristic of a typical nektonic organism: form drag, frictional drag and induced drag. During movement in the semi-submerged state (Sauropterygia, Pinnipedia, etc.) there arises also wave drag, but this kind of resistance is not characteristic of nektonic animals, as they swim either completely submerged all the time (Cephalopoda, fishes, etc.) or completely submerged most of the time (Sauropterygia, Sphenisciformes, Cetacea, Pinnipedia, etc.), and in the semi-submerged state merely rest or move slowly with their propulsive mechanism functioning below capacity. In any case, the general morphological organization of nektonic organisms is adapted to motion in the completely submerged state and is, therefore, conducive to reducing form drag, frictional drag and induced drag. We do not find in the morphology of nektonic organisms any special adaptations immediately aimed at reducing wave drag, which fully corresponds to the ecological factors outlined above.

Thus one can assume with sufficient accuracy that the total drag for nektonic animals is

$$C_D = C_{Dp} + C_{Df} + C_{Di}, \tag{2}$$

where C_{Dp} is the form-drag coefficient, C_{Df} the frictional-drag coefficient and C_{Di} the induced-drag coefficient.

Form drag depends on the difference between the dynamic pressures at the front and rear ends of the moving body. The basic cause of form drag is vortices (Prandtl 1951, and others). It is precisely these vortices that prevent confluence of the flow behind a streamlined body and determine the asymmetric distribution of pressure over the body surface; besides, their formation requires a constant expenditure of energy. The magnitude of form drag is determined primarily by the general shape of the moving body.

The appearance of frictional drag is connected with the friction between the flow and the surface of the moving body. The magnitude of frictional drag depends primarily on whether the boundary layer is laminar or turbulent, since during boundary-layer transition frictional drag increases steeply (Prandtl 1951; Schlichting 1956, and others). Since the state of the boundary layer is to a considerable degree determined by the character of the body surface, the magnitude of frictional drag depends, above all, on the character and area of the body surface.

Induced drag (significant when there is hydrodynamic lift) is due to the flow of fluid over the lateral edges of the body and the associated trailing-vortex wake (Patrashev 1953), and its magnitude depends mainly on the aspect ratio of the body, increasing as the latter diminishes.

To grasp the differences between plankton and nekton it is necessary to examine in greater detail the reasons for and characteristics of the differences in their form drag and frictional drag, expressed by C_{Dp} and C_{Df} respectively. In the case of bodies with a more-or-less streamlined shape, typically possessed by all nektonic animals, the total drag is made up predominantly of frictional drag. The general body shape of nektonic organisms usually approximates well-streamlined bodies of revolution, whose frictional drag is known (Patrashev 1953) to constitute about 0.9 of the total drag.

It is also known (Prandtl 1951; Patrashev 1953; Schlichting 1956, and others) that the resistance of bodies of revolution varies considerably with changes in the Reynolds number *Re*, which expresses the ratio of inertia forces to friction forces:

$$Re = l V/\nu, \tag{3}$$

where l is the length or width of the stream or of the moving body (here the animal's absolute length), V is the velocity of the stream or the speed of the moving body in some frame (here the velocity at which the animal is moving relative to the surrounding water) and ν is the kinematic viscosity of the water. A low Reynolds number means the viscous forces dominate the flow; conversely, a high Reynolds number implies that the main role in the flow is played by inertia forces. Therefore, inertia forces will be decisive for large and rapidly swimming animals (high Reynolds numbers) and viscous forces for small and slowly swimming animals (low Reynolds numbers) (Aleyev 1959b, 1963a).

It has been reported (Prandtl 1951; Schlichting 1956) that in various cases the boundary layer may remain laminar up to *Re* values between 9.0×10^4 and 3.0×10^6, and with further increases in *Re* a transition from laminar to turbulent flow takes place, most often occurring at $Re \approx 5.0\times10^5$. Transition of the boundary layer from laminar to turbulent occurs earlier or later in the *Re* range 9.0×10^4 to 3.0×10^6 according to the intensity of the initial turbulence in the flow, the shape of the body and the nature of its surface – the degree of its roughness and some other factors. Re_{cr} is the value at which the boundary layer in a given flow changes from laminar to turbulent, and is called the critical value. The laminar state of the boundary layer in the majority of cases is of greater advantage energywise than the turbulent state, and pelagic animals whose movement is characterized by critical and supercritical *Re* values usually have special adaptations aiding laminarization of the boundary layer, i.e. prevention of transition to the turbulent state. Owing to such adaptations, the boundary layer on nektonic animals may remain laminar even at $Re > 3.0\times10^6$, and sometimes, possibly, until $Re = 10^7$. This helps to preserve a lower total drag and is examined in greater detail in chapter V.

It follows from the above that at different Reynolds numbers the adaptations associated with the form of movement in an aquatic environment will be essentially different, to a degree proportional to the difference in *Re*. Apparently, at low *Re*, the animal must be adapted to movement under conditions predominantly influenced by viscous forces, i.e. to movement in a relatively viscous medium with comparatively little momentum produced by the propulsive process. On the other hand, at higher *Re*, the animal must be adapted to movement predominantly due to momentum, i.e.

to movement in a relatively inviscid medium with comparatively high momentum developed by the propulsive process. It is these different directions taken by the complex of adaptations associated with locomotion that make up the basic differences between plankton and nekton.

One may mention, for example, that the nektonic body shape is characterized by a comparatively large slenderness ratio, an overall smoothness of contours and the absence of protruding details, which ensures low resistance when moving in water (squids, pelagic fishes, cetaceans, etc.). The planktonic body shape, on the other hand, is characterized by a discontinuous general contour an abundance of protruding details and poor streamlining, ensuring a good parachute effect and, at the same time, creating comparatively high resistance to forward movement (Entomostraca, Schizopoda, jellyfish, larvae of Polychaeta, etc.).

A more accurate analysis is required to find a quantitative criterion for distinguishing between plankton and nekton according to these characteristics. It is known from hydromechanics (Prandtl 1951; Martynov 1958; Voitkunsky 1964, and others) that the drag during movement of completely immersed bodies depends to a considerable degree on their slenderness ratio, i.e. on the ratio of their length to the diameter of a circle of area equal to their maximum cross-sectional area. In our case the unknown thickness ratio U of the animal's body may be determined as the ratio of the diameter D of a circle of area equal to the maximum cross-sectional area of the animal's body to its effective length L_c (Aleyev 1972a):

$$U = D/L_c. \tag{4}$$

Here and below the effective length L_c is assumed to be the length of the compact part of the animal's body without thread-like and leaf-like appendages, since the latter do not determine the general picture of the flow over the body. Thus, for example, in the case of pelagic Cephalopoda the effective length L_c is the animal's absolute length, including the arms clasped together, forming altogether a compact cone. Conversely, filiform tentacles of the *Cyanea* jellyfish, which can be up to 30 m long, should not be included in the length L_c. The effective length L_c of the Sagittoidea should be measured from the forwardmost point of the body to the base of the median rays of the tail fin. In Agnatha and fishes with a symmetrical tail fin, as well as in sea snakes (Hydrophidae, Acrochordinae, etc.), cetaceans and sirens, L_c is the length from the end of the vertebral column. In cases when the tail is notably heterocercal (Elasmobranchii, Acipenseriformes, **Ichthyosauria**, etc.) L_c corresponds to the length to the point where the body axis bends, turning towards the increased blade of the caudal fin. In Molidae the caudal peduncle is absent, the fin situated at the rear end of the body being very thick and short, while the transition from the body to the fin is extremely smooth, and L_c for representatives of this family is therefore the absolute length, just as in sea turtles (Chelonioidea, Dermochelyoidea, etc.), penguins (Sphenisciformes) and pinnipeds (Pinnipedia). Spear-shaped or ensiform rostrums clearly distinguishable from the remaining part of the head and occurring in certain fishes (Xiphioidae, some Beloniformes), ichthyosaurs

(*Eurhinosaurus*) and cetaceans (*Monodon*), should not be included in the length L_c in a number of cases such rostrums perform an important hydrodynamic function (Aleyev 1970b), however, when analysing the hydrodynamic properties of the compact part of the body, the latter's length, naturally, should not include the rostrum; in all such cases, one should take as the front end of the effective body length the point at the base of the rostrum where the latter begins to widen steeply.

It has been shown experimentally (Martynov 1958; Pavlenko 1963; Devnin 1967, and others) that the drag on ellipsoids begins steeply increasing for $U > 0.50$ and that on fish-like bodies of revolution with a pointed rear end for $U > 0.30$. This gives grounds for it to be sufficiently accurate that nekton, as an ecomorphological type of animal adapted to active swimming in water, are characterized by values of $U \leqslant 0.40$. This, however, does not mean that all plankton have $U > 0.40$: in plankton we find both forms with high U values (for example, for *Aurelia* and *Cyanea* jellyfish $U \approx 1.0$) and elongated forms for which $U < 0.40$. In the latter case the quantitative differences between plankton and nekton may be found with the help of other morphological criteria.

As a rule, the condition $U \leqslant 0.40$ remains true at every stage of post-embryonic development in the ontogeny of nektonic organisms. An exception is found only in the prelarvae of some fishes with a very large yolk sac and, possibly, in the very early planktonic larvae of some nektonic cephalopods; both these cases obviously concern planktonic stages with a length not more then 1.0 cm and $Re < 2.0 \times 10^2$.

The immediate cause of the increase in drag with increasing U lies in the fact that for a streamlined body with a greater thickness ratio the adverse pressure gradients in the posterior region are larger and the processes of vortex formation intensify accordingly. It is well known (Prandtl 1951, and others) that a decrease in the streamwise pressure in the flow enhances acceleration of the flow and hinders vortex shedding, whereas an increase in the streamwise pressure produces the opposite effect: it retards the flow and enhances vortex formation.

Distinct from a laminar boundary layer, the fluid in a turbulent boundary layer mixes, owing to which there is a constant intensive exchange of impulses between the outer fluid and the boundary layer. This circumstance enables a turbulent boundary layer to withstand without separation a considerably greater pressure increase than is possible for a laminar boundary layer, for which reason the point of separation shifts far down stream, towards the rear end of the streamlined body (Schlichting 1956).

When a stream of fluid flows over some well-streamlined body, for instance that of a nektonic animal, two areas differing in their pressure and velocity gradients may be distinguished along the longitudinal axis of the body. The first stretches from the anterior tip of the body to its maximum cross-section; it is characterized by a drop in pressure in the external flow and an increase in the velocity in the streamwise direction, i.e. from the body's nose to its rear end. This area is called the area of contraction, as the streamlines throughout its length converge. The second area, situated behind the streamlined body's maximum cross-section, extends to its rear end and is characterized by a growing pressure in the external flow and diminishing velocity in the streamwise direction. This area is called the area of diffusion, as the streamlines throughout its length diverge.

Since the possibility of boundary-layer separation occurs, as was mentioned above, during streamwise deceleration of the external flow, it is apparent that separation is possible in the diffusion area, i.e. the vortex system, if one occurs, embraces the rear part of the streamlined body, in particular the rear part of a nektonic animal's body.

Being directly proportional to the adverse pressure gradient, the intensity of the vorticity in the contraction area depends mainly on two quantities. First, it depends on the thickness ratio of the body, as the higher this is, the greater is the so-called rear angle of the body formed by the contours of its profile, owing to which adverse pressure gradients grow in the contraction area; in conformity with this, other conditions remaining unchanged, the intensity of vortex formation will be directly proportional to U. Secondly, the vorticity intensity depends on the velocity of the flowing stream, i.e. in this case on the speed of the nektonic organism, and increases in this will cause the adverse pressure gradient to rise in the contraction area. Hence, on the basis of (3) it follows that the intensity of vortex formation is directly proportional to the Reynolds number.

If all other conditions remain unchanged, the frictional drag is directly proportional to the magnitude of the streamlined body's specific surface S_w, which is the ratio of the total surface area S of the body to its volume W, i.e. SW^{-1}. The body possessing the smallest specific surface is a sphere. With an increase in the body's slenderness ratio or, which is the same thing, with a decrease in U, the specific surface increases. For the same body shape, the specific surface will increase with a decrease in the body's linear dimensions, therefore one may assess the influence of the body shape on the magnitude of the specific surface only by comparing the values of S_w for individuals of the same size. Since this is impossible in the majority of cases, however, as the ranges of the linear dimensions of adult individuals of comparable species do not as a rule overlap, in order to perform this task we obtained a reduced specific surface S_0, defined by the formula (Aleyev 1972a)

$$S_0 = S^{0.5} W^{-0.3(3)}. \tag{5}$$

S_0 is the ratio of the side of a square of area equal to the animal's surface area to the side of a cube with volume equal to the animal's volume. S_0 is absolutely independent of the linear dimensions of the object and may therefore be used to compare the specific body surfaces of any object, including any planktonic or nektonic organism.

It should be stressed that true values of the specific surface S_w cannot be used as a criterion for distinguishing between plankton and nekton. This can be seen from the following comparison: the value S_w for a 10 m long whale *Balaenoptera physalus* is, according to our determination, about 3, for a 1 m dolphin *Phocoena* about 28 and for a 1 cm long young mullet *Mugil saliens* about 3449, whereas the S_0 values for these creatures are 2.70, 2.69 and 2.71 respectively. The S_w values for a 10 m

Balaenoptera and a 1 cm young *Mugil saliens* differ by a factor of more than 1000, yet this by no means indicates that the body shape of the young mullet is as many times better adapted to an increased specific surface than the body shape of the whale; this becomes apparent on comparing the S_0 values, which are very close for all three animals.

The values of the reduced specific surface S_0 we obtained for the most elongated and fastest predators, conspicuously nektonic forms of the *Trichiurus* type, do not exceed 4.40, and in most cases, for nektonic animals with the conventional aspect ratio, are considerably less than 4.00. Therefore, taking into account a certain diversity in S_0 values due to individual and age variability of the nektonic organisms' external structure, one may assume with sufficient accuracy that the condition $S_0 \leqslant 4.50$ is characteristic of the nektonic body shape. It does not follow, however, that the condition $S_0 > 4.50$ is always characteristic of plankton. In plankton we meet both forms with $S_0 > 4.50$ and forms with $S_0 \leqslant 4.50$, but in the latter case either $U > 0.40$ or $Re < 5.0 \times 10^3$ (see below), which allows us to distinguish plankton from nekton in this case also.

In the ontogenesis of nektonic organisms the inequality $S_0 \leqslant 4.50$ in some cases remains valid throughout all the stages of post-embryonic development, while in other cases, in planktonic larvae and the young of most fishes, with the development of special parachute systems in the planktonic stages particularly, we may have the inequality $S_0 > 4.50$.

Therefore the morphological criteria U and S_0 enable a quantitative characterization of the body shape of swimming animals in terms of their aspect ratio. As one can see from the above, the body of a nektonic animal must be sufficiently elongated to ensure a sufficiently low form drag (which is achieved at $U \leqslant 0.40$) but at the same time must not be so extended as to increase essentially the frictional drag (which is attained at $S_0 \leqslant 4.50$) (Aleyev 1972a).

Figure 1, presenting U as a function of Re for planktonic and nektonic animals, shows that in plankton the body thickness ratio varies over a very wide range, from $U < 0.10$ to $U > 1.0$, whereas in nektonic organisms it varies over a smaller range, corresponding to the condition $U \leqslant 0.40$. Therefore, in the range $Re \leqslant 5.0 \times 10^5$ movement of animals in water takes place with very diverse body slenderness ratios, whereas for $Re > 5.0 \times 10^5$ movement with $U > 0.40$ becomes energetically wasteful, since the total drag steeply increases owing, in turn, to the transition to turbulence of the boundary layer and the steep intensification of vortex formation, which brings about an increase in the form drag, i.e. in C_{Dp} (Figure 1). This explains the great variety of body shapes among plankton and the comparatively low variety among nektonic organisms: the latter are subject to the rigid limitations of the condition $U \leqslant 0.40$.

However, at very low U propulsion in the $Re > 5.0 \times 10^5$ regime also becomes energetically wasteful owing to the rising frictional drag, i.e. because of the growth of C_{Df} (Figure 1). All 47 species of nektonic animals we investigated had $S_0 < 4.50$, and usually $S_0 < 4.00$. Most nektonic cephalopods, fishes, reptiles, birds and mammals

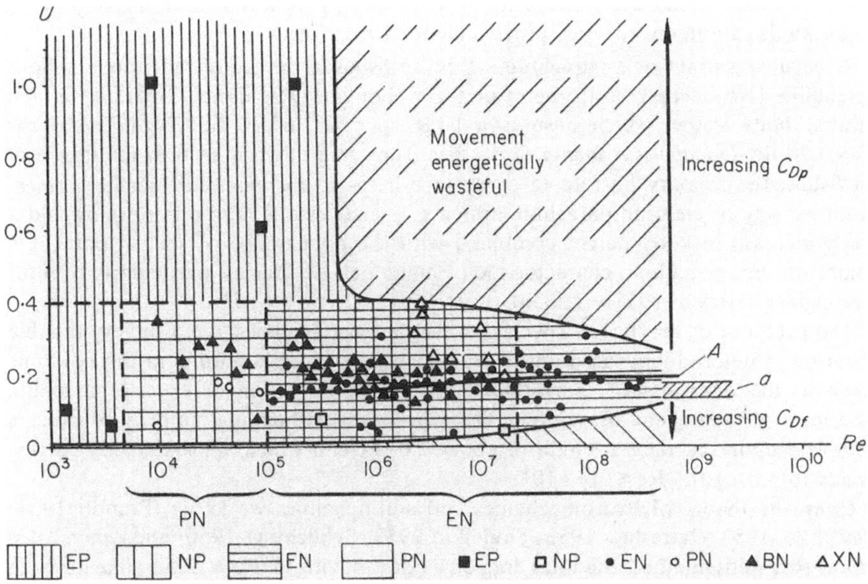

Figure 1. Diagram showing the dependence of *U* upon *Re*. *A*, average *U* values for eunekton; *a*, range of optimal hull elongation for submarines, data from Gerasimov & Droblenkov (1962). EP, euplankton; NP, nektoplankton; BN, benthonekton; EN, eunekton; PN, planktonekton; XN, xeronekton.

have S_0 values in the range of 2.60–3.50 (Table 14); the same range is also characteristic of eunekton, i.e. typical nekton.

Thus some mean *U* value appears to be the most expedient: sufficiently low not to cause a pronounced increase in C_{D_p} but at the same time not so low as to produce a notable increase in C_{Df}. For higher *Re* this range of *U* optimal for nektonic organisms narrows down more and more; this becomes particularly noticeable for $Re > 10^7$ (Figure 1) and is explained by the need for an ever more refined optimization of the nektonic body shape to reduce the hydrodynamic drag encountered by the body in order to deal with the changes in the state of the boundary layer taking place as *Re* increases. The mean *U* for eunekton probably ranges from 0.15 to 0.19, as can be seen from the position of the curve of *U* as a function of *Re* corresponding to eunekton, marked in Figure 1 with the letter A.

An important morphological criterion for distinguishing nekton from plankton is the presence or absence of special parachute systems. Nektonic organisms never have such systems, as they inevitably impair the streamlining of the body. Plankton, on the other hand, nearly always have parachute systems, as their presence enhances passive hovering in the water; they are absent only in very small plankton (less than 1 mm long) and in those which are nearly neutrally buoyant (planktonic eggs of fishes, etc.).

As to structure, parachute systems are diverse, varying from the umbrellas of jellyfish to all kinds of antennae, seta, spines, extended rays, etc.

A peculiar variant of a parachute system in fishes is the alongated ribbon solidus threadlike (*Nemichthys* and others) or the ribbonlike (*Regalecus*, *Trachipterus* and others) body shape, which ensures a large specific surface (in *Regalecus glesne* $S_0 = 4.70$, in *Trachypterus taenia* $S_0 = 4.80$). The same effect is seen in the larvae of all fishes: the primary fin fold in every case helps to increase the specific surface. Another way of creating parachute systems in small fish is lateral flatterning and a steep increase in body height combined with the appearance of special parachute spines on the operculum, characteristic of many pelagic species, particularly of most Carangidae (Aleyev 1957b, 1963a).

The presence or absence of special parachute systems may serve as a very reliable criterion in determining the point of transition from the planktonic to the nektonic stage in the ontogeny of individual species. The reduction of special parachute systems in the ontogeny of nektonic fishes in the overwhelming majority of cases is complete upon reaching a length of about 2.0–5.0 cm, which approximately corresponds to $5.0 \times 10^3 \leqslant Re \leqslant 2.5 \times 10^4$.

From the theory of hydromechanics and aerodynamics we know (Prandtl 1951; Pavlenko 1953; Patrashev 1953; Shuleikin 1953; Schlichting 1956, and others) that at low Reynolds numbers the total drag on a body moving in water or air alike depends very little on its shape, since in this case the drag is determined almost completely by the action of frictional forces, and only when the Reynolds number increases above a certain level does the difference between poorly and well streamlined bodies become significant in this respect. The difference between the drags on a sphere and well-streamlined bodies of revolution becomes noticeable only in the $10^3 \leqslant Re \leqslant 10^4$; it may be assumed with sufficient accuracy that this difference becomes important for $Re > 5.0 \times 10^3$. From this it follows that for $Re \leqslant 5.0 \times 10^3$ there are no prerequisites for the formation of a well-streamlined nektonic body shape, since in this *Re* range frictional drag is dominant whatever the shape of the body. Because of this, $Re = 5.0 \times 10^3$ can be logically taken as nekton's lower limit. With rare exceptions this value of *Re* approximately corresponds to nektonic organisms of absolute length 2–3 cm, from which, in turn, it follows that the absolute length of a nektonic animal is, as a rule, not less than 2 cm.

The transition from laminar to turbulent flow taking place at *Re* values of the order of 10^5 has an important influence on the morphology of nektonic organisms, creating tendencies for the development of special adaptations aimed at preserving laminar flow, which ensures a lower total drag during movement. It has been shown that in fishes (Burdak 1968, 1969a–c, 1970, and others) special laminarizors formed by surface relief due to elasmoid scalation, particularly by ctenii, emerge in ontogeny in the *Re* range 10^4–10^5 and complete their development at $Re \approx 10^5$. It is precisely within this range that (Aleyev & Ovcharov 1969, 1971) areas of turbulent flow appear on a fish body.

Thus, at *Re* values of the order of 10^5 a new function associated with propulsion appears in nektonic organisms: the function of controlling the boundary layer. This limit $Re = 10^5$ determines, therefore, the transition from animals with an uncontrollable boundary layer, which exist in a laminar flow and include the great majority of plankton, to animals with a controllable boundary layer, which dwell in turbulent streams and which include the overwhelming majority of nektonic organisms. Plankton whose Reynolds number is in the range $Re > 10^5$, like nektonic organisms with Reynolds number in the range $Re < 10^5$, are merely exceptions to the general rule.

From the above, it is logical to maintain that $Re = 10^5$ corresponds to the lower limit of eunekton, i.e. typical nekton, whereas the range $5.0 \times 10^3 < Re \leq 10^5$ is characteristic of planktonekton. With rare exceptions $Re = 10^5$ approximately corresponds to a nektonic animal whose absolute length is no greater than 10 cm, from which it follows that the absolute length of eunektonic animals is, as a rule, 10 cm at least.

Thus nekton, as a special ecomorphological type of aquatic animal, may exist theoretically only in the range $Re > 5.0 \times 10^3$, whereas $Re \leq 5.0 \times 10^3$ constitutes the exclusively planktonic range.

There exist also planktonic forms with $Re > 5.0 \times 10^3$, right up to Re values of the order of 2.0×10^7, but for $Re > 2.0 \times 10^7$ there exist nothing but nekton. The upper limit for the existence of euplanktonic forms corresponds to Re values of the order of 5.0×10^5, i.e. approaching the most probable critical Re values. Thus the upper limit on the linear dimensions and swimming speeds of typical planktonic animals is the upper limit of the laminar flow range: not a single euplanktonic form transgresses this limit. This is testimony that the ratio of inertial to frictional forces, characterizing the process of propulsion of a biont and expressed by the Reynolds number Re, constitutes a fundamental ecological factor, which must be taken into account along with such factors as temperature, light, availability of oxygen, etc. Nektoplanktonic species, whose morphology contains some elements of nektonic organization, manifested primarily by satisfaction of the condition $U \leq 0.40$, have $5.0 \times 10^3 < Re \leq 2.0 \times 10^7$, i.e. they move in both subcritical and supercritical regimes. The upper limit for the existence of nektoplankton, corresponding to $Re = 2.0 \times 10^7$, is determined in the main by a decrease in specific body surface with an increase in the animals' linear dimensions, owing to which the provision of an adequate parachute system involves considerable complications.

Thus for $Re \leq 5.0 \times 10^3$ there exist only plankton, in the range $5.0 \times 10^3 < Re \leq 2.0 \times 10^7$ plankton and nekton, and for $Re > 2.0 \times 10^7$ nekton only (Aleyev 1972a).

The definitions of plankton and nekton offered in their day by Hensen (1887) and Haeckel (1890), respectively, lack precise quantitative criteria for discriminating between these types or for distinguishing either of them from benthic organisms. Specific definitions of plankton and nekton taking account of their ecological, morphological and biohydrodynamic peculiarities may be presented as follows (Aleyev 1972a).

Plankton describes an ecomorphological type of biont (plant or animal) which is suspended in the mass of water all or most of the time or swims at the surface (in some cases semi-immersed), which is incapable of sustained active propulsion in the horizontal direction and swims in the regime $Re \leqslant 2.0 \times 10^7$ (as a rule in the regime $Re \leqslant 5.0 \times 10^5$) and whose general body structure is determined on the whole by the development of a complex of adaptations functionally associated with increasing hydrodynamic resistance and the capacity for passive hovering in the mass of water with the minimum expenditure of energy.

Nekton describes an ecomorphological type of biont (animal) which is suspended in the mass of water all or most of the time, which is capable of sustained active propulsion in the horizontal direction and swimming in the regime $Re > 5.0 \times 10^3$ (as a rule in the regime $Re > 10^5$) and whose general body structure is determined on the whole by the development of a complex of adaptations functionally associated with decreasing hydrodynamic resistance and increasing the capacity for active propulsion in the mass of water with the minimum expenditure of energy.

As all the above shows, plankton and nekton are ecological isotopes, i.e. instances of biontic groups which dwell in the same conditions (in this case in a mass of water) but are characterized by alternative directions of development of a complex of adaptations determining the most essential features of their general body structure. It is possible, therefore, to discern two trends in the evolutional development of pelagic animals, corresponding to two forms of fluid flow, i.e. determined by the laws of hydrodynamics. One of these trends – planktonic – is characterized by adaptation to existence in laminar conditions, the other – nektonic – by adaptation to existence in turbulent conditions.

Both planktonic and nektonic animals exhibit a certain variety of forms of locomotion and of general body structure; however, in plankton this variety is incomparably greater than in nekton, where uniformity of functional-morphological organization, the degree of convergence, attains its maximum. The latter is due to the fact that, as the absolute speed of locomotion increases, hydrodynamic resistance grows very rapidly, under an exponential law, and minimizing this drag becomes the organism's most crucial task. In this connection, with rising *Re* (particularly for $Re > 10^6$) one observes a steeply increasing convergence of the functional-morphological spectrum, which results in ever growing uniformity of the general morphological organization of nektonic animals belonging to various systemic groups.

An animal species may have in its ontogeny both nektonic and planktonic stages, as has already been pointed out by Haeckel (1890). All nektonic fishes with planktonic eggs and larvae furnish an example.

Nekton differ from benthic organisms primarily in that the latter are in contact with the bottom or some other immersed solid substratum of inorganic or plant origin, either resting on it or swimming all or most of the time, which in every case has a dominant influence on their ecology and morphology. The distinctions between nektonic animals and burrowing and sessile benthic organisms, like those between nektonic animals and mobile, free-ranging forms moving on the surface of a solid substratum, are obvious enough; certain difficulties arise, however, regarding discrimination between benthonektonic and nektobenthic animals, a matter we shall now take up.

Benthonektonic animals may be clearly distinguished from nektobenthic forms dwelling on the bottom through a whole complex of morphological features associated with camouflage. Usually the shape of the body's cross-section in nektonic forms is either close to a circle or is a dorso-ventrally elongated oval or ellipse, its lower end being somewhat pointed, i.e. the lower part of the body is rounded or pointed like a keel (the nektonic type of camouflage). Yet benthic forms typically have a cross-sectional shape close to either a trapezoid with its broad side downwards, a segment of a circle with its chord downwards or even an isosceles triangle with its base downwards, i.e. the lower part of the body is flattened to some extent (the benthic type of camouflage). In both nektonic and benthic animals these peculiarities of body shape, notwithstanding their different character, serve to eliminate the betraying shadow (see chapter VII).

A characteristic of many nektobenthic animals dwelling on the bottom is dorso-ventral compression of the body (Osteostraci, the benthic Batoidei, Lophiidae and others), which is also aimed at camouflage and in rheophile nektobenthic forms (*Pseudoscaphirhynchus*, etc.) is aimed, in addition, at securing the animal a more stable position at the bottom of a river. Dorso-ventral compression of the body is not, on the whole, characteristic of nektonic animals, though this sometimes occurs in forms ecologically associated with land (xeronektonic Sauropterygia, Placodontia, Testudinata, Pinnipedia, etc.), facilitating their locomotion on land.

Marked lateral compression of the body in benthic animals is as rare an exception as well-pronounced dorso-ventral compression of the body in nektonic ones; all such exceptions do not contradict, however, the general principles of camouflage in benthic and nektonic animals. Thus lateral flattening of the body in the benthic Pleuronectiformes, thanks to the fact that they lie on their sides on the bottom, provides camouflage as successfully as the dorso-ventral flattening of the body in Batoidei, though such a body position is a rare exception for benthic animals and therefore lateral compression of the body is on the whole not characteristic of them. The dorso-ventral flattening of the body in planktonektonic Sagittoidea, itself due to the dorso-ventral character of their locomotive undulations, turns out to be possible owing to the preservation of the planktonic type of camouflage on account of the high degree of the animal's transparency. In eunektonic Mobulidae the dorso-ventral compression of the body, which emerged as a means of camouflage in initial nektobenthic forms, is preserved in the pelagic zone as a morphological means of ensuring a special type of locomotion (see chapter IV), which, given the large linear dimensions of Mobulidae and, moreover, the comparatively small number of dangerous enemies, proves expedient despite certain exposure produced by such a body shape in pelagic conditions.

Many nektobenthic animals dwelling on the bottom or on submarine plants are equipped with all kinds of protective armour, spines, etc.; nektonic organisms as a rule have no structures of this kind.

Benthonektonic forms very clearly differ from nektobenthic ones through their specific hydrostatic features. The buoyancy of nektobenthic forms is, as a rule, markedly negative, whereas that of benthonektonic organisms is neutral or close to neutral.

The above-mentioned features of benthonektonic and nektobenthic animals together allow discrimination between them in every specific case without any particular difficulty.

The distinctions between nekton and terrestrial fauna, seemingly quite obvious, still require some explanation in view of the so-called semiaquatic animals. Serving as a criterion for distinguishing nekton from semiaquatic fauna may be the presence of basic nektonic features in the general body structure and the absence in the diurnal cycle of the necessary ecological connexion with the land or a floating solid substratum in its stead (drifting ice, floating plant material, etc.): in semiaquatic animals this connexion is always maintained (*Amblyrhynchus*, Gavialidae, Crocodylidae, Gaviidae, Podicipedidae, *Nannopterum harrisi*, *Tachyeres brachypterus*, *Ornithorhynchus*, *Desmana*, *Castor*, *Ondatra*, *Lutra*, *Enhydra*, etc.), whereas in nektonic animals it disappears (Chelonioidea, Sphenisciformes, Pinnipedia, etc.). In semiaquatic animals dwelling on the sea-shore, entry into the water is related to the weather conditions: when the weather turns bad (during a storm) the animal remains on land (*Enhydra*, etc.). As for nektonic animals ecologically connected with the land, during certain periods of their life-cycle they stay in the water, remaining afloat whatever the weather (Chelonioidea, Sphenisciformes, Pinnipedia, etc.). One of the criteria of the transition to nekton in birds is the loss of the capacity to fly. Semiaquatic mammals, in distinction from nektonic mammals, are characterized, in particular, by a thick coat of fur which, while the animal is in the water, holds air and plays the role of the main heat-insulating screen; in nektonic animals this furry cover is usually reduced to some extent (Phocidae, Odobenidae) or practically absent (Sirenia, Cetacea), while the functions of a heat-insulating screen are taken over by subcutaneous fatty tissue.

Adaptations to the aquatic mode of life in semiaquatic animals become varied and in some cases very extensive, though in different species one finds these adaptations at every stage of development, beginning from the very first. Among the ecologically diverse semiaquatic forms one may distinguish those whose development tends towards nekton, though they are not yet nektonic. Such, for example, are the *Amblyrhynchus* all the Gavialidae and Crocodylidae among the Reptilia, among the Aves, Gaviidae and Podicipedidae, *Nannopterum harrisi*; among the Mammalia, *Desmana*, *Lutra*, *Enhydra* and many others. To designate this group of animals the term nektoxeron (from the Greek ξηρά, dry, i.e. land, continent) has been proposed (Aleyev 1973a), formed by analogy with the generally accepted term 'nektobenthic'. Nektoxeron are not yet nekton, but terrestrial fauna taking their first steps towards nekton. Accordingly, nekton with a bias towards terrestrial fauna (most often subaquatic nekton, not yet divorced from the land) are designated by the term 'xeronekton'.

In their early ontogeny nektonic animals usually pass through stages which are not nektonic. Later in this book all such stages are described and examined only so far as is necessary for an understanding of the development of nektonic adaptations. Some general considerations regarding the biological development of nektonic organisms may be found in chapter IX.

2. Definition, history and tasks of nektonology

Nektonology (Aleyev 1969d) is the science or study of the general features of the emergence and evolution of nekton as a special ecomorphological type of biont. As an element of the biosphere nekton constitute a global phenomenon and, accordingly, play a rather substantial role both in the life of bodies of water and in the economic activity of man.

The history of nektonology is not very long. Studies of nekton as a definite phenomenon in the life of bodies of water actually began some seventy years after nekton were described by Haeckel (1890), in the 1960's, when the first special investigations into nekton were published (Aleyev 1965c; 1966a–b, 1969a,d, etc.), in distinction from numerous earlier published studies on diverse questions pertaining to the classification, ecology, functional morphology, biohydrodynamics, biohydroacoustics and other aspects of research on individual systemic groups or species of nektonic animals.

The first attempt to reveal and classify certain adaptations common to nekton is contained in work by Abel (1912, 1916), devoted to the paleobiology of vertebrates and cephalopods. This attempt, however, concerned only adaptations associated with forward propulsion and was limited to distinguishing 'adaptational types' of animal based on differences in body shape. The narrow morphological analysis underlying this classification of the phenomenon, carried out without taking into account the biohydrodynamic aspects, naturally led to errors and contradictions, in view of which Abel's attempt to classify what was known about nekton is today of historical interest only.

From around 1930 there have been published a noticeable and ever-growing number of studies of the functional morphology and biohydrodynamics of nektonic animals and from 1950 also of the bioacoustics of nektonic animals, mostly fishes and cetaceans. These investigations made up the 'prehistory', as it were, of nektonology, helping to prepare the ground for strictly nektonological investigations into the general characteristics of nekton.

In 1963 the present author published the book 'Functional Principles of the External Structure of Fishes', in which the characteristics of the functional morphology of fishes were for the first time deliberately treated as a specific case of general nektonic characteristics, the author feeling that his main task was to demonstrate that the general trend in the processes of fish evolution is determined to the highest degree by the affiliation of these or other groups to nekton, plankton or benthos (Aleyev 1963a). Using the example of fishes, and to some extent cetaceans, all basic groups of nektonic adaptations were described, as well as a complex of new nektonological methods which subsequently came to be widely used in studying not only fishes and agnathans, but also the overwhelming majority of other nektonic animals: cetaceans, pinnipeds, sea reptiles and cephalopods.

The formation of a group of nektonologists in 1963 at the Institute of Biology of South Seas, Ukrainian Academy of Science was a landmark in the consolidation of nektonology as a science. Investigations have covered all the main systematic groups of nekton: fishes, cetaceans, pinnipeds, sea reptiles and cephalopods. The first studies of this group, partially reviewed in a separate paper (Aleyev & Mordvinov 1971), include general theoretical nektonological work (Aleyev 1965c, 1966a, 1969a,d, 1970a, 1972a,b, 1973a–c, 1974, and others); studies of the functional morphology and biohydrodynamics of nektonic cephalopods (Zuyev 1966, and others); visualization studies of the flow around fish carried out by using a new technique which avoids disturbing the observed flow (Aleyev & Ovcharov 1969, 1971, 1973a,b) and special miniature turbulence sensors (Varich 1971); a series of studies of the biohydrodynamics of the skin teguments of fishes (Burdak 1968, 1969a–c, 1970, 1972, 1973a,b, and others) and cetaceans (Aleyev 1970a, 1973c,d, 1974); studies of high-speed propulsion of nektonic animals (Aleyev 1965b, 1970b; Ovchinnikov 1966a,b, 1970, and others); studies on a quantitative bases of the locomotion apparatus of nektonic animals, which for the first time allowed an approach to a quantitative assessment of the use made by the propulsive mechanism of the kinetic energy of the boundary layer (Aleyev 1969a, 1973b); studies of the biohydrostatics of nektonic animals (Aleyev 1963a,b, 1966a,b; Mordvinov 1969 and others); studies of the unsteadiness in nektonic animal movement carried out with the help of new instruments operating on the principles of sonic sounding (Komarov 1970, 1971, and others); comparative studies of the hydrodynamics of nektonic organisms (Aleyev 1972b; Aleyev & Kurbatov 1974; Kurbatov 1973), and many others.

Among other investigations of the 1960's one should mention those by Hertel (1963, 1966), Pershin (1965a,b, 1967, 1969a,b) and others into certain aspects of the biohydrodynamics of nektonic organisms.

Nektonology today has two principal tasks: The first and main one is to reveal the basic, fundamental features characterizing nekton as a special type of aquatic biont and to explain the paths of the emergence and evolution of nekton as an element of the biosphere. The performance of this task requires, above all, general functional-morphological and biohydrodynamic investigations into nekton.

The second task is the further development of hydrobionic studies associated with nekton, since hydrobionic studies can only develop properly on the basis of nektonological data, in particular, data on the functional morphology of nektonic animals; this is confirmed by the entire history of the development of hydrobionics, both in the Soviet Union (Aleyev 1965b; A. I. Berg 1965) and abroad (Tavolga 1969; Petrova 1970, and others).

3. Nektonological investigation methods

The principal nektonological methods are 1. functional-morphological analysis, 2. observation of living nektonic organisms and 3. experiment.

1. **Functional-morphological analysis** is the backbone of any study of adaptation. The collection of material necessary for carrying out morphological analysis and also for staging experiments with nektonic organisms requires no special implements or catching techniques; the well-known methods of catching nektonic animals common in the fishing trade may also be used for this purpose.

This author believes the functional morphology of animals to be a branch of zoology whose principal task is to study the dynamics of morphological specializations from the aspects of phylogeny and ontogeny, and to establish the general characteristics pertaining to the evolution of animal forms. According to its content and methods, functional morphology may be evolutional, ecological or experimental. Central among the problems of functional morphology is the question of the interrelations between function and shape, which is superbly reflected in the name 'functional morphology'.

2. **Observing living nektonic organisms** in their natural habitat provides most important primary ecological and other information essential for investigating the general characteristics of nekton. This information may be derived by various methods (underwater and aerial photography and filming, instrumental recordings of various signals from nektonic organisms, underwater observations, etc.).

3. **Experiment** is an integral part of nektonological investigations, since in the overwhelming majority of cases it provides the only possible method of verifying theoretical models.

The mathematical treatment of results is an absolute must for nektonological methodology. When studying the development of adaptations, the morphological feature being investigated and the function associated with it must be represented by dimensionless numerical characteristics suitable for further operations with the aim of carrying out comparative analysis. Whatever the complexity of the mathematical treatment, one may distinguish three fundamentally different levels of quantitative assessment of functional-morphological factors.

The first corresponds to the use for quantitative assessment of absolute values only, expressed in conventional units of measurement, length, volume, weight, etc.; i.e. we receive a number of absolute values A_1, A_2, $A_3, \ldots, A_n$. Only the simplest quantitative determinations are possible at this stage.

The second level corresponds to the use of primary of simple relative values B which are the result of relating one absolute value A_1, characterizing the development of the morphological feature under examination, to another value A_2, playing the role of a base or standard, i.e. $B = A_1 \times A_2^{-1}$. The use of primary relative values makes it possible to study the relationships among individual factors and to describe the simplest causative dependencies. In the overwhelming majority of cases, however, primary relative values prove inadequate for describing more complex causative dependencies, the revelation of which demands comparisons between individual relative values.

The third and highest level corresponds to the construction of secondary or complex relative values C derived from the primary relative values B, i.e. $C = f(B_1, B_2, B_3, \ldots, B_n)$. These secondary or complex relative values allow description of quantitative dependencies of any complexity, since they make it possible to take into account any diversity of factors affecting the process under study.

4. General remarks

In this investigation the author felt that his principal goal was to furnish an all-round characterization of nekton as a whole. Accordingly, he felt it inexpedient to describe in detail the morphology, physiology and ecology of all the representatives of nekton: this would have turned the work into a many volumed textbook of zoology, preventing the reader from concentrating on what makes up the principal features of nekton as a definite ecomorphological type of animal. The author therefore tried to steer clear of well-known zoological material as much as he could, while examining in great detail the characteristics of the development of the principal nektonic adaptations constituting the basis of the convergence of nektonic organisms belonging to different systematic groups.

All the drawings in the book were produced by the author, as were all the photographs, cine records and tables whose sources are not acknowledged. The numerical values of all the morphological indices were derived by the author as a result of direct measurement of the animals.

The author has the pleasant duty of thanking A. P. Andriashev for his advice, given without stint. He also thanks O. P. Ovcharov for the technical assistance he gave throughout the work.

With feelings of particular appreciation the author thanks his wife V. D. Burdak, who all through the many years of the writing of the manuscript was the author's daily advisor, critic and helper, sharing with him all the difficulties involved in accomplishing this work.

PART 1. NEKTON: SYSTEMATICS AND GEOGRAPHICAL DISTRIBUTION

As a global phenomenon in the life of bodies of water, nekton display considerable variety with respect to space and time in their systematic composition, ecomorphological variations and geographical distribution, which together make up Part 1 of this investigation. Without taking into account extinct nektonic forms and groups, a more-or-less complete picture of the basic characteristics of the emergence and development of nektonic adaptations is impossible, therefore in the analysis of the systematic composition of nekton both extant and fossil nektonic fauna were taken into account, which allowed nekton to be covered in their entirety. Nor is it possible to deal with nekton without ecomorphological differentiation. The examination of the systematics and distribution of nekton has, therefore, to be preceded by a chapter on the ecomorphological classification of nekton, but the basic characteristics of the classes are given in Part 3 of the book, after an examination of the most important nektonic adaptations.

I. ECOMORPHOLOGICAL CLASSIFICATION OF NEKTON

As an ecomorphological type of biont nekton include both typical representatives and various deviations from the norm approximating other types to some extent. The names of all the subdivisions of nekton are derived (Aleyev 1972a) from the same root as the name of the type. Typical representatives are united in a class under the name of the type with the prefix 'eu' affixed (from the Greek εὖ, well); the names of all the other classes are formed by adding a prefix to the basic root which indicates the tendency of ecomorphological specialization of the members of the class concerned. Thus, for example, nekton tending towards plankton are united in the class planktonekton (and plankton tending towards nekton in the class nektoplankton). The name of the nektonic class tending towards terrestrial fauna has been derived (Aleyev 1972a) from the Greek root ξηρά, dry, i.e. land, continent; for semiaquatic animals tending towards nekton the name nektoxeron is adopted (see Introduction).

The general ecomorphological classification of nekton is as follows (Aleyev 1973a).

Ecomorphological type: *nekton*

1. Class **benthonekton:** nektonic animals not necessarily ecologically connected with the land or surfaces of floating solid substrata jutting out into the air which are ecologically connected with the bottom or immersed surfaces of floating solid substrata (particularly drifting ice or plant material) and use camouflage of the nektonic type. Examples: *Cyprinus, Boreogadus saida,* Omphalosauridae, *Dugong, Eschrichtius.*
2. Class **planktonekton:** nektonic animals not necessarily ecologically connected with any solid substratum whose movement is characterized by $5.0 \times 10^3 < Re \leq 10^5$. Examples: *Flaccisagitta, Abraliopsis,* Myctophidae, *Leucaspius.*
3. Class **eunekton:** nektonic animals not necessarily ecologically connected with any solid substratum whose movement is characterized by $Re > 10^5$. Examples: Architeuthidae, Scombroidei, Ichthyosauroidei, Balaenopteridae.
4. Class **xeronekton:** nektonic animals ecologically connected with the land or the surfaces of floating solid substrata jutting out into the air (drifting ice, plant material, etc.). Examples: Mesosauria, Chelonioidea, Sphenisciformes, Pinnipedia, *Trichechus.*

The differences between the individual nektonic classes can be seen in Table 1. Since members of the same species frequently belong to different ecomorphological classes or types at different ages, this table is a key to affiliation of the species under examination to one nekton class or another at any age. The differences between nekton and plankton or benthic or terrestrial fauna have already been pointed out in the Introduction. Most difficult, as we have seen, is the distinction of nekton from

Table 1. Key to nekton classes (Aleyev 1973a)

1 (6). Forms either incapable of active movement or actively moving. In the latter case typically $Re \leqslant 5.0 \times 10^3$; for $Re > 5.0 \times 10^3$ usually $U > 0.40$, but when $U \leqslant 0.40$, then $S_0 > 4.50$ and body shape is not of nektonic type .... 1. Type **plankton**.

2 (5). Neither an ecological connection with bottom or immersed surfaces of floating solid substrata (ice, plant material, etc.) not a direct ecological connection with the surface of the water obligatory.

3 (4). Forms incapable of active movement or actively moving. In the latter case $Re \leqslant 5.0 \times 10^5$; for $Re > 5.0 \times 10^3$, $U > 0.40$ .... 1.1. Class **euplankton**.

4(3). Actively moving forms. $Re > 5.0 \times 10^3$, $U \leqslant 0.40$, $S_0 > 4.50$. Body shape not of nektonic type .... 1.2. Class **nektoplankton**.

5 (2). Ecological connection with the bottom or immersed surfaces of floating solid substrata (ice, plant material, etc.) or direct ecological connection with the surface of the water obligatory.. Other groups of plankton.

6 (1). Actively moving animals. $Re > 5.0 \times 10^3$ but usually $Re > 10^5$; for $Re \leqslant 10^5$, $S_0 < 4.50$ and body shape is of nektonic type............................ 2. Type **nekton**.

7 (12). Ecological connection with land or surfaces of floating solid substrata jutting out of the water (ice, plant material, etc.) not obligatory.

8 (9). Ecological connection with the bottom or immersed surfaces of floating solid substrata (ice, plant material, etc.) obligatory. Camouflage of nektonic type .... 2.1. Class **benthonekton**.

9 (8). Ecological connection with the bottom or immersed surfaces of floating solid substrata (ice, plant material, etc.) not obligatory.

10(11). $5.0 \times 10^3 < Re \leqslant 10^5$............................ 2.2. Class **planktonekton**.

11(10). $Re > 10^5$ .... 2.3. Class **eunekton**.

12 (7). Ecological connection with land or surfaces of floating solid substrata (ice, plant material, etc.) jutting out of the water obligatory............. 2.4. Class **xeronekton**.

plankton, and therefore, along with the nekton classes, the key includes some planktonic groups (nektoplankton, euplankton); it should be borne in mind that it is not intended to indicate all planktonic groups.

A detailed characterization of nekton classes and their origins is given in chapter IX, following a description of the principal group of nektonic adaptations (chapters III–VIII), which permits a deeper grasp of the specific features of the different ecomorphological variants of nekton. Chapter IX examines also some general characteristics of the evolution of nekton into an ecomorphological type.

To indicate as briefly and accurately as possible the nektonic class of adult or other age stages, species or other systematic groups of animals, or the close-to-nektonic classes of other ecomorphological types, all the ecomorphological classes occurring in the text are labelled with conventional symbols as follows:

BN, benthonekton;	NB, nektobenthos;
PN, planktonekton;	EP, euplankton;
EN, eunekton;	NP, nektoplankton;
XN, xeronekton;	NX, nektoxeron.

In all the tables and occasionally in the text the Latin names of animals are directly followed by the appropriate label in brackets: for example, *Scomber scombrus* (EN), Balaenopteridae (EN), *Gobius melanostomus* (NB), etc. In comparatively rare instances a species has to be given a double label; these cases reflect a situation where the adult (sexually mature) individuals in certain circumstances may be placed in one of two ecological groups. For example; *Sprattus sprattus phalericus* should be followed by (PN/EN), as the adult individuals of this species may be either planktonekton (when under 10 cm in length) or eunekton (when over 10 cm in length).

II. SYSTEMATIC COMPOSITION, GEOGRAPHIC RANGE AND DISTRIBUTION IN BODIES OF WATER

1. Systematic composition

The very definition of nekton makes it clear that this ecomorphological type is represented exclusively by animals. The nektonic animals known so far are distributed among ten classes: Cephalopoda, Sagittoidea, Monorhina (Cephalaspidomorphi), Placodermi, Acanthodei, Chondrichthyes, Osteichthyes, Reptilia, Aves and Mammalia. Though very probable, so far the presence of nektonic forms in the two classes Diplorhina (Pteraspidomorphi) and Amphibia has not been proved. Thus, to date the presence of nektonic forms has been established or is probable for the following twelve classes of animals:

1. Cephalopoda,
2. Sagittoidea,
3. Diplorhina (Pteraspidomorphi),
4. Monorhina (Cephalaspidomorphi),
5. Placodermi,
6. Acanthodei,
7. Chondrichthyes,
8. Osteichthyes,
9. Amphibia,
10. Reptilia,
11. Aves,
12. Mammalia.

We accept the group Sagittoidea according to Tokioka (1965a), following this author also in the names of individual species, whereas we accept the groups Diplorhina, Monorhina and Placodermi according to Obruchev (1964a,b); the group Acanthodei according to Novitskaya & Obruchev (1964), the group Chondrichthyes according to Glikman (1964) and Obruchev (1964c) and the group Osteichthyes according to Vorobyova & Obruchev (1964) and Kazantseva (1964). Names of species in the group Teuthoidea follow Clarke (1966).

The above-mentioned twelve classes of animals are represented among nekton rather differently. Some of them consist predominantly of nektonic animals (for example Osteichthyes), whereas others contain only a few nektonic individuals (Aves).

Among the **Cephalopoda**, most representatives of Endocochlia are nekton. Primarily, these are the almost entirely nektonic groups Belemnoidea and Teuthoidea of the order Decapoda. The group Teuthoidea contains only a comparatively few planktonic forms, most of which belong to the family Cranchiidae. As for Belemnoidea, judging by the morphology of these extinct molluscs, they were good swimmers capable of rapid locomotion in the water, both in the horizontal and vertical direction, which enables one to place them among nekton (Krymgolts 1958), though undoubtedly there were nekto-benthic and possibly planktonic species among them. The Sepioidea, the third suborder of Decapoda, comprise mainly nektobenthic and planktobenthic forms and only to a small extent nektonic ones. The pelagic Octopoda occur only in the group Cirroteuthoidea; they include both planktonic and nektonic species.

Sagittoidea, constitute a predominantly planktonic class. However the larger forms, with individuals more than 2.0 cm long already belong among planktonekton. The characteristic planktonektonic groups include, for example, the order Flaccisagitta, which includes such species as *F. lyra*, more than 4 cm long, *F. hexaptera* and *F. maxima*, which are 6 cm long or more, and *F. gazellae*, whose length, according to available data (Marshall 1958), reaches 10 cm. Also belonging among planktonekton are representatives of the smaller Sagittoidea species whose length in the state of sexual maturity exceeds 2.0 cm, which corresponds to $Re > 5.0 \times 10^3$. An example is provided by large specimens of Black Sea *Sagitta setosa* 2.1–2.2 cm long, for which the author has chronometrically established a swimming speed in the aquarium of about 0.5–0.6 m/s, which corresponds to an Re of about $1.1 \times 10^4 - 1.2 \times 10^4$; the swimming speed of these Sagitta in natural conditions, and accordingly the Reynolds number, is probably higher.

The presence of nektonic forms in the class **Diplorhina** (Pteraspidomorphi) has not been proved so far. It is, however, thought (Obruchev 1964a) that some Diplorhina, particularly individual representatives of Pteraspidida (*Pteraspis, Podolaspis, Doryaspis* and others) which had an elongated rostrum and a comparatively narrow shell combined with fine elongated oral plates and a comparatively well developed tail fin, were capable of rising into the upper layers to feed on plankton. The general character of the morphological organization of the above-mentioned representatives of Diplorhina provides no grounds, however, for a definite conclusion in this regard. Undoubtedly, such forms as *Pteraspis, Podolaspis* or *Doryaspis* could swim through the water and probably fed, to some extent, on objects suspended in it. However they were hardly nektonic animals: the absence of any fins save the tail fin, the general dorsoventral compression of the body and comparatively undeveloped eyes bear testimony to the fact that they belonged rather among nektobenthic forms that spent most of their time lying on the bottom, a general rule for Diplorhina. It is possible that some of the animals we are examining here did adapt themselves to a more roving mode of life but in that case they were probably no more than the most primitive benthonektonic forms. Though even for the most mobile representatives of Diplorhina affiliation to nekton cannot be regarded as proven, below we describe representatives of this class together with other nektonic animals, as this is of considerable interest both when investigating the process of development of individual nektonic adaptations and also with regard to some general functional-morphological problems concerning these animals and paleontology.

Monorhina (Cephalaspidomorphi) form on the whole a benthic class. The only nekton among them belonged to the subclass Anaspida (Birkeniae), most of which, judging by their morphology, were benthonektonic forms, though in all probability they included nektobenthic species as well. We do not find nektonic species among extant Agnatha, which are placed among Monorhina as a special subclass of Cyclostomi (Marsipobranchii); all these, like lampreys and hag-fishes, are nektobenthic animals.

Among the **Placodermi**, there were doubtless nektonic, specifically benthonektonic or possibly even eunektonic forms in the Pachyosteida group. Such, for example, were the deep-bodied, laterally compressed Synaucheniidae, Oxyosteidae and Leptosteidae, with grinding teeth or other jaws; in some of them (*Oxyosteus*, *Synauchenia* and others) the body cross-section narrowed downwards like a wedge (Obruchev 1964b), i.e. their type of camouflage was undoubtedly nektonic. For another group of Placodermi, the Coccosteida, the inclusion of nektonic forms, though not improbable, has not been proved so far; everything indicates that Coccosteida, whose bodies were flattened on the ventral side, were in the majority of cases merely nektobenthic. There are no grounds to suppose that nektonic forms existed in other Placodermi groups.

Acanthodei is the first pronouncedly nektonic group in the history of vertebrates. Of the four Acanthodei orders nektonic forms clearly predominated in at least two: Ischnacanthida and Acanthodida, whose members, judging by their morphology (Novitskaya & Obruchev 1964), were in the main benthonektonic or euektonic. As for the most primitive early forms, united in the order Diplacanthida, these had a body comparatively broad and flattened beneath and preserved the benthic type of camouflage, i.e. were nektobenthic animals, though the presence of benthonektonic forms among them has not been ruled out. Members of the fourth order, Gyracanthida, were, according to all indications, nektobenthic.

One of the two large groups that make up the class **Chondrichthyes**, the Elasmobranchii, is predominantly nektonic; among the nektonic forms are all the pelagic sharks (including benthopelagic species, such as Triakidae) and some rays (Mobulidae) which only later changed over to the pelagic mode of life. A more detailed examination of the ecological composition of the individual subdivisions (orders) of Elasmobranchii with the purpose of establishing the presence of nektonic forms in them is hardly advisable in view of the fact that there is still no generally accepted approach to the systematics of this group. The second large group, Holosephali, constitutes a nektobenthic variant of the development of Chondrichthyes. Extant members of the Holocephali, taking into account their markedly negative buoyancy (Schmalhausen 1916), the structure of their fins and their mode of locomotion (Herald 1962), should be placed among nektobenthic animals; everything indicates also that there were no nektonic forms among the extinct Holocephali.

The class **Osteichthyes** has nektonic forms in both its subclasses, Sarcopterygii and Actinopterygii. The presence of benthonektonic species among Sarcopterygii is unquestionable both for Crossopterygii and for Dipnoi, however among the former they are apparently much more numerous than among the latter. Nektonic organisms are present in most of the large subdivisions of the Actinopterygii; many large groups being predominantly or exclusively nektonic (Clupeiformes, Beloniformes,

Cyprinodontiformes, Scrombroidei, etc.). But since opinions on the subdivision of the subclass Actinopterygii differ (Kazantseva 1964; Greenwood *et al.* 1966, and others), a more detailed examination of the question of the presence of nektonic forms in its various groups is pointless.

We do not find nektonic species among extant **amphibia**. The most aquatic among them are ecologically closely connected with the bottom or land, have a benthic type of camouflage and spend most of their time in contact with some sort of solid substratum, i.e. these are nektobenthic or xerobenthic. Closest to a benthonektonic state are the larvae (tadpoles) of some Anura, which during daylight sometimes spend much of the time suspended in the water, swimming at $Re > 5.0 \times 10^3$. Moreover, notable is a special nektonic type of Anura larva (Bannikov & Denisova 1969) characterized by strong development of not only the dorsal, but also the ventral part of the caudal fin fold. As examples we could point, particularly, to the larvae of *Bombina bombina*, *Pelobates fuscus* or *Rana ridibunda* in the second half of the larval period. However, even bearing in mind the mobility of these and similar larvae and their prolonged stay in the body of the water, they nevertheless cannot be considered benthonektonic; during the night and during parts of the day all of them rest lying on the bottom or hidden in some crevice or other shelter, i.e. actually remain merely mobile nektobenthic forms. Quite probable, however, is the presence of benthonektonic and xeronektonic forms among extinct Amphibia. Paleontological data (Tatarinov 1964a,c,d; Konzhukova 1964a–d,f–h; Shishkin 1964) prompt the assumption of benthonektonic and xeronektonic species among such groups as Temnospondyli (particularly in view of Ichthyostegalia and Trematosauroidea), Plesiopoda, Anthracosauria (particularly Embolomeri) or Nectridia. However the presence of nektonic forms in the class Amphibia cannot yet be considered as proven. Despite all this, Amphibia will be repeatedly mentioned further in connexion with the development of individual nektonic adaptations.

Among the **reptiles** nektonic forms are comparatively numerous and varied, occurring in most of the subclasses. Thus, of the Synaptosauria, some Araeoscelidia (Pleurosauria[1]) and all Sauropterygia and Placodontia are nektonic. Entirely nektonic groups are the Proganosauria (Mesosauria) and Ichthyopterygia (Ichthyosauria). Of the Testudinata, nektonic are some of the Pleurosternidae (Desmemydinae) and all the Thalassemydidae, Apertotemporalidae, Chelonioidea and Dermochelyoidea. Of the Lepidosauria, the Thalattosauria, Mosasauridae, Acrochordinae, Palaeophidae and Hydrophidae are nektonic. Of the Archosauria, the Teleosauridae and Metriorhynchidae were undoubtedly nektonic. Besides, an analysis of available paleontological material (Romer 1956; Huene 1956; Konzhukova, 1964i; Maleyev 1964a,b,d,e; Rozhdestvensky 1964a; Tatarinov 1964n,r; Chudinov 1964a–d; Shishkin 1964, and others) gives grounds to assume that nektonic (xeronektonic) forms cannot be ruled out in the groups Rhynchocephalia (Claraziidae), Choristodera, Aigialosauridae, Dolichosauridae, Cholophidia,

[1] Pleurosauria are sometimes placed (Romer 1956; Huene 1956) among the Lepidosauria.

Phytosauria and in some other groups of reptiles. For variety of nektonic adaptations the reptiles are comparable only to the Osteichthyes; known among nektonic reptiles are xeronektonic, benthonektonic and eunektonic forms.

The **birds**, unlike the reptiles, contain extremely few nektonic forms. Deep specialization connected with adaptation to flight probably created great difficulties in the way of adaptation to the aquatic mode of life. Judging by available paleontological data (Dementyev 1964), adaptation to aquatic life throughout the entire history of this class reached the nektonic stage in two cases only, resulting in the Cretaceous Hesperornithes and Sphenisciformes (Impennes), which arose in the Oligocene. In both of them the development of nektonic adaptations was associated with loss of the capacity to fly. The non-flying *Pinguinus impennis* (Alciformes), exterminated in the 19th century, was xeronektonic or nektoxeronic; it is impossible to ascertain this more precisely for lack of ecological data.

Of the **mammals,** all representatives of the pinnipeds, cetaceans and sirens are nektonic.

2. Geographic range and distribution in bodies of water

As a special ecomorphological type of pelagic animal, nekton have a global range. In the ocean we come across nektonic organisms everywhere, from the Arctic and Antarctic seas to equatorial waters. Nekton are equally widespread in inland waters, most of which are populated with some nektonic organisms or other at all latitudes.

Only very few bodies of water have no nekton. In the ocean nektonic animals are absent, in fact, only in some very isolated bays with an extremely specific hydrochemical regime, such as the upper reaches of the Sivash, and in areas contaminated by hydrogen sulphide, such as, for example, the depths of the Black Sea. In inland bodies of water nektonic animals are absent only in some oversalted lagoons of large lakes, like the Kara-Bogas-Gol of the Caspian, and in the polyhaline salt lakes, such as Lake Elton, apart from in very small bodies of water, i.e. some small lakes, ponds and marshes, in small rivulets, etc., and in all kinds of temporary bodies of water, i.e. pools, holes, etc. In a number of cases even small inland bodies of water contain certain nektonic inhabitants, consisting of various small fishes: such small nektonic forms as *Dallia, Leucaspius, Carassius carassius m. humilis* or Poeciliidae may occur in most types of small bodies of water – marshes, small ponds, pools, holes, etc. Nektonic organisms are found also in a number of underground bodies of water: such, for example, are the fishes of the Amblyopsidae family in the underground waters of Northern America.

There are three points that must be considered first of all **in distinguishing oceanic (marine) nekton from the nekton of inland bodies of water.**

1. On the whole, the systematic composition of oceanic nekton is much richer than that of the nekton found in inland waters, which corresponds to the greater systematic variety in oceanic fauna in general as compared with the fauna of inland

waters. All twelve classes of animals that contain nekton are represented in the oceanic nekton, whereas the nekton of inland bodies of water consist of members of only nine: completely absent here are the Sagittoidea, Cephalopoda and Aves. Oceanic nekton are richer than the nekton of inland waters not only in the number of animal classes represented, but also with regard to species, which also corresponds to the relation between oceanic and freshwater fauna as a whole.

2. Compared with the nekton of inland waters, oceanic nekton are far more varied in ecomorphological composition. All four nekton classes are fully represented in the ocean. However xeronekton occur in inland waters to a considerably lesser extent than in the ocean. This is due to the smaller pelagic zone of inland waters: the constant proximity of the shore results in subaquatic forms preserving such close day-to-day ecological ties with the land that basic nektonic features have not appeared yet either in the ecology or in the morphology of these animals; xeronektonic forms are in fact replaced here by nektoxeric ones. Extant xeronektonic animals dwelling in inland waters are represented in fact only by seals of the Phocidae family and some *Trichechus* species, all of which have their origin in the sea. We do not find among the extant fauna any xeronektonic forms emerging in inland waters; possibly some extinct Reptilia (particularly Mesosauria) and Amphibia (Embolomeri and others) did have such forms. The variety of eunekton steeply diminishes in the smaller inland bodies of water: the smaller the body of water the fewer eunektonic animals it contains, since as the size of the pelagic zone diminishes, ecological ties with the bottom and land inevitably increase.

3. Nektonic animals in the ocean are on the whole larger than those in inland waters. Inland bodies of water do not contain nektonic forms as large as some occurring in the ocean (whales, etc.). Therefore the smaller size of inland bodies of water influences also the size of their nektonic inhabitants: as the size of the body of water decreases, so does the size of its nektonic inhabitants and, as a consequence, the number of planktonektonic forms among them increases while the number of eunektonic species decreases.

Thus the specific features of nekton in inland bodies of water distinguishing them from oceanic nekton are determined mainly by the small size of inland bodies of water as compared with the ocean. For inland bodies of water of diminishing size we see an increasing predominance of benthonektonic and planktonic forms over all other nektonic classes. In extreme cases this process leads to the complete elimination of nektonic fauna and their replacement by a complex of nektobenthic, nektoplanktonic and nektoxeric forms. Hence it follows that the formation of all classes of nekton requires, in particular, a certain degree of ecological separation of hydrobionts from solid substrata, a question that is examined in greater detail in chapter IX.

Another important characteristic of the nekton of inland waters is an oligohaline quality, which precludes the invasion of many oceanic animals into inland waters. It is precisely for this reason that the nekton of inland waters completely lack Chaetognatha (Sagittoidea) and Cephalopoda.

The specific zonal features of the development of nekton, i.e. the specific features connected with latitude, represent on the whole a particular case of the well-known fact that the qualitative variety in the fauna of marine and inland bodies of water diminishes from the equator towards the poles while the populations of individual species increase in the same directions. In the lower latitudes, where the variety among nekton is considerable, the populations of individual species are comparatively small, whereas in the higher latitudes, with a comparatively small variety of nekton on the whole, the populations of individual species may, on the contrary, be very large; examples are offered by such nektonic animals as *Clupea harengus* or *Gadus morhua*.

The habitats of nektonic animals are extremely varied in type and size. One may point, by way of example, to such habitats as arctic (*Boreogadus saida, Arctogadus* sp., *Monodon*), boreal (*Gonatus fabricii, Gadus morhua, Phoca vitulina*), circumtropical (*Onykia carribaea, Mugil cephalus, Thunnus albacora*), notalian (*Moroteuthis ingens, Gasterochisma melampus, Arctocephalus pusillus*), antarctic (*Pagothenia borchgrevinki, Pleurogramma antarcticum, Aptenodytes forsteri*), bipolar (*Cetorhinus maximus, Hirundichthys rondeletii, Eubalaena glacialis*) and cosmopolitan (*Onychoteuthis banksii, Globicephalus melas, Orcinus orca*). The variety in the size of the regions inhabited by nektonic animals can be seen from a comparison of the cosmopolitan habitats of such forms as *Onychoteuthis banksii* or *Orcinus orca* with the extremely limited habitats of some species occurring in inland bodies of water, in particular *Salmo ischchan*, dwelling in the small Transcaucasian lake of Sevan, or *Clupeonella abrau*, from the Abrau Lake (near Novorossiisk) and Abuliond (Turkey).

In inland waters nektonic animals belonging to different ecomorphological classes on average have practically no marked differences according to their size of habitat. In the ocean, though, benthonektonic and xeronektonic forms never have such large habitats as eupelagic nektonic animals, i.e. planktonekton and, particularly, eunekton. This is due, apparently, to the limited areas of the sea bottom and sea-shore with which benthonektonic and xeronektonic animals are respectively ecologically connected. If we take nekton as a whole, then the regions of occurrence of greatest extent are those of some oceanic eunektonic forms, such as *Onychoteuthis banksii, Ommastrephes bartrami, Xiphias gladius, Balaenoptera musculus* or *Orcinus* orca.

The nektonic fauna of individual zoogeographic regions have their own specific features, which are characteristic of the fauna of the regions concerned and mainly due to the predilection of individual animals for definite climatic zones. An analysis of extant nektonic fauna of individual biogeographic regions naturally includes two independent tasks; i.e. examining the nektonic fauna of 1. the ocean and 2. inland bodies of water.

Only a very general biogeographic characterization of the nekton of the ocean can be given at present and, moreover, mostly for other than deep-dwelling forms. This is mainly due to two facts: 1. the regions of occurrence of most oceanic nektonic animals, deep-dwelling ones especially, have on the whole been inadequately studied and 2. most nektonic animals are very mobile and thus travel independently of oceanic currents within enormous bodies of water, sometimes measuring thousands of kilometres.

It is well known that the latitudinal localization in the distributions of marine animals, due to the individual climatic zones, diminishes as we descend to the deeper layers of the water, where we find maximum uniformity of abyssal fauna, nektonic included, owing essentially to the fact that the temperature profile in the meridional direction becomes more uniform the deeper we go. On the other hand, this localization also decreases as a result of diminishing dependence of the hydrobiont on the temperature of the medium, i.e. with the development of homothermy. The large number of deep-dwelling (particularly Theuthoidea and Osteichthyes) and homothermal forms (Sphenisciformes, Pinnipedia, Cetacea, Sirenia) among oceanic nekton complicates still further the biogeographic regionalization of oceanic nekton as a whole. Therefore one may discuss at this stage, apparently, only a regionalization that invalid for deep-dwelling forms and only partially covers sub-abyssal species, i.e. nektogeographic regionalization of the epipelagic zone of the ocean, taking into account both coastal waters and the open sea. It is within this restriction that the nektonic fauna of the ocean are examined in the following pages, where specific features of the individual biogeographic regions are described.

The biogeographic division of the ocean according to the different ecological groups of organisms presents various degrees of difficulty. For instance, in the case of biogeographic division of the ocean for plankton, we are dealing with minimum, and sometimes zero, self-locomotion of the bionts, i.e. with maximum dependence of their domains on the movement of the water (Beklemishev 1969, and others). At the same time we observe here the maximum dependence of the development of forms on temperature, owing to their poikilothermal nature. Therefore in the case of plankton we see the clearest dependence of the geographical ranges of bionts on climatic factors determined by latitude.

On turning from plankton to poikilothermic oceanic nekton additional difficulties emerge owing to these animals' high capacity for independent movement. Though the boundaries of the domains in this case too are predominantly determined by temperature, biogeographic regionalization based on a study of the habitats of poikilothermic nektonic animals cannot be so detailed as that for planktonic animals, since typical distributions of nektonic animals within definite temperature zones do not in practice depend on the system of currents. Parin (1968) reported this for nektonic fishes on the basis of his experience in ichthyogeographic regionalization of the epipelagic zone of the ocean, and Kondakov (1940) and Akimushkin (1963) – for pelagic Cephalopoda.

The distribution of small poikilothermic planktonektonic forms naturally reveals greater dependence on the currents in the ocean and certain other bodies of water and its general characteristics in the ocean closely resemble those of the distribution of plankton. Thus Parin (1968) points out that the distribution of small oceanic fishes, which he describes as 'planktonic' but which according to our classification (Aleyev 1972a, 1973a) are in most cases planktonektonic (Myctophidae, *Cololabis adocetus* and others), is entirely determined by the systems of oceanic currents, representing only a particular case of a plankton distribution. The same is undoubtedly true of the planktonektonic forms among the Sagittoidea and Cephalopoda. This circumstance demonstrates one of the aspects of the planktonektonic complex which may provide an additional criterion for distinguishing planktonekton from eunekton.

Biogeographic regionalization is still more complicated for homothermal nektonic animals, whose extant oceanic representatives include Sphenisciformes, Pinnipedia, Cetacea and Sirenia. A high capacity for independent movement is augmented here by the relative eurythermism of these nektonic animals, on the whole always more pronounced than in poikilothermic organisms, owing to which the spread of homothermal nektonic organisms in the ocean depends comparatively little on temperature. This is expressed primarily by the greater size of their habitats and the higher incidence of cosmopolitanism. The entire set of homothermal nekton, represented by benthonekton, xeronekton and eunekton, is subdivided in this respect into two groups. The first comprises nekton ecologically connected with a solid substratum, i.e. benthonekton (*Dugong, Eschrichtius*) and xeronekton (Sphenisciformes, Pinnipedia, and from Sirenia, *Trichechus* and *Hydrodamalis*); these forms preserve in great measure a predilection for definite climatic zones, determined by the presence of ecological connections with definite areas of the littoral region (Sirenia, *Eschrichtius*) or sea-shore (Sphenisciformes, Pinnipedia). The second group comprises eunektonic forms, consisting of eunektonic Cetacea; on the whole these are characterized by a lesser predilection for definite climatic zones and the highest percentage of cosmopolitan species. Thus, of the 22 eunektonic Cetacea species occurring in the seas of the Soviet Union, ten, i.e. more than 45% of all the forms,[1] are cosmopolitan or nearly so. Among the factors determining the geographic domain of cetaceans, apparently one of the first places goes to the distribution of food organisms, whereas temperature has a smaller direct bearing on their distribution, its effect, according to every indication, being felt primarily as a factor determining the boundaries of suitable feeding grounds. The predilection of, for example, *Physeter* for the tropics is obviously connected with the presence in these latitudes of the bulk of Teuthoidea above all (Tomilin 1957). No less obvious is the predilection of Mystacoceti, which is a straining apparatus for catching food, for arcto-boreal and antarcto-notalian waters with their higher numbers of planktonic invertebrates and pelagic fishes (Tomilin 1970, and others). The fish-eating Odontoceti, which are predators of the pursuing type, reveal on the whole the least predilection for definite climatic zones and furnish the most vivid examples of complete cosmopolitanism (*Delphinus delphis, Globicephalus melas, Orcinus orca,* etc.).

Taking all the above into account, it is advisable to examine biogeographic regionalization as applied to contemporary oceanic nekton only within the

[1] Calculated by the author from data by Tomilin (1957).

framework of a system of units of the largest biogeographical rank, i.e. proceeding from the basic generally accepted (Parin 1968; Beklemishev 1969) pelagic biogeographic areas the arcto-boreal, tropical and antarcto-notalian regions.

Comparing the geographical distribution of nekton belonging to most diverse systematical groups, we find a coincidence of the areas they inhabit at the borders of the tropical region with the arcto-boreal and antarcto-notalian regions. For example, the northern and southern boundaries of the area inhabited by the flying fish of the Exocoetidae family and the sea turtles of the family Cheloniidae practically coincide; the same boundaries in the northern and southern hemispheres determine the bipolar distribution of the squids of the Gonatidae family, and also of penguins (Sphenisciformes) in the southern hemisphere, which confirms the objective significance of these boundaries as nektogeographic divisions (Figure 2). We accept, following Parin (1968), as the specific northern and southern boundaries of the tropical region, separating it from the arcto-boreal and antarcto-notalian regions, the northern and southern boundaries of the area inhabited by the fish *Hirundichthys rondeletii* (Figure 2).

Below, when describing the non-deep-water nekton of individual biogeographical regions of the ocean, we use data pertaining predominantly to those forms and groups for whose geographic distribution our data are most complete. We shall concern ourselves mainly with the general features of these regions, principally the incidence of large groups, and to a lesser degree with individual forms of nektonic animal. It goes without saying that it is impossible to enumerate in the present work all the nektonic animals of any region and we therefore have to discuss typical examples only, mentioning only the most characteristic groups and forms. When describing the geographic distribution of individual nektonic species and groups we use data contained in the following faunistic works, which are in most cases not cited again to avoid cluttering up the text: Günther 1877; Verrill 1879–1882; Allen 1880; Jordan & Evermann 1896–1900; Jordan, Tanaka & Snyder 1913; Jordan, Evermann & Clark 1930; Chun 1900, 1910, 1913; Weber & Beaufort 1911, 1913, 1916, 1922, 1929, 1931, 1936, 1940, 1951; Murray & Hjort 1912; Murray-Levick 1914, 1915; Naef 1923; Dammerman 1924; Evermann & Shaw 1927; Foerste 1928; Fowler 1928, 1936, 1938; McCulloch 1929; Murphy 1915, 1936; Normann 1935, 1937; Thiel 1938; Kondakov 1940, 1941, 1948; Whitley 1940, 1956; Hilderbrand 1946; Nybelin 1947, 1951; Thomson 1947; Bigelow & Schroeder 1948, 1953a,b; Fraser-Brunner 1950; Svetovidov 1948, 1952; Berlioz 1950b; Schmidt 1950; Pierce 1951, 1953, 1958, 1962; Tokioka 1952, 1956, 1959, 1965a,b; Vannucci & Hosoe 1952; Ekman 1953; Herre 1953; Andriashev 1954, 1964, 1968, 1970, 1971; Bourdelle & Grasse 1955; David 1955, 1958; Frechkop 1955; Okada 1955; Petit 1955; Suares Caabro 1955; Suares Caabro & Madruga 1960; Tortonese 1956; Bieri 1957; Furnestin 1957, 1959, 1962a,b; Tomilin 1957; N. R. Marshall 1958; T. C. Marshall 1965; Scheffer 1958; Lindberg & Legeza 1959, 1965; Lindberg & Krasyukova 1969; Gosline & Brock 1960; McAllister 1960; Norris & Prescott 1961; Smith 1961; Smith & Smith 1963; Terentyev 1961; Akimushkin 1963, 1970;

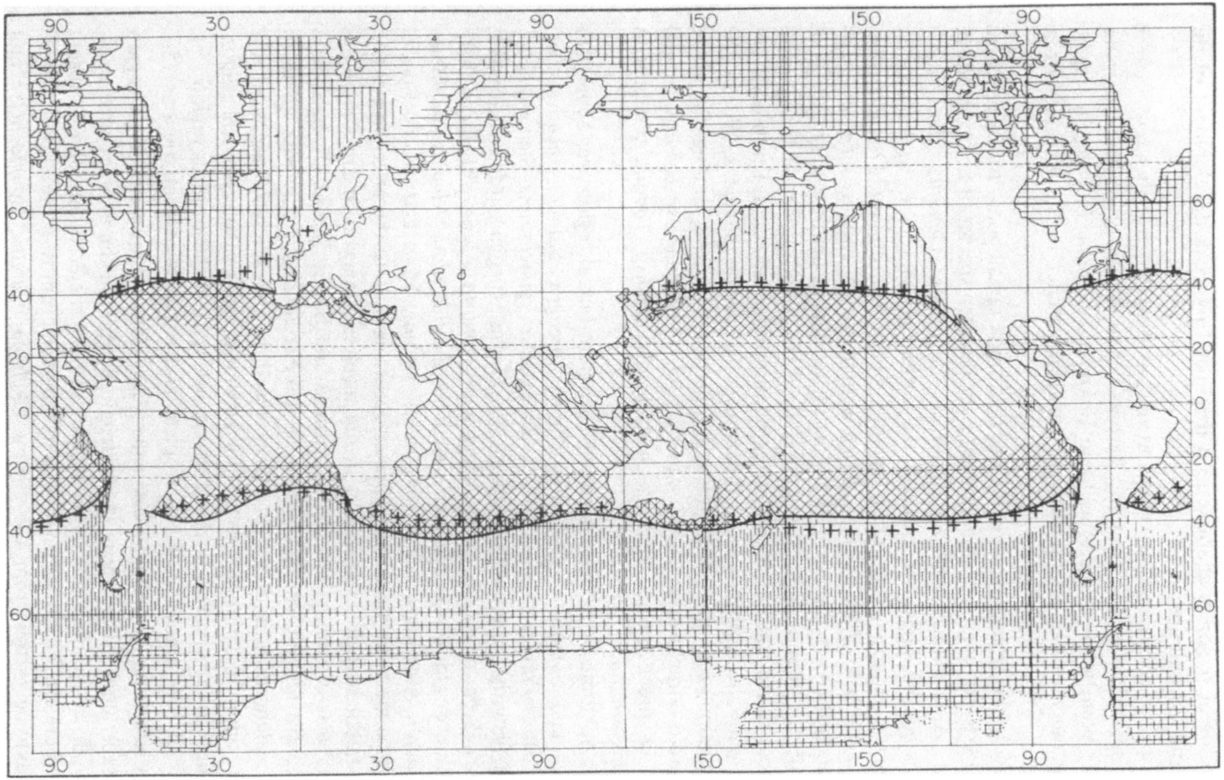

Figure 2. Geographic distribution of some nektonic animals in the ocean. ////, *Hirundichthys rondeletii* (Val.) (Parin 1968); \\\\\\\\, Exocoetidae (Parin 1968); ++, Cheloniidae (Terentyev 1961); ≡≡≡, *Boreogadus saida* (Lepechin) (northern hemisphere; Svetovidov 1948) and *Pleuragramma antarciticum* Boul. (southern hemisphere; Andriashev 1964); ||||, *Gonatidae* (Clarke 1966; Nesis 1973; ¦¦¦, Sphenisciformes (Bobrinsky 1951). Heavy lines, northern and southern boundaries of area where *Hirundichthys rondeletti* (Val.) is found, taken as the borders between the tropical and the arcto-boreal and antarcto-notalian biogeographic regions respectively.

Alvariño 1963, 1966; Chapski 1963; King 1964; Clarke 1966; Nishimura 1968; Parin 1968; Pilleri 1972; Nesis 1973, 1974.

The arcto-boreal region is characterized first of all by weak development of the Sagittoidea and Cephalopoda.

Of the 58 species of Sagittoidea so far known (Tokioka 1965a), only a few occur in the northern seas of the USSR, from the Barents Sea to the Sea of Chukotka; among these we may point to the planktonektonic *Eukrohnia hamata, Parasagitta elegans* and *Flaccisagitta maxima.* Sagittoidea are somewhat richer in the warmer boreal regions.

Equally few in the arcto-boreal region are nektonic Cephalopoda. In the northern seas of the USSR, for example, there occur fewer than ten species of nektonic squid, in particular *Architeuthis princeps, Todarodes sagittatus, Gonatus fabricii* and *Onychoteuthis banksii.* These are joined in the warmer boreal regions by *Moroteuthis robusta, Gonatus magister, Gonatopsis borealis, Galiteuthis armata, Ommastrephes caroli, Illex illecebrosus, Illex coindeti, Brachioteuthis riissei, Taonius megalops, Taonius pavo* and others. In the north-west part of the Pacific, about twenty species of Teuthoidea occur, amounting to less than 10% of the total number of this group's known species; at the same time some nektonic representatives of Teuthoidea are found in quantity here, including, for example, *Todarodes pacificus.* Within the region in question, one may distinguish Teuthoidea belonging among the more cold-loving arcto-boreal forms, like *Gonatus fabricii* and *Gonatopsis borealis,* as well as more widely spread species, such as *Onychoteuthis banksii, Todarodes sagittatus* and *Galiteuthis armata.*

Among the nektonic fishes characteristic of the arcto-boreal region are: *Somniosus microcephalus* and *Somniosus pacificus, Clupea harengus, Salmo, Oncorhynchus, Mallotus villosus, Paralepis coregonoides, Myctophum punctatum, Stenobrachius, Gadidae* (particularly *Boreogadus, Arctogadus, Gadus, Eleginus, Melanogrammus* and *Theragra*), *Sebastes marinus* and *Anoplopoma fimbria.* Along with these there occur many more widely spread forms, such as *Carcharodon, Alopias vulphinus, Sphyrna zygaena, Alepisaurus, Mugil cephalus, Xiphias, Mola* and others. Special mention must be made of the bipolar species occurring in both the arcto-boreal and the antarcto-notalian region but absent in the tropics; among them are *Cetorhinus maximus, Anotopterus pharao, Scomberesox saurus* and *Brama brama.* Of the species with a bipolar distribution we ought to name also *Sardinops sagax,* whose northern forms (*S.s. melanosticta* and *S.s. coerulea*) dwell in the southern areas of the arcto-boreal region and partially in the northernmost areas of the tropical region, while the southern forms dwell in the southermost parts of the tropical region (*S.s. ocellata*) and at the boundary of the tropical and antarcto-notalian regions (*S.s. sagax* and *S.s. neopilchardus*). On the whole the ichthyofauna of the arcto-boreal region are incomparably poorer than those of the tropical region in the number of species.

Absolutely absent in the arcto-boreal region are nektonic Aves and almost absent are Reptilia; there are only occasional visits to the southernmost (boreal) parts of this region by nektonic turtles from the genera *Chelonia, Caretta, Eretmochelys* and

Dermochelys apart from those by sea snakes of the Hydrophidae group to the southernmost parts of the arcto-boreal region in the Sea of Japan.

The mammals are represented among the nekton of the arcto-boreal region by many pinnipeds and cetaceans and one species of siren, now extinct. Of the pinnipeds, particularly characteristic are *Eumetopias, Callorhinus, Odobenus, Erignatus, Phoca vitulina* and *Phoca hispida, Pagophoca, Histriophoca* and *Cystophoca.* Among the Cetacea limited to the arcto-boreal region are *Balaena, Eschrichtius, Hyperoodon ampullatus, Berardius bairdii, Mesoplodon bidens, Lissodelphis borealis, Lagenorhynchus acutus, Lagenorhynchus albirostris, Phocoenoides dalli, Phocoena phocoena, Delphinapterus* and *Monodon.* Besides, common in the arcto-boreal region are species of cetaceans with a very extensive, cosmopolitan range, particularly such genera as *Balaenoptera, Megaptera, Physeter, Ziphius, Delphinus, Tursiops, Globicephalus, Pseudorca* and *Orcinus.* Of the sirens the endemic *Hydrodamalis gigas* used to inhabit the northern Pacific before it became extinct in the 18th century.

The tropical region is characterized by a general diversity of nekton both as regards the large groups represented and as regards the number of species.

Very rich in the tropics are the nekton among the Sagittoidea and Cephalopoda. It is here also that we find the great majority of Sagittoidea. The Indo-Pacific waters are the centre of this group's range (Tokioka 1965a,b): 51 out of the 58 known Sagittoidea species, which amount to about 88%, occur in the Indo-Pacific region.

No less characteristic of the tropical region are nektonic Cephalopoda. Between the latitudes 40°N and 40°S the Cephalopoda group as a whole reaches its greatest numbers, both with regard to the number of species, 88% of the species of this group being present, and also with regard to the numbers within each species (Kondakov 1940; Akimushkin 1970), which is equally true of nektonic Teuthoidea. The abundant teuthifauna of the tropics attract teuthiphages from other systematic groups of nekton, particularly Cetacea (see below). Among the groups and forms of Cephalopoda characteristic of the tropical nekton one should note the Loliginidae and most of the Ommastrephidae (*Ommastrephes bartrami, Ommastrephes pteropus, Symplectoteuthis oualaniensis, Ornithoteuthis antillarum, Todaropsis eblanae, Todarodes sagittatus,* etc.), Onychoteuthidae (*Onychoteuthis banksii, Chaunoteuthis mollis, Onykia carribaea, Tetronychoteuthis dussumieri,* etc.), Brachioteuthidae (*Brachioteuthis riissei*), Bathyteuthidae (*Bathyteuthis abissicola*), Thysanoteuthidae, Cycloteuthidae, Enoploteuthidae (*Abralia, Abraliopsis, Enoploteuthis, Pterygioteuthis, Pyroteuthis, Ancistrocheirus, Thelidioteuthis*), Lycoteuthidae, Chiroteuthidae, Lepidoteuthidae, Grimalditeuthidae, etc.

The nektonic fishes of the tropics are extremely varied. One may point out a whole range of entirely tropical groups, of which the most characteristic are the Rhincodontidae, Mobulidae, Megalopidae, Elopidae Albulidae, Chanidae, Dussumeriidae, Bregmacerotidae, Exocoedae, Pomatomidae, Coryphaenidae, Istiophoridae, Nomeidae, Lutianidae, Pomadasyidae, Chaetodontidae, Echeneidae, Acanthuridae, etc. Almost entirely tropical are the Gempylidae, Trichiuridae, Scombridae,

Xiphiidae, Carangidae, Sciaenidae, Serranidae, Zeidae, Balistidae and Molidae; only individual representatives of these groups occur in summer in the warmest areas of the arcto-boreal and antarcto-notalian regions. A characteristic element of the oceanic nekton of the tropical region is constituted by Clupeidae, most of whom are tropical sea fishes (Svetovidov 1952). Of the individual genera and species of tropical fishes outside the above-mentioned groups, one should note in particular *Pterolamiops longimanus, Isistius brasiliensis, Prionace glauca, Carcharodon carcharias, Galeocerdo cuvieri, Carcharhinus falciformes, Sphyrna zygaena, Euprotomicrus bispinatus, Thrissocles, Stolephorus, Anchoa, Cetengraulis, Engraulis ringens, Oxyporhamphus micropterus, Euleptorhamphus viridis, Cololabis adocetus, Sphyraena barracuda*, etc.

Also most numerous in the tropics, as compared with other regions, are nektonic reptiles. Here occur all the Chelonioidea, Dermochelyoidea, Acrochordinae and Hydrophidae (Figure 2). Representatives of the Chelonioidea (with the exception of *Lepidochelys*), Dermochelyoidea and Hydrophidae occasionally occur beyond the tropical region, but only in the warmest parts of the arcto-boreal and antarcto-notalian regions, directly adjacent to it. The range of the Acrochordinae is restricted to the waters of the north-eastern part of the Indian Ocean and the Indo-Malayan Archipelago, i.e. is entirely within the tropical region.

Nektonic Aves are occasionally represented in the tropical region by species of Sphenisciformes. On the whole this group is characteristic of the antarcto-notalian region, though some species are found near the coast of South Africa, i.e. within the tropical region as well as near the coast of South America, arising along its Pacific shores up to the equator itself, where one species, *Spheniscus mendiculus*, dwells near the Galapagos Islands (Figure 2).

Of the nektonic Mammalia we find in the tropical region Pinnipedia, Cetacea and Sirenia. Pinnipedia are on the whole characteristic of the arcto-boreal and antarcto-notalian regions though, only a few forms occurring in the tropics. Most characteristic of the tropical region are *Monachus* and *Mirounga*, of the Phocidae, and *Zalophus*, of the Otariidae. In addition, *Arctocephalus australis* dwells in the notalian and part of the tropical waters around South America (northwards up to Rio-de-Janeiro in the Atlantic), and also around the Galapagos Islands.[1] In the northernmost areas of the tropical region we meet also *Eumetopias, Callorhinus* and *Phoca vitulina*, and in the southernmost regions *Otaria* (*O. jubata*, along the eastern coast of the Pacific north of Peru), *Arctocephalus* and *Hydrurga*. The cetaceans characteristic of the tropical region include *Sotalia, Steno, Neomeris, Grampus* and *Kogia;* these do not occur outside the tropics, without exception. The tropics are the habitat too of representatives of the more widespread forms of Cetacea, many of whom are cosmopolitans; we may note among these froms *Balaenoptera, Megaptera, Physeter, Ziphius, Delphinus, Stenella caeruleoalbus, Tursiops, Globicephalus, Pseudorca* and *Orcinus*. Besides, the bipolar *Eubalaena*, absent in equatorial waters, occurs in the

[1] Quite possibly, it is not *A. australis* but the endemic *A. galapagoensis* that dwells near the Galapagos Islands (Frechkop 1955).

northern and southern parts of the tropical region. Of the Sirenia, only *Dugong* and *Trichechus* dwell exclusively in the tropical region.

The antarcto-notalian region, like the arcto-boreal region is distinguished on the whole by a comparatively poor selection of Sagittoidea and Teuthoidea.

Of the 58 known Sagittoidea species only nine appear in antarctic waters (Tokioka 1965b). Of the planktonektonic species we may note here, particularly, *Caecosagitta macrocephala, Eukrohnia hamata, Flaccisagitta maxima* and *Flaccisagitta gazellae*, the latter being found exclusively in the antarcto-notalian region.

The Teuthoidea are also fewer here, the number of species diminishing from the equator southwards: in the high latitudes of the Antarctic the entire Cephalopoda group is represented by no more than about twenty species, while nektonic Teuthoidea are still fewer. Noteworthy among the characteristic antarcto-notalian Teuthoidea are *Moroteuthis ingens, Kondakovia, Gonatus antarcticus, Psychroteuthis glacialis, Alluroteuthis antarcticus, Mesonychoteuthis hamiltoni*, Batoteuthidae (*Batoteuthis*), as well as the more widespread *Onychoteuthis banksii, Bathytheuthis abissicola, Taonius pavo, Galiteuthis armata* and *Brachioteuthis riissei*, all of which have a predilection for this region (Figure 2).

Nektonic fishes of the antarcto-notalian region are comparatively few as regards the number of species and have been inadequately studied so far. Noteworthy among them are *Clupea bentincki, Clupea fuegensis*,[1] three species of *Sprattus* (*S. bassensis, S. muelleri* and *S. arcuatus*), *Ethmidium chilcae*, two species of *Engraulis* (*E. australis* and *E. anchoita*), *Notolepis coatsi*, three species of *Merluccius* (*M. hubbsi, M. gayi* and *M. australis*), *Paradiplospinus antarcticus, Gasterochisma*, some pelagic forms of Nototheniidae (*Pagothenia borchgrevinki, Pagothenia brachysoma, Trematomus newnesi, Dissostichus, Pleuragramma* and others) and Chaenichthyidae (*Cryodraco antarcticus, Pagetopsis macropterus, Neopagetopsis ionah, Champsocephalus gunnari, Pseudochaenichthys georgianus*, etc.). Worth noting among the high-latitude dwellers is the cryophilic benthonektonic *Pagothenia borchgrevinki*, ecologically connected with immersed surfaces of ice, and similar in this respect to the arctic *Boreogadus* (Figure 2); the young of *Pagothenia* make use of crevices and cavities in the ice as a refuge in which to hide when danger approaches (Andriashev 1968, 1970, 1971). The warm regions of the antarcto-notalian region are visited by the more widespread fishes from the tropical region. These include, for example, *Xiphias* and *Allothunnus falli*, common near the coast of Chile. Also present in the antarcto-notalian region, as in the arcto-boreal region, are fishes with bipolar habitats: dwelling here, for example, are the southern populations of *Cetorhinus maximus, Anotopterus pharao, Scomberesox saurus* and *Brama brama*.

There are no nektonic reptiles in the antarcto-notalian region with the exception of rare visits by nektonic turtles from the groups Chelonioidea and Dermochelyoidea to the northernmost parts.

[1] *C. bentincki* and *C. fuegensis* may possibly belong to the genus *Sprattus*, or to a separate, new genus (Svetovidov 1949, 1952).

Nektonic Aves are represented here by numerous Sphenisciformes species, this, indeed, being the centre of their range.

Nektonic mammals are represented in the antarcto-notalian region by pinnipeds and cetaceans. Characteristic pinnipeds of this region are *Otaria, Arctocephalus, Lobodon, Leptonychotes, Hydrurga, Ommatophoca* and *Mirounga leonina*. Of the cetaceans, *Hyperoodon planifrons, Berardius arnouxii, Mesoplodon grayi, Lissodelphis peronii* and *Cephalorhynchus* all have a predilection for antarcto-notalian waters; along with these forms, equally common here are the very widespread species, such as *Eubalaena glacialis, Balaenoptera, Megaptera, Physeter, Ziphius, Delphinus, Tursiops, Globicephalus, Pseudorca* and *Orcinus*.

A biogeographic characterization of the nekton of inland bodies of water, like that of the nekton of the ocean, may be given at present only in units of the largest biogeographical rank, i.e. in terms of the generally accepted divisions of the continents into ecological regions (Bobrinsky 1946): the arctogean, paleogean, neogean and notogean regions (Figure 3). The extant nekton of inland waters include members of only four classes of animals: the Chondrichthyes, Osteichthyes, Reptilia and Mammalia. Since nektonic Chondrichthyes, Reptilia and Mammalia are rather rare in inland waters and do not decisively affect the general aspects of nekton anywhere, the characteristics of the nekton of individual regions are in most cases almost entirely restricted to those of the nekton of the Osteichthyes, which, as regards the present investigation, are of limited interest only. A more general and important aspect of this matter is the fact that the extant nekton of inland bodies of water are made up almost without exception of fishes of the class Osteichthyes. Accordingly, below we shall concentrate our attention on the range and distribution of large groups of nektonic animals, giving much less attention to the distribution of small groups and individual representatives. When describing the range of animals in inland waters we draw on the following general and specific works, which will not be cited in every instance (Allen 1880; Day 1885; Jordan & Evermann 1896–1900; Boulenger 1905; Pellegrin 1911, 1933; Günther 1920; Bertin 1948, 1951; Bertin & Arambourge 1958b; Berg 1948, 1949a,b; Hora 1937; Myers 1937; Okada & Matsubara 1938; Hardenberg 1941; Nichols 1943; Nichols & Griscom 1917; Monod 1950; Poll 1950; Blanc 1954; Bourdelle & Grasse 1955; Frechkop 1955; Petit 1955; Tomilin 1957; Scheffer 1958; Chapski 1963; King 1964; Drozdov 1969).

The nekton of the inland waters of arctogean land masses consist of Chondrichthyes, Osteichthyes and Mammalia.

Of the Chondrichthyes we note the sharks that enter rivers (the Mississippi, Euphrates and Indus), such as *Carcharhinus leucas, Galeocerdo cuvieri* and *Negaprion brevirostris*. Characteristic of the nektonic Osteichthyes here are the Lepisosteiformes, Amiiformes, Acipenseriformes, *Alosa, Clupeonella, Signalosa*, Salmonidae, Thymallidae, Esocidae, Umbridae, Dalliidae, Hiodontidae, Cyprinidae, Catostomidae, Centrarchidae and Percidae (Figure 3).

The mammals represented among the arctogean nekton include pinnipeds, cetaceans and sirens, but the most characteristic of them are the pinnipeds. Very

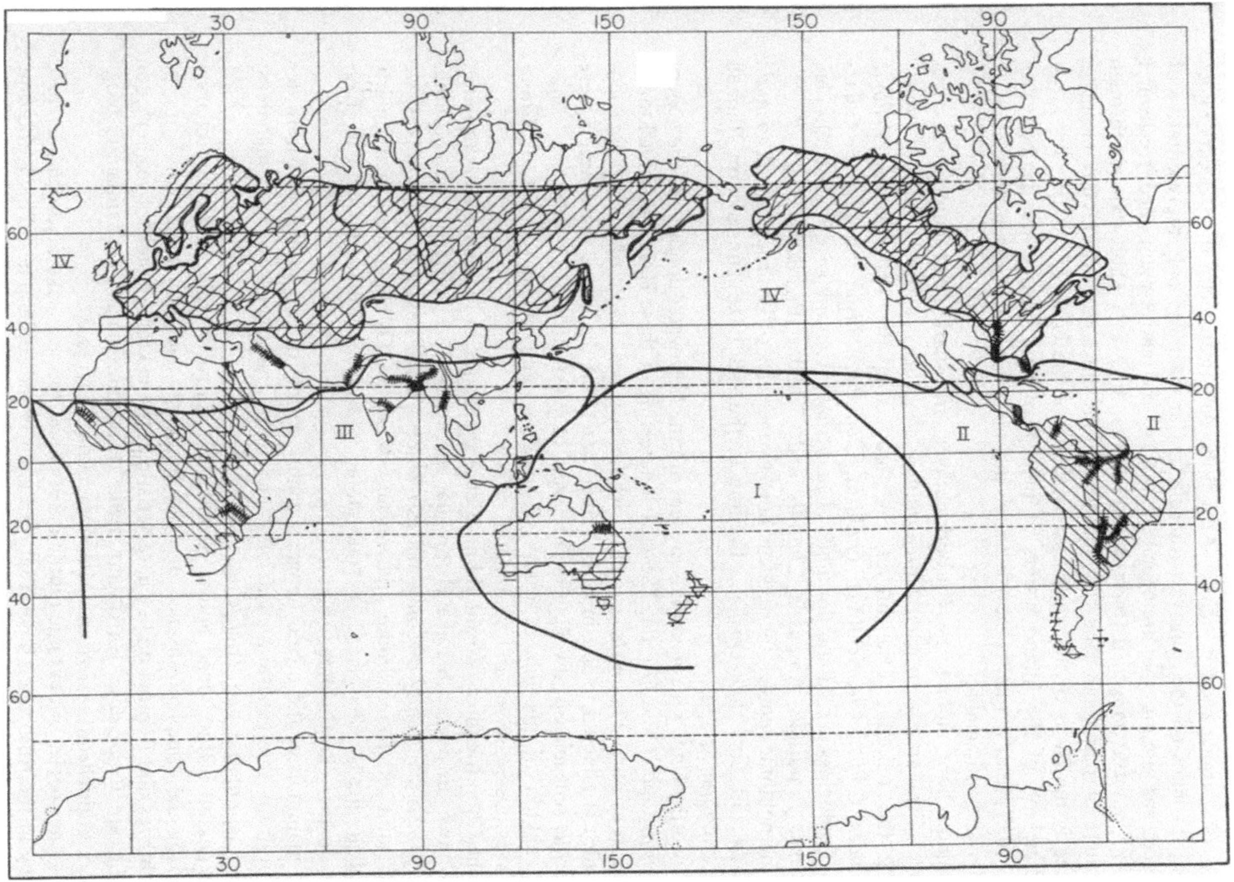

Figure 3. Geographic distribution of some nektonic animals in inland waters. Heavy lines, borders of zoogeographic regions, (I) notogean, (II) neogean, (III) paleogean and (IV) arctogean (Bobrinsky 1946); ≡, Galaxiidae (Rass 1971, with modifications); \\\, Characidae (Ilyin & Rass 1971); ///, *Esox* (Berg 1948; Spanovskaya 1971); ++++, sharks appearing in fresh water (Scheuring 1929–1930).

common among the pinnipeds here are three species of the genus *Phoca* (*Pusa*): *Ph. hispida*, *Ph. caspica* and *Ph. sibirica*. Also in the American sector there occurs *Phoca vitulina* (in lakes on the Ungava peninsula in Canada). The only representative of the cetaceans among the arctogean nekton is the above-mentioned freshwater dolphin *Platanista gangetica* (Indus), however its range as a whole extends through the waters of paleogean land. Of the sirens one must note *Trichechus manatus*, dwelling in the rivers of the south-eastern states of the USA.

The nekton of the inland waters of paleogean land masses are distinctive in having the greatest qualitative diversity, including Chondrichthyes, Osteichthyes, Reptilia and Mammalia.

Common here among the Chrondrichthyes are sharks that enter rivers, for instance *Carcharhinus leucas*, *Carcharhinus gangeticus* and *Galeocerdo cuvieri*. Among the nektonic Osteichthyes one may note the Pellonulinae (*Pellona*, *Clupeoides*, *Corica*, *Clupeichthys*), *Gudusia*, *Gonialosa*, *Ilisha macrogaster*, Galaxiidae, Osteoglossidae, Pantodontidae, Characidae, Cyprinidae, Phallostethoidei, *Polynemus paradiseus*, *Ambassis*, *Chandra* and Cichlidae (Figure 3).

Of the nektonic reptiles there occur here two species of Hydrophidae: *Laticauda laticauda* and *Laticauda crocheri*, dwelling in the salt Lake Tungano on Rannell Island, Solomon Islands.

Noteworthy among the mammals present are the freshwater dolphins of the genera *Platanista*, *Lipotes* and *Orcella*. The most interesting of these is the benthonektonic, benthivorous blind *Platanista gangetica*, whose eyes are devoid of a crystalline lens and which is adapted to life in turbid river water. Of the sirens there occurs here *Trichechus senegalensis*, dwelling in the rivers of West Africa.

The nekton of the inland waters of neogean land masses consist of Chondrichthyes, Osteichthyes and Mammalia.

Of the Chondrichthyes we should note here the sharks of the genus *Carcharhinus*, which move up rivers; in Lake Nicaragua, there is even a 'resident' *Carcharhinus nicaraguensis*, constantly dwelling in fresh water. Of the nektonic Osteichthyes, we should point out *Rhinosardina amazonica*, Galaxiidae, Osteoglossidae (particularly *Arapaima gigas*), Characidae, Serrasalmidae, Gasteropelecidae, Poeciliidae, *Cestraeus*, *Joturus* and Cichlidae (Figure 3).

The mammals of the inland neogean waters consist of two species of freshwater dolphin from the genera *Stenodelphis* and *Inia* and two species of manatees: *Trichechus manatus* and *Trichechus inunguis*.

The nekton of the inland bodies of water of notogaean land masses are characterized by the complete absence of Chondrichthyes and Mammalia, consisting of Osteichthyes and Reptilia only. Characteristic of the nektonic Osteichthyes here, for example, are *Potamalosa*, *Harengula tawilis* (Hawaian Islands), Galaxiidae, Osteoglossidae, *Aldrichetta forsteri* and Melanotaeniidae (Figure 3).

Only one species of nektonic reptile occurs here: the sea snake *Hydrophis semperi*, dwelling in the freshwater Lake Taal on Luzon Island, Philippines. This is the only case of Hydrophidae adapting to freshwater life.

On the whole the notogaean nekton are distinguished by a general lack of variety, which can most probably be explained by the relatively small size of these inland bodies of water and their complete isolation from the inland waters of Afro-Eurasia and America.

The distribution of nekton in bodies of water is determined by the ecology of the individual nektonic species.

In the ocean benthonektonic species range from the shore level to depths of thousands of metres. Planktonektonic and eunektonic forms spread from the coastal strip to areas far away from any land, and from the surface of the water to the greatest oceanic depths. The greatest depths are reached by planktonektonic and eunektonic fishes and cephalopods. The range air-breathing subaquatic eunektonic species is on the whole restricted to the upper 1000 m of the water; a record with regard to depth of sinking among extant subaquatic eunektonic forms is set, probably by *Physeter*, which is capable (Tomilin 1957, 1962a; Berzin 1971, and others) of going down to a depth of 2000 m. The oceanic xeronektonic forms (Chelonioidea, Dermochelyoidea, Sphenisciformes and Pinnipedia) are confined mainly to the upper 100 m of the water, though individual species descend much deeper. The most eurybathic among extant oceanic xeronektonic forms, for example, are some pinnipeds: it is known, for instance (Kooyman 1966), that *Leptonychotes weddelli* is capable of descending to a depth of 600 m.

Benthonektonic species are as a rule distributed all over the bottom in inland waters, from the shore level to the greatest depths. Planktonektonic and eunektonic species also occur on the whole from the surface of the water to the greatest depths. Xeronektonic forms (in general uncharacteristic of inland bodies of water), as in the ocean, show a predilection for the upper layers of water, however in inland bodies of water, owing to their small size and, particularly, small depth, this specific feature is in most cases not so marked as in the ocean.

The ecological diversity of nekton as a whole is extremely great. By preserving basic nektonic features and remaining dwellers in water, nektonic animals populate the most diverse biotopes and exist in extremely varying conditions: in the ocean and in inland waters, in high latitudes and at the equator, in the sun-drenched surface layers of water and in the pitch-dark oceanic depths and underground water, in hot springs and sea-water at subzero temperatures, in water with a high salt content and the nearly salt-free water in peat marshes, in direct proximity to the shore and in the vast expanses of the oceanic pelagic zone, among floating icebergs and among coral, in crystal-clear oceanic and lake waters and in the very opaque, turbic water of some rivers, in stagnant water and in rushing mountain streams, etc.

It is obvious that such a wide distribution of nektonic animals in the ocean and in inland waters and their existence in such widely differing conditions have become possible owing to the development, in the course of the evolution of individual animal groups, of a complex of most essential nektonic adaptations, the development of which is the foundation of the convergence of nektonic animals belonging to different systematic groups. These adaptations are examined in Part 2 of this book.

PART 2. FUNDAMENTAL NEKTONIC ADAPTATIONS

Adaptation to active propulsion in water undoubtedly had the most fundamental influence on formation of the general features of nektonic animals which characterizes them as a definite ecomorphological type. Nothing else facilitated to such an extent the strong convergence of animals from different, and at times very remote, systematic groups during their transition to the nektonic mode of life. If we arrange the individual groups of adaptations in order of diminishing influence on the general morphological organization of nektonic animals, we obtain the following series.

1. Adaptations associated with maintaining the body suspended in the water and providing forward propulsion.
2. Adaptations associated with camouflage.
3. All other adaptations.

Accordingly, below we concentrate our attention on analysis of the complex of adaptations associated with movement and maintaining the body suspended in the water (chapters III–VI). Adaptations associated with camouflage are also examined in detail (chapter VII). Other adaptations are examined in less detail (chapter VIII) since their influence on the general morphological organization of nektonic animals is, on the whole, much weaker than that of the adaptations mentioned above.

III. MAINTAINING THE BODY SUSPENDED IN THE WATER

1. Statodynamic types of nektonic animals

One of the most characteristic features of nektonic animals, as of aquatic animals in general, is their capacity to keep themselves suspended in the water. This capacity is ensured first of all hydrostatically, by approximation of the animal's mean density to that of the water, either fresh or salt, in which it dwells. The animal's buoyancy Δ, which is the difference between the density ρ_1 of the water and its own mean density ρ (Aleyev 1963a,b),

$$\Delta = \pm(\rho_1 - \rho), \tag{6}$$

may be either negative (when $\rho_1 < \rho$), positive (when $\rho_1 > \rho$) or close to neutral (when $\rho_1 \approx \rho$)[1]. According to which is true, the movements aimed at maintaining the body in a state of suspension will be completely different. We may thus distinguish three fundamentally different types of relationship between the density ρ_1 of the medium that forms the habitat (air or water) and the mean density ρ of the animal. Since it has been shown (Aleyev 1963a; 1965a–c; 1966a,b) that for every particular type of statics there is a specific type of dynamics i.e. there are specific ways of creating vertical forces and specific means of locomotion, the types under investigation should be called statodynamics (Aleyev 1966b).

First statodynamic type: $\Delta < -0.03$. Well developed in this case are morphological adaptations aimed at creating hydrodynamic or aerodynamic lift (swimming and flying animals) or adaptations ensuring locomotion with pressure against a solid substratum (benthic and terrestrial fauna). Belonging to this type are all aerobionts, and of the hydrobionts, all benthic animals dwelling on the surface of the ground and the young of some nektonic animals. Though water is a hypogravitational medium (Korzhuev 1965; Korzhuev & Glazova 1965, 1969), the buoyancy of the majority of benthic forms is strongly negative.

Second statodynamic type: $-0.03 \leqslant \Delta \leqslant +0.03$. In this case special morphological adaptations aimed at creating lift or sinking forces are weakly developed or absent. Belonging to this type are the majority of hydrobionts: all benthic animals dwelling underground as well as the overwhelming majority of planktonic and nektonic forms.

Third statodynamic type: $\Delta > +0.03$. In this case morphological adaptations are usually developed for the purpose of neutralizing the effect of the buoyancy force,

[1] All densities are taken here to be measured relative to that of pure water (i.e. in g/c.c.) and thus are strictly speaking specific gravities.

i.e. aimed at creating sinking forces. Belonging to this type are the adults of some aquatic insects during their stay in the body of the water, as well as some Cetacea (Balaenidae, *Physeter* and others), Pinnipedia and Sphenisciformes when they are in the surface layers of the water (see below) and, possibly, some other aquatic animals.

Since the great majority of nektonic animals are of the second statodynamic type they require as a rule a very insignificant hydrodynamic correction to their buoyancy. The need for a more substantial correction appears only in the young of some nektonic animals, whose buoyancy differs markedly from zero.

2. Density and buoyancy

The static equilibrium of a nektonic animal swimming in water depends on the equality of two forces: the force G of gravity, acting at the centre of gravity c and directed downwards, and the force Q exerted by the hydrostatic pressure of the water, acting at the geometrical centre of the animal, the centre of pressure q, and directed upwards (Figure 4). When $G = Q$ the animal is neutrally buoyant, when $G > Q$ its buoyancy is negative and when $G < Q$ its buoyancy is positive. The moment of the couple due to Grand Q is very insignificant, since the centre of gravity c and the centre of pressure q are very close together in nektonic animals: the distance between these two points projected onto a horizontal plane comes to no more than 3–5% of the animal's absolute length, as has been reported earlier for fishes (Magnan & Sainte-Lagüe 1929a; Kozyrev 1950) and in the present work for fishes, cetaceans and pinnipeds.

We determined the centre of gravity on the longitudinal axis of the body of nektonic animals in air by piercing them with a steel rod (for inflexible animals) or by placing them on a board with a transverse beam of triangular or circular cross-section underneath (for flexible animals). The centre of pressure was found as the centre of gravity of a model of the animal made of a homogeneous material (wood). The

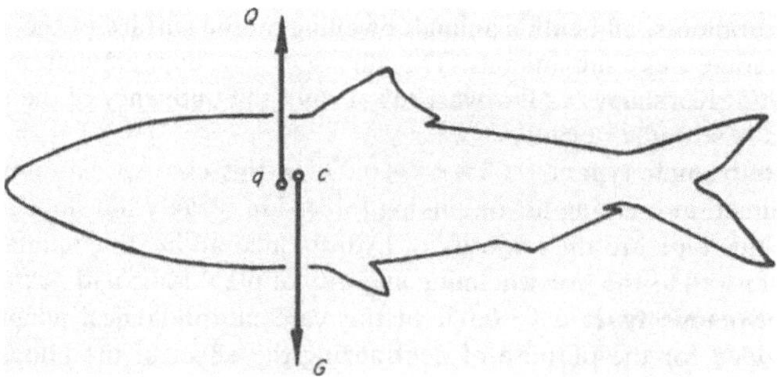

Figure 4. Points of action of forces due to gravity (G) and hydrostatic pressure (Q). c, centre of gravity; q, centre of hydrostatic pressure.

projection of the distance between the points *c* and *q* onto a horizontal plane was 0.4–4.0% for the 28 fish species listed in Table 17. Usually it was 0.5–2.5% of the body's absolute length; in *Delphinus delphis* and *Phocoena phocoena* it was 0.5–2.0% and in *Otaria jubata* 1.0–2.0%.

The data from the following sources on the density and buoyancy of nektonic animals have been obtained by different methods, which prevents them from being comparable: cephalopods (Abel 1916; Denton 1961; Denton, Shaw & Gilpin-Brown 1958; Denton & Gilpin-Brown 1959, 1960, 1961a–c, 1966; Denton, Gilpin-Brown, Howarth 1961; Clarke 1962; Zuyev 1963, 1966; and others), fishes (Moreau 1876; Popta 1910; Magnan & Lariboisiere, 1912; Schmalhausen 1916; Taylor 1921; Magnan & Sainte-Laguë 1929a; Grove & Newell 1939; Tester 1940; Plattner 1941; Stas 1941; Lowndes 1942, 1955, 1956; Andriashev 1944; Black 1948; Affleck 1950; Jones 1951, 1952, 1957; Jones & Marshall 1953; Alexander 1959b,d, 1966a, 1967; Aleyev 1963a,b; Satchell 1966; Bone & Roberts 1969, and others) and mammals (Smirnov 1914, 1929; Ognev 1935; Nikulin 1937; Sleptsov 1940, 1952; A. S. Sokolov 1955; Tikhomirov 1964; Babenko & Morozov 1968; Mordvinov 1969; Morozov & Tomilin 1970 and others). To obtain comparable data we determined the density of nektonic Sagittoidea, Osteichthyes, Reptilia, Aves and Mammalia. The density of Sagittoidea was found with the aid of a series of NaCl solutions of different concentrations and that of all other nektonic animals by the volumetric-weight method (Aleyev 1963a), i.e. by finding the animal's weight and volume (by the method of water displacement). All the density determinations were carried out with freshly killed animals without any fixation.

From the literature we have borrowed data on the density of cephalopods and pinnipeds obtained (Zuyev 1963; Mordvinov 1969) by our methods (Aleyev 1963a), as well as data on the density of pinnipeds obtained (A. S. Sokolov 1955) by similar methods, and also Sleptsov's data (1952) on the density of whales (Table 2).

To assess the density and buoyancy of nektonic animals we frequently resorted to the observation of living Sagittoidea, Chondrichthyes, Osteichthyes, Chelonioidea, Hydrophidae, Sphenisciformes, Pinnipedia and Cetacea.

The determination of the density of nektonic animals which contain gas-filled cavities (the sepion of Sepioidea, the swim-bladder of fishes or the lungs and special air sacs in reptiles, birds and mammals) is considerably more difficult owing to the fact that when the animal sinks from a certain level into deeper water its buoyancy diminishes, since all the volumes of gas get compressed by the increasing external hydrostatic pressure, while, conversely, during ascent its buoyancy increases, as all the volumes of gas expand with the decrease in external pressure. Therefore, if the buoyancy of an animal with gas-filled cavities becomes zero at a certain depth, this means that it is in a state of unstable equilibrium, since its buoyancy inevitably becomes positive when it moves upwards and negative when it sinks (Aleyev 1963a). We determined the buoyancy of fishes with a swim-bladder from specimens caught at depths not greater than 10–15 m, which, within the accepted accuracy of the determination (0.01), practically eliminates the above-mentioned undesirable effect

Table 2. Density ρ and buoyancy Δ of nektonic animals

Species and nektonic types	For vertebrates, length L to end of vertebral column; for invertebrates absolute length L_a (cm) Variation	Mean	Number of specimens examined n	Density of animal ρ (g/c.c.) Variation	Mean	Density of water in which it was caught ρ_1 (g/c.c.)	Buoyancy Δ	Gaseous inclusions in body	Source
1	2	3	4	5	6	7	8	9	10
Phoca caspica Gmel. (XN)	—	90.5	12	0.90–0.99	—	1.01	+0.11	Present	Mordvinov (1969)
Phoca hispida ladogensis Nordq. (XN)	—	—	—	0.9–1.05	—	1.00	+0.10	”	Sokolov, A. S. (1955)
Balaenidae (EN)	—	—	—	—	0.95	1.03	+0.08	”	Sleptsov (1952)
Physeter catodon L. (EN)	—	—	—	—	0.95	1.02	+0.07	”	Sleptsov (1952)
Eudyptes chrysolophus Brandt (XN)	—	61.1	1	—	0.97	1.03	+0.06	”	Author's data
Delphinus delphis ponticus Barab. (EN)	120.0–168.5	145.4	4	0.96–1.00	0.97	1.01	+0.04	”	”
Phocoena phocoena (L.) (EN)	101.0–128.8	114.3	6	0.97–1.01	0.98	1.01	+0.03	”	”
Balaenopteridae (EN)	—	—	—	—	1.00	1.03	+0.03	”	Sleptsov (1952)
Squalus acanthias (L.) (EN)	124.1–130.0	126.2	6	1.01–1.02	1.01	1.01	0.00	Absent	Author's data
Acipenser güldenstädti colchicus V. Marti (BN)	89.5–109.9	94.0	6	1.01–1.02	1.01	1.01	0.00	Present	”
Sprattus sprattus phalericus (Risso) (PN/EN)	7.1–9.0	8.3	10	1.01–1.02	1.01	1.01	0.00	”	”
Clupeonella delicatula delicatula (Nordm.) (PN/EN)	6.0–8.2	6.8	10	1.01–1.02	1.01	1.01	0.00	”	”
Alosa kessleri pontica (Eichw.) (EN)	18.0–20.8	20.1	10	1.01–1.02	1.01	1.01	0.00	”	”

Leuciscus cephalus (L.) (BN)	22.0–27.1	25.4	10	1.00–1.01	1.00	1.00	0.00	”	”
Chalcalburnus chalcoides derjugini (Berg) (EN)	8.4–14.1	10.7	10	1.00–1.01	1.00	1.00	0.00	”	”
Alburnoides bipunctatus fasciatus (Nordm.) (EN)	9.9–10.8	10.1	10	1.00–1.01	1.00	1.00	0.00	”	”
Vimba vimba tenella (Nordm.) (BN)	10.7–14.8	13.5	10	1.00–1.01	1.00	1.00	0.00	”	”
Barbus tauricus Kessl. (BN)	19.2–24.7	21.8	10	1.00–1.01	1.00	1.00	0.00	Present	Author's data
Cyprinus carpio carpio L. (BN)	27.3–32.5	30.3	9	1.00–1.01	1.00	1.00	0.00	”	”
Carassius carassius (L.) (BN)	12.7–13.6	12.9	10	1.00–1.01	1.00	1.00	0.00	”	”
Anguilla anguilla (L.) (EN)	92.0–110.0	99.0	4	1.01–1.02	1.01	1.01	0.00	”	”
Belone belone euxini Günth. (EN)	31.2–46.0	33.8	10	1.00–1.02	1.01	1.01	0.00	”	”
Odontogadus merlangus euxinus (Nordm.) (BN)	15.3–17.5	16.6	10	1.01–1.02	1.01	1.01	0.00	”	”
Gasterosteus aculeatus L. (BN)	4.5–5.9	5.5	10	1.00–1.02	1.01	1.01	0.00	”	”
Gambusia affinis holbrooki Girard (BN)	2.3–5.1	3.8	10	1.00–1.01	1.00	1.00	0.00	”	”
Atherina mochon pontica Eichw. (EN)	9.1–11.4	10.5	10	1.01–1.02	1.01	1.01	0.00	”	”
Mugil auratus Risso (BN)	28.3–32.0	30.4	10	1.01–1.02	1.01	1.01	0.00	”	”
Sphyraena barracuda (Walb.) (EN)	79.0–99.6	95.2	3	1.02–1.02	1.02	1.02	0.00	”	”
Serranus scriba (L.) (BN)	11.2–13.5	12.9	10	1.00–1.02	1.01	1.01	0.00	”	”
Percarina demidoffi maeotica Kuzn. (BN)	5.6–8.5	7.0	7	1.01–1.02	1.01	1.01	0.00	Present	Author's data
Stizostedion lucioperca (L.) (BN)	45.0–55.1	49.4	8	1.01–1.02	1.01	1.01	0.00	”	”
Spicara smaris (L.) (EN)	11.8–14.5	13.1	10	1.01–1.02	1.01	1.01	0.00	”	”
Trachurus mediterraneus ponticus Aleev (EN)	38.7–50.1	43.0	10	1.01–1.02	1.01	1.01	0.00	”	”

Table 2 (contd.)

Species and nektonic types	For vertebrates, length L to end of vertebral column; for invertebrates absolute length L_a (cm)		Number of specimens examined n	Density of animal ρ (g/c.c.)		Density of water in which it was caught ρ_1 (g/c.c.)	Buoyancy Δ	Gaseous inclusions in body	Source
	Variation	Mean		Variation	Mean				
1	2	3	4	5	6	7	8	9	10
Diplodus annularis (L.) (BN)	6.2–8.9	8.0	10	1.00–1.02	1.01	1.01	0.00	,,	,,
Umbrina cirrosa (L.) (BN)	30.5–36.7	32.4	10	1.01–1.02	1.01	1.01	0.00	,,	,,
Chromis chromis (L.) (BN)	6.7–8.9	8.0	10	1.01–1.02	1.01	1.01	0.00	,,	,,
Crenilabrus ocellatus Forsk. (BN)	10.1–12.0	11.4	10	1.01–1.02	1.01	1.01	0.00	,,	,,
Crenilabrus tinca (L.) (BN)	12.6–18.1	15.4	10	1.01–1.02	1.01	1.01	0.00	,,	,,
Scomber scombrus (L.) (EN)	28.9–37.0	36.1	10	1.01–1.03	1.02	1.02	0.00	,,	,,
Gobius ophiocephalus Pall. (BN)	12.8–15.2	14.2	10	1.01–1.02	1.01	1.01	0.00	,,	,,
Enhydrina schistosa (Daud.) (EN)	89.5–97.0	92.2	3	1.01–1.03	1.02	1.02	0.00	,,	,,
Caretta caretta (L.) (XN)	—	114.5	1	—	1.02	1.02	0.00	Present	Author's data
Chelonia mydas (L.) (XN)	86.0–102.0	94.0	2	1.01–1.03	1.02	1.02	0.00	,,	,,
Eretmochelys imbricata (L.) (XN)	19.0–19.4	19.2	2	1.01–1.03	1.02	1.02	0.00	,,	,,
Sagitta setosa Müll. (PN)	2.0–2.2	2.1	7	1.02–1.03	1.02	1.01	−0.01	Absent	,,
Oxynotus centrina (L.) (EN)	—	65.0	1	—	1.03	1.02	−0.01	,,	,,
Acipenser stellatus Pall. (BN)	94.3–110.0	100.5	8	1.02–1.03	1.02	1.01	−0.01	Present	,,
Engraulis encrasicholus maeoticus Rus. (EN)	8.5–11.0	10.1	30	1.01–1.03	1.02	1.01	−0.01	Scarce or	,,

Gymnammodytes cicerellus (Raf.) (BN)	9.6–11.2	10.3	10	1.01–1.03	1.02	1.01	−0.01	Absent	”
Crenilabrus quinquemaculatus (Bl.) (BN)	7.0–9.1	8.1	10	1.01–1.03	1.02	1.01	−0.01	Present	Author's data
Pomatomus saltatrix (L.) (EN)	56.6–69.0	64.5	10	1.02–1.03	1.02	1.01	−0.01	Scarce or absent	”
Sarda sarda (B1.) (EN)	60.4–78.9	66.9	10	1.01–1.03	1.02	1.01	−0.01	Absent	”
Auxis thazard (Lac.) (EN)	42.0–46.0	44.2	6	1.02–1.03	1.03	1.02	−0.01	”	”
Remora remora (L.) (EN)	—	55.0	1	—	1.03	1.02	−0.01	Absent	Author's data
Loligo vulgaris Lam. (EN)	25.1–30.0	27.0	6	1.03–1.07	1.05	1.03	−0.02	Absent	Zuyev (1966)
Loligo forbesi (Steen.) (EN)	—	33.0	1	—	1.05	1.03	−0.02	”	”
Loligo edulis Hoyle (EN)	—	12.3	1	—	1.05	1.03	−0.02	”	”
Acroteuthis media (L.) (EN)	—	14.0	1	—	1.05	1.03	−0.02	”	”
Ommastrephes pteropus (Steen.) (EN)	—	14.0	1	—	1.05	1.03	−0.02	”	”
Symplectoteuthis oualaniensis (Less.) (EN)	50.1–55.0	—	3	1.05–1.05	1.05	1.03	−0.02	”	”
Illex coindeti (Ver.) (EN)	25.1–30.0	26.1	3	1.04–1.06	1.05	1.03	−0.02	Absent	Zuyev (1966)
Todarodes sagittatus (Lam.) (EN)	—	50.8	1	—	1.05	1.03	−0.02	”	”
Todarodes pacificus (Steen.) (EN)	—	40.2	1	—	1.05	1.03	−0.02	”	”
Onychoteuthis banksii (Leach) (EN)	—	17.1	1	—	1.05	1.03	−0.02	”	”
Ancistroteuthis lichtensteini (D'Orb.) (EN)	20.1–25.0	—	2	1.04–1.06	1.05	1.03	−0.02	”	”
Gonatus fabricii (Licht.) (EN)	—	24.0	1	—	1.05	1.03	−0.02	”	”
Sphyrna zygaena (L.) (EN)	125.6–220.5	173.0	2	1.03–1.05	1.04	1.02	−0.02	”	Author's data
Mustelus mustelus (L.) (BN)	122.0–128.0	125.0	3	1.03–1.05	1.04	1.02	−0.02	”	”
Engraulis encrasicholus ponticus Alex. (EN)	9.5–10.9	10.3	20	1.03–1.05	1.04	1.01	−0.03	Scarce or absent	”

(Aleyev 1963a). We accept as the mean buoyancy of all nektonic animals with lungs (reptiles, birds and mammals) the buoyancy they possess in the surface layers of the water, where they spend most of their time (Table 2).

To determine the characteristic buoyancy of various species of nektonic animals we used adult, sexually mature specimens caught in a period of maximum fatness, i.e. the values of ρ given in Table 2 are minima, while the values of Δ are maxima. These and values of the buoyancy close to them are characteristic of nektonic animals throughout the greater part of the year and should therefore be considered as normal, whereas noticeable decreases in buoyancy are in most cases relatively short-lived and should be regarded as deviations from the norm. Moreover, such deviations are characteristic almost entirely of bladderless fishes and are observed predominantly at the high-latitude borders of their domains where the thermal regime is no longer conducive to sustaining adequately intensive nourishment throughout the year (Aleyev 1956b, 1958d) and where, because of this, the hydrostatic parameters for fat deposition are upset from their optimal values during some seasons of the year. A reduction in buoyancy in fishes without swim-bladders is observed during intensive fat expenditure in the breeding period (Aleyev 1963a).

On the basis of studies of the bouyancy of benthonektonic, planktonektonic, eunektonic and xeronektonic forms (Table 2), taking into account the variation in buoyancy with age (Table 3) and the season, we may conclude that the density ρ of the nektonic animals examined generally fluctuates from 0.90 to 1.09 and their buoyancy Δ from +0.11 to −0.08. It may be noted that the buoyancy of nektonic animals varies within approximately the range +0.1 to −0.1.

As may be seen from Table 2, in which the species are presented in order of diminishing maximum mean buoyancy Δ, a buoyancy close to neutral is the general rule for nektonic animals: the mean buoyancy Δ of most of the species examined lies between +0.03 and −0.03, thanks to which the expenditure of energy to maintain the body in a state of suspension is reduced to a minimum.

Only nektonic birds (*Eudyptes*) and some nektonic mammals (*Phoca*, Balaenidae, *Physeter* and *Delphinus*), which breathe air and dive more or less deeply in pursuit of food, have $\Delta > +0.03$. This 'safety margin' of positive buoyancy compensates to some extent for the hydrostatic changes experienced by these animals during deep diving.

Reports of high density and negative buoyancy for sharks and Acipenseridae (Schmalhausen 1916; Staš 1941; Lowndes 1955, and others) refer either to the nektobenthic forms (*Scyliorhinus*, etc.) or to young, sexually immature individuals. Adult nektonic sharks and Acipenseridae have neutral or nearly neutral buoyancy (Table 2), while the amount of body fat in *Cetorhinus* is so great (Grove & Newell 1939) that it allows this shark to swim at the surface of the water.

In the majority of nektonic animals changes in buoyancy with age are practically absent or hardly perceptible; this is characteristic of Sagittoidea, Cephalopoda, fishes with a normally developed swim-bladder, reptiles, birds and mammals (Table 3). However the buoyancy of some benthonektonic fishes with a comparatively well developed swim-bladder may be markedly negative when they are young because of the great relative weight of the ossifications of the cutaneous covering (Table 3, *Acipenser*). The buoyancy of fishes with a rudimentary swim-bladder (Table 3; *Pomatomus, Scomber* and *Sarda*) or none at all (Table 3: *Squalus* and *Mustelus*) in youth is, as a rule, negative, but in maturity becomes approximately neutral on account of accumulation of fat (Aleyev 1963a).

Seasonal changes in the buoyancy of most nektonic animals are very slight or practically absent, since the mechanisms controlling the level of buoyancy in the majority of cases do not actually depend on the regime of exogenic feeding (Sagittoidea, Cephalopoda, fishes with a swim-bladder, Reptilia, Aves and Mammalia).

Only bladderless fishes, and those for whom the quantity of gas in the bladder is very small, have a buoyancy which notably diminishes during some seasons of the year (*Gymnammodytes, Scomber, Sarda, Auxis, Pomatomus* and *Engraulis*). This is explained by the body's diminishing fat content, as is well demonstrated by the examples of *Scomber scombrus* and *Sarda sarda*, whose fat content in the period after spawning increases from 2–3% to 20–23% (Dragunov 1950; Tolgay 1957) and whose buoyancy in the same period increases from between −0.07 and −0.05 to −0.01 (Aleyev 1963a). For *Auxis thazard* we have established seasonal changes in buoyancy from −0.05 to −0.01.

3. Modes of ensuring neutral buoyancy

The neutral or nearly neutral buoyancy of nektonic animals is mainly due to the presence in the body of inclusions lighter than water, such as gases and fat, and also hydration of body tissues and diminution of the heavy-element content of the skeleton.

Hydration of tissues and diminution of the heavy-element content of the skeleton as a means of increasing buoyancy is observed in Sagittoidea, which have no rigid skeletal formations, and in Teuthoidea, whose bodies are more than 90% water. It has been reported (Denton, Shaw & Gilpin-Brown 1958) that some Teuthoidea, particularly the planktonektonic *Verrilliteuthis hyperborea* and the eunektonic *Galiteuthis armata*, have cavities in the body occupying up to two-thirds of the animal's entire volume and filled with fluid with a relatively high ammonium and low sodium content, which accounts for its density of about 0.01, i.e. much less than that of sea water. These cavities containing light fluid ensure neutral or nearly neutral buoyancy. Such a mechanism is not really acceptable, however, in mobile eunektonic species and occurs very seldom among them, since reservoirs of light fluid occupy much space and thus reduce the effective volume of the mantle cavity, which has an adverse effect on the locomotion capacity of cephalopods.

Among fishes, hydration of tissues and a lighter skeleton owing to decalcification occur in deep-water species.

Table 3. Variation in density ρ and buoyancy Δ of nektonic animals with age

Species and nektonic type	For fishes, length to end of vertebral column L; for cephalopods, absolute length L_a (cm) Variation	Mean	Number of specimens examined n	Density of animal ρ (g/c.c.) Variation	Mean	Density of water in which it was caught ρ_1 (g/c.c)	Buoyancy Δ	Gaseous inclusions in body	Source
1	2	3	4	5	6	7	8	9	10
Squalus acanthias	19.0–36.0	23.3	3	1.05–1.05	1.05	1.01	−0.04	Absent	Author's data
L. (EN)	40.0–49.6	44.2	10	1.03–1.05	1.04	1.01	−0.03	”	”
	80.0–85.0	83.0	9	1.02–1.04	1.03	1.01	−0.02	”	”
	92.0–109.0	97.0	10	1.01–1.04	1.02	1.01	−0.01	”	”
	124.1–130.0	126.2	6	1.01–1.02	1.01	1.01	0.00	”	”
Acipenser gülden-	31.0–41.6	34.0	10	1.03–1.05	1.04	1.01	−0.03	Present	”
städti colchicus	89.5–109.9	94.0	6	1.01–1.02	1.01	1.01	0.00	”	”
V. Marti (BN)									
Sprattus sprattus	5.0–7.0	5.9	10	1.01–1.02	1.01	1.01	0.00	”	”
phalericus (Risso)	7.1–9.0	8.3	10	1.01–1.02	1.01	1.01	0.00	”	”
(PN/EN)									
Alosa kessleri pon-	9.7–11.9	10.2	10	1.01–1.03	1.01	1.01	0.00	”	”
tica (Eichw.) (EN)	18.0–20.8	20.1	10	1.01–1.02	1.01	1.01	0.00	”	”
Leuciscus cephalus	5.0–6.5	5.9	10	1.00–1.01	1.00	1.00	0.00	”	”
(L.) (BN)	22.0–27.1	25.4	10	1.00–1.01	1.00	1.00	0.00	”	”
Alburnoides bipuncta-	5.6–6.3	6.0	10	1.00–1.01	1.00	1.00	0.00	Present	Author's data
tus fasciatus (Nordm.)	9.9–10.8	10.1	10	1.00–1.01	1.00	1.00	0.00	”	”
(EN)									
Vimba vimba tenella	5.9–7.5	7.0	10	1.00–1.01	1.00	1.00	0.00	”	”
(Nordm.) (BN)	10.7–14.8	13.5	10	1.00–1.01	1.00	1.00	0.00	”	”
Barbus tauricus	3.9–5.1	4.4	10	1.00–1.01	1.00	1.00	0.00	”	”
Kessl. (BN)	19.2–24.7	21.8	10	1.00–1.01	1.00	1.00	0.00	”	”
Odontogadus merlangus	6.0–7.1	6.5	10	1.01–1.02	1.01	1.01	0.00	”	”
euxinus (Nordm.) (BN)	15.3–17.5	16.6	10	1.01–1.02	1.01	1.01	0.00	”	”
Atherina mochon	4.8–7.1	6.4	10	1.01–1.02	1.01	1.01	0.00	”	”
pontica Eichw. (EN)	9.1–11.4	10.5	10	1.01–1.02	1.01	1.01	0.00	”	”
Mugil auratus Risso	2.5–4.0	3.1	10	1.01–1.02	1.01	1.01	0.00	”	”
(BN)	28.3–32.0	30.4	10	1.01–1.02	1.01	1.01	0.00	”	”

Stizostedion lucio-	15.2–20.0	18.4	10	1.01–1.02	1.01	1.01	0.00	”	”
perca (L.) (BN)	45.0–55.1	49.4	8	1.01–1.02	1.01	1.01	0.00	”	”
Spicara smaris	5.1–5.8	5.3	10	1.01–1.02	1.01	1.01	0.00	”	”
(L.) (EN)	11.8–14.5	13.1	10	1.01–1.02	1.01	1.01	0.00	”	”
Trachurus mediterraneus	6.2–7.2	6.8	10	1.01–1.02	1.01	1.01	0.00	”	”
ponticus Aleev (EN)	38.7–50.1	43.0	10	1.01–1.02	1.01	1.01	0.00	”	”
Crenilabrus tinca	4.0–5.4	4.7	10	1.01–1.02	1.02	1.01	−0.01	”	”
(L.) (BN)	12.6–18.1	15.4	10	1.01–1.02	1.01	1.01	0.00	”	”
Sagitta setosa	1.0–1.2	1.1	10	1.02–1.03	1.02	1.01	−0.01	”	”
Müll. (PN)	2.0–2.2	2.1	7	1.02–1.03	1.02	1.01	−0.01	”	”
Acipenser stellatus	—	2.04	1	—	1.06	1.01	−0.05	Present	Author’s data
Pall. (BN)	4.7–8.2	6.4	2	1.05–1.05	1.05	1.01	−0.04	”	”
	31.0–55.0	41.6	4	1.04–1.05	1.04	1.01	−0.03	”	”
	—	75.0	1	—	1.03	1.01	−0.02	”	”
	94.3–110.0	100.5	8	1.02–1.03	1.02	1.01	−0.01	”	”
Scomber scombrus	16.0–17.1	16.6	10	1.05–1.07	1.06	1.01	−0.05	Absent	”
L. (EN)	17.9–19.5	18.7	10	1.03–1.05	1.04	1.01	−0.03	”	”
	20.8–23.0	22.1	10	1.01–1.03	1.02	1.01	−0.01	”	”
Pomatomus saltatrix	4.4–6.9	5.5	20	1.02–1.05	1.03	1.01	−0.02	Present	”
(L.) (EN)	15.0–18.1	16.3	16	1.04–1.07	1.06	1.01	−0.05 }	Present	”
	23.1–26.0	24.5	10	1.02–1.05	1.03	1.01	−0.02 }	or	”
	56.6–69.0	64.5	10	1.02–1.03	1.02	1.01	−0.01 }	absent	”
Sarda sarda (B1.)	16.5–19.2	18.0	10	1.07–1.09	1.08	1.01	−0.07	Absent	”
(EN)	64.4–78.9	66.9	10	1.01–1.03	1.02	1.01	−0.01	”	”
Loligo vulgaris	0.1–5.0	—	50	1.05–1.06	1.06	1.03	−0.03	”	Zuyev (1966)
Lam. (EN)	5.1–10.0	—	12	1.03–1.07	1.05	1.03	−0.02	”	”
	20.1–30.0	—	15	1.03–1.07	1.05	1.03	−0.02	”	”
Symplectoteuthis	0.1–5.0	—	23	1.05–1.07	1.06	1.03	−0.03	”	”
oualaniensis (Less.)	5.1–10.0	—	5	1.03–1.07	1.05	1.03	−0.02	”	”
(EN)	40.1–55.0	—	16	1.03–1.07	1.05	1.03	−0.02	”	”
Illex coindeti	0.1–5.0	—	2	1.04–1.06	1.05	1.03	−0.02	Absent	”
(Ver.) (EN)	20.1–30.0	—	6	1.03–1.07	1.05	1.03	−0.02	”	”
Ancistroteuthis	0.1–5.0	—	9	1.05–1.06	1.06	1.03	−0.03	Absent	Zuyev (1966)
lichtensteini	5.1–15.0	—	11	1.03–1.07	1.05	1.03	−0.02	”	”
(D’Orb.) (EN)	15.1–25.0	—	5	1.04–1.06	1.05	1.03	−0.02	”	”
Mustelus mustelus	—	49.0	1	—	1.06	1.02	−0.04	”	Author’s data
(L.) (BN)	122.0–128.0	125.0	3	1.03–1.05	1.04	1.02	−0.02	”	”

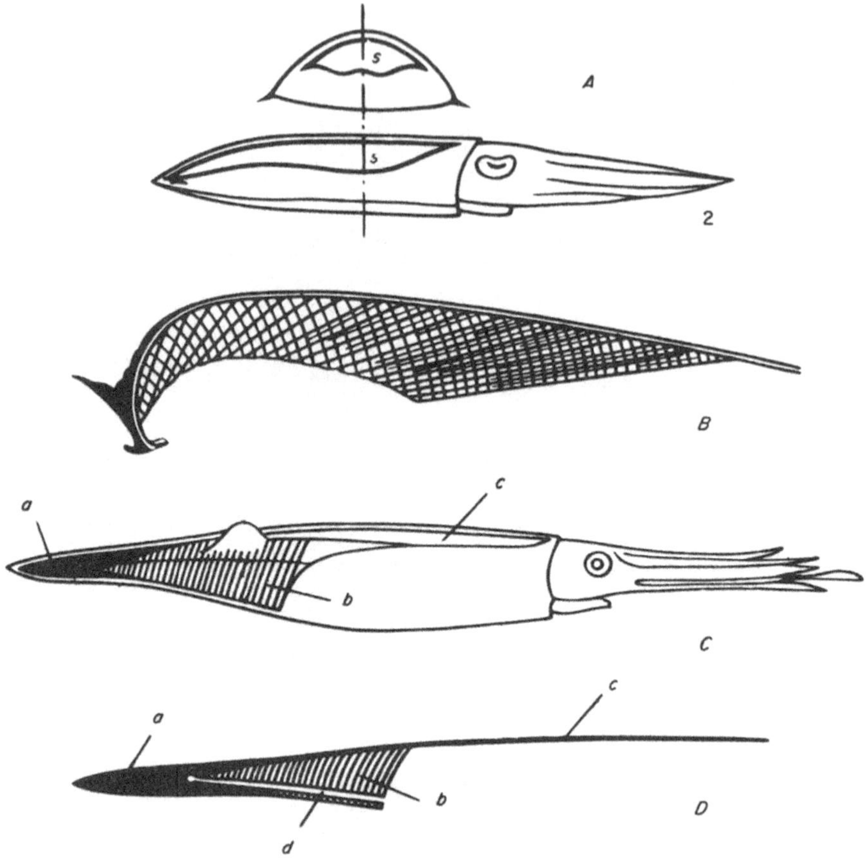

Figure 5. Hydrostatic apparatus of nektonic invertebrates. *A*, *Sepia officinalis* L., (1) cross-section and (2) right-hand view; *B*, longitudinal section of shell of *Sepia orbignyana* Fer. (after Naef 1922); *C*, reconstruction of *Megateuthis gigantea* (Schloth) (after Naef 1922, with modifications); *D*, schematic longitudinal section of shell of *Belemnites* (after Lang 1900; see Kondakov 1940, modified). *a*, roster; *b*, phragmocone; *c*, proostracum; *d*, siphon; *s*, shell.

Many nektonic animals contain **gas-filled cavities**, which likewise help to bring their buoyancy to the neutral level: Belemnoidea, Sepioidea, most fishes, and reptiles, all birds and all mammals.

In Belemnoidea and Sepioidea the gaseous inclusions take the form of gas-filled chambers in the shell (Davitashvili 1949; Krymgolts 1958; Luppov, Kiparisova & Krymgolts 1958; Shimansky 1962). In Sepioidea the rudimentary shell, the sepion (Figures 5A,B), has become a hydrostatic apparatus consisting of a large number of autonomous rigid calcified chambers with semipermeable walls filled with liquid and gas; the total volume of the sepion is about 9% of that of the animal (Denton 1961; Denton & Gilpin-Brown 1960, 1961a,b; Denton, Gilpin-Brown & Howarth 1961). When the external pressure changes the volume of the gas in the chambers changes too since liquid is forced through the semi-permeable walls from the blood stream into the chambers and vice versa, i.e. the sepion holds a variable volume of gas. In extinct Belemnoidea, most of which were nektonic (Abel 1916; Krymgolts 1958), the function of a hydrostatic gas apparatus was fulfilled by the middle part of the inner shell, the phragmocone (Abel 1916), which was divided by septa into individual chambers permeated by a siphon (Figures 5C,D). An examination of the hydrostatic apparatus of extant *Nautilus* (Bidder 1962) has shown that the relative volume of gas and liquid in the chambers is changed with the help of the siphon. Apparently all cephalopods provided with a shell containing gas chambers are capable of osmotically changing the amount of gas in these chambers (Joysey 1961), thus controlling their buoyancy. In Belemnoidea the massive, heavy rostrum at the caudal end of the body (Figures 5C,D) acts as a counterweight, balancing the negative buoyancy of the cranial part (Kondakov 1940; Akimushkin 1963, and others).

The hydrostatic gas apparatus of fish is the swim-bladder (Figures 6A,B), whose volume is usually 5–10% of that of the whole fish (Evans & Damant 1928; R. M. Alexander 1959a,c, 1961). While physostomous fishes can easily correct their buoyancy by swallowing atmospheric air or expelling some of the gas in the swim-bladder into the surroundings, physoclistic fishes can change the volume of their bladder only by the relatively slow process of gas secretion and reabsorption, which makes fairly intensive muscular exertion necessary for extensive vertical motion, as has been reported (Kanwisher & Ebeling 1957), for instance, for bathypelagic fishes which make diurnal vertical migrations over distances of up to 400 m. This, however, cannot serve as grounds for the conclusion (Konstantinov 1965) that a physoclistic swim-bladder has no hydrostatic functions of any importance; on the contrary, it should be stressed that any gas-filled cavity in the animal's body inevitably increases its buoyancy. Everything indicates that, phylogenetically, the physoclistic bladder emerged in fishes precisely to preserve the hydrostatic function of the swim-bladder: at depths of hundreds or thousands of metres in the sea the gas in the swim-bladder would otherwise simply be pressed out into the surroundings through the *ductus pneumaticus.*

Fishes having a bladder divided into two portions, anterior and posterior (Figure 6B), by transferring gas from one part to the other, can eliminate the unnecessary difference or, on the other hand, create it. Fishes whose bladder is not divided (Figure 6A) cannot do this.

In the fastest pelagic physoclistic fishes, which continually make considerable vertical movements, the presence of a swim-bladder destroys neutral buoyancy (Jones 1949, 1951, 1952; Denton & Marshal 1958). Therefore, in the process of evolution such fishes, mostly pelagic predators (*Auxis*, *Sarda*, *Scomber*, etc.), lost the swim-bladder. A bladderless fish of neutral buoyancy, irrespective of its vertical

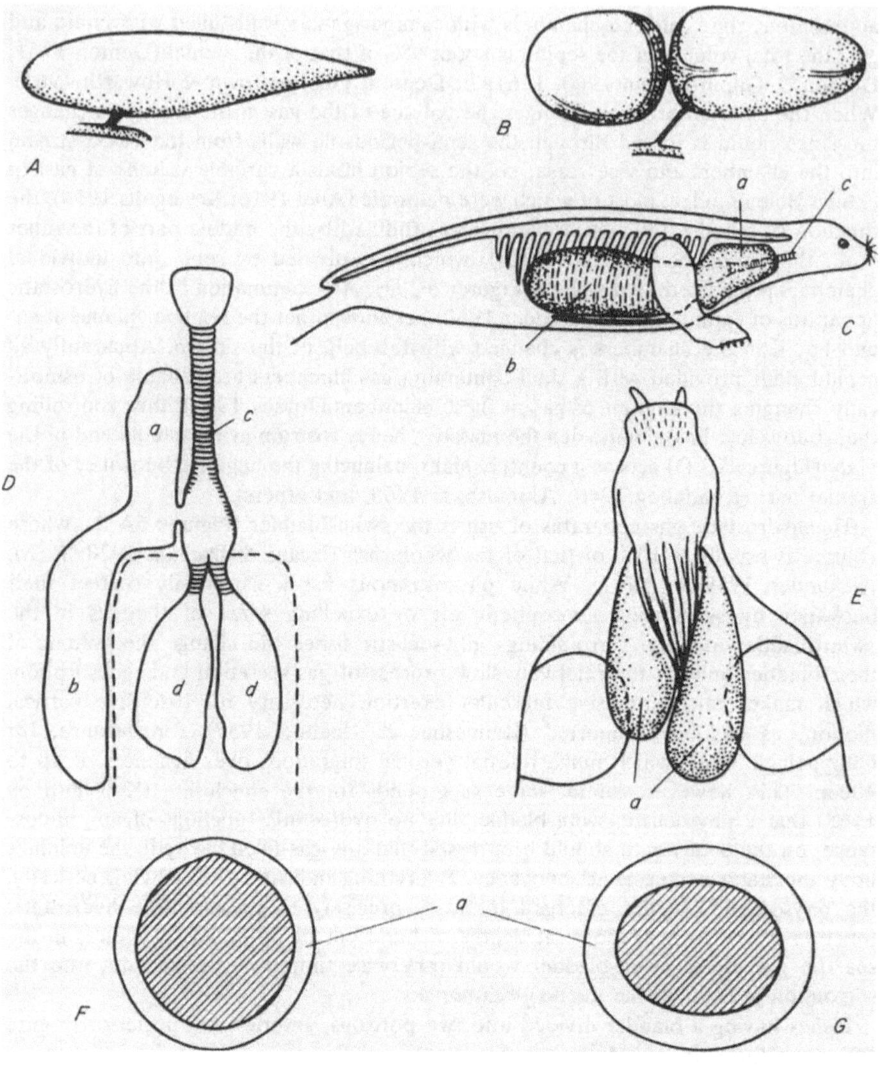

Figure 6. Hydrostatic apparatus of nektonic vertebrates. *A*, swim-bladder of *Acipenser güldenstädti colchicus* V. Marti; *B*, swim-bladder of *Cyprinus carpio carpio* L.; *C*, air sac of *Histriophoca fasciata* Zimm. (*a*, cervical portion of air sac; *b*, trunk portion; *c*, trachea; after Sleptsov 1940); *D*, top view of respiratory system of *Histriophoca fasciata* Zimn., (*a*, cervical portion of air sac; *b*, trunk portion; *c*, trachea; *d*, lungs; after Sleptsov 1940); *E*, air sacs of *Odobenus rosmarus* (L.) (a) (after Sleptsov 1940, with modifications); *F*, cross-section of midbody of *Delphinus delphis ponticus* Barab. (*a*, subcutaneous blubber); *G*, cross-section of midbody of *Pagophoca groenlandica* (Erxl.) (*a*, subcutaneous blubber).

migrations, is always in a state of stable equilibrium, since it is practically 'incompressible' (Aleyev 1963a).

In all nektonic animals with lungs (reptiles, birds and mammals), the hydrostatic gas apparatus is provided by the lungs themselves and sometimes also by special air sacs. In Sphenisciformes, in addition the air trapped among the feathers, is of some hydrostatic importance, while the absence of pneumatic bones in this group (Ognev 1941) should be viewed as an adaptation aimed at reducing excessive buoyancy. All nektonic animals with lungs can easily regulate their buoyancy by changing the quantity of air in their lungs or air sacs.

The structure of the respiratory organs and related systems of cetaceans and sirenians has a number of very peculiar features (Petit 1955; Bourdelle & Grasse 1955; Tomilin 1957, 1962a; Yablokov 1961, 1965, and others) owing to their profound, allround adaptation to the aquatic mode of life: very few sternal ribs, relatively large and elongated lungs, as well as very elastic alveoli, whose walls in cetaceans are supplied with a well-developed musculature. Thanks to these special features the volume of the thoracic cavity can diminish considerably during the animal's descent into deep water without harm to the thorax and lungs, while the latter's relatively large volume prolongs the respiratory pause. The slanted position of the diaphragm determines the dorsal arrangement of the lungs and tends to shift the animal's centre of gravity downwards, thus increasing its static stability. Both in cetaceans and in sirenians the respiratory passages are supplied with a system of cartillaginous rings, valves and sphincters, and in cetaceans also with a system of air sacs to ensure reliable plugging of the passages during immersion of the animal. The cetaceans' characteristic of having the nostrils on top of the head makes for maximum ease in contacting the air while the isolation of the alimentary and respiratory passages prevents water from getting into the respiratory passages when the mouth is opened under water.

In the pinnipeds we find a peculiar hydrostatic gas apparatus in the form of special air sacs connected to the trachea or to the oesophagus (Figures 6C–E). These may be filled to some extent with air from the respiratory passages (when the sac is connected to the trachea, as in *Histriophoca*) or by swallowing air (when the sac is connected to the oesophagus, as in *Odobenus*), and serve as an adaptation regulating buoyancy (Sleptsov 1940; Fay 1960; Schevill, Watkins & Ray 1966; Sokolov, Kosygin & Tikhomirov 1966; Sokolov, Kosygin & Shustov 1968). Partial reduction of the cartilaginous rings of the trachea, characteristic of Pinnipedia, and the presence in it of an elastic membrane which deflects during a dive ensure reliable closure of the trachea. This adaptation is better developed in those species whose members dive to greater depths, particularly *Leptonychotes* (Sokolov, Kosygin & Shustov 1968; A. S. Sokolov 1969b, and others).

Fat, being lighter than water, plays an important role in ensuring neutral buoyancy in fishes and nektonic mammals.

There is some fat in the bodies of all fishes, but in many cases the fat content is so great as to appear to be one of the most important factors determining the general degree of buoyancy; this is characteristic particularly of pelagic sharks and many Clupeidae, Engraulidae, Scombridae, etc. Among the bladderless fishes we have examined, neutral or nearly neutral buoyancy is due to an increased quantity of body fat in *Squalus, Oxynotus, Gymnammodytes, Pomatomus, Scomber, Sarda, Auxis* and *Remora* (Table 2). The distribution of the fat in the body varies. In some cases fat accumulates predominantly in the muscles, in the subcutaneous layer and in the abdominal cavity (*Scomber, Sarda, Auxis, Trachurus, Engraulis, Pomatomus* and *Sphyrna*); in other cases the fat deposits are of a more local character. Thus, for example, in some Gonostomidae and Stomiatidae which make considerable vertical migrations, after metamorphosis the swim-bladder stops secreting gas and fills up with fatty tissue (Marshall 1960); consequently in this case the bladder's hydrostatic function is preserved, though it is no longer a gas-filled cavity, but a functionally improved fat deposit. In many sharks (*Lamna, Somniosus, Squalus, Oxynotus,* etc) it is the liver that becomes the main fat depository, and along with its normal functions, it acquires that of a hydrostatic organ (Aleyev 1963a). Moreover, owing to its high fat content the density of the liver considerably diminishes ($\rho = 0.94$), thanks to which its buoyancy becomes strongly positive, while its weight relative to that of the whole body considerably increases, as can be seen from the following comparison: in predators whose liver has negative buoyancy and fulfils no hydrostatic functions, including sharks, for example *Sphyrna, Mustelus, Sarda, Trachurus* or *Pomatomus,* the liver's weight is 1–5% of that the whole fish, whereas in *Squalus* and *Oxynotus* it comes to 23–25%. The relative size of the liver and its fat content increases in *Squalus* with age, while its density noticeably drops, which points to its increasing hydrostatic function (Aleyev 1963a).

In nektonic mammals, especially cetaceans and pinnipeds, hydrostatic fat deposits take the form of a thick subcutaneous layer (Figures 6F,G): in *Delphinapterus,* for example, its thickness reaches 20 cm and in *Balaenoptera* 15–35 cm (Bourdelle & Grasse 1955). Along with its hydrostatic function, the mammals' subcutaneous fat also provides heat insulation (Tomilin 1940; Kanwisher & Sundnes 1966, and others); in Sphenisciformes a similar function is fulfilled by a thick covering of feathers.

4. Hydrodynamic correction of buoyancy

Nektonic organisms use adaptations of both the passive and active type for hydrodynamic correction of buoyancy. **Passive adaptations** can be effective only during forward movement and are connected with the formation of structures acting as lifting surfaces; these adaptations are not intended to perform any special supporting movements. **Active adaptations**, which are connected with the performance of special supporting movements aimed at creating hydrodynamic lift, may act both during forward movement and in its absence. Characteristic of the Agnatha, Placodermi, Acanthodei, Elasmobranchii and primitive Osteichthyes (Acipenseriformes, etc.) are only passive adaptations, while Cephalopoda and the great majority of extant fishes (including nearly all the Actinopterygii) and reptiles have both types of adaptation.

1. Passive adaptations: lifting surfaces

The creation of hydrodynamic lift by means of lifting surfaces is characteristic to some degree of the great majority of nektonic animals except birds and mammals, for which $\Delta > 0$ at every stage of adaptation to the nektonic mode of life.

In primary aquatic vertebrates negative buoyancy is as a rule combined with an asymmetric shape of the caudal fin, which directly creates vertical hydrodynamic forces. In the case of a hypocercal tail fin the lower lobe, which includes the axis of the body, is always less flexible than the upper lobe. Therefore, during transverse movements of the fin, the more flexible upper lobe, because of resistance on the part of the water, always lags behind the lower lobe, owing to which the entire blade of the tail fin is inclined at a negative angle of attack and inevitably creates a force directed not exactly forwards, but with a vertical component also (Figure 7A). An epicercal fin, having a more flexible upper lobe, is inclined during transverse movements at a positive angle of attack (Figure 15) and creates, accordingly, and upward force (Figure 7C).

In some nektobenthic forms of Diplorhina (Phlebolepidida, Heterostraci, etc.) and benthonektonic Anaspida, the caudal fin used to be hypocercal, creating a downward force during movement.

The Heterostraci had no way of neutralizing this force. Having no fins but the caudal one (Figure 8A), when swimming actively the animal inevitably moved in an upward arc until the longitudinal axis of the body was at a certain positive angle of attack $\alpha > 0$. Then the propulsive mechanism was 'switched off' and the animal began to dive. During this dive the anterior part of the body, which was more compact and thus had a lower specific surface area than the tail, and, moreover, was covered by heavy armour, sank faster than the tail. As a result the front part of the body soon became inclined downwards, whereupon either the propulsive mechanism was 'switched on' again to repeat the active part of the cycle, i.e. movement in an upward arc, or the animal settled on the bottom (Figure 9). Thus, in the absence of pectoral fins, hypocercy in the Heterostraci was a means of ensuring the condition $\alpha > 0$, under which the entire body acted as a lifting surface.

The origin and primary function of hypocercy were similar in Phlebolepidida and Anaspida, neither of which had pectoral fins. However, hypocercy in members of these groups was combined with the presence of an anal fin (Figure 8B), which during undulations of the body for the purpose of locomotion created lift on the same principle as the lower lobe of an epicercal tail. This force partially or fully neutralized the downward force created by the hypocercal tail fin and was most probable for Anaspida, in which the anal fin was well developed (Figure 8B). Because of this, in Anaspida the body was apparently no longer inclined during

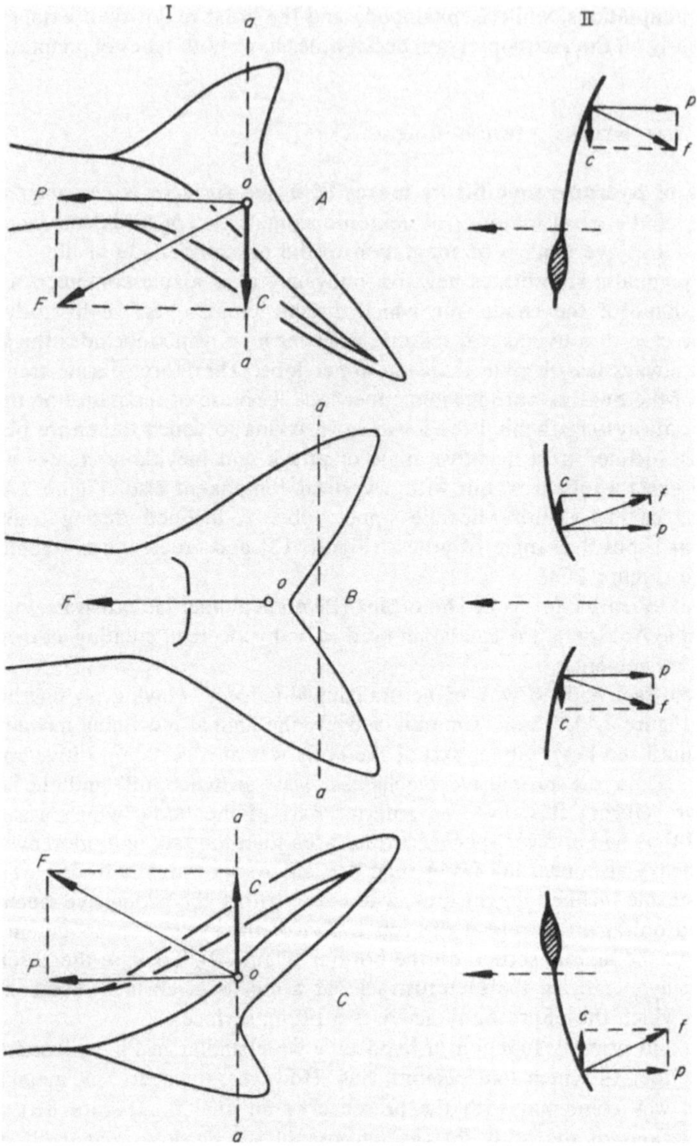

Figure 7. Diagram showing operation of caudal fins of different shapes as lifting surfaces. I, side view; II, rear view (section at *aa*). *A*, hypocercy; *B*, isocercy; *C*, epicercy. *o*, centre of fin blade; *F*, *f*, propulsive force created by fin; *P*, *p*, *C*, *c*, its components; ←– – –, direction of transverse fin movement. See text for further explanations.

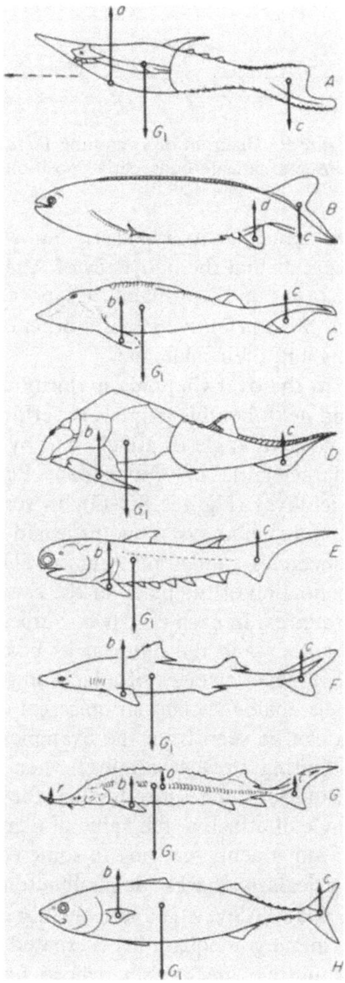

Figure 8. Points of action of basic vertical hydrodynamic forces created by lifting surfaces in nektonic (*B*, *E-H*) and nektobenthic animals (*A*, *C*, *D*). *A*, *Pteraspis rostrata toombsi* White (after White 1935, modified); *B*, *Pharyngolepis oblongus* Kiaer (after Kiaer 1924, modified); *C*, *Hemicyclaspis murchisoni* (Egert.) (after Stensio 1932, modified); *D*, *Pterichthys milleri* Agass. (after Watson 1935; see Obruchev 1964b, modified); *E*, *Climatius reticulatus* Agass. (after Watson 1937, modified); *F*, *Squalus acanthias* L.; *G*, *Acipenser stellatus* Pall.; *H*, *Scomber scombrus* L. *a*, force acting on body's lower surface; *b*, force created by pectoral fins and their analogues; *c*, vertical component of propulsive force created by caudal fin; *d*, force created by anal fin; *o*, force created by trunk at $\alpha = 0$; *r*, force created by rostrum; G_1, force due to residual weight.

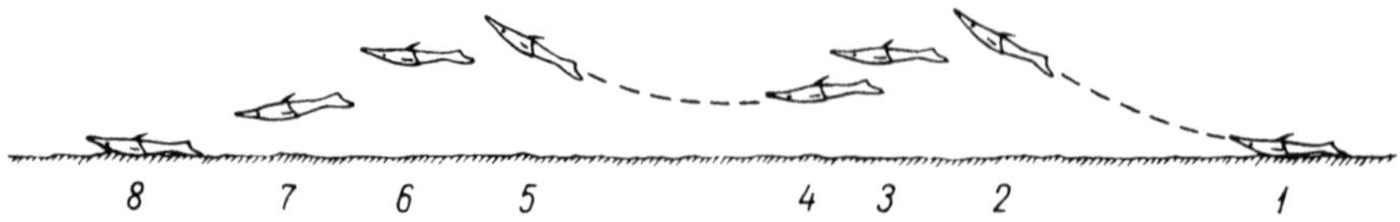

Figure 9. Diagram of swimming *Pteraspis*. Dashed line indicates stretches of active swimming; numbers indicate consecutive positions of animal.

swimming as in *Pteraspis*, but was horizontal, as in all nektonic animals, which suggests that the buoyancy of Anaspida approached the neutral level characteristic of nekton. In all probability hypocercy in primary aquatic vertebrates corresponds to the most primitive general organization, particularly the complete absence of paired fins and their analogues.

In the overwhelming majority of cases the function of lifting surfaces in nektonic and nektobenthic animals is performed by pectoral fins or their analogues, situated at a positive angle of attack, and by an epicercal tail fin. This system of creating lift is characteristic of Osteostraci, Placodermi, Acanthodei, Chondrichthyes and Osteichthyes (Figures 8C–G). In some higher Osteichthyes (Scombridae, Carangidae, etc.) a similar system is furnished by the pectoral fins and a completely symmetrical, (isocercal) caudal fin (Figure 8H); in this case the fish can actively change the inclination of the blade of the caudal fin and, accordingly, the magnitude of the lift it produces. In each case two sources of lift are created, one situated at the front end of the body and the other at its posterior end (Figures 8B–H); even in the very early phylogenetic stages of adaptation to pelagic life both these sources, pectoral fins or their analogues and an epicercal caudal fin, usually emerged almost simultaneously, as can be seen from the examples of Osteostraci and Placodermi (Figures 8C,D).

During slow swimming, when the action of the fins as lifting surfaces proves insufficient, the entire body of the nektonic animal becomes inclined at some positive angle of attack α, the value of α increasing in proportion to the decrease in the speed of movement, reaching in some cases +30°. This is demonstrated not only in fishes, particularly sharks (Schmalhausen 1916; Aleyev 1963a) and the young of Acipenseridae (Aleyev 1963a), but also in Teuthoidea (Zuyev 1965, 1966); during rapid swimming adequate lift is created in Teuthoidea at $\alpha = 0$ on account of the hydrojet propulsive mechanism, whose funnel is always directed somewhat downwards. A similar pattern of lift forces was undoubtedly characteristic also of Belemnoidea, most of which (Krymgolts 1958) were nektonic.

All the above-mentioned animals, the Belemnoidea, Teuthoidea, sharks and Acipenseridae, have a comparatively broad body with a height (H) to width (I) ratio close to or less than one (Table 4), which ensures sufficiently good lifting-surface properties. In the ontogeny of squids and fishes alike the properties of the body as a lifting-surface change according to the value of Δ: in squids, with the preservation of a slightly negative buoyancy the ratio H/I diminishes (Table 4: *Loligo vulgaris*, *Symplectoteuthis*), which improves the lifting-surface properties of the body, while in

Table 4. Maximum height (H) to width (I) ratio of the body and buoyancy Δ of nektonic animals whose bodies act as a lifting surface

Species and nektonic type	For vertebrates, length L to end of vertebral column; for invertebrates, absolute length L_a (cm)	Number of specimens examined n	H/I	Δ	Source
Sagitta setosa Müll. (PN)	2.2–2.2	6	0.75	−0.01	Author's data
Loligo edulis Hoyle (EN)	12.3	1	1.00	−0.02	Zuyev (1966)
Symplectoteuthis	0.1–3.0	1	1.00	−0.03	”
oualaniensis	3.1–15.0	8	0.96	−0.02	”
(Less.) (EN)	15.1–27.0	17	0.95	−0.02	”
	27.1–39.0	9	0.95	−0.02	”
	39.1–48.0	5	0.92	−0.02	”
	48.1–54.0	2	0.92	−0.02	”
Onychoteuthis banksii (Leach) (EN)	16.5	1	0.87	−0.02	”
Ancistroteuthis lichtensteini (D'Orb) (EN)	20.5	1	0.85	−0.02	”
Todarodes pacificus (Steenstr.) (EN)	40.2	1	0.79	−0.02	”
Loligo vulgaris	0.1–3.0	50	1.00	−0.03	”
Lam. (EN)	3.1–12.0	12	0.95	−0.02	”
	12.1–21.0	6	0.90	−0.02	”
	21.1–33.0	7	0.83	−0.02	”
	35.1–42.0	3	0.79	−0.02	”
Mustelus mustelus (L.) (BN)	49.0	1	0.95	−0.04	Author's data
Squalus acanthias L. (EN)	19.0–36.0	3	0.85	−0.04	”
	40.0–49.6	10	0.88	−0.03	”
	57.5–66.0	2	0.91	−0.02	”
	90.0–109.0	10	0.91	−0.01	”
	114.0–130.0	7	0.91	0.00	”
Acipenser ruthenus L. (BN)	48.0	1	1.02	−0.01	Author's data
Acipenser stellatus	2.0	1	1.00	−0.05	”
Pall. (BN)	4.7–8.2	2	0.98	−0.04	”
	31.0–55.0	4	1.00	−0.03	”
	75.0	1	1.03	−0.02	”
	94.3–110.0	8	1.07	−0.01	”
Acipenser güldenstädti colchicus V. Marti (BN)	56.3	1	1.00	−0.02	”
Acipenser sturio L. (BN)	55.4	1	1.00	−0.03	”
Huso huso (L.) (BN)	53.0	1	1.00	−0.03	”
Scaphirhynchus platorhynchus (Raf.) (BN)	55.0	1	0.67	—	”
Eretmochelys imbricata (L.) (XN)	20.9	1	0.47	0.00	”
Caretta caretta (L.) (XN)	114.5	1	0.44	0.00	”
Chelonia mydas (L.) (XN)	144.0	1	0.40	0.00	”

some nektonic sharks and the sturgeon family this ratio increases with higher buoyancy (Table 4: *Squalus*, *Acipenser stellatus*), as a result of which the lifting-surface properties of the body worsen.

Owing to the presence of lungs in all secondary aquatic nektonic animals, reptiles, birds and mammals, their buoyancy was close to neutral at all stages of adaptation to nektonic life; therefore lift due to lifting surfaces is not on the whole characteristic of them. Hypocercy in nektonic reptiles (Ichthyosauria, Metriorhynchidae. Mosasauridae, etc.) is due to the presence of nektoxeric and xeronektonic stages in the phylogeny of these animals: while they were crawling on land the hypochordal lobe of the tail fin was inevitably injured, owing to which it was the epichordal lobe that underwent preferential development, i.e. a hypocercal type of fin took shape (Figures 10A, 11A, 12A). Also conducive to the development of hypocercy in nektonic reptiles was the considerable dorsoventral asymmetry of the body in the nektoxeric and xeronektonic stages of their phylogeny. The low position of the limbs with comparatively weak specialization inevitably increased the resistance of the ventral side of the body and during swimming created a parasitic moment on body making the animal's head rotate downwards; this moment could be neutralized, in particular, by the creation of a reverse moment by a hypocercal tail, rotating the head upwards. During the eunektonic stage ecological contact with the land was lost, dorso-ventral body asymmetry diminished, buoyancy approached the neutral level and the tail fin became nearly isocercal (Figures 10D,E, 11C–E, 12A). The same process of decreasing tail-fin hypocercy and approximation to the isocercal type can be observed in the ontogeny of ichthyosaurs (Figures 11C–E), indicating reduced dorso-ventral asymmetry and increasing buoyancy with age.

It has not been ruled out that in the initial stages of adaptation to the nektonic mode of life hypocercy in reptiles had a certain significance also in ensuring the condition $\alpha > 0$, under which the entire body acted as a lifting surface. Such a method of creating lift can be observed in the example of terrestrial lizards (*Lacerta*, etc.), which, when swimming, always assume an inclined position with a positive angle of attack usually amounting to 15–20° (Figure 13).

In Sagittoidea with slightly negative buoyancy (Table 2, *Sagitta*) the body is compressed dorso-ventrally (Table 4, *Sagitta*) and is fitted with horizontal lateral fins, which improve its properties as a lifting surface.

As one can see from the above, the function of lifting surfaces in nektonic animals is fulfilled mainly by fins: the pectoral and the caudal fins.

Pectoral fins predominantly acting as lifting surfaces and frontal rudders usually turn into isolated well-streamlined rigid planes (Figure 14), with a low total drag and comparatively restricted vertical mobility (sharks, Acipenseridae, Ichthyosauria, Cetacea, etc.) which makes it easier to maintain them in the horizontal position (Figures 15, 26 and 31). Moreover, the mobility of the skeletal elements of the limb forming the fin also diminishes, and in secondary aquatic nektonic animals, particularly ichthyosaurs, cetaceans and sirenians, one usually observes a shortening of its proximal and an extension of its distal elements, accompanied, as a rule, by

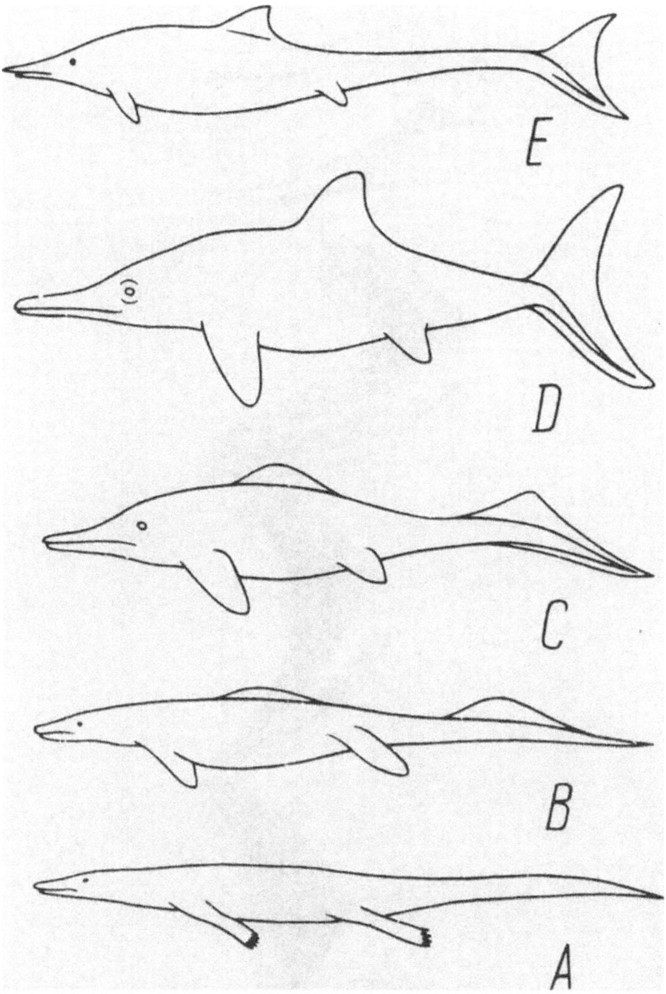

Figure 10. Development of body shape and caudal fin in nektonic reptiles. *A, Mesosaurus; B, Cymbospondylus; C, Mixosaurus; D, Stenopterygius; E, Nannopterygius* (after Kuhn 1937, with modifications).

hyperphalangism and, more seldom, by hyperdactylism (Figure 14). A reduction in the vertical mobility of the fin is achieved by restricting its mobility where its proximal elements join the bones of the pectoral girdle; in some cases (Acipenseriformes, etc.) the vertical mobility of the pectoral fins is practically non-existent because of this.

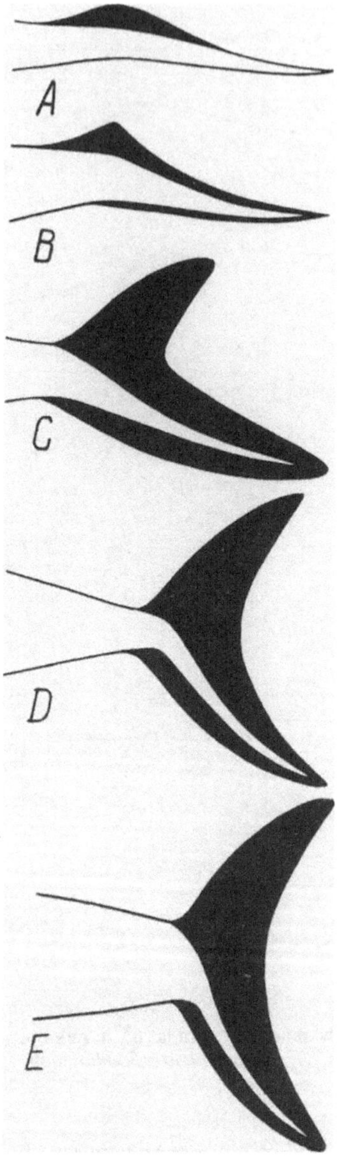

Figure 11. Caudal-fin development in Ichthyosauria. *A*, *Cymbospondylus* sp.; *B*, *Mixosaurus nordenskjoldi* Wiman; *C*, *D*, *Stenopterygius quadriscissus* Quenst., young and adult specimens respectively; *E*, *Macropterygius trigonus* (Owen) (*A*, after Kuhn 1937; *B*-*E* after Abel 1912, with modifications).

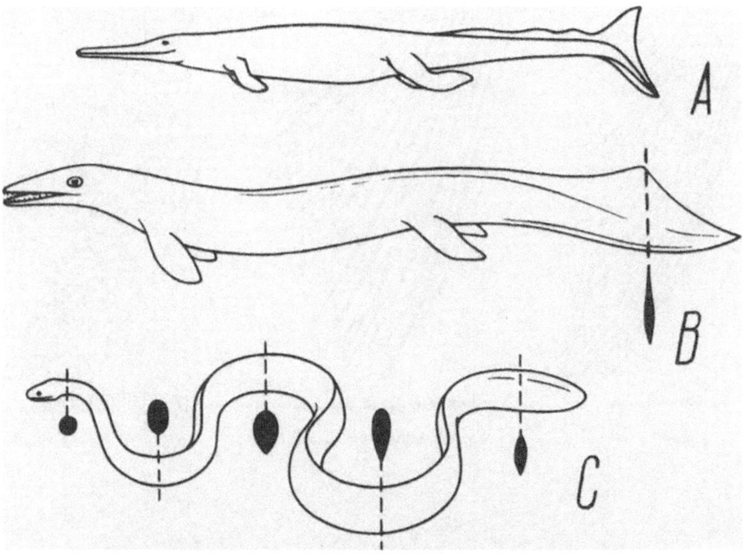

Figure 12. A, *Metriorhynchus jaekeli* Schmidt (after Abel 1907); B, *Clidastes velox* Marsh, with cross-section of caudal fin, author's reconstruction based on Merriam's material (1897; see Khozatsky & Yuryev 1964); C, *Enhydrina schistosa* (Daud.), with body cross-sections.

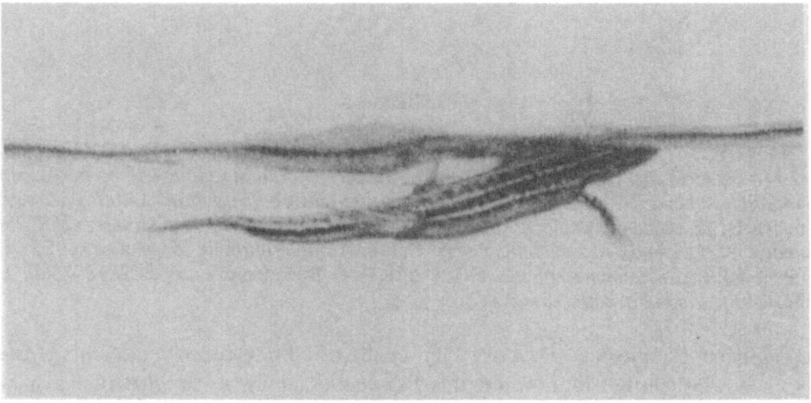

Figure 13. Swimming *Lacerta vivipara.* Photo by the author.

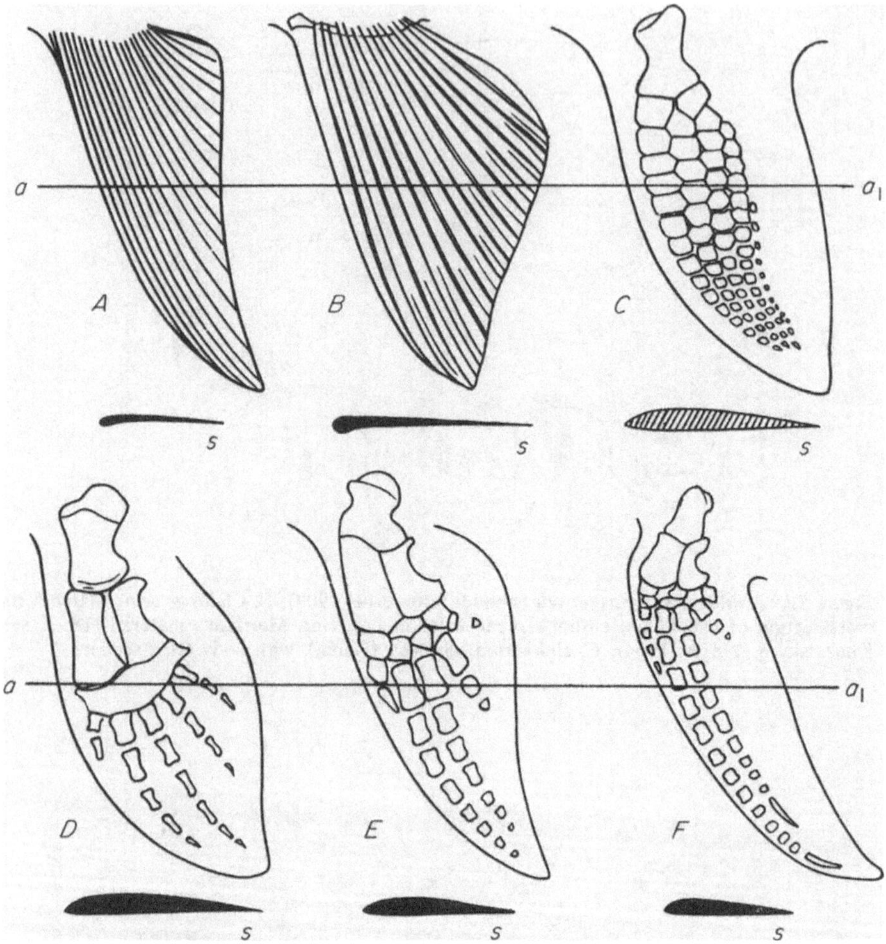

Figure 14. Pectoral fins functioning as lifting surfaces. *A, Prionace glauca* (L.); *B, Acipenser güldenstädti colchicus* V. Marti; *C, Stenopterygius quadriscissus* Quenst. (after Abel 1912, modified); *D, Eubalaena glacialis* (Bonnat.) (after Van Beneden & Gervais 1868–1880, modified); *E, Delphinus delphinis ponticus* Barab. (after Bourdelle & Grasse 1955, with modifications); *F, Balaenoptera physalus* (L.) (after Van Beneden & Gervais 1868–1880, with modifications). *s*, cross-section along aa_1.

It has been established (Harris 1936, 1938) using a gypsum model of *Mustelus canis* in a wind tunnel that, when this fish moves along its longitudinal axis with $\alpha = 0$, about 86% of the total lift force is created by its pectoral fins; similar relationships apparently hold also for other sharks and Acipenseridae.

The heterocercal type of caudal fin (sharks, Acipenseridae, Ichthyosauria, etc.)

Figure 15. Swimming *Squalus acanthias* (L.) Pectoral fins extended, caudal fin and elongated lobule of second dorsal fin situated at positive angle to transverse body movement. Cine-record.

always operates in the same way, inevitably and purely mechanically creating downward (Figure 7A) or upward (Figure 7C) forces. The fact that a heterocercal tail fin creates vertical forces has been noted by many investigators (Schulze 1894; Ahlborn 1895; Whitehouse 1918; Breder 1926; Harris 1936, 1937a, 1938, 1953; Grove & Newell 1936, 1939; Affleck 1950; Gray 1953a,b; Kobi & Pristovsek 1959; Aleyev 1959b, 1963a, R. S. Alexander 1965, 1966b, 1968, and others).

The isocercal tail fin of the higher Actinopterygii, depending on the direction in which undulations pass along it, is capable of creating differently directed forces. Thus, if the plane of the fin's displacements is parallel to the body's longitudinal axis and the blade of the fin is perpendicular to this plane, then the force created by the fin is also directed along the body's longitudinal axis (Figure 16A), as was observed in most Actinopterygii with neutral buoyancy during propulsive movements (Figures 25B, 27A, 28). If the plane of bending is parallel to the longitudinal axis of the fish and the blade of the fin is inclined at an acute positive angle to the direction of the transverse sweep of the fin, the force it creates is directed forwards and upwards (Figure 16B), which may be observed, particularly, in sharks (Figures 15, 26A), in Acipenseridae (Figure 26B) and some Actinopterygii with a high density (*Scomber, Sarda, Mullus*, etc.; see Figures 27B, 29B). If the plane of bending is situated at an angle to the longitudinal axis of the fish, the force created by the fin always has a longitudinal and a vertical component (Figures 16C), the latter increasing the greater this angle. Propulsive movements of this type are a particular feature of fishes characterized by swimming in an upward or downward arc; in these one usually observes a pronounced separation of the lobes of the caudal fin, since such a manoeuvre is always carried out mainly by the action of a single lobe (*Pelecus, Abramis, Cyprinus, Mugil*, etc.). If this angle increases to 90° the longitudinal component disappears and the fin creates a vertical force only (Figure 16D), which according to the direction of propagation of the undulations, may be directed either upwards or downwards.

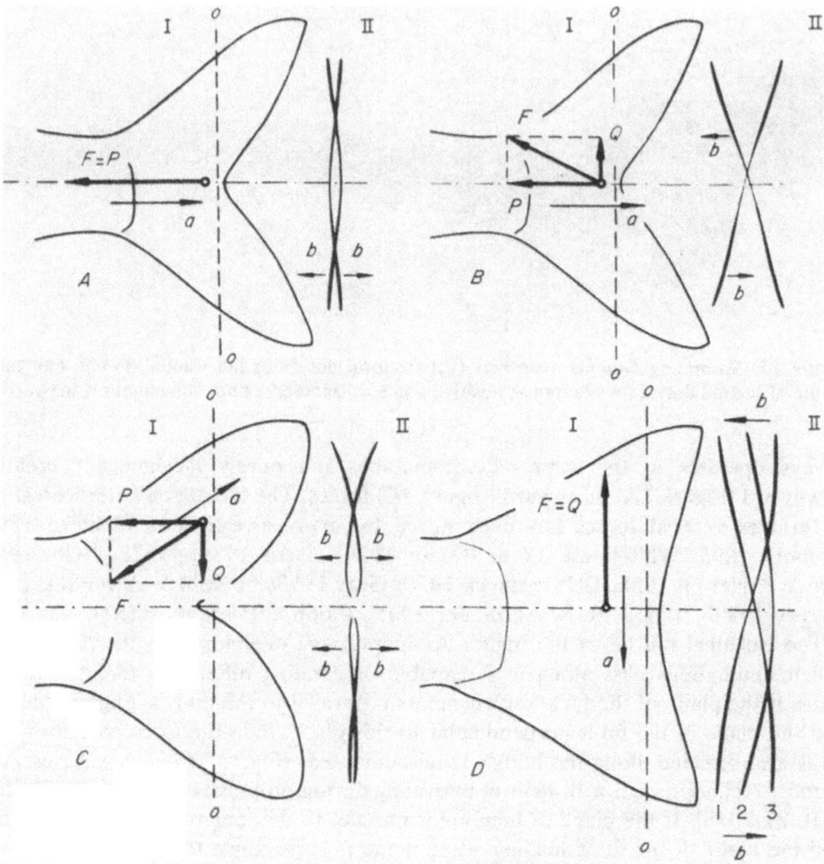

Figure 16. Diagram of caudal-fin performance for different directions of propulsive undulations. *a*, direction of wave movement; *b*, direction of transverse fin movement; *F*, force created by fin; *P,Q*, its components. I, side view, II, section along *oo*; *A,B*, plane of fin undulation parallel to longitudinal axis of fish's body; *A*, lobe membrane vertical; *B*, lobe membrance at positive angle to direction of transverse movement; *C*, plane of fin undulation parallel to longitudinal axis of fin's upper lobe (upper lobe only is active); *D*, plane of fin undulation perpendicular to longitudinal axis of fish.

It has been shown (Le Mare 1936; Aleyev 1963a) that in some sharks a certain amount of lift is created by the elongated posterior lobes of the dorsal fins, which during the fish's flexural movements are at a positive angle of attack (Figure 15); however the relative magnitude of this force is negligibly small. A more substantial lift force is generated in a similar way by the anal fin of some fishes in which this fin has a long base (some Siluroidei, etc.).

When the body of a nektonic animal has a comparatively low aspect ratio and for one reason or another is more convex dorsally than ventrally, it becomes capable during purely longitudinal motion, i.e. at $\alpha = 0$, of creating hydrodynamic lift according to Bernoulli's law on account of the difference between the speeds of the streams flowing over and under it and between the dynamic pressures on its upper and lower surfaces. In most cases, however, this force proves to be either negligibly small or actually zero. The reasons for this lie in the characteristics of a low aspect ratio wing, as represented by the body of the nektonic animal, and the large variation in the body shape of most nektonic animals.

It has been shown (Aleyev 1963a) that, according to the theory of aerodynamics (Prandtl 1951; Arzhanikov & Maltsev 1952; Karafoli 1956; Martynov 1958, and others), the properties of a **nektonic body as a lifting surface** improve (1) with an increase in this surface's slenderness ratio, i.e. in this case with an increase in the ratio λ of the square of body width I to the area S of the horizontal projection of the lifting surface formed by the body, (2) with an increase in the relative curvature f of the body profile and (3) with a decrease in the relative body height H. From this, the index N_0 of the lifting capacity of the nektonic body for the case $\alpha = 0$ can be expressed as

$$N_0 = \lambda \frac{f}{L_b} \frac{L_b}{H} = \frac{\lambda f}{H} = \lambda f H^{-1}, \tag{7}$$

where L_b is the length of the chord of the profile of the lifting surface formed by the body and f is the maximum distance between this profile's chord and its median epichordal line.

In view of the particular curvature of the profile of the body when its upper side is more convex than its lower side, in addition to the lift at $\alpha = 0$, a parasitic pitching moment M_0 is also created (Aleyev 1963a, 1965a–c, 1966a,b; Ohlmer 1964, and others); the effect of this moment is countered by the action of the pectoral fins (Aleyev 1963a).

It has been established experimentally using a gypsum model in a wind tunnel (Harris 1936, 1938) that the lift force created by the body of *Mustelus canis* at $\alpha = 0$ comes to about 4.6% of the total lift force created by the model. In *Mustelus* $N_0 = 0.047$ (Table 5). Since among nektonic sharks the body of Triakidae (*Mustelus, Triakis*, etc.) has maximum dorso-ventral asymmetry, the great majority of eunektonic sharks being much more symmetrical, it should be assumed that the relative value of the lift force created by the body of *Mustelus* at $\alpha = 0$, equal to 0.046, is close to the maximum value for sharks; this value should thus undoubtedly be regarded as negligibly small. Bearing this in mind, from the aspect concerning us here we should examine only those nektonic animals for which $N_0 > 0.030$ (Table 5).

The unsteadiness in body shape in nektonic animals, emerging as a result of the continuously changing shape of the body during propulsive and respiratory movements and various other manoeuvres, considerably reduces the already not very high quality of the body as a lifting surface capable of creating lift at $\alpha = 0$.

Thus, on the basis of analysis of vast numbers of underwater photographs and ciné films of cetaceans, both those taken by the author (1973b) in the Black Sea and those available in the literature (Essapian 1955, and others), and also on the basis of frame-by-frame examination of numerous ciné records, the author has come to the conclusion that in cetaceans, just as in pinnipeds and sirenians, during swimming the body has on the whole a nearly symmetrical but also rather changeable profile, owing to which it is incapable of creating appreciable lift in any way at $\alpha = 0$. The at first glance essential dorso-ventral asymmetry of Balaenopteridae and plankton-eating sharks (*Rhincodon* and *Cetorhinus*) is practically eliminated when these animals swim with their mouths open (Figure 17), which for straining animals, among which they belong, is usual.

When Teuthoidea are swimming, at different moments of the propulsive/respiratory cycle, first the dorsal then the ventral side of the body becomes more convex (Zuyev 1966). Therefore one may assume with sufficient accuracy that the body of Teuthoidea at $\alpha = 0$ does not create any hydrodynamic lift, owing to which Teuthoidea (and cephalopods on the whole) are not further examined from this aspect.

Thus we come to the conclusion that creation of hydrodynamic lift by the body of a nektonic animal under the condition $\alpha = 0$ is extremely rare. In more or less developed form it is characteristic of only a few Chondrichthyes (benthonektonic Triakidae and the young of some nektonic sharks), Osteichthyes (benthonektonic Acipenseridae) and reptiles (xeronektonic Testudinata); it also probably took place among turtle-like xeronektonic Placodontia (Cyamodontoidei, Henodontoidei, etc.). Characteristically, all these animals are of either benthonektonic or xeronektonic type and have a body which is more convex on the upper than on the lower side. This is a side-effect of the development of a number of adaptations having nothing to do with hydrodynamic correction of buoyancy. Such a body shape in benthonektonic animals, combined with the low position of the mouth (Figure 8G), makes it easier to grab food off the bottom, while reducing the betraying shadow, and is still more prominent among nektobenthic Sepiidae, Osteostraci, Antiarchi, Orectolobidae, Scyliorhinidae, Triglidae, etc. The flattening and straightening out of the entire ventral surface in xeronektonic animals, combined with considerably greater convexity of the dorsal surface, make it easier to travel on land; these peculiarities of body shape are pronounced in all Sauropterygia, Placodontia and Testudinata.

We do not find examples where the body creates hydrodynamic sinking forces under the condition $\alpha = 0$ among nektonic animals. Considerable dorso-ventral asymmetry of the body with a more convex ventral contour was characteristic of some Heterostraci (Figure 8A), yet they were incapable of rectilinear movement along their longitudinal axis, owing to which no sinking force was generated while they were swimming. This is why the Heterostraci are not examined from this aspect.

We shall now examine **the mechanism whereby an animal's body creates hydrodynamic lift under the condition $\alpha = 0$** using the examples of some sharks (*Mustelus*, *Squalus* and *Scyliorhinus*), Acipenseridae (*Acipenser*, *Huso* and *Scaphirhynchus*)

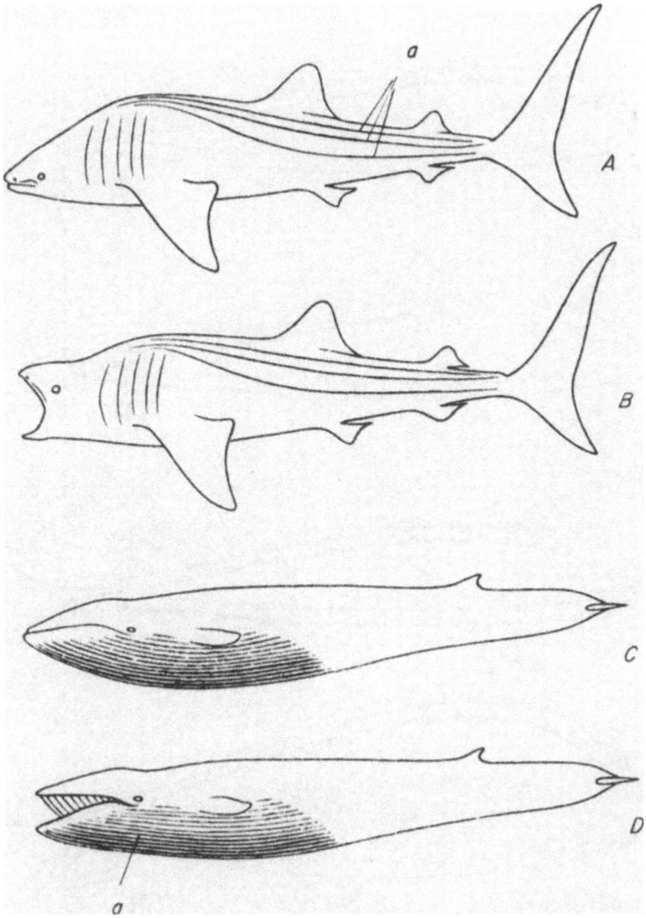

Figure 17. A,B, Rhincodon typus Smith; C.D, *Balaenoptera physalus* (L.). *a,* longitudinal keels (*Rhincodon*) or folds (*Balaenoptera*) on body surface. See explanations in text.

and Chelonioidea (*Chelonia, Caretta* and *Eretmochelys*). As has already been shown (Aleyev 1963a, 1965a, 1966a,b) the logitudinal profile of all these animals has a noticeable similarity to the profiles of some aeroplane wings (intended to create a high lift force), particularly asymmetric profiles of the TsAGI 'B' series (Figure 18).

Table 5, which shows nektonic animals arranged in order of increasing N_0, demonstrates that among the animals examined, sharks, Acipenseridae and Chelonioidea, the greatest lift capacity of the body is exhibited, as one should expect, by sea turtles. This lift created by the body possibly assumes for turtles a still greater

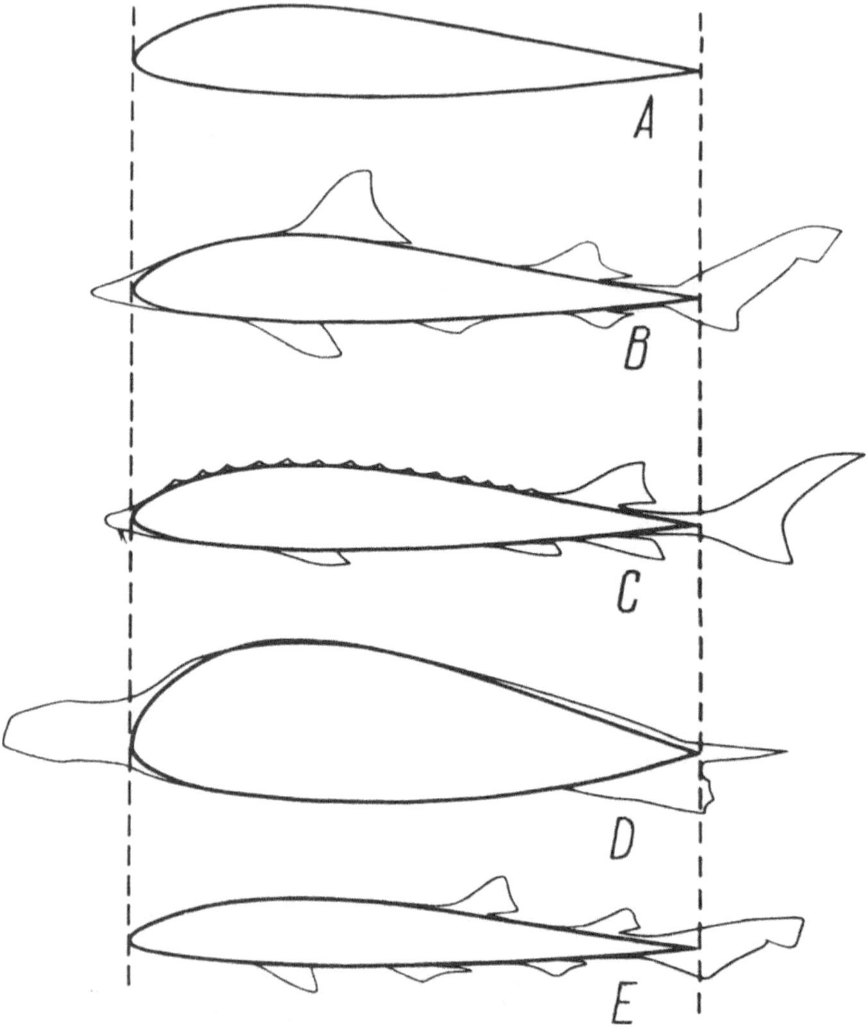

Figure 18. *A*, wing profile, "B" series, TsAGI* (after Martynov 1958); *B*, profile of *Mustelus mustelus* (L); *C*, *Acipenser güldenstädti colchicus* V. Marti; *D*, *Chelonia mydas* (L.); *E*, *Scyliorhinus canicula* (L.). Inscribed in each animal's profile is the TsAGI "B" series profile of corresponding thickness.

* Central Aerohydrodynamic Institute, Moscow.

Table 5. Length L_b of chord of profile of lifting surface formed by body, rostrum development index R_0, lift capacity N_0 of body and buoyancy Δ for nektonic animals

Species and nektonic type	Length L of animal to end of vertebral column (cm)	L_b (cm)	R_0	N_0	Δ
Acipenser stellatus Pall. (BN)	2.0	1.3	0.07	0.021	−0.05
	4.7	3.1	0.11	0.028	−0.04
	8.2	5.4	0.12	0.032	−0.04
	31.0	18.0	0.13	0.040	−0.04
	34.1	19.8	0.13	0.040	−0.03
	46.2	27.0	0.13	0.039	−0.03
	55.0	33.0	0.12	0.036	−0.03
	75.0	46.5	0.11	0.031	−0.02
	98.0	64.9	0.09	0.025	−0.01
	110.0	76.5	0.07	0.023	−0.01
Acipenser sturio L. (BN)	55.4	38.8	0.08	0.039	−0.03
Acipenser güldenstädti colchicus V. Marti (BN)	56.3	39.6	0.06	0.041	−0.02
Squalus acanthias L. (EN)	45.7	33.8	0.11	0.042	−0.04
Mustelus mustelus (L.) (BN)	49.0	36.3	0.08	0.047	−0.04
Acipenser ruthenus L. (BN)	48.0	32.7	0.08	0.056	−0.01
Huso huso (L.) (BN)	53.0	31.4	0.12	0.066	−0.03
Scaphirhynchus platorhynchus (Raf.) (BN)	55.0	35.3	0.16	0.085	—
Eretmochelys imbricata (L.) (XN)	20.9	15.3	—	0.110	0.00
Chelonia mydas (L.) (XN)	144.0	96.5	—	0.118	0.00
Caretta caretta (L.) (XN)	114.5	87.0	—	0.133	0.00

importance when they dive to a depth where their buoyancy should diminish.

In *Scaphirhynchus, Pseudoscaphirhynchus* and *Acipenser stellatus* the rostrum, which is compressed dorso-ventrally and bent upwards slightly, has a specific hydrodynamic function, creating an upward force on the front end of the body and partially compensating for the action of the body moment M_0 (Figure 8G); this function of the rostrum is analogous to the similar function of pectoral fins.

According to the morphological interpretations of the term, a rostrum is found in all sharks and Acipenseridae. In sharks it fulfils the function of a fairing, masking a very wide mouth slit. In the comparatively immobile benthonektonic Acipenseridae the rostrum acts mainly as a well-streamlined 'bracket', carrying the tendrils with tactile receptors in front of the mouth (Figure 8G); the end of the rostrum is usually raised to some degree, which reduces the probability of its being obstructed by the roughness in the bottom topography.

In the hydrodynamic sense, the rostrum forms a special structure in cases where the centre-line of the profile of the fish is convex downwards near the front (Aleyev 1963a) (Figure 18B,C). The centre-line of a wing profile creating a lift force at $\alpha = 0$ has an arched frontal part convex upwards (Prandtl 1951, and others). This gives grounds to believe that the rostrum cannot be included within the lifting surface formed by the body and constitutes a separate hydrodynamic structure. It is precisely for this reason that, in drawing an analogy between the body of a fish and a lifting surface, one should take as the length L_b of the chord of this surface's profile the length of the fish without the tail fin and without the rostrum if the rostrum is of hydrodynamic significance. Whereas in the case where the rostrum is not of hydrodynamic significance, the part of the head anterior to the mouth slit should be regarded as part of the lifting surface formed by the body (Figure 18E).

To assess the hydrodynamic significance of the rostrum, the index R_0 of rostrum development may be suggested (Aleyev 1963a):

$$R_0 = S_r^{1/2} L_b^{-1}, \tag{8}$$

where S_r is the area of the horizontal projection of the part of the rostrum of hydrodynamic significance. An increase in R_0 corresponds to growing hydrodynamic significance of the rostrum and vice versa. One can easily trace the functional relations among R_0, N_0 and Δ, particularly in the ontogeny of *Acipenser stellatus*. Table 5 and Figure 19 show that in its early ontogeny, before the fish reaches a length of 30 cm and while $\Delta < -0.03$, the load capacity of the body develops progressively (i.e. N_0 increases), owing to which the body moment M_0 inevitably increases, as do the dimensions of the rostrum (i.e. R_0 increases), which creates a force neutralizing this parasitic moment. The buoyancy increases at a later stage to values $\Delta > -0.03$ causes a reduction in the body's lift capacity (N_0 diminishes), owing to which the body moment M_0 and the rostrum neutralizing it, naturally, diminish too (R_0 diminishes); the curves $R_0 = f(L)$ and $N_0 = f(L)$ are practically coincident (Figure 19).

For Acipenseridae and Chelonioidea the fact that the body creates a lift force at $\alpha = 0$ has been demonstrated experimentally with towed models of *Acipenser stellatus* (Aleyev 1963a) and of *Chelonia mydas* (Aleyev 1965c, 1966a,b). The models were made of wood and had a slightly negative buoyancy. When modelling the sturgeon a specimen 55.0 cm long was taken as the original; the model was 158.0 cm long and was made without paired fins in order to guarantee a lift force in the experiment due to the body only. When making the model of a turtle shell we chose a specimen of *Chelonia mydas* with a shell of length 94.5 cm for the original; the model was 100.0 cm long. In the experiment the model had only one degree of freedom: it could move in the dorso-ventral direction, rotating around a relatively immobile longitudinal axis of the towing system, to which it was joined by means of hinges connected to brass rods rigidly fixed laterally in the model and ensuring that the condition $\alpha = 0$ was observed during towing. The towing speed was 1.4 m/s. With the apparatus stationary and during towing alike, when the hinged device was fixed in the position corresponding to a state of immobility (Figures 20 and 21, frame 1), the models, because of their negative buoyancy and the lateral position of the supporting rods, were placed in the water on their sides, so that their profiles could be observed from above. During free towing the unconstrained models gradually rotated around the longitudinal axis of the towing system under the action of the

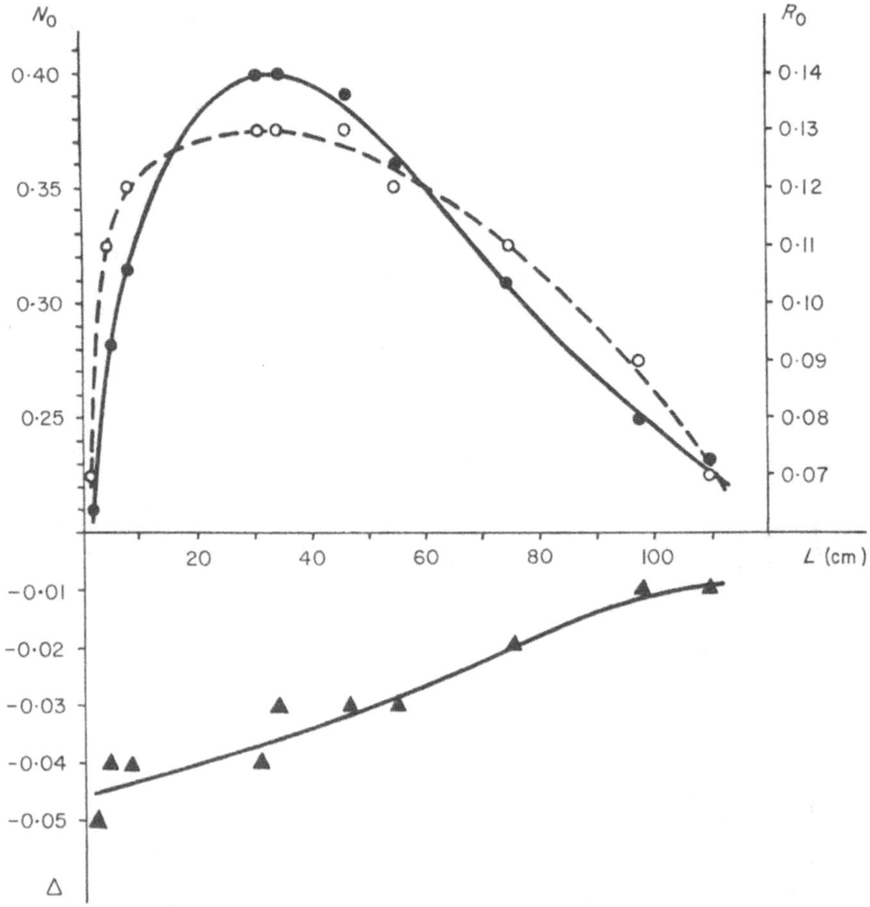

Figure 19. Variation of R_0, N_0 and Δ in the ontogeny of *Acipenser stellatus* Pall. —●—, N_0; - -○- -, R_0; —▲—, Δ.

hydrodynamic lift force, and assumed the normal position characteristic of the original, so that they were observable from above no longer in plan but dorsally (Figures 20 and 21).

2. Active adaptations: special supporting movements

Creation of hydrodynamic life by means of special supporting movements is a property of all cephalopods, some Chondrichthyes (Mobulidae and benthic rays) and the majority of Osteichthyes (with the exception of some primitive groups, such as

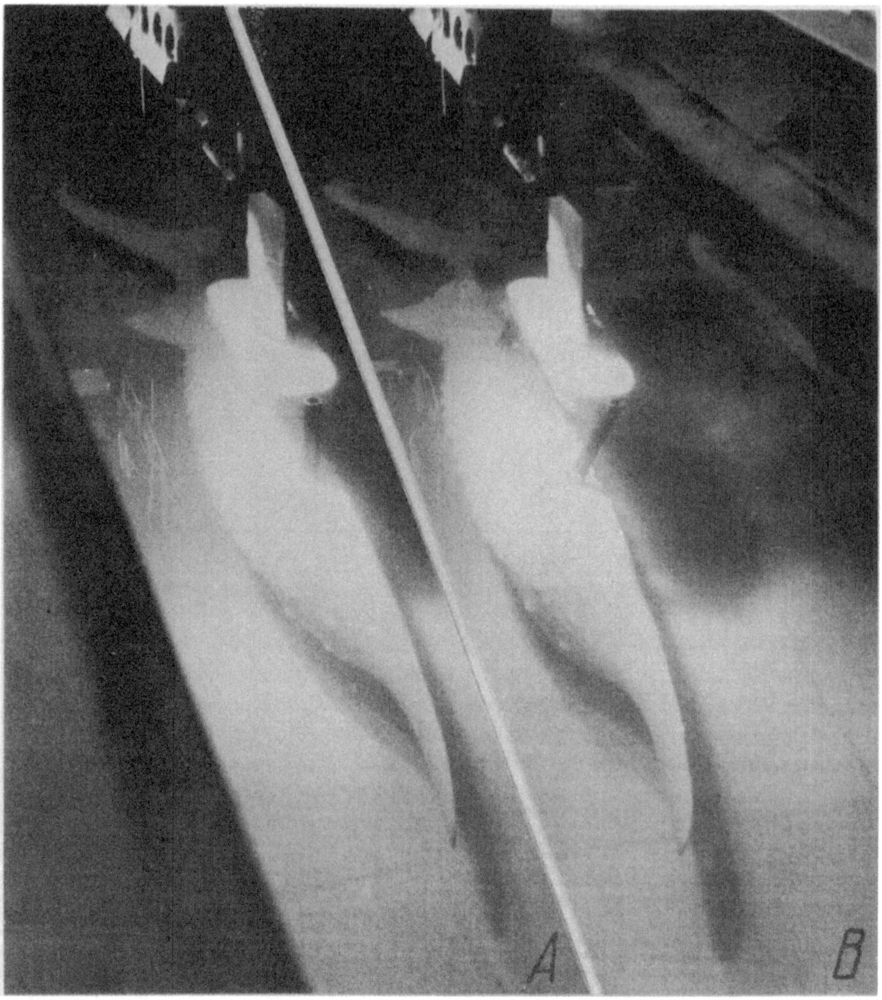

Figure 20. Model of *Acipenser stellatus* Pall. in biohydrodynamic channel. *A*, immobile; *B*, during towing (showing the dorsal shift).

Acipenseriformes, etc.); it is not characteristic of reptiles, birds and mammals since the buoyancy of these animals is always close to neutral.

No special supporting movements have been observed in Sagittoidea; these, with their slightly negative buoyancy, maintain the body suspended through conventional propulsive movements.

In some Teuthoidea, which usually have a slightly negative buoyancy, lift may be

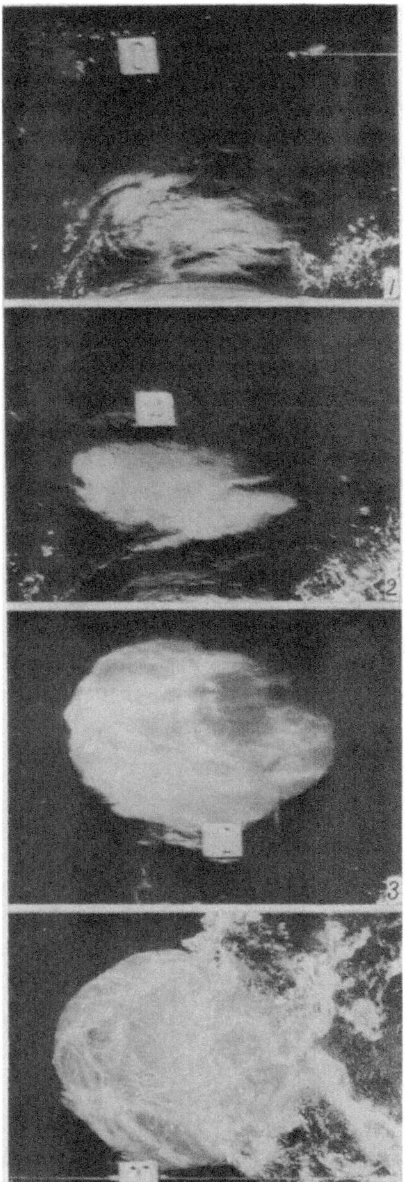

Figure 21. Consecutive photos of a moving model of a turtle shell. See explanation in text.

created on account of undulation of the mantle fins or the action of the hydrojet propulsor. When a squid is 'hanging' motionless in the water in a somewhat inclined position, lift forces are created by both the undulating fins and the action of the hydrojet propulsive mechanism, whose funnel is directed steeply ventrally (Zuyev 1965, 1966). Here the horizontal components of the forces and their moments balance (Figure 22). During rapid swimming the force due to the residual weight is effectively neutralized entirely by the action of the hydrojet propulsive mechanism, whose funnel is always directed slightly downwards.

Among fishes, active adaptations for hydrodynamic purposes are developed to the greatest degree in all species which usually stay in a suspended state in the absence of forward movement (Poeciliidae, Serranidae, etc.). Supporting movements are most often carried out by pectoral fins and the caudal fin, which make up an integral

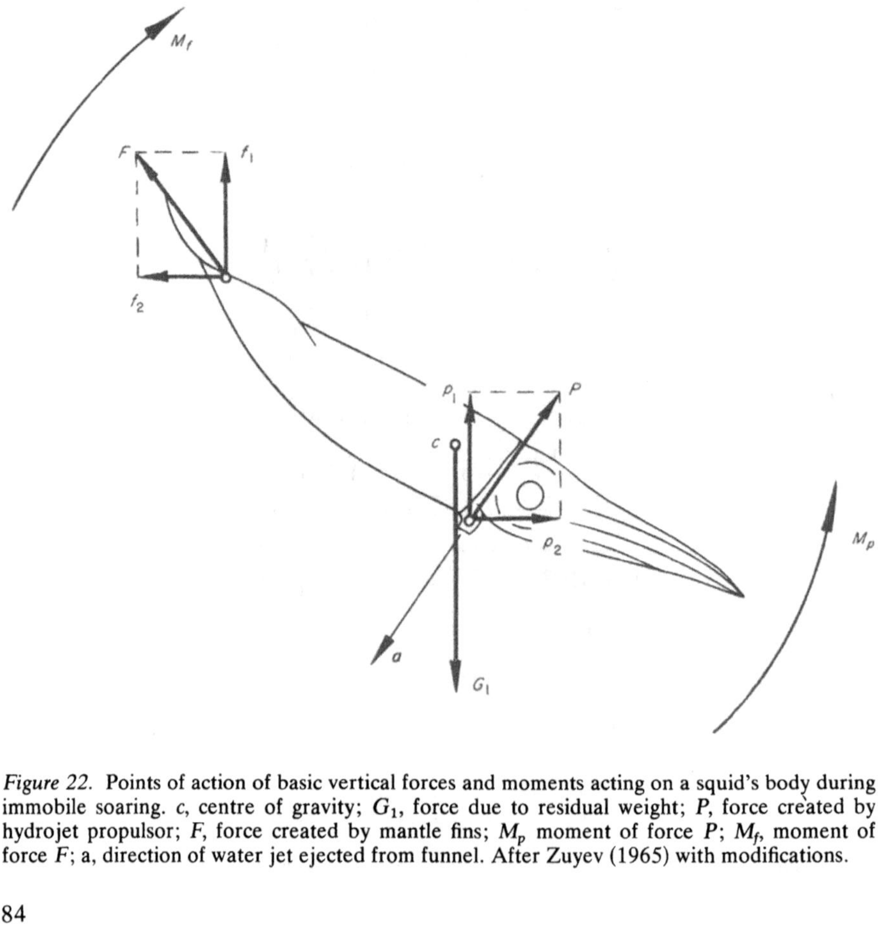

Figure 22. Points of action of basic vertical forces and moments acting on a squid's body during immobile soaring. *c*, centre of gravity; G_1, force due to residual weight; *P*, force created by hydrojet propulsor; *F*, force created by mantle fins; M_p moment of force *P*; M_f, moment of force *F*; a, direction of water jet ejected from funnel. After Zuyev (1965) with modifications.

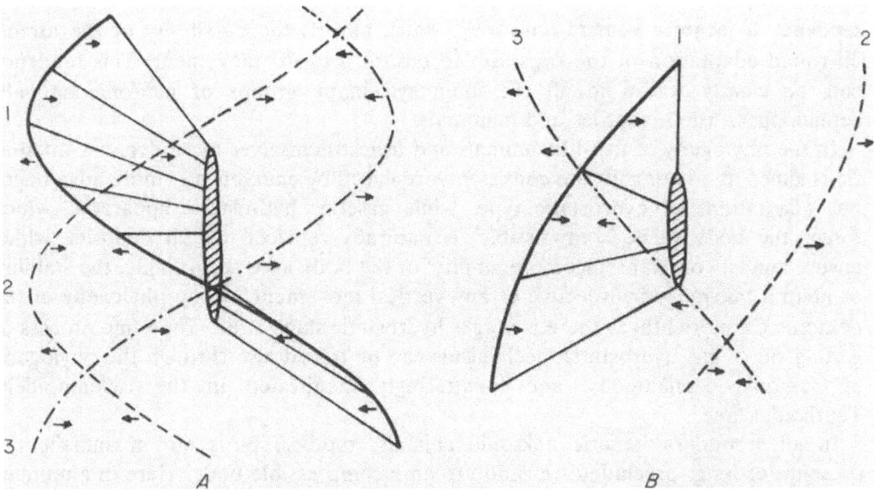

Figure 23. Undulations of pectoral fin when $(A)\frac{2}{3}$ and $(B)\frac{1}{4}$ of a wavelength can be fitted into the fins width. Numbers designate successive positions of fin; arrows indicate direction of fin blade's movement in each position; hatched oval areas show base of fin (after Harris 1937a).

system, ensuring the fish adequate stability; the undulations travel vertically downwards across these fins. Thus the fish seems to be 'suspended' in the body of the water at three points and may even rise, like a helicopter, as has been observed for Poeciliidae (Harris 1937; Aleyev 1963a) and Serranidae (Aleyev 1963a), for example. By changing the mode of action of the pectoral and caudal fins, the fish can easily eliminate roll or pitch or, conversely, create them (Magnan & Sainte-Laguë 1929a; Aleyev 1963a). Various fishes may have different numbers of wavelengths spanning the width of the pectoral fin: sometimes, for example in *Sphaeroides*, up to three-quarters of a wavelength (Figure 23A) and in other cases, for example in Serranidae, only $\frac{1}{4}$–$\frac{1}{6}$ of a wavelength (Figure 23B). The wider the pectoral and caudal fins, the more vertical wavelengths can be fitted into their width and, consequently, the stronger is the reaction force of the water supporting the fish. Therefore, particularly in those fishes in which the motions of the pectoral fins play an important role, these fins are usually wide and rounded, i.e. fan shaped (many Serranidae, Gasterosteidae, Labridae, Poeciliidae, etc.).

5. Phylogeny of adaptations associated with supporting the body in a suspended state

One of the most general trends manifested in the course of adaptation to the nektonic mode of life in the phylogeny of various systematic groups is undoubtedly **the**

tendency to acquire neutral buoyancy, which affords the possibility of the utmost all-round adaptation of the organism to ensure forward movement. This tendency can be clearly traced in all the main systematic groups of nektonic animals: cephalopods, fishes, reptiles and mammals.

In the phylogeny of fish-like animals and fishes themselves hydrodynamic adaptations aimed at creating lift are consistently replaced by energetically more advantageous adaptations of hydrostatic type, while gaseous hydrostatic apparatus, which forces the body to be compressible, is naturally replaced by fat deposits, which ensure almost complete 'incompressibility' of the body and, accordingly, the stability of neutral buoyancy irrespective of any vertical movement. In the phylogeny of the nektonic Chondrichthyes there is no gas hydrostatic stage at all. The same process of getting rid of gas hydrostatic mechanisms can be traced also through the phylogeny of nektonic cephalopods, and is strikingly manifested in the Belemnoidea-Teuthoidea line.

In all secondary aquatic nektonic animals, reptiles, birds and mammals, the presence of lungs precluded transition to an incompressible body. Here in a number of cases adaptations to help the thoracic cavity to withstand considerable pressures appeared. Despite this, however, the presence of gas inclusions in the body and the connection with the surface of the water owing to breathing air are undoubtedly essential limiting factors in the evolution of secondary aquatic nektonic animals.

IV. CREATION OF PROPULSIVE FORCE (LOCOMOTION)

1. General

Forward movement of any animal in any medium can be effected by two methods: first, by attachment to some other moving object, and second, by creation of a propulsive force of its own. The first method is very rare among nektonic animals and is never found without the second (Echeneidae, etc.). Of greater interest is the second method, being more characteristic of nektonic animals and ensured by the most subtle complex of adaptations.

The only mode of locomotion for the great majority of benthonektonic, planktonektonic and eunektonic forms is swimming in the body of the water. Only a very few eunektonic and planktonektonic species, some cephalopods and, among fishes, some Exocoetidae, Hemirhamphidae ex p., Pantodontidae and Gasteropelecidae, are capable not only of swimming, but also of flying; such flight, however, in most cases is but a leap out of the water with passive gliding, and only in still rarer cases is active flight, as in Gasteropelecidae. All xeronektonic species, in addition to swimming, can also travel over solid substrata, either crawling or walking.

In the majority of cases nektonic animals have their back upwards, their head forwards and their longitudinal axis directed along the path of movement. In cephalopods, during swimming the head usually points backwards though they all retain the capacity to move with the head forwards. Many fishes and sea snakes (Hydrophidae) are capable of 'backing up', but none of them uses this mode of locomotion as the principal one, or even frequently. Among special types of locomotion one should point to *Synodonthis batensoda*, which normally swims belly upwards (Heuman, Jacobsohn & Vilter 1941; Daget 1948). Swimming back downwards is also habitual for Phocidae (Ray 1963).

The **propulsive or locomotive force**, imparting forward movement to the animal, is created in all cases as a result of special movements by certain parts of the animal's body. All body parts mechanically interacting with the surrounding medium with the aim of creating a propulsive force are henceforth referred to collectively as the propulsor. The propulsor comprises working propulsive surfaces, which, when the propulsive mechanism is active, directly press against the surrounding medium (water, air or solid substrata).

The **propulsive mechanisms of nektonic animals** fall into three groups.

I. Aquatic propulsors, intended to create a propulsive force during swimming, may be grouped in three main classes: 1. undulatory, 2. paddling and 3. hydrojet.

II. Aerial propulsors, intended to create a propulsive force during flight.
III. Tactile propulsors, intended to create propulsive thrust during motion over the surface of solid substrata (bottom, coast, ice cover, etc.).

The efficiency η of a nektonic animal's propulsive mechanism can be found from the formula

$$\eta = W_m/(W_t - W_0), \qquad (9)$$

where W_m is the energy expended actually converted into thrust, W_t is the total energy expended by the organism within the same period and W_0 is the rest exchange energy. Since various authors (Shuleikin 1934, 1953, 1965, 1966, 1968; Kovalevskaya 1953; Bainbridge 1958a, 1960; Pyatetski 1970; Pyatetski & Kayan 1971; Matyukhin 1973; Matyukhin *et al.* 1973, and others) have suggested numerous modifications of the method of finding the efficiency of the propulsors of swimming animals, the η values obtained by different investigators are, as a rule, not really comparable, owing to which one should approach their assessment with great caution. Usually for fishes $\eta > 0.30$, with η increasing for bigger fishes and higher swimming speeds. Of the greatest interest, however, in determining the properties typical of nekton are not the absolute values of η, which are always subjective to some extent, but its trends with the development of locomotive or other nektonic adaptations. In this connection it should be noted that, in animals with a propulsor undergoing axial undulations, the value of η, all other conditions remaining unchanged, diminishes with increasing frequency of the flexural body movements as has been reported for fishes (Shuleikin 1934, 1953; Pyatetsky & Kayan 1971, and others).

η may be increased or reduced through the development of various morphological adaptations, but a decrease in η due to a less advantageous morphology of the propulsive mechanism is always compensated for by definite adaptations of a physiological nature. Further on, when speaking of increasing or diminishing efficiency of the propulsor in connection with some morphological change or other during ontogeny or phylogeny, we shall assume that the propulsor becomes structurally more-or-less adapted to ensuring high efficiency.

2. Swimming. Undulatory propulsors

For nektonic animals, swimming in the body of the water by means of wavelike motions of the body or fins is the most typical mode of locomotion. This mode of locomotion is characteristic in some measure of nearly all the large systematic groups (classes) of nektonic animals: Sagittoidea, Cephalopoda, Diplorhina, Monorhina, Placodermi, Acanthodei, Chondrichthyes, Osteichthyes, Reptilia and Mammalia. Also, for most of these it is the principal or even the only mode (Sagittoidea, Diplorhina, Monorhina, Placodermi, Acanthodei, Chondrichthyes, most of the Osteichthyes and nektonic Reptilia, and among Mammalia-Sirenia, Cetacea and some

Pinnipedia). Locomotion by means of wavelike body motions is characteristic also of many euplanktonic and nektoplanktonic animals (pelagic Polychaeta, larvae and young fishes, etc.), as well as of many semi-aquatic forms (all the nektoxeric Urodela, the nektobenthic larvae of all Anura and some nektoxeric mammals: Desmanidae, *Enhydra, Castor* and others).

Propulsors acting on the principle of creating wavelike deflexions of the working surfaces are called undulatory propulsors (Aleyev 1969a, 1973b). When a propulsive wave passes along an undulatory propulsor the oblique leading surfaces of the wave constantly encounter resistance on the part of the water, as a result of which the propulsor is subjected to some force P directed contrary to the direction of travel of the propulsive wave. This force P is the propulsive force effecting the animal's forward movement. It is obvious that when the propulsive wave travels caudad along the propulsor the animal will move with its anterior end forward and when the propulsive wave travels craniad the animal will be moving with its posterior end forward. An undulatory propulsor is always bilateral, i.e., when in operation, its two sides alternately act as working surfaces, subjecting themselves to forces of resistance (Figure 24). This is the fundamental distinction between an undulatory propulsor and a paddling one (see below).

During the operation of an undulatory propulsor, the resistance force f acting on each elementary section of its working surface at each moment is directed against the displacement of that element of the working surface relative to the surrounding medium (Figure 24). The propulsive or locomotive force P created by the propulsor is, as Figure 24 shows, the sum of the components p of the elementary forces f along the animal's longitudinal axis or in other words, since the body's longitudinal axis is parallel to the path of progression, along the path of progression. Meanwhile, it follows from Figure 24 that the magnitude of p is directly proportional to the sine of the angle α between the path of progression TT_1 and the tangent DD_1 to the working surface at the point of application of f, i.e. $p = f \sin \alpha$. Therefore

$$P = \sum_{i=1}^{n} (f_i \sin i\alpha), \qquad (10)$$

where n is the number of elementary sections making up the propulsor.

One may distinguish three types of undulatory propulsor: axial, pseudo-axial and peripheral.

1. Axial Undulatory Propulsors

A. General: Principle of action and operation

An undulatory propulsor of the axial type (Aleyev 1969a, 1973b) comprises elements of the axial skeleton and elements of their direct extension, the caudal fin, more or less rigidly mechanically connected to it. During the operation of an axial propulsor the wavelike locomotive deflexions always involve the actual body of the animal in some measure, i.e. the longitudinal axis itself bends. These flexural

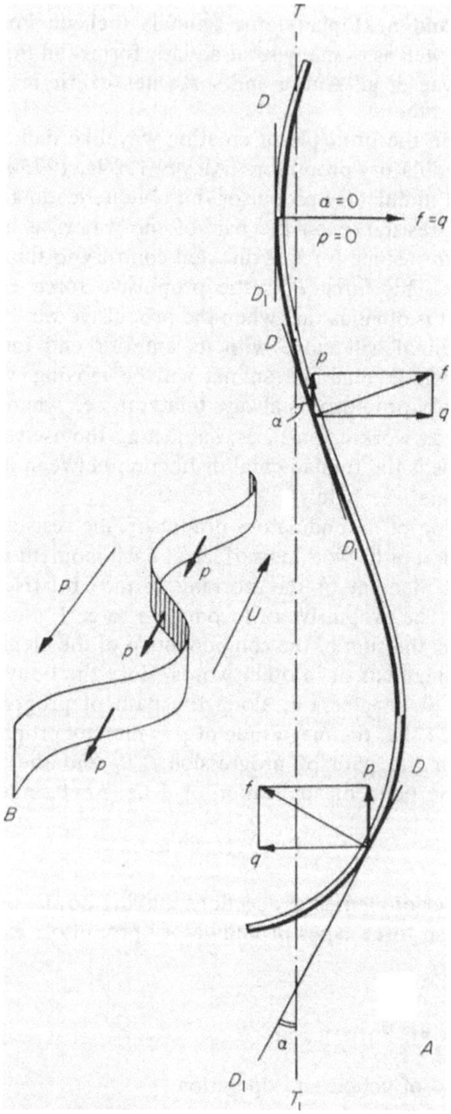

Figure 24. Schematic diagram of (*A*) performance and (*B*) sterometric model of undulatory propulsor. *U*, direction of propulsive wave; *P*, propulsive force; *f*, elementary resistance forces of medium acting on working surfaces of propulsor; *p*, *q*, components of *f*. Heavy lines indicate working surfaces; *TT*, trajectory; *DD*, tangent to working surface at point of action of elementary force *f*. See text for further explanations.

movements are more often than not lateral, since these are more easily ensured on account of the bilateral body symmetry inherent in all nektonic animals: lateral locomotive flexural movements are characteristic of all agnathians, fishes and reptiles (Figures 25A,B, 26–29). More seldom they are dorsoventral, as in Sagittoidea, Sirenia and Cetacea (Figure 25C).

In Sagittoidea the dorsoventral character of the body's locomotive flexural movements and its dorsoventral compression are most probably due to the body having turned into a lifting surface.

The dorsoventral character of the locomotive flexural movements in Sirenia and Cetacea is due to the necessity of making bending the median plane, frequently performed during respiration, as easy as possible. Reptiles, with their much lower level of metabolism, need to rise to the surface of the water much more seldom than mammals, which was of decisive significance in determining the paths of morphological adaptation of members of these groups when transferring from the terrestrial to the nektonic mode of life. Reptiles preserved here their original type of locomotion based on lateral body deflexions, inherent in the majority of them not only in water but also on land, whereas sirenians and cetaceans were compelled to go over to a fundamentally different type of locomotion based on dorsoventral body deflexions.

The axial undulatory propulsor is one of the most common nektonic propulsive mechanisms. It is inherent in Sagittoidea, Agnatha, the great majority of fishes (including the fastest among them), all the variants of reptiles most completely adapted to the aquatic mode of life (including Ichthyosauria, Mosasauridae and Metriorhynchidae) and in all sirenians and cetaceans.

A typical propulsor of the axial undulatory type is characteristic of **Sagittoidea** (Aleyev 1973b). It is well known (Burfield 1927; Ghirardelli 1952, 1968; Rose 1957; Beauchamp 1960; Reeve 1966, and others) that Sagittoidea swim on account of dorso-ventral body deflexions effected by the alternate contraction and relaxation of the longitudinal musculature of the body; the horizontally situated flippers of these animals are incapable of independent movement and function merely as passive elastic planes. According to this author's observations, the propulsive mechanism of a *Sagitta setosa* about 2 cm long involves its entire body, and the amplitude of the propulsive wave noticeably increases caudad, reaching at the rearmost point of the body about 20% of the body's total length. About half a propulsive wavelength can be fitted into the entire length of the body, and flexural movements are made at a frequency of not less than 10 c/s. The main part of the propulsive mechanism is the rear half of the body, which carries lateral fins and the caudal fin. In smaller forms (*Sagitta setosa*, etc.) the caudal fin is rounded at the end, while in larger species (*Flaccisagitta hexaptera*, etc.) it is sometimes divided into two lobes (Murray & Hjort 1912; Kuhl 1938; Hyman 1959; Ghirardelli 1968).

An axial undulatory type of propulsor is very characteristic of all **the agnathians** and the great majority of **fishes**. That fishes move by bending their body was known in India in the 4th century BC (Hora 1935). Later on, the same was noted by

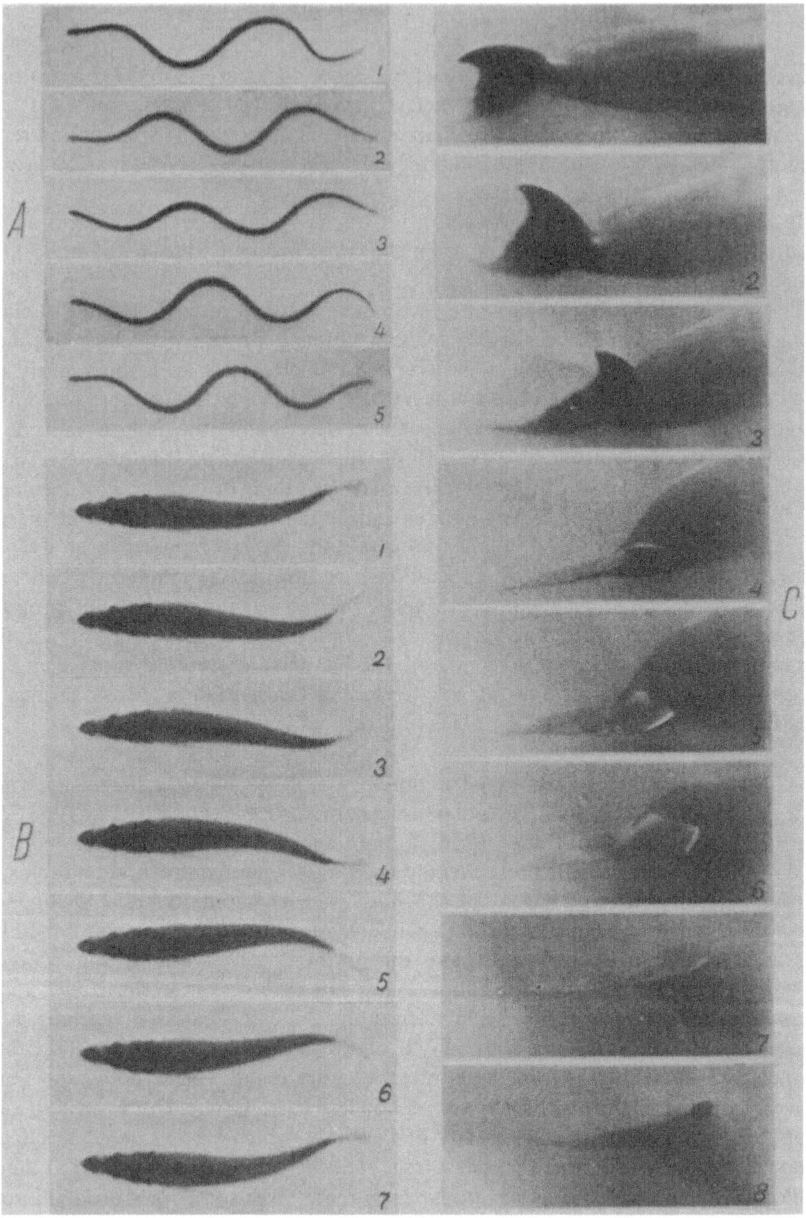

Figure 25. Axial undulatory propulsor. *A*, *Enhydrina schistosa* (Daud.), anguilliform propulsor; *B*, *Trachurus mediterraneus ponticus* Aleev, scombroid propulsor based on lateral bending; *C*, *Tursiops truncatus* (Montagu), scombroid propulsor based on dorsoventral bending. Cinerecord.

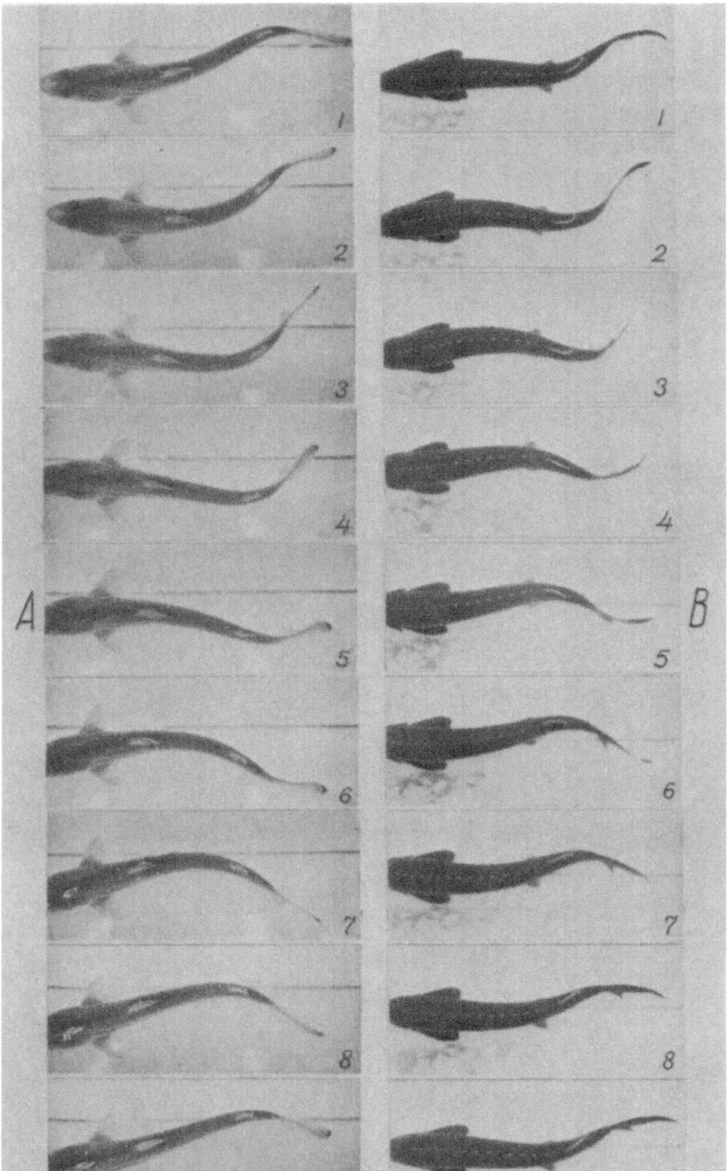

Figure 26. Axial undulatory propulsor based on lateral bending. *A*, *Squalus acanthias* L.; *B*, *Acipenser güldenstädti colchicus* V. Marti. Cine-record.

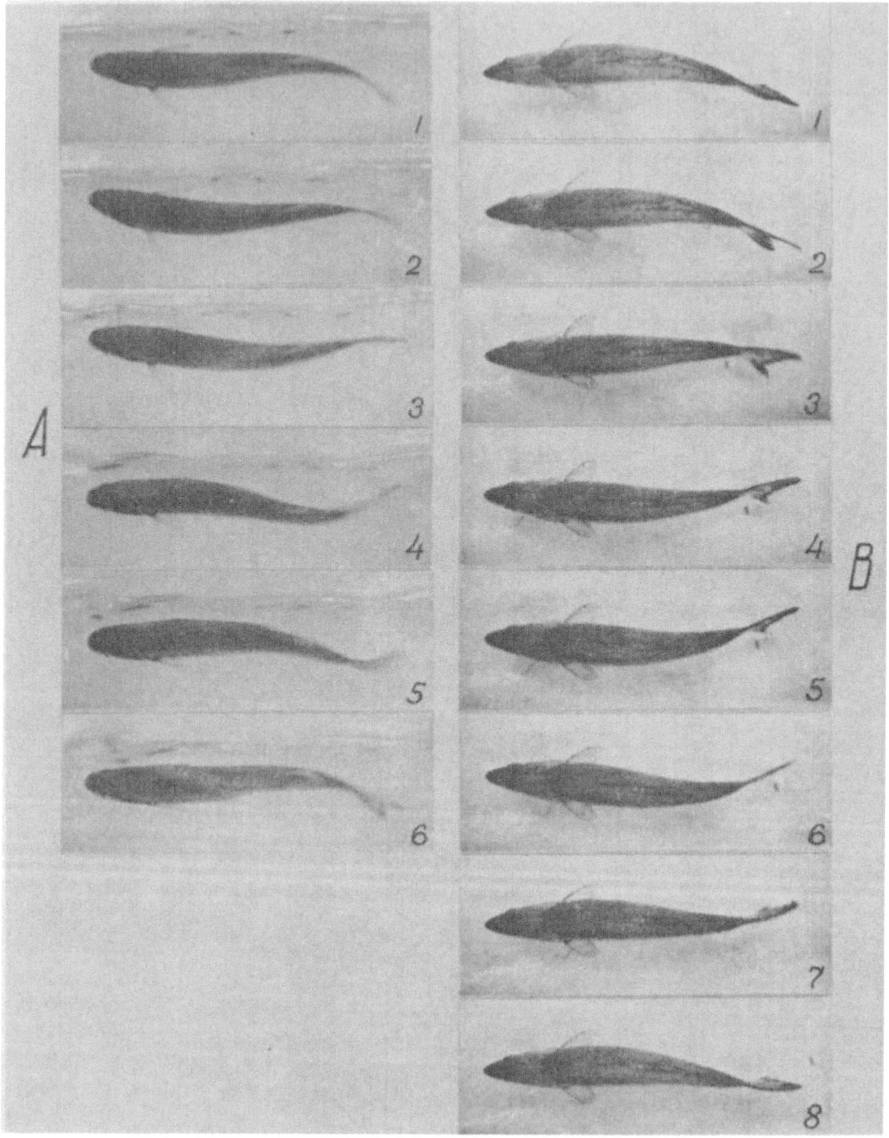

Figure 27. Axial undulatory propulsor based on lateral bending. *A*, *Mugil auratus* Risso; *B*, *Pomatomus saltatrix* (L.). Cine-record.

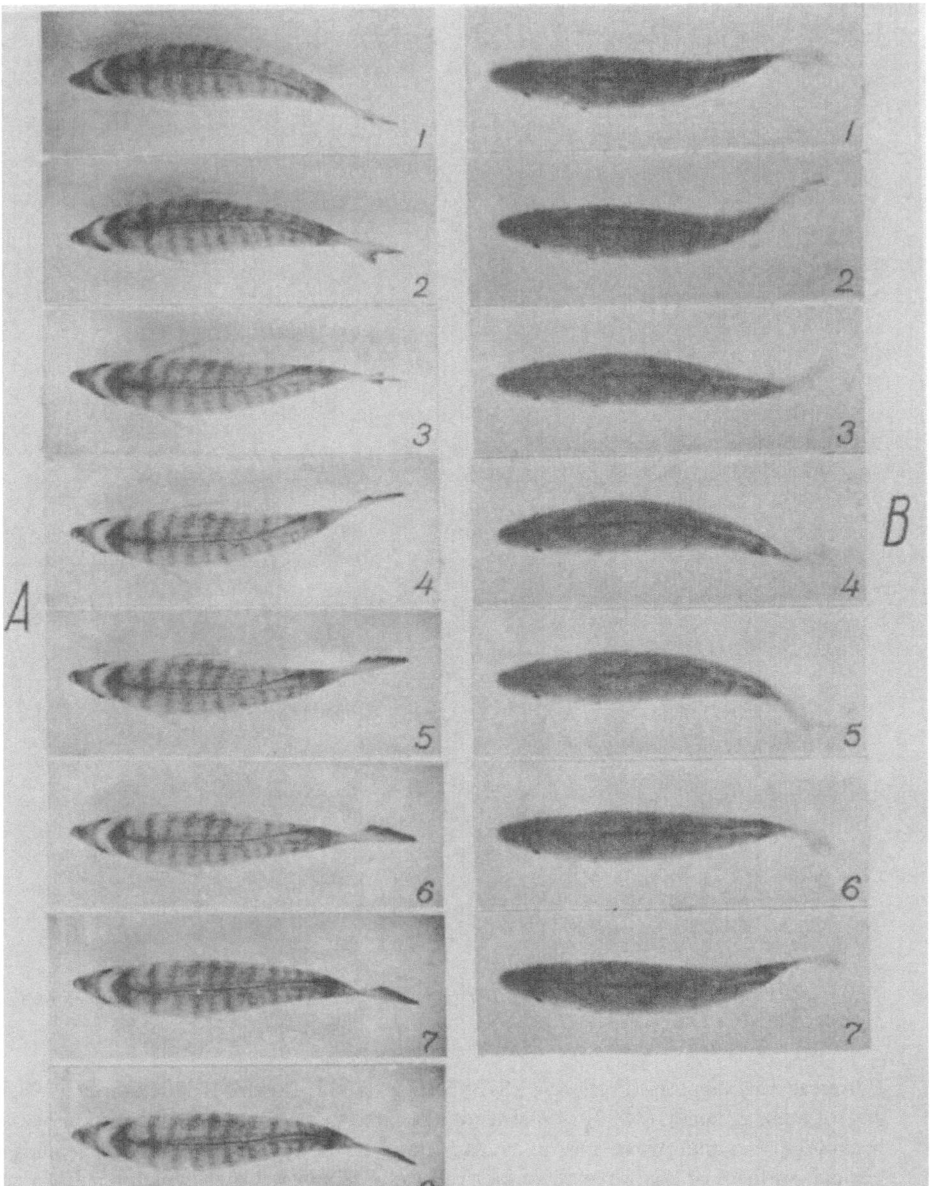

Figure 28. Axial undulatory propulsor based on lateral bending. *A*, *Puntazzo puntazzo* (Cetti); *B*, *Diplodus annularis* (L.). Cine-record.

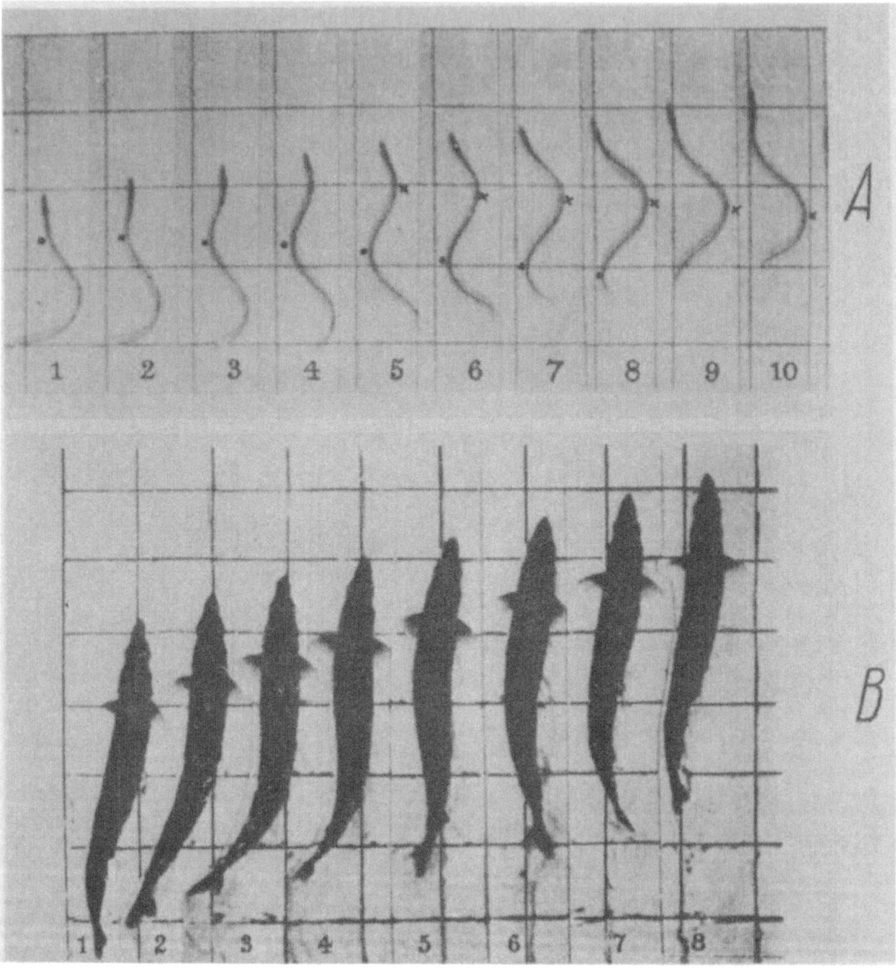

Figure 29. Propulsive movements of (*A*) *Anguilla anguilla* (L.) and (*B*) *Scomber scombrus* L. Cine-record (after Gray 1933c).

European investigators (Pettigrew 1873; Strasser 1882; Amans 1888; Marey 1894, and others). Strasser (1882) was one of the first to point out that a fish moves because a muscular wave passes caudad along its body and the oblique leading (caudal) surfaces of this wave encounter resistance from the water, which results in a forward force being continuously imparted to the fish's body. At a later period this mode of fish locomotion was examined theoretically in greater detail as an example of oscillating movement connected with the propagation along the body of waves of definite wavelength and amplitude (Gray 1930, 1933a–e; Shuleikin 1934, 1953, 1968; Lighthill 1960, 1969, 1971; Kelly 1961; Kelly, Rentz & Siekmann 1964; Smith & Stone 1961; Bonthron & Fejer 1962; Siekmann 1962a,b; 1963a,b, 1965, 1966; Aleyev 1963a, 1965b, 1969a, 1973b; Gadd 1963; Pao & Siekmann 1964; Reece, Uldrick & Siekmann 1965; Fierstine 1966; Magnusson 1966; Magnusson & Prescott 1966; Lund 1967; Logvinovich 1970, 1973; Pershin 1970, and others).

The shorter the relative length of the fish's body, the fewer myomeres it contains, the less flexible it is and the greater is the degree to which propulsive bending movements are localized in the rear part of the body and vice versa (Houssay 1909a,b, 1912; Breder 1926; Aleyev 1963a, 1969a). Long flexible fishes, such as eels, generate a propulsive wave whose amplitude is large and remains nearly constant throughout the length of the body, only slightly increasing caudad, whereas in fishes with the conventional form of body the wave amplitude is smaller, notable increasing caudad (Figures 25–29). The speed of propulsive waves, like the speed of the fish's forward movement, is found to be lowest in long anguilliform fishes and highest in fishes with a short dense body, such as Scombridae (Magnan 1929, 1930; Gray 1930, 1933a–e, 1936a, 1949, 1953a,b, 1957, 1968; Shuleikin 1934, 1949, 1953, 1968; Harris 1934; Langton 1949; Bainbridge 1958a,b, 1960; Nursall 1956, 1958a,b; Oehmichen 1958; E. Kramer 1959, 1960; Aleyev 1963a, 1969a, 1973b; Bohun & Winn 1966; Hertel 1967a). It has been shown (Boddeke, Slijper & van der Stelt 1959) that short muscular fibres are capable of more rapid contractions than long ones. The three-dimensionally curved myomeres of fish-like organisms and fishes themselves make for maximum effectiveness of the performance of the muscular apparatus during flexural motions of the animal's body (Chevrel 1913; Nursall 1956, 1962; Le Danois 1958). The significance of the caudal fin as a locomotive organ increases at lower aspect ratios (Gray 1933e).

The two-lobe shape of the caudal fin characteristic of the majority of nektonic fishes, many eunektonic reptiles, many eunektonic mammals and some Sagittoidea should be regarded above all (Aleyev 1959b, 1963a, 1965b) as an adaptation aimed at preventing vortex formation on the caudal fin and using the kinetic energy of the boundary layer. It has been shown (Aleyev 1959b, 1963a, 1973b) that, in nektonic animals with greater linear dimensions, greater absolute and relative speeds of movement and higher *Re* values, the propulsive movements are increasingly concentrated at the rear end of the body, which leads in extreme cases to the caudal fin turning into the only propulsive element and its shape becoming ever more concave and elongated transversely, resulting in the improvement of its qualities as a propulsor.

Compared with anguilliform fishes, which have poorly streamlined bodies, fishes with a short, well-streamlined body have a lower η. Shuleikin (1934) explains this by the fact that in the latter case horizontal transverse forces acting upon the front and rear ends of the fish body during swimming cause oscillations of the body to the right and left of its course, whereas in the former case, of anguilliform fishes, these are only slight. It should be borne in mind, however, that during fast movement of such fishes as *Scomber* the amplitude of the propulsive wave is considerably less than during slow movement, as in the ciné records of *Scomber* movements made by Gray (1933c) and used by Shuleikin (1934) for assessing the magnitude of the above-mentioned transverse forces. Therefore the parasitic role of these transverse forces, even in short-bodied fishes, such as *Scomber*, is probably not very great.

In the opinion of this writer (Aleyev 1963a), the lower η in fishes with a short, well-streamlined body, as compared with anguilliform fishes, is explained first of all by the decrease in the relative magnitude of the area of the working surfaces of the propulsive mechanism: the propulsor of an eel is its entire body, whereas in such fishes as the tuna group it is caudal fin only, owing to which the relative area of the working surfaces of their propulsor is about 10 times less than that in anguilliform fishes (see below).

The reports (Lowndes 1955) that the propulsive wave of some flexible fishes (*Scyliorhinus* and *Conger*) is not a plane wave but a spiral wave have not been confirmed by the results of special filming (Aleyev 1963a). In the nektobenthic *Myxine glutinosa* a helical rotation of the body is sometimes observed (Hans 1960), most probably because of the ability of hag-fishes to burrow into the body of their prey like a borer (Aleyev 1963a).

In the phylogeny of aquatic **reptiles** an axial undulatory propulsor develops more often in cases where the primary terrestrial form had a somewhat elongated body and tail and the mode of locomotion on land was close to that of snakes. Such are the Pleurosauria, Mesosauria, Ichthyosauria, Thalattosauria, Aigialosauridae, Mosasauridae, Dolichosauriae, Acrochordinae, Palaeophidae, Hydrophidae, Teleosauridae, Metriorhynchidae, Cholophidia, Phytosauria, Claraziidae, Choristodera, etc. (Bauer 1898; Merriam 1902; Abel 1907, 1912, 1922; Huene 1916, 1923, 1956; Kuhn 1937; Oehmichen 1938; Romer 1956; Vyushkov 1964; Konzhukova 1964d; Maleyev 1964a–e; Rozhdestvensky 1964a,b; Tatarinov 1964g,k–t; Khozatski & Yuryev 1964; Chudinov, 1964a–d, and others). All these animals have as their propulsors a flexible body together with a laterally compressed paddle-like tail (Abel 1907, 1912, 1922; Kuhn 1937; Huene 1956; Vyushkov 1964; Konzhukova 1964i; Rozhdestvensky 1964a; Tatarinov 1964k,l; Khozatski & Yuryev 1964; Carpenter 1966; Bellairs 1969). Moreover, the propulsive role of the tail is all the more significant the shorter and thicker the body. In some cases there appears on the tail a clearly demarcated two-lobe fin (Metriorhynchidae and Ichthyosauria). In early Triassic Ichthyosauria, which had a relatively elongated body and a faintly demarcated tail fin (Figures 10B,C, 11A,B), the function of a propulsive mechanism was fulfilled by the entire body together with the tail, yet in later Jurassic and Cretaceous forms the body became shorter, thicker and less flexible, while propulsive motions were performed almost entirely by the tail fin, which assumed the two-lobe shape characteristic of all fast-swimming nektonic animals (Figures 10D,E 11D,E, 98C) (Bauer 1898; Merriam 1902; Fraas 1911; Abel 1912; Heune 1916, 1923, 1956; Kuhn 1937; Oehmichen 1938; Kripp 1954; Romer 1956; Tatarinov 1964k; Bellairs 1969).

In aquatic Ophidia propulsive movements are performed by the whole body (Figure 25A). In the most specialized eunektonic Hydrophidae the entire body, except the frontal part, is considerably laterally compressed (Figure 12C).

The propulsor of **sirenians** and **cetaceans** is the tail flipper (Murie 1880; Dexler & Freund 1906; Abel 1912; Kellogg 1928, 1938; Gray 1936b, 1968; Slijper 1936, 1958, 1961; Gray & Parry 1948; Narkhov 1939; Parry 1949a,b; Petit 1955; Tomilin 1957, 1962a–c, 1965; Kruger 1958; Aleyev 1959b, 1963a, 1965b; Backhouse 1960; Borkhvardt 1965; Harrison & King 1965; Pershin, Sokolov & Tomilin 1970; Yablokov, Belkovich & Borisov 1972, and others). This acts on the same principle as the caudal fin of fishes (Aleyev 1959b, 1963a, 1965b, 1973b; E. Kramer 1960), the only difference being that it is not vertical, as in fishes and aquatic reptiles, but horizontal, and its oscillations are, accordingly, not horizontal but vertical (Figure 25C).

Rotating movements of the caudal peduncle allegedly performed during swimming by dolphins (Shuleikin 1935, 1949, 1953, 1968; Stas 1939a,b) have not been confirmed (Parry 1949b, Sleptsov 1952; Rosen 1961; Tomilin 1957, 1965; Aleyev 1959b, 1963a, 1965b, 1973b; Yablokov, Belkovich & Borisov, 1972). A moving dolphin *Phocoena phocoena* was specially filmed in a biohydrodynamic channel 21 m long with the aid of a 'Scopa' automatic scanning system, built on the principle of photoelectric scanning (see chapter V). This quite definitely showed that during the performance of propulsive movements the tail fluke moves rhythmically in the dorso-ventral plane without any lateral displacements whatever and without any rotation around the animal's logitudinal axis: throughout the propulsive cycle, the centre of the tail fluke (the dorsal crest of the caudal peduncle) is seen in the frame at the same distance from the longitudinal black lines drawn on the bottom of the channel, along which the dolphin is moving (Figure 30). What one sees is only a decrease in size of the image of the tail fluke at the moment of maximum downward deflexion, when it is farthest from the cine camera (Figure 30, frames 1, 9), and conversely, an increase in the image size at the moment of maximum upward deflexion, when the fluke is closest to the camera (Figure 30, frame 5), which was situated above the dolphin on the platform of a self-propelled automatic scanning system which followed the dolphin along a monorail track running above the water (Figure 31).

The assertion that '. . . the body of fishes and dolphins performs rotating movements around the longitudinal axis synchronous to the performance of the tail fluke' and that the presence of such movements '. . . has been proved by filming materials both in fishes (Magnan & Sainte-Lague, 1929; Gray 1933) and in dolphins (Shuleikin 1953, 1965) . . .' (Kudryashov 1969, p. 62) is groundless since these sources (Magnan & Sainte-Lague 1929a; Gray 1933c; Shuleikin 1953, 1965) do not show the presence of such rotating movements by direct observation. The rotating movements performed by the tail of a dolphin immobilized on a special dry stand in the air (Shuleikin 1935, 1953) cannot be associated with the normal propulsive movements of a dolphin in natural surroundings. Nor are the above-mentioned rotatory movements proved by material obtained (Kudryashov 1969) in a narrow annular pool of

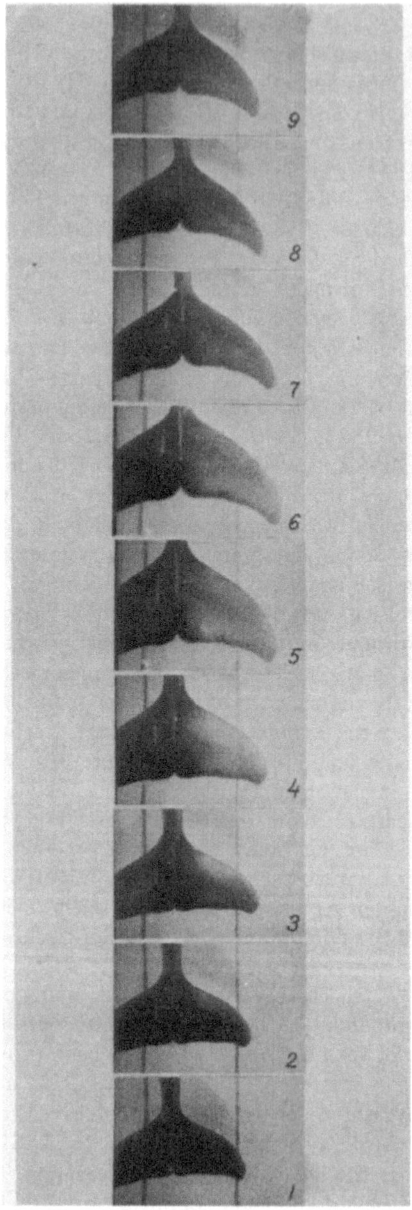

Figure 30. Plan view of propulsive movements of tail fin of *Phocoena phocoena* (L.). Filmed in biohydrodynamic channel by means of automatic scanning system, moving along channel, following dolphin along monorail; lines are longitudinal lines drawn on channel bottom.

Figure 31. Phocoena phocoena (L.) in biohydrodynamic channel. Moving over dolphin on monorail is the automatic scanning system with a cine-camera which filmed the cine-record shown in figure 30.

the Hydrophysical Marine Institute, Ukrainian SSR Academy of Sciences, where dolphins were compelled to swim in an arc, continuously turning laterally.

The axial undulatory propulsor is inherent to some degree also in a number of nektoxeric animals: Amphibia (Urodela and larvae of Anura), Reptilia (Crocodylidae, *Amblyrhynchus*, etc.) and Mammalia (*Enhydra*, *Castor*, *Ondatra*, Desmanidae, etc.)

B. The propulsive action of the body

It is clear from the above that the action of the animal's body as an axial undulatory propulsor necessitates, from the morphological point of view in particular, the satisfaction of three basic conditions (Aleyev 1973b): 1. the presence of a certain body flexibility, permitting the body to perform lateral or dorso-ventral undulations of adequate amplitude; 2. the presence of a definite muscular apparatus to ensure the propulsive undulations of the body; 3. the presence of a definite lateral (or dorso-ventral) surface which, as the animal performs swimming movements, directly presses against the surrounding water.

When examining the propulsive action of the body it is advisable to take into account the distribution of individual morphological peculiarities along the longitudinal axis of the nektonic animal's body by dividing it into five equal elementary sections, each $0.2L$ wide. Division of the body into more than five sections is undesirable, since when examining small animals, say 5–8 cm long, the absolute values of the measurements obtained are very small, which considerably decreases their accuracy (Aleyev 1963a).

We always determined transverse **body flexibility** in the plane of the propulsive bendings of the body: for fishes and reptiles (snakes) the lateral plane, for cetaceans the dorso-ventral. Upon bending the animal in this plane such that its body forms an arc of angle 45–90°, we can measure the chords of the arcs formed by each elementary section to obtain five values: d_1, d_2, d_3, d_4 and d_5 (corresponding to each section numbered from the head). By subtracting each d_n found from the length of an elementary section ($0.2L$) and dividing the differences obtained by their sum, we shall find the index E_n of transverse flexibility of each of the sections (Aleyev 1963a):

$$E_n = \frac{0.2L - d_n}{\sum_{i=1}^{5} (0.2L - d_i)}, \qquad (11)$$

where n is the number of section under examination.

Only freshly killed animals after the disappearance of rigor mortis were used for this study. The frontal point of the animal, lying on the table, was fixed with a special peg, and a weighted cord was attached to the end of the vertebral column. By means of a pulley the weight was dropped over the edge of the table so that it exerted a pull on the posterior point of the vertebral column, bending the body (Figure 32). The axle of the pulley could be adjusted vertically, which in every case ensured that the

Figure 32. Instrument for determining body flexibility of nektonic animals.

pull was horizontal. When the body had assumed its final position we twice found the magnitude of d_n for each of the sections by measuring the distance between steel needles that marked the borders between them, for fishes and reptiles when bending right and left and for cetaceans, when bending upwards and downwards; the final d_n values were taken as the means of these measurements.

When measuring the d_n for particularly flexible animals (snakes and eels), we had to bend the individual sections one after another with a standard load and sometimes reduce their length to $0.05L$, obtain for each of the $0.2L$ long sections four quantities of the form $0.05L - d_n$ and then find their average. Since the right-hand side of (11) is the ratio of a quantity of the form $0.2L - d_n$ to their sum, this change of method does not affect the final results.

The method described is not applicable to Sagittoidea, since their bodies have too great a flexibility. We assessed the flexibility of Sagittoidea by visual observation of the amplitude of the propulsive wave.

The E_n values we obtained for 37 nektonic species are given in Table 6, where the animals are arranged in order of decreasing E_1, E_2, E_3, etc., i.e. in order of caudally diminishing body flexibility, as well as in Figure 33 in the form of the curves $E_n = f(L)$. Table 6 and Figure 33 show that in the overwhelming majority of cases nektonic animals exhibit increasing body flexibility from head to tail; in elongated, snake-like and anguilliform animals this increase is mild (Table 6; Figure 33, *Anguilla*) and in animals with a shorter and denser body more pronounced (Table 6; Figure 33, *Sprattus-Pomatomus*). In the fastest fishes we observe a secondary decrease in flexibility in the tail portion of the body (Table 6; Figure 33, *Sarda-Istiophorus*). All these peculiarities in the distribution of flexibility along the body's longitudinal axis are associated with adaptation to various swimming speeds and a definite mode of forward movement.

Body flexibility in nektonic animals depends on many factors. Of the most general significance in this regard are **the size and shape of the cross-sections of the body.** The relative magnitude Q_n of the cross-sectional area q_n of the body in section n may be expressed as the ratio of the side of a square of equal area to the animal's effective length L_c:

$$Q_n = q_n^{1/2} L_c^{-1}. \tag{12}$$

The ratio G_n of the diameter k_n of this cross-section parallel to the propulsive flexural movements to the diameter l_n perpendicular to k_n will serve as the relative magnitude of the shape of the cross-section of the body in section n:

$$G_n = k_n l_n^{-1}. \tag{13}$$

A decrease in Q_n and G_n corresponds to an increase in body flexibility in the direction of propulsive flexural movements and vice versa. For fishes and Hydrophidae usually $G_n < 1$ (Table 6, Figures 12C, 96–98); the lateral compression of the body of fishes accompanying its conversion into a propulsive mechanism has been noted by many authors (Schlesinger 1911a; Schmalhausen 1916; Breder 1926; Gray 1933c,d; Aleyev 1963a, and others). The inequality $G_n < 1$ is equally characteristic of Sagittoidea; for a 2.2 cm long *Sagitta setosa* we obtained $G_1 = 0.70$, $G_2 = 0.78$, $G_3 = 0.80$, $G_4 = 0.90$ and $G_5 = 0.95$.

The second primary factor influencing body flexibility is the **number of flexible links in the axial skeleton** i.e. the number of vertebrae. The effect of this factor on flexibility, noted earlier for fishes (Barsukov 1959a; Yakovlev 1966 and others) and for cetaceans (Pershin 1969a), is characteristic in some measure of all vertebrates except forms with a very low degree of axial-skeleton calcification (sharks, etc.). We

Table 6. Indices of transverse flexibility E_n, cross-sectional area ratios Q_n, cross-sectional shape indices G_n and numbers of elastic joints in axial skeleton V_n for nektonic animals.

Species, nektonic type and V ($\equiv \sum V_n$)	Length L to end of vertebral column (cm)	Index	Number n of body section: 1	2	3	4	5
Enhydrina schistosa (Daud.) (EN), $V=203$	90.0	E_n	0.29	0.17	0.11	0.14	0.29
		Q_n	0.01	0.02	0.02	0.02	0.01
		G_n	1.20	0.61	0.51	0.34	0.40
		V_n	46	34	30	35	58
Anguilla anguilla (L.) (EN), $V=113$	91.0	E_n	0.15	0.16	0.18	0.21	0.30
		Q_n	0.05	0.05	0.05	0.04	0.03
		G_n	0.92	0.83	0.83	0.63	0.40
		V_n	16	20	21	23	33
Gymnammodytes cicerellus (Risso) (BN), $V=64$	10.3	E_n	0.02	0.23	0.24	0.24	0.27
		Q_n	0.05	0.06	0.06	0.06	0.03
		G_n	0.80	0.75	0.75	0.69	0.85
		V_n	2	15	15	15	17
Trichiurus lepturus L. (EN), $V=175$	124.0	E_n	0.02	0.09	0.12	0.22	0.55
		Q_n	0.03	0.03	0.03	0.02	0.01
		G_n	0.57	0.62	0.60	0.50	0.40
		V_n	16	30	36	43	50
Sprattus sprattus phalericus (Risso) (*PN/EN*), $V=48$	10.0	E_n	0.00	0.20	0.22	0.25	0.33
		Q_n	0.08	0.11	0.10	0.08	0.05
		G_n	0.51	0.44	0.42	0.41	0.33
		V_n	2	14	12	11	9
Engraulis encrasicholus ponticus Alex. (EN), $V=47$	10.0	E_n	0.00	0.20	0.21	0.24	0.35
		Q_n	0.09	0.12	0.14	0.11	0.06
		G_n	0.82	0.68	0.69	0.67	0.58
		V_n	3	10	10	12	12
Clupeonella delicatula delicatula (Nordm.) (PN/EN), $V=42$	7.0	E_n	0.00	0.19	0.21	0.25	0.35
		Q_n	0.10	0.15	0.15	0.12	0.06
		G_n	0.77	0.43	0.42	0.43	0.31
		V_n	0	10	11	11	10
Alosa kessleri pontica (Eichw.) (EN), $V=49$	25.0	E_n	0.00	0.18	0.22	0.26	0.34
		Q_n	0.08	0.12	0.13	0.10	0.06
		G_n	0.43	0.48	0.43	0.44	0.40
		V_n	0	14	12	12	11
Belone belone euxini Günth. (EN), $V=75$	50.4	E_n	0.00	0.16	0.39	0.33	0.12
		Q_n	0.01	0.04	0.04	0.04	0.03
		G_n	1.61	0.67	0.67	0.63	0.90
		V_n	0	13	20	21	21
Spicara smaris (L.) (EN), $V=24$	15.5	E_n	0.00	0.15	0.22	0.27	0.36
		Q_n	0.11	0.16	0.18	0.14	0.07
		G_n	0.77	0.43	0.42	0.43	0.31
		V_n	0	7	6	6	5
Sciaena umbra L. (BN), $V=25$	27.1	E_n	0.00	0.13	0.18	0.25	0.44
		Q_n	0.12	0.20	0.20	0.15	0.07
		G_n	0.65	0.52	0.52	0.50	0.55
		V_n	1	7	6	6	5

Table 6 (*contd.*)

Species, nektonic type and V ($\equiv \sum V_n$)	Length L to end of vertebral column (cm)	Index	Number n of body section				
			1	2	3	4	5
Squalus acanthias L. (EN)	112.0	E_n	0.00	0.06	0.13	0.23	0.58
		Q_n	0.07	0.10	0.06	0.04	0.02
		G_n	1.66	0.91	0.89	1.00	1.00
Tetrapturus belone Raf. (En), $V=24$	149.5	E_n	0.00	0.06	0.47	0.35	0.12
		Q_n	0.01	0.06	0.06	0.05	0.04
		G_n	1.00	0.48	0.53	0.48	0.65
		V_n	0	6	6	6	6
Pomatomus saltatrix (L.) (EN), $V=26$	60.2	E_n	0.00	0.06	0.19	0.36	0.39
		Q_n	0.10	0.15	0.14	0.12	0.06
		G_n	0.55	0.58	0.57	0.53	0.75
		V_n	0	7	7	6	6
Sarda sarda (Bl.) (EN), $V=53$	48.8	E_n	0.00	0.05	0.23	0.38	0.34
		Q_n	0.08	0.15	0.16	0.11	0.04
		G_n	0.75	0.68	0.68	0.55	1.49
		V_n	1	14	13	12	13
Trachurus mediterraneus ponticus Aleev (EN), $V=24$	43.3	E_n	0.00	0.05	0.20	0.43	0.32
		Q_n	0.10	0.15	0.16	0.12	0.06
		G_n	0.74	0.67	0.64	0.83	1.51
		V_n	0	8	6	5	5
Scomberomorus commersoni (Lac.) (EN), $V=49$	59.7	E_n	0.00	0.05	0.13	0.42	0.40
		Q_n	0.08	0.13	0.13	0.10	0.05
		G_n	0.90	0.69	0.57	0.64	0.97
		V_n	3	14	11	11	10
Lebistes reticulatus Peters (BN), $V=30$	1.6	E_n	0.00	0.05	0.12	0.37	0.46
		Q_n	0.15	0.21	0.19	0.13	0.10
		G_n	1.43	0.88	0.75	0.50	0.30
		V_n	0	6	8	8	8
Gambusia affinis holbrooki (Girard) (BN), $V=32$	2.0	E_n	0.00	0.05	0.12	0.37	0.46
		Q_n	0.10	0.17	0.17	0.12	0.08
		G_n	1.00	0.68	0.58	0.53	0.33
		V_n	0	9	8	8	7
Odontogadus merlangus euxinus (Nordm.) (BN), $V=53$	41.6	E_n	0.00	0.05	0.11	0.25	0.59
		Q_n	0.12	0.16	0.15	0.10	0.04
		G_n	1.00	0.66	0.62	0.68	0.67
		V_n	0	12	11	11	19
Istiophorus platypterus (Show & Nodder) (EN), $V=24$	152.0	E_n	0.00	0.04	0.48	0.36	0.12
		Q_n	0.01	0.06	0.07	0.06	0.04
		G_n	1.25	0.46	0.42	0.50	0.50
		V_n	0	4	7	6	7
Thunnus alalunga (Bonn.), (EN), $V=39$	67.0	E_n	0.00	0.04	0.13	0.50	0.33
		Q_n	0.11	0.19	0.19	0.13	0.03
		G_n	0.86	0.72	0.69	0.98	3.45
		V_n	1	11	10	8	9

Table 6 (*contd.*)

Species, nektonic type and V ($\equiv \sum V_n$)	Length L to end of vertebral column (cm)	Index	Number n of body section: 1	2	3	4	5
Mugil auratus Risso (BN), $V=24$	34.2	E_n	0.00	0.04	0.13	0.28	0.55
		Q_n	0.12	0.16	0.17	0.12	0.07
		G_n	0.86	0.71	0.67	0.65	0.60
		V_n	1	6	5	6	6
Abramis brama (L.) (BN), $V=41$	50.5	E_n	0.00	0.04	0.12	0.37	0.47
		Q_n	0.11	0.18	0.19	0.14	0.07
		G_n	0.56	0.33	0.31	0.37	0.55
		V_n	0	8	12	11	10
Sphyraena barracuda (Walb.) (EN), $V=24$	93.8	E_n	0.00	0.04	0.09	0.26	0.61
		Q_n	0.06	0.10	0.11	0.11	0.06
		G_n	0.75	0.69	0.71	0.62	0.75
		V_n	0	4	6	7	7
Acipenser güldenstädti colchicus V. Marti (BN), $V=100$	120.0	E_n	0.00	0.04	0.07	0.18	0.71
		Q_n	0.08	0.11	0.07	0.04	0.02
		G_n	1.00	1.00	1.00	1.00	0.50
		V_n	0	16	16	18	50
Scomber scombrus L. (EN), $V=31$	23.1	E_n	0.00	0.03	0.15	0.30	0.52
		Q_n	0.09	0.15	0.16	0.12	0.04
		G_n	0.82	0.65	0.76	0.75	1.52
		V_n	0	8	7	8	8
Auxis thazard (Lac.) (EN), $V=40$	38.9	E_n	0.00	0.03	0.14	0.50	0.33
		Q_n	0.11	0.17	0.17	0.13	0.03
		G_n	0.88	0.77	0.76	0.70	2.44
		V_n	3	10	9	8	10
Xiphophorus helleri Heck. (BN), $V=28$	4.7	E_n	0.00	0.03	0.12	0.27	0.58
		Q_n	0.12	0.18	0.20	0.15	0.11
		G_n	1.22	0.63	0.50	0.44	0.27
		V_n	0	7	7	8	6
Hirundichthys rondeletii (Cuv. et Val.) (EN), $V=47$	18.0	E_n	0.00	0.02	0.12	0.27	0.59
		Q_n	0.10	0.13	0.13	0.11	0.06
		G_n	0.83	0.93	0.78	0.67	1.00
		V_n	1	11	10	12	13
Coryphaena hippurus L. (EN), $V=31$	64.0	E_n	0.00	0.02	0.06	0.17	0.75
		Q_n	0.12	0.13	0.12	0.09	0.04
		G_n	0.50	0.48	0.47	0.47	0.67
		V_n	2	8	7	7	7
Prionace glauca (L.) (EN)	180.0	E_n	0.00	0.01	0.04	0.14	0.81
		Q_n	0.07	0.18	0.08	0.03	0.02
		G_n	1.70	1.09	0.67	1.10	0.50
Xiphias gladius L. (EN), $V=26$	165.0	E_n	0.00	0.00	0.10	0.57	0.33
		Q_n	0.02	0.02	0.10	0.09	0.05
		G_n	2.00	2.00	0.62	0.65	1.09
		V_n	0	0	7	11	8

Table 6 (*contd.*)

Species, nektonic type and $V(\equiv\sum V_n)$	Length L to end of vertebral column (cm)	Index	Number n of body section 1	2	3	4	5
Chaetodon striatus L. (BN), $V=24$	11.0	E_n	0.00	0.00	0.04	0.18	0.78
		Q_n	0.05	0.05	0.05	0.04	0.03
		G_n	0.47	0.27	0.18	0.15	0.32
		V_n	0	6	6	6	6
Phocoena phocoena (L.) (EN), $V=62$	128.8	E_n	0.00	0.00	0.02	0.17	0.81
		Q_n	0.11	0.18	0.18	0.11	0.01
		G_n	1.00	1.00	1.00	1.12	1.75
		V_n	1	11	14	12	24
Delphinus delphis ponticus Barab. (EN), $V=67$	178.5	E_n	0.00	0.00	0.02	0.13	0.85
		Q_n	0.08	0.17	0.19	0.11	0.01
		G_n	0.90	1.08	1.06	1.25	4.20
		V_n	0	10	17	19	21
Tursiops truncatus (Montagu) (EN) $V=54$	223.0	E_n	0.00	0.00	0.01	0.10	0.89
		Q_n	0.11	0.16	0.17	0.12	0.04
		G_n	0.96	0.89	1.08	1.50	4.40
		V_n	0	10	13	14	17

determined the number of flexible links in the vertebral column (V) and within each elementary section (V_n); in cases of vertebral fusion the associated intervertebral link was disregarded (Table 6). In nektonic animals in which the entire body serves as the propulsor, usually $V > 100$, which ensures high body flexibility throughout almost its whole length; such are Trichiuroidei, Anguilliformes, Mesosauria, Mosasauridae, Hydrophidae, etc. (Table 6, *Enhydrina, Anguilla* and *Trichiurus*). When the propulsive action is concentrated in the rear portion of the body, usually $V < 100$, and most frequently $V < 70$, as in most of nektonic fishes, cetaceans and sirenians (Table 6, species beginning from *Sprattus*). In Ichthyosauria, in which the total number of vertebrae is about 120–160 (Fraas 1891; Huene 1956; Romer 1956), the number within the trunk up to the point where the body axis begins to descend into the lower tail lobe amounts to approximately half, i.e. is within the range characteristic of the majority of fishes, cetaceans and sirenians.

Yet in particularly fast fishes the decrease in body flexibility owing to fewer articulated links in the vertebral column proves inadequate. In this case either special transverse growths appear on the vertebrae in the caudal part of the body (Figure 34A), usually occurring, for example, in the majority of Scombridae (Aleyev 1963a), or enlarged zygapophyses embrace the spinal processes of adjacent vertebrae (Figure 34C), as one may observe in Istiophoridae (Gregory & Conrad 1937; Ueyanagi & Watanabe 1965). These adaptations both lead to an additional reduction in body flexibility (Figure 35C).

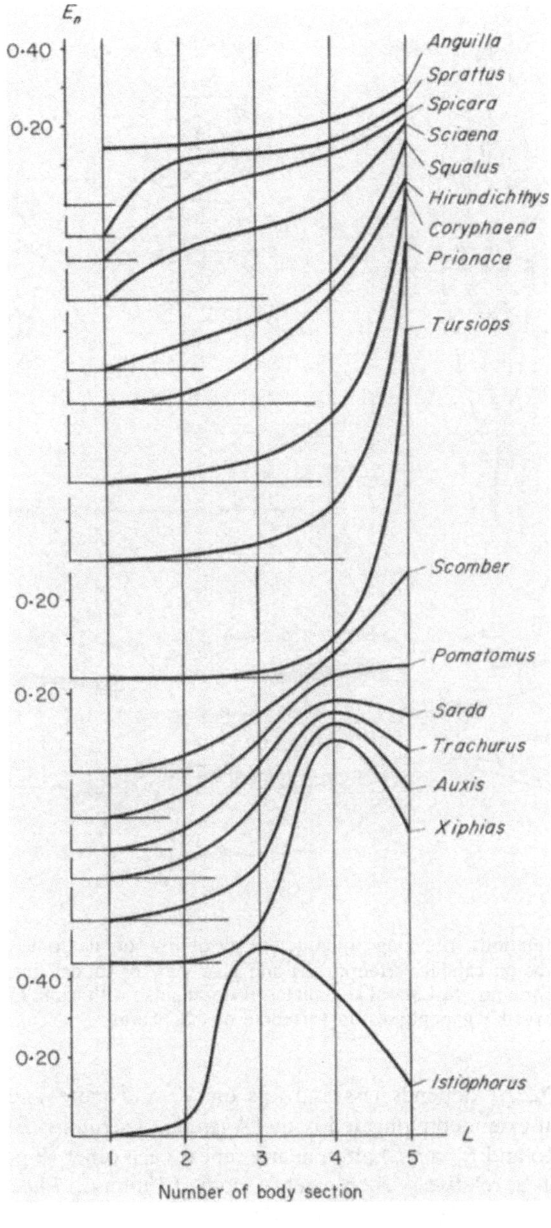

Figure 33. E_n versus L for different nektonic animals.

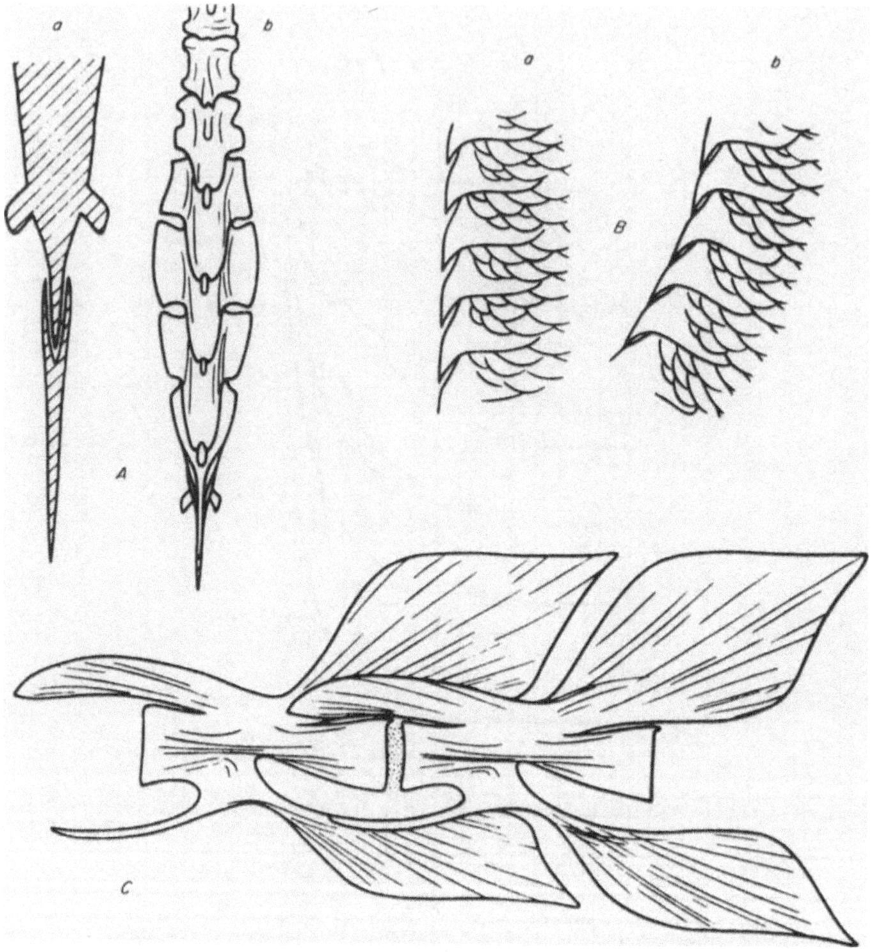

Figure 34. Adaptations reducing longitudinal flexibility of nektonic animal's body. *A*, transverse growths on caudal vertebrae (*a*) and plan view of caudal fin rays rigidly fixed to hypurale (*b*) in *Thunnus*; *B*, Lateral keel plates of Carangidae with trunk of fish straight (*a*) and bent (*b*); *C*, enlarged zygapophyses on vertebrae of *Istiophorus*.

At higher Q_n, E_n depends less and less on V_n and more and more on Q_n. In animals with an extended elongated body (*Anguilla, Trichiurus, Enhydrina*, etc.) the curves $V_n = f(L)$ and $E_n = f(L)$ often nearly repeat each other (Figure 35A), whereas in the case of a relatively short, dense body (*Thunnus, Phocoena*, etc.) E_n is determined mainly by Q_n (Figures 35C,D).

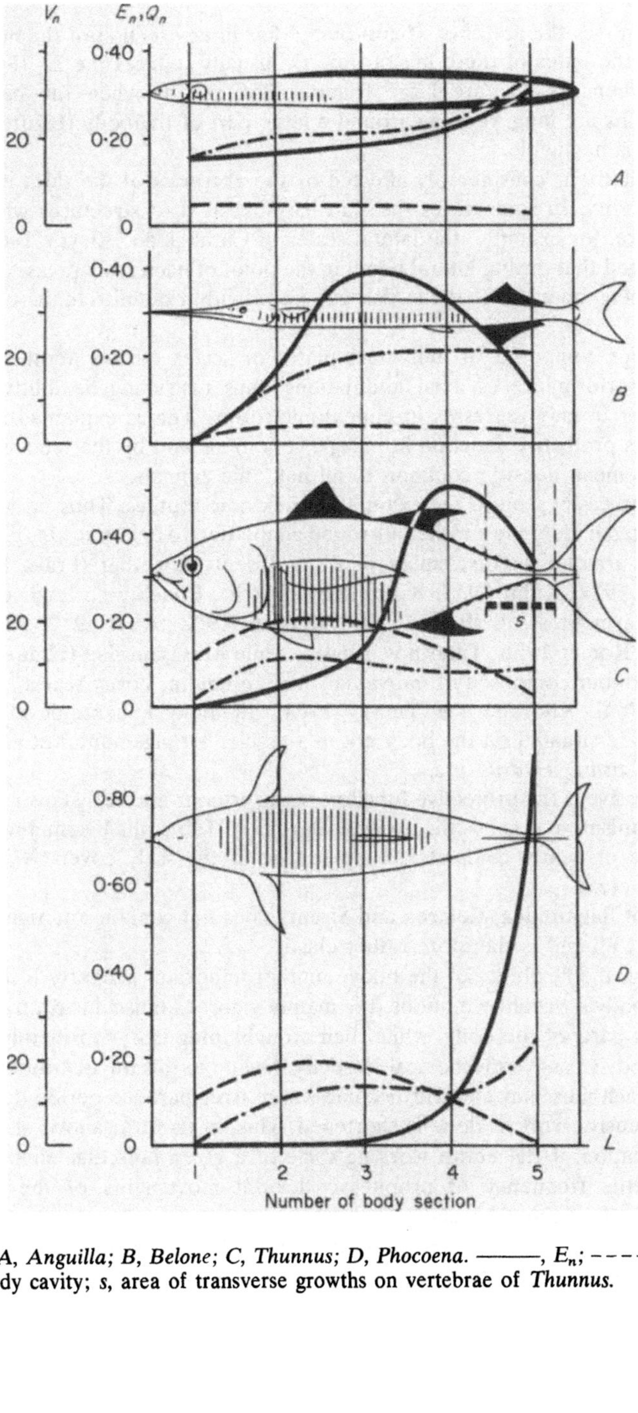

Figure 35. *A, Anguilla; B, Belone; C, Thunnus; D, Phocoena.* ———, E_n; – – – –, Q_n; —·—·, V_n; ||||||, body cavity; *s*, area of transverse growths on vertebrae of *Thunnus*.

In Osteichthyes **the presence of non-paired fins** in any section of the body, i.e. the presence of the bases of these fins, at low Q_n usually reduces the E_n (Figure 35B). However, when the Q_n are large (Figures 35C,D), or when the bases of the non-paired fins are long, running around a large part of the body (Figure 35A), this effect remains negligible.

Body flexibility is considerably affected by **the character of the skin:** its flexibility and distensibility. In some cases the skin has special rigid structures which restrict flexibility, like, for example, the lateral scales of Carangidae (Aleyev 1963a), which are so arranged that during lateral bending the point of each scale presses against the base of its neighbour, so that the fish can bend within definite limits only (Figure 34B).

Any armour composed of individual plates or scales creates additional friction during the performance of axial undulations, thus restricting flexibility, owing to which armour usually regresses in eunektonic forms. This is explained also by the atrophy of its protective function in pelagic conditions and by the tendency towards reducing the mean density, common to all nektonic animals.

This can be easily seen in the example of nektonic reptiles. Thus, in Metriorhynchidae who had undergone a rather all-round adaptation to nektonic life (Figure 12A), the external armour was extremely weak or absent altogether (Fraas 1902; Abel 1907, 1912, 1922; Kuhn 1937; Konzhukova 1964i). Eunektonic Ichthyosauria had no external armour at all (Bauer 1898; Merriam 1902; Abel 1912; Huene 1916, 1923, 1956; Romer 1956; Tatarinov 1964k), while Mosasauridae (Figure 12B), had only weak armour composed of individual small, granular, horny scales (Abel 1912; Tatarinov 1964l; Khozatsky & Yuryev 1964). In many eunektonic Hydrophidae these scales are situated on the body not in a tegular arrangement, but adjacent one to another (*Distira*, *Hydrus*, etc.).

Also regressive is the protective function of the armour and scaly cover in nektonic fish-like organisms and fishes themselves (Burdak 1973b), which sometimes leads to the complete or nearly complete disappearance of the scaly cover (*Xiphias*, some Scombridae, etc.).

The skin of Sagittoidea, Cetacea and Sirenia does not contain any rigid structural formations at all and is therefore rather elastic.

Along with the flexibility of the body, another important property is its **elasticity**. During the body's wavelike motions it is mainly vigorous muscular contractions that bend various parts of the body, while their straightening is in considerable measure due to the body's passive mechanical elasticity. Owing to this the distribution of work in time for each muscular element becomes more irregular: the period during which the most intensive work is done is shortened. This, in its turn, allows shortening of the total duration of the entire working cycle of a given muscular element, i.e. an increase in the frequency of propulsive flexural movements of the body and, consequently, the animal's speed.

The further the muscular fibre or skin is from the body's longitudinal axis, the more it is extended during bending of the body and the greater is the effectiveness of the elastic forces at the moment when the body curvature begins to decrease. Body elasticity, therefore, increases at higher G_n. For the fastest nektonic animals usually $G_5 > 1$ (Table 6, *Sarda, Trachurus, Thunnus, Scomber, Auxis, Xiphias, Phocoena, Delphinus* and *Tursiops*; Figure 46), which is partially connected with ensuring higher elasticity at the rear end of the body.

In Osteichthyes additional factors in body elasticity are the scales (Barsukov 1960b; Aleyev 1963a) and intermuscular bonelets. In Istiophoridae body flexibility and elasticity are determined to a considerable extent by the flexibility of the zygapophyses; it is interesting that in *Istiophorus* and *Tetrapturus*, which do have these rigid vertebral links, the body is comparatively thinner than in representatives of related genera, *Makaira* and *Xiphias*, in which they are developed to a lesser degree (Figures 96–98). An adaptation similarly aimed at reducing the flexibility and increasing the elasticity of the body are the forklets at the proximal ends of the caudal fin rays, which embrace the hypurale (Figure 34A), as in the case of Scombridae.

The muscular apparatus of the axial undulatory propulsor consists in all nektonic animals of a complex of longitudinal body muscles. One may distinguish two types of propulsive muscular apparatus, corresponding to the two types of propulsive body deflexions: lateral and dorso-ventral.

The lateral muscular apparatus is characterized by muscular complexes situated to the right and left of the axis of the body and acting as symmetric antagonists to produce the body's propulsive flexural motions (agnathians, fishes, amphibians and reptiles; see Figure 36A).

In Cyclostomi each of the lateral muscular bands is mainly formed by *m. parietalis, m. obliquus* and *m. rectus abdominalis* (Suvorov 1948; Fontaine 1958).

In fishes the lateral muscular bands are composed mainly of the very powerful *m. lateralis magnus*, but also include the system *m. obliquus* and *m. rectus abdominalis*

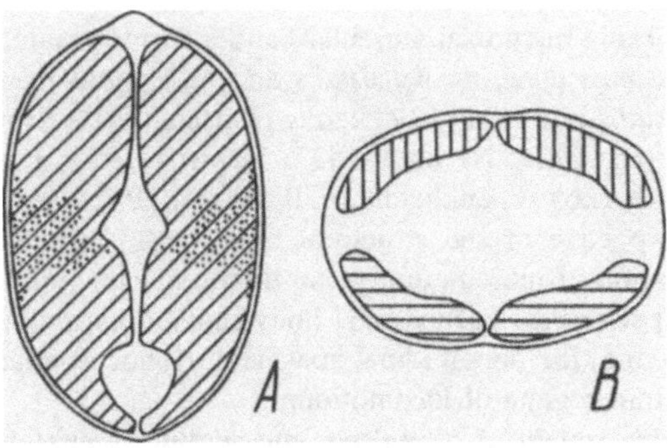

Figure 36. Body cross-section of (*A*) *Thunnus thynnus* (L.) *and* (*B*) *Sagitta setosa* Müll. ////, lateral muscular bands; ||||||, dorsal muscular bands; ≡, ventral muscular bands; ∴∴, red musculature.

(Wiedersheim 1906; Shimkevich 1912; Suvorov 1948; Le Danois 1958). In Actinopterygii the *m. lateralis magnus* is in addition covered by the longitudinal *m. rectus lateralis* (*s. m. lateralis superficialis*) (Greene 1913; Ihle *et al.* 1927).

The distinctions between groups of fishes in the degree of development of the so-called 'red musculature', i.e. the vascular plexuses in the lateral muscles, appear as a result of different degrees of adaptation to rapid swimming. Functionally, these vascular plexuses, connected by a powerful vascular system in the skin, are a means of ensuring an adequate level of the biochemical processes in the intensively performing muscles, and also, perhaps, a means of shedding excessive heat generated in the muscles into the external medium, which resembles a similar process in cetaceans and sirenians (aided by complex vessels), representing a universal thermoregulatory mechanism (Fawcett 1942; Tomilin 1950, 1951; Scholander & Schevill 1955; van Utrecht 1958). It is known (Portier 1903; Legendre 1934; Morrow & Mauro 1950; Van Oosten 1957) that in tunas and some related fast fishes the body temperature during movement may be several degrees (in *Thunnus alalunga* up to 9°, in *Th. thynnus* up to 12°C) above that of the water. Cooling off through lateral body surfaces, as Walters (1962) believes, has also a hydrodynamic importance, which is examined below.

The lateral muscular bands in reptiles are composed mainly of a powerful system of *m. iliocostalis*, but also contain *m. longissimus dorsi*, *m. obliquus*, *m. rectus abdominalis*, *m. caudi-femoralis* and *m. ilio-ischiocaudalis* (Butschli 1910; Ihle *et al.* 1927).

The dorso-ventral muscular apparatus is characterized by muscular complexes situated dorsally and ventrally with respect to the axis of the body and acting as symmetrical antagonists to produce propulsive flexural movements of the body (Sagittoidea, Cetacea and Sirenia; Figure 36B).

In Sagittoidea the dorsal and ventral muscular bands are both formed by paired longitudinal muscles stretching from the head to the end of the tail (Burfield 1927); see Figure 36B.

In Cetacea and Sirenia the dorsal muscular band is formed mainly by the powerful system of *m. longissimus dorsi*, *m. iliocostalis* and *m. spinalis dorsi* and the ventral mainly by *m. hypaxialis* and *m. ischiocaudalis* (Butschli 1910; Howell 1927; Ihle *et al.* 1927; Slijper 1936, 1961; Narkhov 1937; Sleptsov 1952; Bourdelle & Grasse 1955; Petit 1955; Agarkov & Lukhanin 1970; Berzin 1971).

In any case, irrespective of the structural type of the propulsive muscular apparatus, the distribution of musculature along the body's longitudinal axis is merely the result of those peculiarities of external body shape determining the distribution of body volume along the longitudinal axis and is not connected per se with adaptation to a certain regime of locomotion.

The overall **relative weight of propulsive musculature** is nektonic animals within the individual systematic groups increases with the measure of adaptation to active forward movement; this may be clearly traced in the example of Osteichthyes. Thus, according to our data, in *Abramis brama* with a length to the end of the vertebral column of $L = 38.2$ cm, the total weight of the propulsive musculature ensuring the operation of the axial undulatory propulsor makes up 52% of the animal's total weight, while in *Belone belone euxini* with $L = 68.2$ cm this figure is 58%, in *Sphyraena sphyraena* with $L = 30.8$ cm, 60%, in *Alosa kessleri pontica* with $L = 25.0$ cm, 62%, in *Acipenser güldenstädti colchicus* with $L = 120.0$ cm, 65%, in *Sarda sarda* with $L = 58.8$ cm, 67%, and in *Anguilla anguilla* with $L = 91.0$ cm, 69%. According to Bainbridge (1960, 1962) the figure for *Carassius auratus* is 42%, for *Leuciscus leuciscus* 60% and for *Salmo gairdneri irideus* 65%, while for *Sphyraena barracuda*, according to Walters (1966), it is 66–70%. Similar figures for cetaceans, according to our data, are: in *Delphinus delphis ponticus* with $L = 178.5$ cm, 22%, in *Tursiops truncatus* with $L = 223.0$ cm, 33% and in *Balaenoptera musculus* (after Slijper 1961), 40%. Thus, in fishes the relative weight of propulsive musculature is on the whole higher (42–70%) than in cetaceans (22–40%). The explanation is that in the latter subcutaneous fat accounts for a considerable part of the total weight of the animal, owing to which the relative weight of all the other body components diminishes; thus, according to our data, the weight of subcutaneous fat in *Tursiops truncatus* with $L = 223.0$ cm makes up 34.0% of the animal's total weight. With diminishing absolute dimensions the relative weight of the subcutaneous fat in cetaceans, which fulfils the function of heat insulation, increases while the relative weight of the propulsive musculature diminishes accordingly, as can be seen when comparing corresponding figures for *Balaenoptera*, *Tursiops* and *Delphinus*.

For the trunk to function as an axial undulatory propulsor, of essential importance given lateral propulsive bending of the body is the area of its vertical longitudinal projection (i.e. its projection onto the median plane), and given dorso-ventral bending, the area of its horizontal longitudinal projection (i.e. its projection onto a horizontal plane). We shall henceforth refer to these projections as **working propulsive surfaces**.

The effectiveness of any section of the body as an element of an axial undulatory propulsor, determined by the magnitude of the propulsive force its creates increases 1. with an increase in its propulsive surface area and 2. with a greater degree of flattening, since a flat surface set perpendicular to the direction of the flow encounters much greater resistance than a convex one. Accordingly the elements of an axial undulatory propulsor are characterized by $G_n \ll 1$ (Figures 12B,C, 136, 137).

The relative area S_n of the working propulsive surface of a given elementary section n can be found from the relationship

$$S_n = s_n \left(\sum_{i=1}^{5} s_i \right)^{-1}, \tag{14}$$

where s_n is the absolute area of the working propulsive surface of section n. The degree of flatness of the propulsive surface of any section of the body can be quantitatively expressed through the thickness ratio B_n of that section in the direction of propulsive movements, since the degree of flatness of the working propulsive surface is always inversely proportional to the body's thickness ratio in the given section:

$$B_n = k_n s_n^{-(1/2)} \tag{15}$$

where k_n is the average diameter of section n in the direction of propulsive bending. The value of k_n for each elementary section should be found separately for the trunk and for fins, since their degrees of flatness differ sharply. For the trunk we shall obtain

$$k_{cn} = \tfrac{1}{3}(k_1 + k_2 + k_3), \tag{16}$$

where k_1, k_2 and k_3 are respectively the body diameters in the direction of propulsive bending at points at the beginning, middle and end of the given elementary section. For the fins we shall find

$$k_{pn} = \tfrac{1}{2}(k_1 + k_2), \tag{17}$$

where k_1 and k_2 are now the fin's thickness at its base and at its apex respectively. Since, however, k_2 is always zero here,

$$k_{pn} = \tfrac{1}{2}k_1. \tag{18}$$

The values of S_n and B_n are essential for obtaining the index C_n of the distribution of the propulsive action along the animal's longitudinal axis (see below).

C. The caudal fin as an element of an axial undulatory propulsor

The caudal fin, an attribute of the overwhelming majority of nektonic forms swimming by means of wavelike body undulations, is an essential specialized element of an axial undulatory propulsor. The caudal fin has diverse functions.

1. The caudal fin participates in the general propulsive movements of the body, creating a propulsive force, this usually being its basic function.
2. Functioning as a passive lifting surface, a heterocercal caudal fin creates a vertical force either lifting or sinking the posterior part of the animal's body.
3. During inert movement the caudal fin acts as a stabilizer and partially a vertical rudder.
4. In most of the Actinopterygii the caudal fin is capable of performing a number of complicated movements causing the emergence of forces acting in a vertical plane at an angle to the body's longitudinal axis, which helps to change the inclination of the direction of movement to the vertical.

The first three functions may be described as passive since when performing them the caudal fin acts simply as an elastic plane more or less rigidly fixed to the end of the caudal peduncle. The last function, however, is active, since when performing it the fin carries out a number of complex movements of its own, acting not simply as a single plane, but as a complex system consisting of individual elements: – rays – articulately joined to the end of the vertebral column.

The movements of the caudal fin aimed at creating vertical forces have been examined in chapter III, while its functions as a stabilizer and rudder will be examined below, in chapter VI. Here, we shall dwell predominantly on the action of the caudal fin as a propulsor.

The structure and functions of the caudal fin in different nektonic animals, mainly fishes and cetaceans, are examined in a number of studies (Murie 1880; Strasser 1882; Schulze 1894; Ahlborn 1895; Dexler & Freund 1906; Osburn 1906; Regan 1910a,b; Whitehouse 1910a,b, 1918; Abel 1907, 1912; Schmalhausen 1913, 1916; Huene 1916, 1923, 1956; Magnan & Sainte-Lague 1929a; Gray 1933è,f, 1953a,b; Graham-Smith 1936; Grove & Newell 1936; Harris 1934, 1936, 1937a, 1938, 1953; Narkhov 1939; Vasnetsov 1941, 1948; Kermack 1943; Chabanaud 1944; Kripp 1954; Parry 1949a,b; Affleck 1950; Gero 1952; Petit 1955; Romer 1956; Aleyev 1957b, 1959b, 1963a, 1973b, 1974; Tomilin 1957, 1962a–c, 1965; Nursall 1958a, 1962, 1963a,b; Barsukov 1959a,b; Kobi & Pristovsek 1959; E. Kramer 1959, 1960; Walters 1962; Bainbridge 1963; Tatarinov 1964k; Pershin 1969b; Cutchen 1970, and others). However, the question of what gave rise to one of the crucial features of the nektonic caudal fin, its two-lobe shape, has become the subject of special studies only recently (Nursall 1958a; Aleyev 1959b, 1963a; Walters 1962, and others), though many authors had earlier examined the action of the separate lobes of the caudal fin (Schulze 1894; Ahlborn 1895; Schmalhausen 1913, 1916; Grove & Newell 1936; Kermack 1943, and others). The question of the significance of the fin's asymmetric forms has been elaborated in greater detail (Schulze 1894; Ahlborn 1895; Schmalhausen 1913, 1916; Grove & Newell 1936; Harris 1936, 1938, 1953; Kripp 1954; Vasnetsov 1948; Aleyev 1959b, 1963a; R. S. Alexander 1965, 1966a, 1967, 1968, and others).

One of the reasons for the emergence of the vertical two-lobe tail is to ensure autonomous movement of its upper and lower halves. However, a **more general cause of the division of the caudal fin into two lobes** is to be found in the specific features of the counterflow over the bodies of nekton (Aleyev 1959b, 1963a). In this context one should primarily bear in mind two points: 1. the position of the tail fin relative to the boundary layer and the wake flow and 2. its position relative to the potential vortex formation zone.

Being situated at the posterior end of the body, the caudal fin of all nektonic animals is to some degree within the area occupied by the added mass formed by the thick boundary layer flowing off the body, or in other words, within the wake flow (Figures 37, 72–75). This facilitates its higher effectiveness as a propulsor since use can be made of the kinetic energy of the boundary layer: pressing against the water forming the boundary layer, i.e. moving in the same direction as the animal, is more advantageous than pressing against a relatively immobile potential stream (Aleyev 1965b). In practice, however, in the overwhelming majority of nektonic animals the tail fin cannot be situated wholly within the wake flow: this is prevented by the **threat of vortex formation on the caudal fin**, which is the central factor determining its shape.

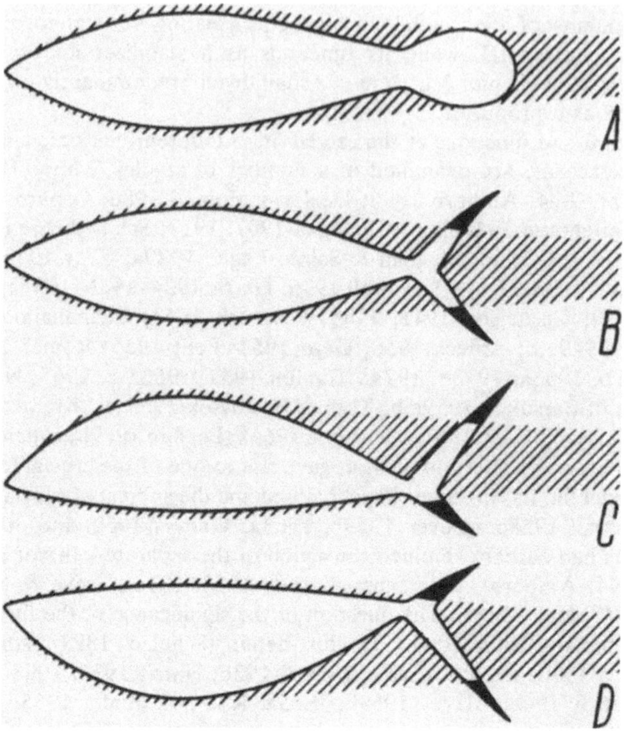

Figure 37. Position of caudal fins of different forms relative to boundary layer and hydrodynamic wake for same degree of development of vortex formation. *A*, rounded caudal fin; *B-D*, tail fin of transversely elongated shape. *B*, isocercal; *C*, epicercal; *D*, hypocercal. Boundary layer and hydrodynamic wake shown in hatched. See explanation in text.

Since boundary-layer separation over the tail can only occur at a certain distance from its leading edge, the shorter the tail fin and the greater its span, the smaller is its portion, all other conditions remaining unchanged, within the vortex zone, and vice versa. It is known (Schlichting 1956) that moving a surface against the stream is a means of preventing boundary-layer separation, and therefore no separation can

---

Figure 38. Flow over trailing edge of tail fin. *A*, *Cyprinus carpio* L., silk streamer indicating convergent character of flow (cine-record); *B-D*, location of zone of potential vortex formation (horizontal hatching) on trailing side of tail fins of different shapes for the same degree of development of vortex formation. *B*, rounded tail fin; *C*, emarginate, for low speed and strong convergence of flow; *D*, emarginate, for high speed and weak convergence of flow. Zone of potential vortex formation over tail fin designated by shaded hatching. Equal vectors a indicate direction of flow and distance from leading edge of fin to frontal border of zone of potential boundary-layer separation. Further explained in text.

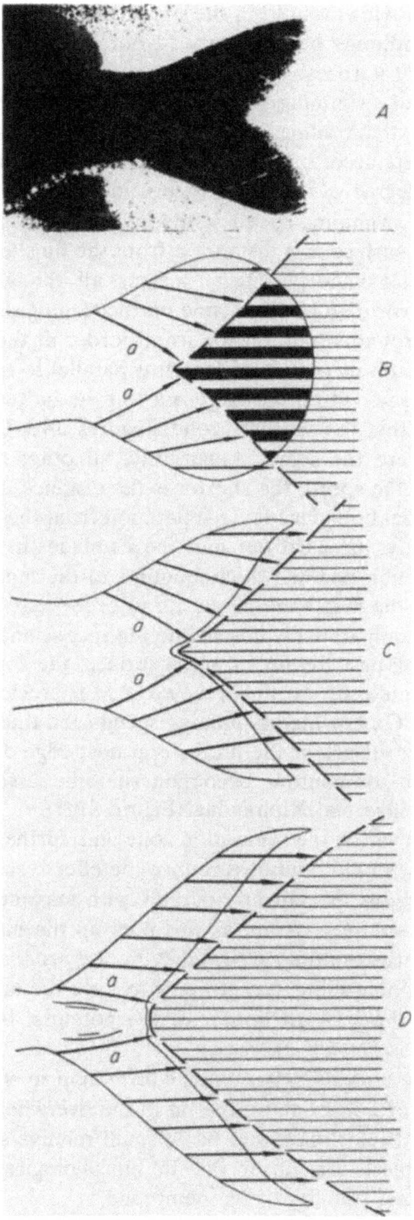
A
a
a
B
a
a
C
a
a
D

take place on the surface of the tail facing the stream. On the other (trailing) surface, however, which is moving away from the stream, if the fin's transverse sweep exceeds a certain speed conditions become very favourable for boundary-layer separation and the formation of vortices.

Films of the tail of swimming fishes and dolphins which we made specially show that the flow over the trailing surface of the tail fin is always of a somewhat convergent character, according to the position of silk streamers fixed to the fin (Figure 38A), the degree of this convergence increasing with decreasing speed. Let us assume that at a swimming speed V the boundary layer begins to separate from the tail fin's trailing surface at a distance a from the fin's leading edge, i.e. at a local Reynolds number $Re = aV\nu^{-1}$. Then, bearing all the above in mind, one may imagine a potential vortex formation zone on the trailing surface of the fin, as shown in Figure 38: for a rounded tail fin, the front border of the potential vorticity zone, drawn through the ends of vectors of length a parallel to the flow and with origin at the leading edge, would cut off the larger part of the fin blade, i.e. the larger part of the fin would be within the vorticity zone (Figures 38B–D), if the development of vortex formation were the same. Apparently, all other conditions remaining unchanged, the higher the speed, the shorter is the distance a separating the border of the region of potential boundary-layer separation from the fin's leading edge (Figure 38), and consequently, the narrower must be its blades in order to keep in front of this potential separation line or, which amounts to the same thing, the greater must be the concavity of the rear edge of the fin in order that its blades are outside the vorticity zone. For comparatively low swimming speeds and thus strong convergence of the stream flowing past the fin's trailing surface, the concavity forms the wedge-like notch in the middle of the fin's rear edge so characteristic of many fishes and cetaceans (Figure 38C). For high swimming speeds and thus little convergence of the flow past the trailing surface of the fin, its rearmost edge does not have such a deep notch in the middle, its contour becoming smoother, which is characteristic, for example, of Scombridae and Xiphioidae (Figure 38D).

Having the tail fin within the separation zone and, further downstream, in the area of detached vortices, would doubtless reduce the effectiveness of its performance. In fulfilling all its functions the tail fin operates with maximum efficiency if it presses against a dense vortex-free stream, as this puts up the most resistance to the fin's transverse movements, resulting, accordingly, in the production of the most effective thrust. The most rational thing, therefore, is to have the tail fin outside the vorticity zone, i.e. ahead of the forward border of the potential boundary-layer separation zone, which indeed is actually the case.

In slow-swimming animals, when vortex formation in the water flowing past the tail would be weak or non-existent, the fin in the overwhelming majority of cases is rounded, not divided into lobes, and has a small relative span (Figure 39A), which represents the simplest variant of tail-fin morphological organization, ensuring, moreover, the greatest stability of its membrane.

Higher swimming speeds and, consequently, higher frequencies of the tail fin's

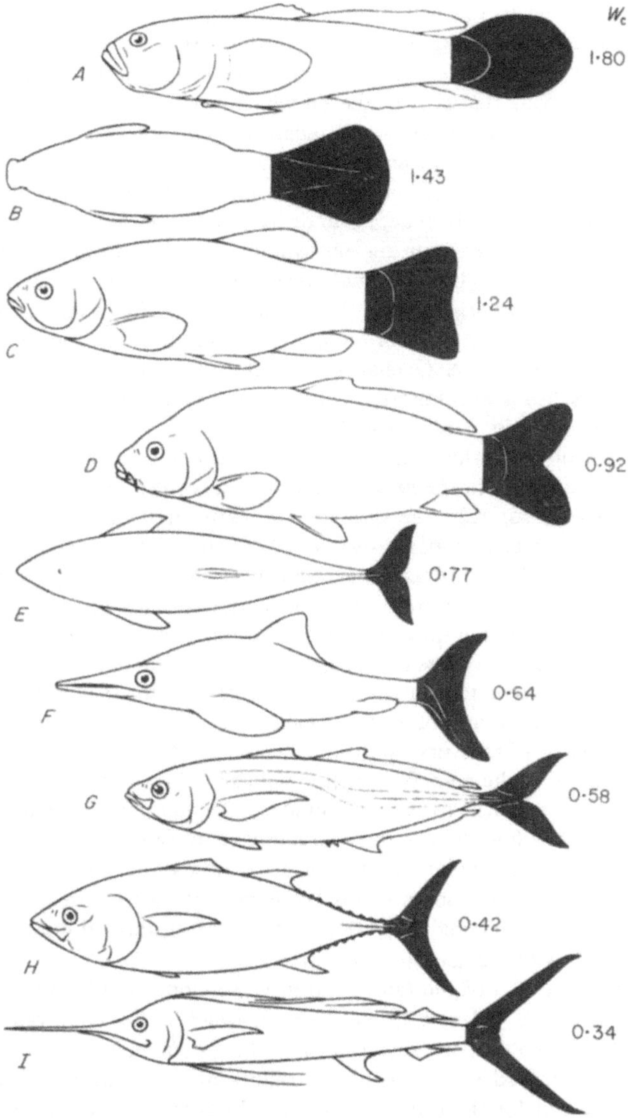

Figure 39. Shape of tail fin in various nektonic animals. *A*, *Gobius ophiocephalus* Pall.; *B*, *Trichechus manatus* L.; *C*, *Tinca tinca* (L.); *D*, *Cyprinus carpio carpio* L.; *E*, *Phocoena phocoena* (L.); *F*, *Macropterygius trigonus* (Owen); *G*, *Trachurus mediterraneus ponticus* Aleev; *H*, *Thunnus thynnus* (L.); *I*, *Istiophorus platypterus* (Show and Nodder). See text.

transverse propulsive movements introduce the possibility of boundary-layer separation, which, according to the above, must start at the middle of the tail fin's rear edge, since a rounded fin is longest along the median line. Hence the need now arises to counter the adverse effects stemming from the fin's being in the vortex zone. This is achieved by removing that part of the fin which in the given mode of locomotion could become a zone of boundary-layer separation and the origin of vortices. The first stage of this process is reduction of the curvature of the fin's rear edge: it becomes less convex (Figure 39B). Later, there appears a concavity in the rear edge of the tail fin (Figure 39C), which begins to deepen as the swimming speed grows, i.e. with the expansion of the zone on the fin where vortices might appear. In the long run this results in the formation of a two-lobe tail fin with well-distinguished lobes (Figure 39D). With the enlargement of the tail fin's span its lobes increasingly extend beyond the boundaries of the wake of the body. This has been proved experimentally for fish (Gero 1952; Aleyev & Ovcharov 1973a,b): the turbulent zone which forms behind the moving fish is encountered mainly by the middle of the tail fin, whereas the tips of its lobes are in a less turbulent flow (Figures 73–75).

The formation of a two-lobe fin is enhanced also by the fact that during the formation of the concavity in the fin its length near its centre-line is reduced, and, in order to compensate for this, the area of the parts of it most distant from the centre-line increases, i.e. its lobes grow away from the centre-line of the body; in other words they grow longer. Besides, an increase in the tail fin's aspect ratio, which can be found according to the ordinary aerodynamic formula for a wing plan form ($\lambda_c = l^2 S^{-1}$, where l is the tail fin's span and S its area), improves its properties as a lifting surface and its effectiveness as a propulsive mechanism (Nursall 1958a; Aleyev 1963a); which also stimulates transverse growth of the fin. In extreme cases this process leads to the formation of a very concave tail fin indeed: its lobes gradually assume an increasingly transverse position, eventually turning into narrow blades almost perpendicular to the animal's longitudinal axis. This is observed, for example, in the fastest fishes, such as Coryphaenidae, Scombridae, Xiphiidae and Istiophoridae, whose tail fins all have the characteristic transversely elongated crescent or bifurcated shape. A similar tail-fin shape is found in the most specialized ichthyosauras, cetaceans and some sirens (*Dugong*; see also Figures 39E–I). Thus, with increasing speed a rounded single-lobe tail fin is transformed during phylogeny into a two-lobe fin with elongated lobes.

Thus the functional aim of the tail fin's transverse elongation is to place its lobes outside the vortex zone (Aleyev 1959a, 1963a), which is the main reason for the formation of the two-lobe shape of tail fin in nektonic animals.

A sweptback leading edge and a concave trailing edge are advantageous not only for the caudal fin, but for the other fins as well. In every case fins of such a shape create large transverse forces, which in the various cases may be important as propulsive, lift or equilibrium-maintaining forces. Fins of this type occur in fast and sturdy swimmers (Scombroidei, Carangidae, Cetacea, etc.). At the same time a deeply concave tail fin is not suited to providing rapid acceleration: when a sudden

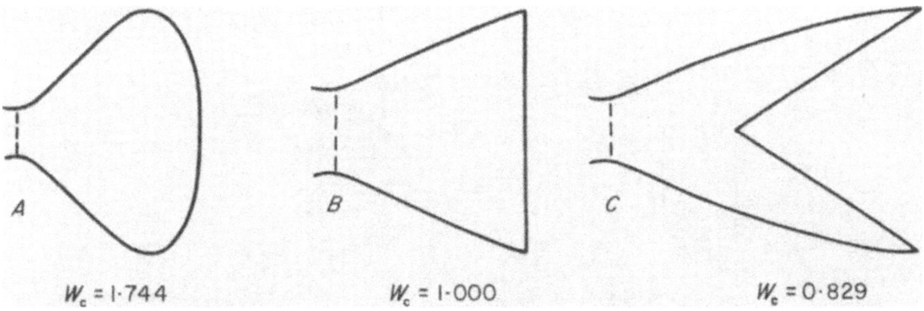

Figure 40. Variation in tail-fin shape at λ_c = constant = 1.950. *A*, rounded; *B*, truncated; *C*, emarginate. See text.

forward spurt has to be made, a steeply increasing angle of attack results in loss of speed. Therefore animals which often make sharp spurts from rest (Sagittoidea, *Salmo*, *Esox* and *Sphyraena*) have a less emarginate, more fan-like tail fin; with such fin shape there is no threat of losing speed even at comparatively large angles of attack.

The determination of the tail fin's transverse slenderness ratio λ_c, though allowing assessment of its degree of transverse elongation, gives no idea of its shape, since at exactly the same value of λ_c its shape can vary from being rounded to having two well-pronounced lobes (Figure 40). The index W_c of the tail-fin shape can be found by comparing the areas of the fin's proximal and distal parts. Let us assume that we understand as the proximal part of the tail fin the entire area between the outer contour of the fin, its base and the straight line aa_1 through the points of the fin furthest from the animal's longitudinal axis; this area represents, therefore, the sum of two areas: those of the parts of the fin proper and of its concavity situated forward of the line aa_1 (Figure 41, horizontal hatching). If the tail fin has elongated ribbon-like appendages along the margins serving mainly for camouflage, which are as a rule without any hydrodynamic function (Figure 41, shaded areas), the line aa_1 should be drawn taking account only of the shape of the basic fin blade and ignoring such appendages, whose area, accordingly, must not be included in the area of the fin's distal part (Figures 41F–H,K). The presence of ribbon-like or thread-like appendages on the tail fin of nektonic animals corresponds, as a rule, to cases when the fin is not a propulsor (Tetrodontiformes) or functions as a propulsor only rarely (Labridae, Scaridae, Pomacentridae, Zeiformes, etc.).

Let us find the area *A* of the fin's proximal part, the area *B* of its distal part and the area *C* of the concavity in the fin's proximal part situated forward of the line aa_1 (Figure 41, inclined hatching). The required index W_c of the tail-fin shape may be defined as follows:

$$W_c = \frac{A}{A} + \frac{B}{A} - \frac{C}{A}. \qquad (19)$$

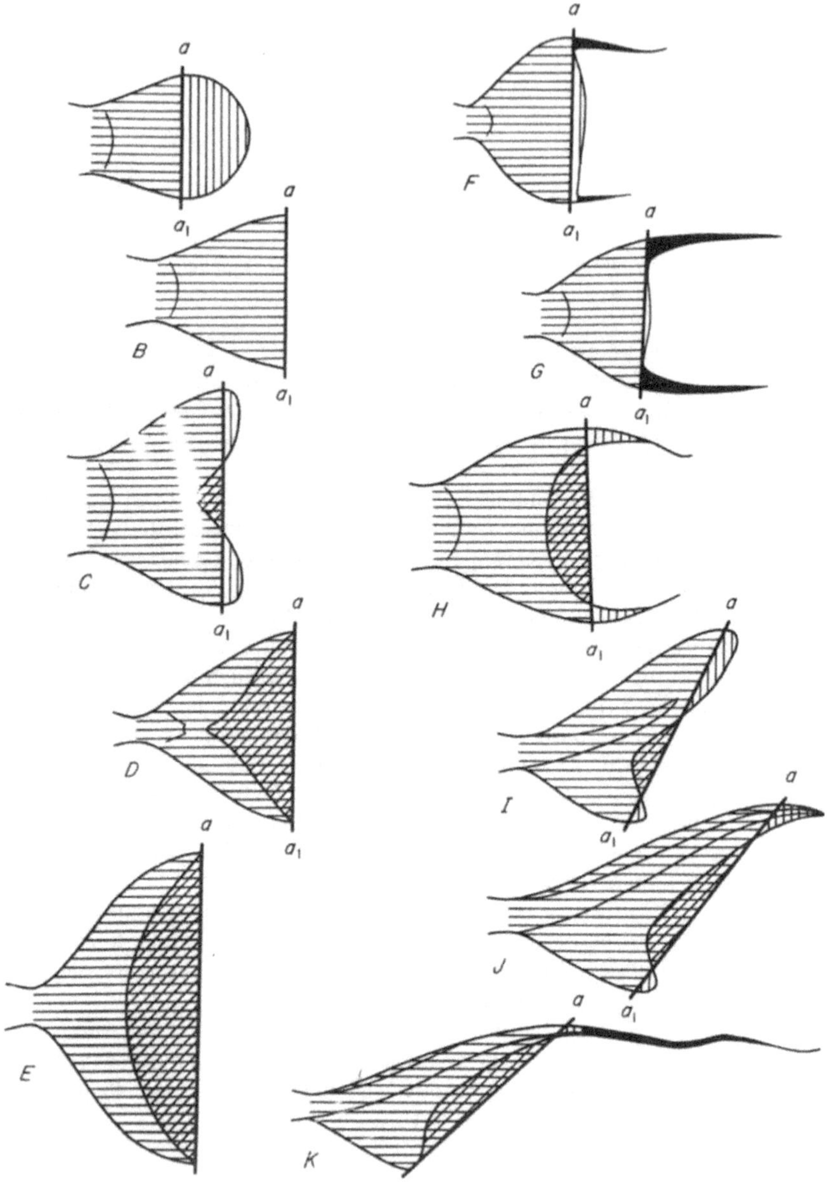
a
a₁
B
C
D
E
F
G
H
I
J
K

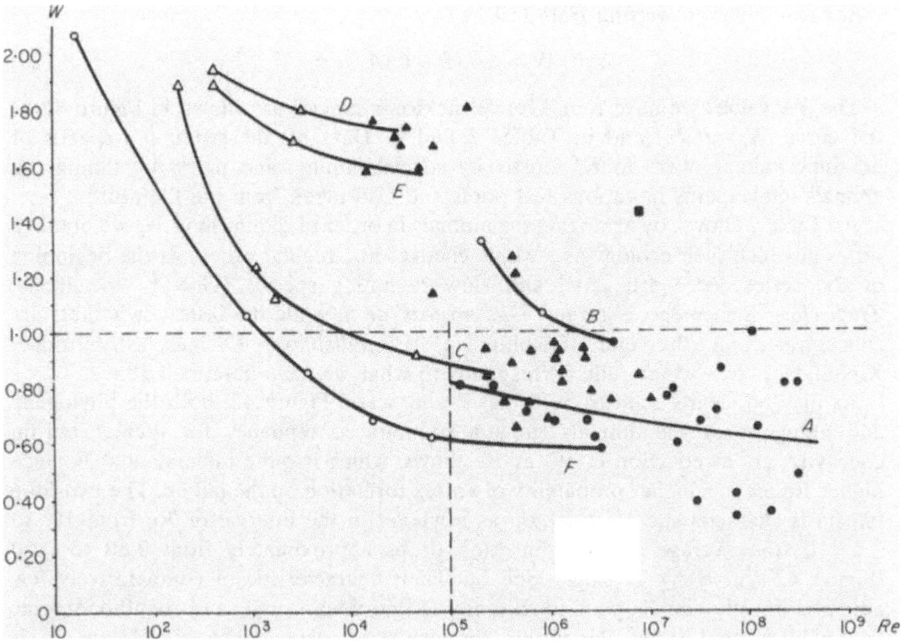

Figure 42. W_c versus *Re.* Points on graph correspond to individual species. *A*, eunekton, phylogeny; *B-F*, ontogeny: *B*, *Squalus acanthias* L.; *C*, *Carassius auratus* (L.); *D*, *Lebistes reticulatus* Peters; *E*, *Gambusia affinis holbrooki* (Girard); *F*, *Trachurus mediterraneus ponticus* Aleev. Filled symbols, sexually mature individuals; open symbols, young fry. ▲, △, benthonekton; ●, ○, eunekton; ■, xeronekton.

Obviously, the first term on the right-hand side will always be equal to one; the second term is the relative area of the fin's distal part and the third the relative area of its concavity.

In the case of a rounded fin with a convex trailing edge we shall always have $W_c > 1$, since $(C/A) = 0$ and $(B/A) > 0$, with W_c increasing with greater convexity of the fin's trailing edge (Figure 41A). In the case of a fin truncated at the rear along a straight line $W_c = 1$, since $B = 0$ and $C = 0$ (Figure 41B). In the case of a concave fin, whether symmetrical or not, we shall always have $W_c < 1$, since always $B \ll C$ and frequently $B = 0$; the more concave the fin, the smaller is W_c (Figures 41C–E,H–K).

Figure 41. Diagram for measuring areas of proximal (horizontal hatching) and distal (vertical hatching) parts of tail fin and concavity of its proximal part (slanted hatching). *A*, *Gobius ophiocephalus* Pall.; *B*, *Trachinus draco* L.; *C*, *Cyprinus carpio carpio* L.; *D*, *Trachurus mediterraneus ponticus* Aleev; *E*, *Xiphias gladius* L.; *F*, *Cyphomycter eoume* (Lesson); *G*, *Hipposcarus harid vexillus* Smith; *H*, *Anthias squamipinnis* (Peters); *I*, *Squalus acanthias* L.; *J*, *Acipenser stellatus* Pall.; *K*, *Pseudoscaphirhynchus kaufmanni* (Bogd.).

After manipulation we find from (19)

$$W_c = 1 + (B - C)A^{-1}. \tag{20}$$

The W_c values we have found for 52 nektonic species are shown in Figure 42 as the curve $W_c = f(Re)$ and in Tables 7 and 8. Data on the swimming speeds of nektonic animals were found mostly by special filming, and partly by timing the animals' movements in various test pools and also taken from the literature.

As Table 7 shows, by arranging the animals in order of diminishing W_c we obtain a series in which their ecology as a whole changes in a regular order. At the beginning of the series there are small and slow-swimming species (*Gobius*, Poeciliidae, *Trichechus*, Sciaenidae, *Serranus*, *Gasterosteus*, etc.), while the fastest swimmers are concentrated at the end (Delphinidae, *Hirundichthys*, *Orcinus*, Scombridae, Xiphioidae, etc.), which fully corresponds to what we have described above.

As may be clearly seen from Tables 7 and 8 and Figure 42, both the phylogeny and ontogeny of the animals show a pronounced tendency for greater tail-fin concavity, i.e. a reduction in W_c as *Re* grows, which is quite understandable, since higher *Re* mean a higher probability of vortex formation on the tail fin. The two-lobe tail fin is characteristic of eunekton as a whole; in the interval of *Re* from 10^5 to 3.0×10^8 the average W_c for eunekton drops approximately from 0.80 to 0.60 (Figure 42, curve A). A single-lobe tail fin is characteristic of comparatively few animals, mostly small planktonektonic (Sagittoidea, etc.) and benthonektonic (Poeciliidae, etc.) forms. Naturally, the described dependence of W_c on *Re* is preserved only in those animals in whom the tail fin is the main propulsor (Tables 7 and 8) and is absent when it does not function as a propulsor or serves as only an additional propulsor and only rarely, as, for instance, in many deep-bodied fishes of the coral reefs.

The shape of the tail fin in the ontogeny of nektonic animals changes from a more rounded to a less rounded or concave shape; which can in all cases be traced in the reducing W_c values, these changes being all the more pronounced the greater the values of *Re* (Figure 42).

A tail fin situated in a horizontal plane (i.e. in a plane perpendicular to the median one) is always symmetrical (Sagittoidea, Cetacea and Sirenia). A fin situated in a vertical (median) plane (agnathians, fishes and nektonic reptiles) may, however, be either symmetrical or asymmetrical. We have already dealt in chapter III with the factors determining the asymmetry of the tail fin of nektonic animals connected with the development of adaptations aimed at maintaining the animal in a state of suspension in the body of water (creating vertical forces). We shall dwell here, therefore, only on those aspects of tail-fin asymmetry that relate to its performance as a propulsor.

Earlier investigators (Schulze 1894; Ahlborn 1895; Schmalhausen 1913, 1916; Harris 1936, 1937; Grove & Newell 1936; Kermack 1943; Gray 1953, and others), when considering the functional significance of asymmetric tail-fin shapes, proceeded from the assumption that tail-fin asymmetry inevitably entails the emergence of

Table 7. Maximum absolute (V_m) and relative (V_r) speed of propagation, Reynolds number *Re* and tail-fin shape index W_c for nektonic animals.

Species and nektonic type	Absolute length L_a of animal (cm)	Maximum speed V_m (cm/s)	V_r (L_a/s)	Source	*Re*	W_c
Gobius ophiocephalus Pall. (BN)	15.0	100	6.6	Author's data	1.5×10^5	1.80
Lebistes reticulatus Peters (BN)						
male	2.5	60	24.0	,,	1.5×10^4	1.75
female	4.0	70	17.5	,,	2.8×10^4	1.70
Mollienisia velifera Regan (BN)	4.4	75	17.0	,,	3.3×10^4	1.72
Xiphophorus maculatus Günth.(BN)	4.0	70	17.5	,,	2.8×10^4	1.68
Xiphophorus helleri Heck (BN)	7.0	100	14.3	,,	7.0×10^4	1.66
Gambusia affinis holbrooki (Girard) (BN)						
male	2.5	60	24.0	,,	1.5×10^4	1.58
female	5.0	80	16.0	,,	4.0×10^4	1.58
Trichechus manatus (L.) (XN)	300.0	300	1.0	*	9.0×10^6	1.43
Sciaena umbra L. (BN)	28.0	150	5.4	,,	4.2×10^5	1.27
Serranus scriba (L.) (BN)	22.0	200	9.1	,,	4.4×10^5	1.20
Gasterosteus aculeatus L. (BN)	9.0	70	7.8	,,	6.3×10^4	1.13
Physeter catodon L. (EN)	1800.0	600	0.3	Tomilin (1962a)	1.1×10^8	1.00
Umbrina cirrosa (L.) (BN)	42.0	250	5.9	Author's data	10^6	0.96
Squalus acanthias L. (EN)	120.0	350	2.9	,,	4.2×10^6	0.96
Odontogadus merlangus euxinus (Nordm.) (BN)	26.0	200	7.7	,,	5.2×10^5	0.94
Stizostedion lucioperca (L.) (BN)	60.0	300	5.0	,,	1.8×10^6	0.94
Carassius carassius (L.) (BN)	20.0	100	5.0	,,	2.0×10^5	0.94
Gadus morhua macrocephalus Til. (BN)	72.0	250	3.5	Chestnoi (1961)	1.8×10^6	0.93
Salmo trutta m. fario L. (BN)	25.0	485	19.4	Denil (1937) (See Bainbridge 1958a)	1.2×10^6	0.93

Table 7 (contd.)

Species and nektonic type	Absolute length L_a of animal (cm)	Maximum speed			Re	W_c
		V_m (cm/s)	V_r (L_a/s)	Source		
Cyprinus carpio carpio L. (BN)	50.0	200	4.0	Author's data	10^6	0.92
Perca fluviatilis L. (BN)	24.0	165	6.9	Radakov & Protasov (1964)	4.0×10^5	0.90
Globicephalus melas (Traill) (EN)	450.0	1130	2.5	Johannessen & Harder (1960)	5.1×10^7	0.87
Salmo trutta L. (BN)	30.5	366	12.0	Bainbridge (1958)	1.1×10^6	0.86
Carassius auratus (L.) (BN)	12.5	159	12.7	Blaxter & Dickson (1959)	2.0×10^5	0.85
Huso huso (L.) (BN)	200.0	400	2.0	Author's data	8.0×10^6	0.85
Balaenoptera physalus (L.) (EN)	2500.0	830	0.3	Tomilin (1962a)	2.1×10^8	0.82
Balaenoptera musculus (L.) (EN)	2750.0	970	0.4	Tomilin (1962a)	2.7×10^8	0.82
Esox lucius L. (BN)	44.0	279	6.3	Ohlmer & Schwartzkopff (1959)	1.2×10^6	0.82
Engraulis encrasicholus ponticus Alex. (EN)	14.0	180	12.8	Author's data	2.5×10^5	0.81
Sprattus sprattus phalericus (Risso) (EN)	12.0	100	8.4	,,	1.2×10^5	0.81
Mugil cephalus L. (BN)	55.0	400	7.3	,,	2.2×10^6	0.79
Sphyraena barracuda (Walb.) (EN)	130.0	1210	9.3	Gero (1952)	1.6×10^7	0.79
Phocoena phocoena (L.) (EN)	128.0	1080	8.4	Johannessen & Harder (1960)	1.4×10^7	0.77
Salmo salar L. (BN)	91.5	1116	12.2	Gray (1953b)	10^7	0.76
Acipenser güldenstädti colchicus V. Marti (BN)	150.0	300	2.0	Author's data	4.5×10^6	0.76
Scardinius erythrophthalmus (L.) (BN)	22.0	130	5.9	*Gray* (1953b)	2.9×10^5	0.75
Leuciscus leuciscus (L.) (BN)	20.2	296	13.3	Gray (1957)	5.5×10^5	0.74
Tursiops truncatus (Montagu) (EN)	350.0	1400	3.2	Backhouse (1960)	4.9×10^7	0.72
Clupea harengus pallasi Val. (EN)	28.0	200	7.1	Chestnoi (1961)	5.6×10^5	0.71
Delphinus delphis ponticus Barab. (EN)	178.5	1400	7.8	Tomilin (1962a)	2.5×10^7	0.68
Scomber scombrus L. (EN)	38.0	300	7.9	Blaxter & Dickson (1959)	1.1×10^6	0.68

Table 7 (contd.)

Species and nektonic type	Absolute length L_a of animal (cm)	Maximum speed V_m (cm/s)	V_r (L_a/s)	Source	*Re*	W_c
Hirundichthys rondeletii (Cuv. et Val.) (EN)	24.0	833	34.7	W. M., (1953)	2.0×10^6	0.68
Abramis brama (L.) (BN)	60.0	150	2.5	Author's data	9.0×10^5	0.68
Rutilus rutilus rutilus (L.) (BN)	24.0	200	8.3	,,	4.8×10^5	0.66
Orcinus orca (L.) (EN)	800.0	1500	1.9	Tomilin (1962a)	1.2×10^8	0.66
Pomatomus saltatrix (L.) (EN)	55.0	500	9.1	Author's data	2.7×10^6	0.62
Acanthocybium solandri (Cuv. et Val.) (EN)	150.0	2140	14.3	Walters & Fierstine (1964)	1.9×10^7	0.60
Trachurus mediterraneus ponticus Aleev (EN)	56.0	600	10.7	Author's data	3.4×10^6	0.58
Xiphias gladius L. (EN)	300.0	3610	12.0	Nursall (1962)	3.0×10^8	0.49
Thunnus thynnus (L.) (EN)	300.0	2500	8.3	Nursall (1962)	7.5×10^7	0.42
Thunnus albacora (Lowe) (EN)	150.0	2070	13.8	Walters & Fierstine (1964)	3.1×10^7	0.40
Makaira indica (Cuv. et Val.) (EN)	450.0	3610	8.0	Nursall (1962)	1.6×10^8	0.36
Istiophorus platypterus (Show & Nodder) (EN)	250.0	3048	12.2	Walford (1937)	7.6×10^7	0.34

* For *Dugong* we give (Tomilin 1971) a swimming speed of 5.0 m/s, adopting, therefore, for the slower *Trichechus* a speed of 3.0 m/s.

vertical forces either supporting or sinking the posterior end of the body and is directly aimed at creating such forces. It has been recently shown (Aleyev 1959b, 1963a) that tail-fin asymmetry in nektonic animals, apart from creating vertical forces, is aimed also at keeping the propulsive force symmetrical with respect to the animal's body and the direction of its movement. Tail-fin asymmetry often emerges also in animals with neutral buoyancy, when there is clearly no need to create vertical forces supporting or sinking the posterior part of the body; in such cases this asymmetry is merely a consequence of the vertical asymmetry of the nektonic animal's body.

The functional importance of heterocercy is revealed by analysing the conditions for dynamic equilibrium of a nektonic animal. Such equilibrium is dependent, particularly, on two things.

Table 8. Variation with age of the maximum absolute (V_m) and relative (V_r) swimming speed, Reynolds number (*Re*) and caudal-fin shape index W_c in nektonic animals.

Species and nektonic type	Absolute length L_a; of animal (cm)	Maximum swimming speed V_m (cm/s)	Maximum swimming speed V_r (L_a/s)	*Re*	W_c
Squalus acanthias L. (EN)	20.0	100	5	2.0×10^5	1.32
	47.9	190	4	9.0×10^5	1.06
	120.0	350	3	4.2×10^6	0.96
Alosa kessleri pontica	8.0	—	—	—	0.72
(Eichw.) (EN)	25.0	—	—	—	0.69
Salmo trutta labrax m.	7.0	—	—	—	0.95
fario L. (BN)	15.4	—	—	—	0.88
	25.0	485	19.4	1.2×10^6	0.93
Carassius auratus (L.) (BN)	0.7	2.5	3.6	1.7×10^2	1.88
	1.2	9.0	7.5	1.1×10^3	1.23
	1.5	13	9	1.9×10^3	1.12
	7.0	90	16	6.3×10^4	0.92
	12.5	159	13	2.0×10^5	0.85
Odontogadus merlangus	5.2	—	—	—	0.98
euxinus (Nordm.) (BN)	26.0	200	7.7	5.2×10^5	0.94
Atherina mochon pontica	1.6	—	—	—	0.79
Eichw. (EN)	8.1	—	—	—	0.77
Mugil auratus Risso (BN)	1.8	—	—	—	0.85
	34.2	—	—	—	0.68
Lebistes reticulatus	0.8	5	6.3	4.0×10^2	1.94
Peters (BN)	1.6	25	14	4.0×10^3	1.79
	2.5	60	24	1.5×10^4	1.75
	4.0	70	17	2.8×10^4	1.70
Gambusia affinis holbrooki	0.8	5	6.3	4.0×10^2	1.88
(Girard) (BN)	1.3	23	18	3.0×10^3	1.67
	2.5	60	24	1.5×10^4	1.58
	5.0	80	16	4.0×10^4	1.58
Spicara smaris (L.) (EN)	5.1	—	—	—	0.82
	15.1	—	—	—	0.74
Pomatomus saltatrix (L.) (EN)	4.8	—	—	—	0.86
	10.5	—	—	—	0.74
	20.8	—	—	—	0.69
	55.0	500	9.1	2.7×10^6	0.62
Trachurus mediterraneus	0.2	0.7	3	1.5×10	2.07
ponticus Aleev (EN)	0.9	10.0	11	9.0×10^2	1.05
	2.0	38	19	7.6×10^3	0.86
	10.0	140	14	1.4×10^5	0.68
	21.5	300	14	6.5×10^5	0.62
	56.0	600	11	3.4×10^6	0.58

Table 8 (*contd.*)

Species and nektonic type	Absolute length L_a of animal (cm)	Maximum swimming speed		*Re*	W_c
		V_m (cm/s)	V_r (L_a/s)		
Megalaspis cordyla (L.) (EN)	7.0	—	—	—	0.71
	18.4	—	—	—	0.63
	38.0	—	—	—	0.58
Scomber scombrus L. (EN)	12.1	—	—	—	0.73
	38.0	300	7.9	1.1×10^6	0.68
Xiphias gladius L. (EN)	49.5	—	—	—	0.54
	318.0	3610	12	3.0×10^8	0.51
Delphinus delphis L. (EN)	90.0	—	—	—	0.83
	178.5	1400	7.8	2.5×10^7	0.68

First, the point of application of the drag force and the point of application of the propulsive force created by the tail fin, the centre of the tail fin's blade,[1] must lie on a straight line passing through the centre of gravity of the whole animal and parallel to its path, which for brevity we shall henceforth call the trajectory of the centre of gravity (Figure 43). Here there are no rotational moments about the centre of gravity and the animal moves in a straight line.

Second, the areas of the tail fin's upper and lower lobes must be positioned similarly relative to the axis of the wake flow and its upper and lower boundaries. This makes the propulsive forces created by each of the fin's lobes equal and makes the point of application of the total propulsive force *P* created by the tail fin lie on the axis of the wake flow, i.e. makes this force symmetrical relative to the overall flow over the nektonic animal (Figure 37). It also ensures the absence of rotational movements and creates prerequisites for rectilinear movement.

Both these factors affecting tail-fin shape operate in one and the same direction. When the animal's body has a more-or-less symmetrical profile, the point where the tail fin is attached to the vertebral column[2] and the centre of this fin's blade are on the trajectory of the centre of gravity and at the same time on the axis of the wake flow. In this case the fin is symmetrical (Figures 37A,B, 43A). If for some reason, the upper contour of the body's profile becomes in the course of phylogeny more convex than the lower profile, then the point of attachment of the tail fin is usually below the centre of gravity and, at the same time, below the axis of the wake flow. In this case the tail fin grows upwards, i.e. becomes epicercal, owing to which the centre of its

[1] This point was found as the centre of gravity of a cardboard tail-fin template of a definite scale.

[2] Here and below we take this point to be the mid-point of the part of tail stem lying on a vertical line through the beginning of the tail fin (either its upper or lower lobe).

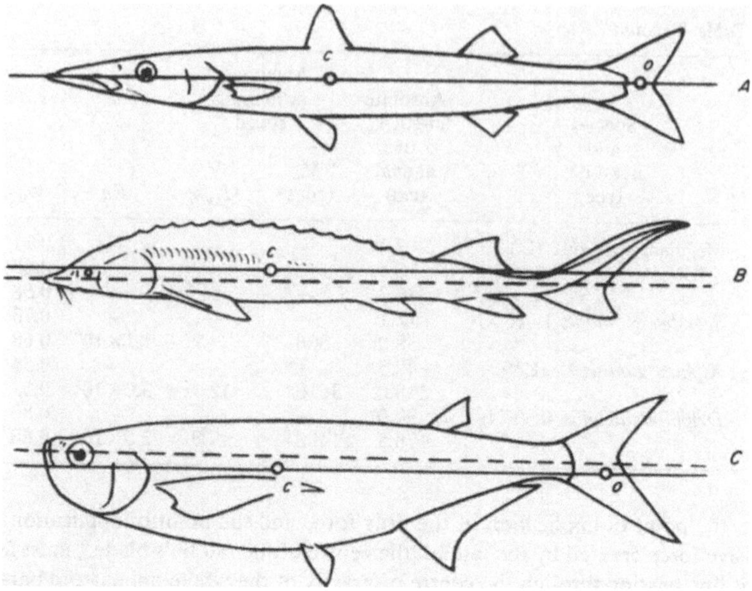

Figure 43. Trajectory of centre of gravity (continuous line) and longitudinal body axis (dashed line) of fish. *c*, centre of gravity; *o*, centre of blade of tail fin. *A*, *Sphyraena sphyraena* (L.), absolute length = 44.5 cm; *B*, *Acipenser güldenstädti colchicus* V. Marti, 102.0 cm; *C*, *Pelecus cultratus* (L.), 29.1 cm.

blade, the point of application of the propulsive force it creates, moves upwards, closer to the trajectory of the centre of gravity and also to the axis of the wake flow (Figures 37C, 43B). When the animal's body has a more convex lower profile, the point of attachment of the tail fin is above both the centre of gravity and the axis of the wake flow. Accordingly, in this case the tail fin grows downwards, i.e. becomes hypocercal, owing to which the point of application of the propulsive force created by the fin again moves closer to the trajectory of the centre of gravity and the axis of the wake flow (Figures 37D, 43C).

Numerous examples bear testimony to the fact that epicercy emerges precisely when the upper contour of the body's profile is more convex than the lower (Figures 8C–G, 44A,B) or when the body is wider dorsally and the centre of gravity is noticeably displaced dorsally (Figures 44C,D), the degree of tail-fin asymmetry and the degree of vertical body asymmetry being as a rule directly related (Figure 44), as can be seen when comparing various species and in the ontogeny of any one species (Aleyev 1963a). Sometimes, with a more convex upper contour of the body's profile the tail fin remains symmetrical; this always corresponds to cases of slow movement (*Carassius*, *Cyprinus*, etc.; Figure 39D).

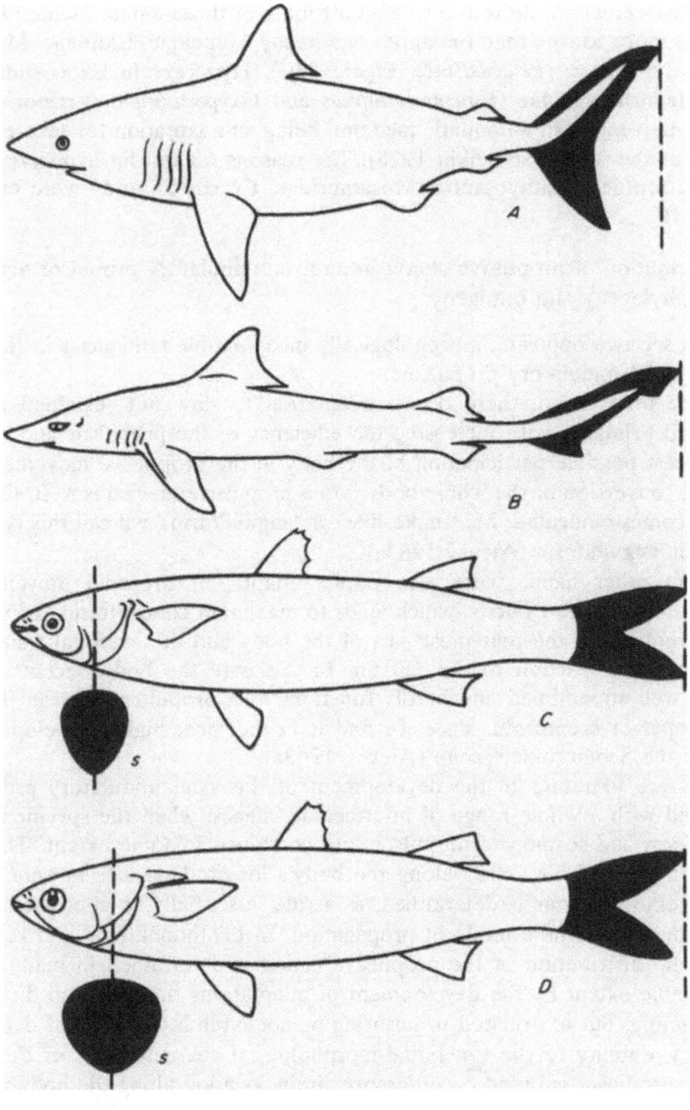

Figure 44. Dependence of degree of tail-fin asymmetry on dorsoventral body asymmetry in (*A*, *B*) sharks and (*C*, *D*) mullets. *A*, *Isurus nasus* (Bonnat.), body profile and tail fin nearly symmetrical; *B*, *Mustelus canis* (Mitch.), body profile and tail fin sharply asymmetric; *C*, *Mugil saliens* Risso, body slightly broadened dorsally, tail nearly symmetrical; *D*, *Mugil cephalus* L., anterior part of body steeply broadened dorsally, tail fin notably epicercal. *C*, *D*, after Burdak (1957), with modifications, showing body cross-section (*s*).

The hypocercal tail fin is as a rule an attribute of those animals whose lower body profile is more convex than the upper one (many Clupeidae, Elopidae, Megalopidae, of the Cyprinidae – *Pelecus*, etc.: Figure 43C). Hypocery in Exocoetidae and the flying Hemirhamphidae (*Euleptorhamphus* and *Oxyporhamphus*) is not directly related to movement in an aquatic medium, being an adaptation for take-off from the surface of the water (Shuleikin 1928). The reasons for tail-fin hypocercy in extinct nektonic reptiles (Ichthyosauria, Mosasauridae, Crocodilia, etc.) were examined in chapter III.

D. Distribution of propulsive action in an axial undulatory propulsor and its variation in phylogeny and ontogeny

One can see two opposite, morphologically incompatible tendencies in the development of axial undulatory propulsors.

On the other hand, there occurs adaptation to slow but sustained movement, concerned primarily with increasing the efficiency of the propulsor and resulting in the greatest possible participation of the body in the propulsive movement and the eventual conversion of the whole body into a propulsive mechanism. In this case the body becomes elongated, i.e. 'snake-like' or 'anguilliform;' we call this type of axial propulsor **anguilliform** (Aleyev 1963a).

On the other hand, there also occurs adaptation to rapid movement, both sustained and in short bursts, which leads to maximum concentration of the propulsive movements in the rear-most part of the body and the eventual transfer of the entire propulsive action to the tail fin. In this case the body becomes relatively shorter, well streamlined and hardly functions as a propulsor; we call this type of axial propulsor **scombroid**, since we find it in the most highly developed form in fishes of the Scombroidei group (Aleyev 1963a).[1]

These two extremes in the development of the axial undulatory propulsor are connected with a whole range of intermediate stages, when the specific features of anguilliform and scombroid propulsors are combined to some extent. The distribution of the propulsive action along the body's longitudinal axis in eunektonic and planktonektonic forms is determined, as a rule, essentially or almost essentially by adaptation to a definite mode of propagation. In benthonektonic and xeronektonic species the distribution of the propulsive action is determined in many cases to a considerable extent by the development of adaptations that have no direct bearing on swimming, but are related to ensuring manoeuvrability, means of defence, etc.

Since variability of the functional-morphological characteristics of the axial undulatory propulsor is based on differences in its location along the body's longitudinal axis, a quantitative assessment of the distribution of propulsive action along the longitudinal axis seems of great interest. Such an assessment has been attempted by Bainbridge (1963), but his method fails to account for the shape of the body's

[1] This type of propulsor has also been called *carangiform*, after the family Carangidae (see Breder 1926, and others).

cross-section, which constitutes, as has been shown above, one of the crucial factors determining the characteristics of a given elementary body section as a propulsor. Failure to take this factor into account resulted in underestimation by Bainbridge of the role of the tail fin in the locomotion of the fishes studied.

Our method of assessing the quantitative distribution of the propulsive action along the longitudinal axis of eunektonic animals (Aleyev 1969a) is based on assessing the area of the working propulsive surface, the degree of its curvature and the amplitude of the propulsive wave in an elementary section n of the body. Taking all these factors into account, an expression for the index C_n of the distribution of the propulsive action of an axial undulatory propulsor along the body's longitudinal axis may be written as follows:

$$C_n = P_n\left(\sum_{i=1}^{5} P_i\right)^{-1}, \tag{21}$$

where

$$P_n = S_n A_n B_n^{-1}. \tag{22}$$

S_n and B_n are indices characterizing, according to formulae (14) and (15), the area and degree of curvature respectively of the working propulsive surface in section n. A_n is an index characterizing the amplitude of propulsive body undulations in the given section. Its magnitude (Aleyev 1963a) is directly proportional to the speed of the transverse movements of the given section of the body:

$$A_n = a_n\left(\sum_{i=1}^{5} a_i\right)^{-1}, \tag{23}$$

where a_n is the amplitude of the propulsive wave in the middle of elementary section n according to ciné records.

By substituting the expressions (14) and (15) for S_n and B_n in (22) we obtain

$$P_n = \frac{s_n\left(\sum_{i=1}^{5} s_i\right)^{-1}}{k_n s_n^{-(1/2)}} A_n = s_n^{3/2}\left(k_n \sum_{i=1}^{5} s_i\right)^{-1} A_n. \tag{24}$$

The right-hand side of (24) is an expression characterizing the contribution of elementary section n to the total propulsive force created during the operation of the axial undulatory propulsor. Since the contributions to k_n from the trunk and the fins will always differ, two expressions similar to the right-hand part of (24) must be obtained for each elementary section: one for the trunk and the other for the fins. Accordingly, instead of (24) we shall have

$$P_n = \left[s_{cn}^{3/2}\left(k_{cn} \sum_{i=1}^{5} s_i\right)^{-1} + s_{pn}^{3/2}\left(k_{pn} \sum_{i=1}^{5} s_i\right)^{-1}\right] A_n, \tag{25}$$

where the subscript c refers to the trunk and the subscript p to the fins. The expression (25) is the formula for the P_n on the basis of which, using (21), the required indices C_n of the longitudinal distribution of propulsive action of an axial undulatory propulsor should be obtained.

A simultaneous increase in C_5 and decrease in C_1, C_2, C_3 and C_4 corresponds to greater concentration of the propulsive action in the rear part of the body and vice versa. A functional-morphological analysis shows that nektonic animals with an anguilliform propulsor always have $C_5 \leq 0.50$, whereas species with a scombroid propulsor always have $C_5 > 0.50$, owing to which the value of C_5 can serve as a criterion for distinguishing between these two types of propulsive mechanism.

By arranging in Table 9 eunektonic and planktonektonic species in order of increasing C_5, we obtained a series in which both the ecology and the length of the animal change in a regular manner. At the beginning of the series appear the smallest and slower swimming nektonic animals, in particular all the planktonektonic species, whereas at its end, for $C_5 > 0.90$, the largest and fastest forms are concentrated, including large Scombridae, *Istiophorus* and dolphins. The ecological significance of this series is, therefore, that the growth in the animal's average speed is accompanied by a higher degree of localization of propulsive action in the rear part of the body: from *Enhydrina* to *Tursiops* we observe a gradual transition from typical anguilliform propulsors to typical scombroid ones, as shown graphically in Figure 45 in the form of curves of C_n versus longitudinal position along the body.

The tendency towards concentrating the propulsive action in the rearmost part of the body observed in nektonic animals at higher speeds of locomotion is determined by the steadily growing need to limit the energy expended during locomotion.

In the case of adaptation to sustained but slow movement, the optimal propulsor for eunektonic and planktonektonic forms is the anguilliform propulsor, which provides the highest efficiency. The tendency towards increasing to the utmost the propulsor's efficiency develops most frequently in nektonic animals in connection with adaptation to life in the oligotrophic conditions of the oceanic depths, where elongated, anguilliform shapes, particularly among fishes, are relatively more numerous than in the surface layers of the ocean and land-locked bodies of water (Aleyev 1963a). The morphological provision of an anguilliform propulsor requires that the entire body be turned into a propulsive mechanism, as is indicated by the low C_5 values ($C_5 < 0.50$) for such nektonic animals as *Enhydrina* and *Anguilla* (Table 9). Because of the more-or-less uniform distribution of propulsive action along the body's longitudinal axis (Figure 45, *Enhydrina, Anguilla*), the body has a ribbon-like or anguilliform structure (Figures 12C, 25A, 29A, 35A), with comparatively slight variation in the size Q_n and shape G_n of its cross-sections and its transverse flexibility E_n along the longitudinal axis (Table 6, *Trichiurus, Enhydrina, Anguilla*). The elongated body of animals with an anguilliform propulsor performs with maximum effectiveness the functions of both a propulsive mechanism and a rudder. Therefore, characteristic of them is a certain reduction of paired limbs, culminating in their complete disappearance in the most specialized forms (Figure 25A). Characteristic of Osteichthyes with an anguilliform propulsor is also reduction of the scaly cover: since the body flexibility of nektonic animals is inversely proportional to the

Table 9. Index C_n of distribution of propulsive action of axial undulatory propulsors in eunektonic and planktonektonic animals.

Species and nektonic type	Length L of animal to end of vertebral column (cm)	C_n				
		Number of body section				
		1	2	3	4	5
Enhydrina schistosa (Daud.) (EN)	90.0	0.06	0.09	0.17	0.25	0.43
Anguilla anguilla (L.) (EN)	91.0	0.05	0.07	0.16	0.23	0.44
Sagitta setosa Müll. (PN)	2.2*	0.00	0.01	0.08	0.21	0.70
Clupeonella delicatula delicatula (Nordm.) (PN)	7.0	0.03	0.05	0.10	0.10	0.72
Sprattus sprattus phalericus (Risso) (PN/EN)	10.0	0.03	0.04	0.07	0.14	0.72
Alosa kessleri pontica (Eichw.) (EN)	25.0	0.03	0.04	0.08	0.09	0.76
Spicara smaris (L.) (EN)	15.5	0.02	0.05	0.07	0.08	0.78
Scomber scombrus L. (EN)	23.1	0.00	0.01	0.04	0.16	0.79
Pomatomus saltatrix L. (EN)	60.2	0.00	0.01	0.02	0.18	0.79
Engraulis encrasicholus ponticus Alex. (EN)	10.8	0.03	0.03	0.05	0.09	0.80
Trachurus mediterraneus ponticus Aleev (EN)	43.3	0.00	0.01	0.04	0.11	0.84
Squalus acanthias L. (EN)	112.0	0.00	0.02	0.03	0.06	0.89
Belone belone euxini Günth. (EN)	50.4	0.00	0.01	0.02	0.08	0.89
Auxis thazard (Lac.) (EN)	38.9	0.00	0.00	0.02	0.08	0.90
Sphyraena barracuda (Walb.) (EN)	93.8	0.00	0.00	0.01	0.08	0.91
Sarda sarda (Bl.) (EN)	58.1	0.00	0.00	0.01	0.06	0.93
Hirundichthys rondeletii (Cuv. et Val.) (EN)	18.0	0.00	0.00	0.01	0.04	0.95
Thunnus alalunga (Bonnat.) (EN)	89.0	0.00	0.00	0.01	0.04	0.95
Thunnus thynnus (L.) (EN)	195.0	0.00	0.00	0.01	0.03	0.96
Istiophorus platypterus (Show & Nodder) (EN)	162.0	0.00	0.00	0.00	0.02	0.98
Phocoena phocoena (L.) (EN)	128.8	0.00	0.00	0.00	0.02	0.98
Delphinus delphis ponticus Barab. (EN)	178.5	0.00	0.00	0.00	0.01	0.99
Tursiops truncatus (Montagu) (EN)	223.0	0.00	0.00	0.00	0.01	0.99

* Body length to base of median rays of tail fin.

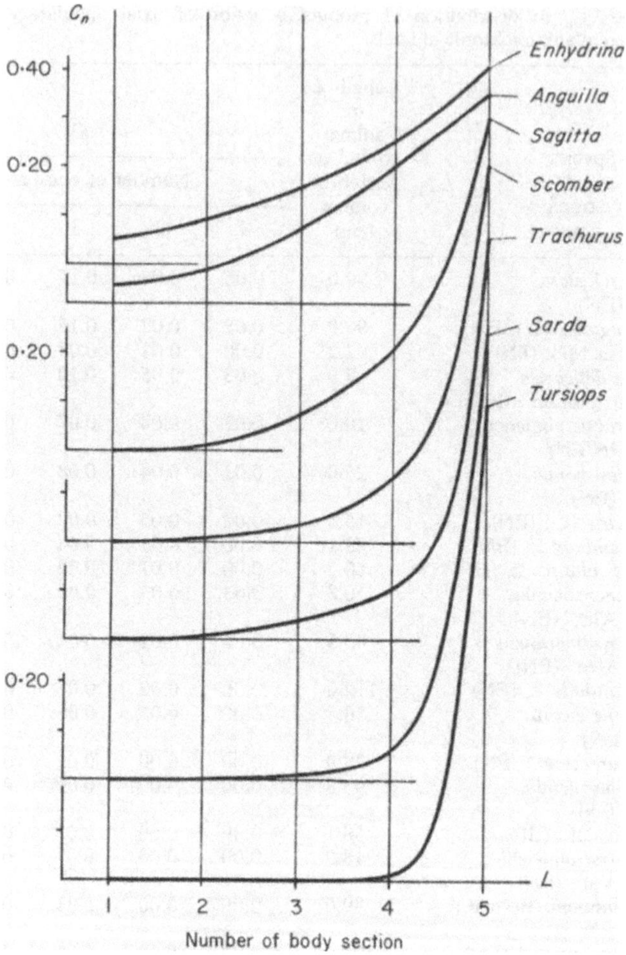

Figure 45. C_n versus longitudinal position along the body for nektonic animals.

size of its scales (Grinberg 1950; Aleyev 1963a), the conversion of the whole body into a propulsive mechanism inevitably leads to a reduction in their size and even their complete disappearance (Saccopharyngiformes, etc.).

With increased swimming speeds there is an increasing tendency towards improving the general streamlining of the trunk, reducing the amplitude of its propulsive undulations and increasing their frequency (Shuleikin 1934; Nursall 1958a; E. Kramer 1960; Aleyev 1963a). This in turn is known (Strasser 1882; Nursall 1958a) to require a stronger axial skeleton and a less flexible body, which is most often achieved by increasing the areas Q_n of the body's cross-sections, particularly Q_2 or Q_3 (Table 6, species beginning from *Sprattus*). All this leads to the foremost part of the body being the first to be excluded from propulsive work, followed by the more posterior parts, and results in the concentration of propulsive action more and more in the rearmost part of the body, the area occupied by the tail fin, as is clear from the values of C_5 (Table 9, species from *Sagitta* to *Tursiops*; Figure 45). Eventually the trunk assumes the most-streamlined shape possible and during movement its anterior half remains absolutely straight, propulsive undulations involving its rear half only, the caudal stem predominantly (Figure 25B,C); the function of propulsor is transferred almost completely to the tail fin (Figure 45, *Sarda–Tursiops*). This creates optimal conditions for the secondary utilization of the kinetic energy of the boundary layer, since the tail fin, being situated at the rear end of the body, where the boundary layer is thickest and contains maximum kinetic energy, can make most effective use of this energy (Aleyev 1965b).

The specialization of the scombroid propulsor goes furthest **in the fastest swimmers,** in particular in most of the Scombridae, Xiphiidae and Istiophoridae and in many Carangidae. The high frequency and sharp changes in the direction of the transverse movements of the tail part of the body demand a considerable reduction in the flexibility of this part of the animal's body in order to preserve adequately rhythmic propulsive performance. This can be seen from the changing values of E_4 and E_5: whereas in most nektonic animals $E_4 < E_5$, in the fastest swimmers, *Trachurus, Sarda, Scomberomorus, Auxis, Thunnus, Xiphias, Tetrapturus* and *Istiophorus*, $E_4 > E_5$ (Table 6, Figure 33). The thinner caudal stem in this case usually turns into a rigid beam with the tail fin at its end and set in motion by the contraction of the lateral muscles (Bertin & Arambourg 1958; E. Kramer 1960; Aleyev 1963a). The necessary transverse rigidity of the caudal stem is achieved in the animals under examination by raising the general elasticity of the vertebral column (Rockwell, Evans & Pheasant 1938) and thanks to the development of special tail keels (Figure 46), which in various groups of nektonic animals have a different morphological basis (Aleyev 1963a). In Scombridae these keels are due to transverse growth of the caudal vertebrae (Priol 1939–1943; see also Figure 34A). The reduction in E_5 in Carangidae is attained at the expense of keel-like plates situated on the sides of the body (Figure 34B), whose action has been described above. Cetacea, *Xiphias* and some sharks (*Carcharias*, etc.) have tail keels of connective tissue and partially cutaneous origin. Istiophoridae have no tail keels along the median line of the body, but strongly enlarged neural and hemal zygapophyses embrace the spinal processes of neighbouring vertebrae (Figure 34C), which also diminishes the flexibility of the vertebral column. The degree of caudal keel development is directly proportional to the animal's speed, as is clearly evident in the examples of Carangidae and sharks (Aleyev 1955, 1957b,c, 1963a). These keels, apart from helping to reduce the lateral flexibility of the caudal stem, also serve as stabilizers (Houssay 1912; Aleyev 1955; Bertin 1958b) and improve the streamlining of the cross-sections of the caudal stem during the latter's transverse propulsive movements (E. Kramer 1960; Aleyev

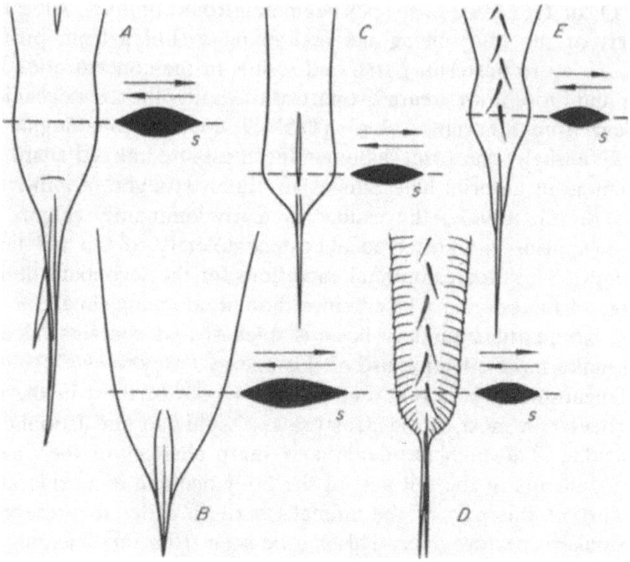

Figure 46. Caudal keels of nektonic animals. *s*, body cross-section, arrows indicating direction of transverse propulsive movements of caudal peduncle. *A*, *Carcharodon carcharias* (L.); *B*, *Tursiops truncatus* (Montagu); *C*, *Xiphias gladius* L.; *D*, *Megalaspis cordyla* (L.); *E*, *Thunnus alalunga* (Bonnat.).

1963a, 1964a). They thus prevent the emergence of parasitic cavitation over the caudal stem, which is important for reduction of total hydrodynamic resistance (see chapter V). When the keels contain a bony skeleton (Scombridae) they probably (Nursall 1958a) function as levers to which the lateral muscles are attached.

In all the cases of extreme specialization of the scombroid propulsor described (Scombridae, Xiphiidae, Istiophoridae, Cetacea, etc.) we observe maximum concentration of the propulsive action in the rearmost part of the body (Table 9, Figure 45). Thus, for example, in *Trachurus* $C_5 = 0.84$, in *Thunnus* $C_5 = 0.96$ and in *Tursiops* $C_5 = 0.99$. Here the relative area of the total propulsive surface of the scombroid propulsor is reduced, inevitably leading to a drop in its efficiency (Aleyev 1963a): while, for example, in the case of the anguilliform propulsor of *Anguilla* 0.9 of the propulsive force is created by a part of the body surface comprising 0.86 of the total lateral surface area $(S_2 + S_3 + S_4 + S_5)$, in the case of the scombroid propulsor of *Tursiops* the propulsive surface consists of only 0.09 of the total lateral surface area (S_{p5}), i.e. ten times less. Accordingly, in latter case the tail fin becomes the only propulsor.

The assertion (Pershin 1969b) that the elongation of the caudal stem in cetaceans helps to raise swimming speeds at the expense of increasing the amplitude of the tail's oscillations contradicts firmly established facts: it is well known (Gray 1933c; Harris 1934; Shuleikin 1934, 1968; Aleyev 1959b, 1963a, 1965b, and others) that adaptation to rapid movement leads to the reduction of the amplitude of transverse tail-fin movements in cetaceans, and also in fishes. During rapid swimming we observe in both fishes and cetaceans examples of transverse tail-fin movements of the smallest amplitude; the gain in speed is obtained in this case at the expense only of a greater frequency of the transverse fin movements.

With higher speeds the scaly cover of fishes also undergoes considerable changes. Since, as we have seen, scales in general make lateral bending of the body more difficult (the bigger the scales the more so), at higher frequencies of propulsive undulations first there takes place a reduction in the size of the scales and then their complete disappearance (Scombridae, Xiphiidae, etc.), the scales first disappearing over the most flexible, rear part of the body, as can be seen, for example, in Scombridae (Figures 96A,B; see also Fraser-Brunner 1950; Aleyev 1963a). Thus both cases of extreme specialization of the propulsive mechanism, towards increasing its efficiency and towards increasing swimming speeds, demand a certain reduction of the scaly cover or even its complete elimination (Aleyev 1963a). However, structural changes in the scaly cover in fishes are connected not only with the characteristics of the structure and action of the propulsor, but also with a number of other factors: primarily the development of adaptations aimed at reducing resistance to movement (see chapter V). Of great interest in this context is the appearance in various groups of particularly fast fishes of elongated, needle-like scales, characteristic in particular of Istiophoridae and some Carangidae (Figure 142); during the body's flexural movements such small scales produce hardly any friction by rubbing against each other, while creating longitudinal relief on the body's surface, which proves useful for controlling the boundary layer (see chapter V).

In accordance with the comparatively low efficiency of the scombroid propulsor and the considerable amount of work thus performed by the muscles per unit time, the fastest nektonic animals develop compensating adaptations of a physiological character. The intensity of their metabolism is comparatively very high and is usually combined with great voracity, which is equally characteristic of fast fishes and cetaceans. In fishes this is reflected in an increased volume and number of erythrocytes and higher haemoglobin content in the blood, in a considerably higher oxygen capacity of the blood (Korzhuev 1965; Bulatova & Korzhuev 1952) and in a somewhat higher body temperature in such fast fishes as the tuna group and *Makaira* (Portier 1903; Legendre 1934; Morrow & Mauro 1950). In cetaceans, owing to adaptation to fast movement under water and continuous diving, and in pinnipeds, only in connection with diving, the intermittent type of respiration emerges and rather large oxygen deposits develop in the form of muscular haemoglobin (Scholander 1940; Tawara 1950, and others): in *Delphinus delphis*, for example, the fraction of myohaemoglobin comes to about 45% and in *Phoca* (*Pusa*) *caspica* to more than 60% of the total amount of haemoglobin in the blood and muscles (Korzhuev, Balabanova, Evstratova & Moderatova 1965; Korzhuev & Glazova 1971).

The phylogenetic development of the axial undulatory propulsor **in benthonektonic and xeronektonic animals** follows the same pattern as has been described above for eunekton. However, this pattern is in most cases concealed to some extent by the development of other adaptations not functionally associated with the creation of propulsive forces but dependent on the particular form of contact with the bottom or land.

Among benthonektonic species we come across both forms with an elongated body and an anguilliform propulsor (*Gymnammodytes*, etc.) and deep-bodied forms with a typical scombroid propulsor (many coral fishes: of the Cyprinidae – *Abramis*, *Blicca*, *Carassius*, etc.; see also Table 10). The formation of both of these is connected with the development of adaptations which increase the animal's manoeuvrability and stability (see chapter VI) and is due to dwelling near the bottom, in the midst of a labyrinth of obstacles formed by the benthic relief: plant growths, corals, rocks, etc. In some benthonektonic forms concentration of the propulsive action in the rear of the body is due to the development of protective armour (*Ostraction*; etc.).

In xeronektonic and nektoxeric forms with an axial undulatory propulsor (Mesosauria (XN), Urodela (NX), Crocodylidae (NX), *Desmana* (NX), *Castor* (NX), etc.), the development of adaptations not connected with propulsion in water

Table 10. Index C_n of distribution of propulsive action of axial undulatory propulsors in benthonektonic and xeronektonic animals.

Species and nektonic type	Length L of animal to end of vertebral column (cm)	C_n				
		Numbers of body section				
		1	2	3	4	5
Gymnammodytes cicerellus (Raf.) (BN)	10.3	0.05	0.06	0.13	0.28	0.48
Acipenser güldenstädti colchicus V. Marti (BN)	120.0	0.00	0.01	0.02	0.27	0.70
Sciaena umbra L. (BN)	27.1	0.00	0.02	0.04	0.20	0.74
Odontogadus merlangus euxinus (Nordm.) (BN)	41.6	0.00	0.00	0.10	0.16	0.74
Gambusia affinis holbrooki (Girard) (BN)	2.0	0.00	0.01	0.06	0.06	0.87
Abramis brama (L.) (BN)	50.5	0.00	0.00	0.02	0.08	0.90
Mugil auratus Risso (BN)	34.2	0.00	0.00	0.01	0.08	0.91
Xiphophorus helleri Heck. (BN)	4.7	0.00	0.01	0.02	0.05	0.92
Lebistes reticulatus Peters (BN)	1.6	0.00	0.00	0.03	0.03	0.94
Pagophoca groenlandica (Erxl.) (XN)	178.0	0.00	0.00	0.01	0.04	0.95

Table 11. Variation with age of the transverse body flexibility E_n, body cross-sectional area ratio Q_n, body cross-sectional shape index G_n, number of flexible connections in the axial skeleton V_n and longitudinal distribution of propulsive action C_n in nektonic animals.

Species and nektonic type	Length L of animal to end of vertebral column (cm)	Index	Numbers of body sections				
			1	2	3	4	5
Sprattus sprattus phalericus (Risso) (PN/EN)	6.0	E_n	0.00	0.24	0.24	0.24	0.28
		Q_n	0.08	0.10	0.10	0.09	0.05
		G_n	0.50	0.39	0.34	0.36	0.39
		V_n	1	14	13	11	9
		C_n	0.05	0.06	0.09	0.10	0.70
	10.0	E_n	0.00	0.20	0.22	0.25	0.33
		Q_n	0.08	0.11	0.10	0.08	0.05
		G_n	0.51	0.44	0.42	0.41	0.33
		V_n	2	14	12	11	9
		C_n	0.03	0.04	0.07	0.14	0.72
Odontogadus merlangus euxinus (Nordm.) (BN)	5.2	E_n	0.00	0.18	0.22	0.25	0.35
		Q_n	0.11	0.14	0.11	0.07	0.04
		G_n	0.65	0.60	0.50	0.55	0.50
		V_n	0	12	12	12	17
		C_n	0.03	0.03	0.12	0.18	0.64
	41.6	E_n	0.00	0.05	0.11	0.25	0.59
		Q_n	0.12	0.16	0.15	0.10	0.04
		G_n	1.00	0.66	0.62	0.68	0.67
		V_n	0	12	11	11	19
		C_n	0.00	0.00	0.10	0.16	0.74
Alosa kessleri pontica (Eichw.) (EN)	8.0	E_n	0.00	0.21	0.23	0.25	0.31
		Q_n	0.08	0.12	0.13	0.10	0.06
		G_n	0.46	0.44	0.38	0.41	0.40
		V_n	0	13	13	12	11
		C_n	0.04	0.06	0.08	0.09	0.73
	25.0	E_n	0.00	0.18	0.22	0.26	0.34
		Q_n	0.08	0.12	0.13	0.10	0.05
		G_n	0.43	0.48	0.43	0.44	0.40
		V_n	0	14	12	12	11
		C_n	0.03	0.04	0.08	0.09	0.76
Pomatomus saltatrix (L.) (EN)	5.0	E_n	0.00	0.21	0.24	0.25	0.30
		Q_n	0.09	0.14	0.14	0.11	0.06
		G_n	0.53	0.40	0.40	0.36	0.50
		V_n	0	6	7	7	6
		C_n	0.03	0.04	0.11	0.16	0.66
	60.2	E_n	0.00	0.06	0.19	0.36	0.39
		Q_n	0.10	0.15	0.14	0.12	0.06
		G_n	0.55	0.58	0.57	0.53	0.75
		V_n	0	7	7	6	6
		C_n	0.00	0.01	0.02	0.18	0.79

Table 11. (*contd.*)

Species and nektonic type	Length L of animal to end of vertebral column (cm)	Index	Numbers of body sections				
			1	2	3	4	5
Trachurus mediterraneus ponticus Aleev (EN)	9.3	E_n	0.00	0.22	0.24	0.28	0.26
		Q_n	0.09	0.13	0.13	0.10	0.05
		G_n	0.53	0.56	0.56	0.61	1.00
		V_n	0	6	7	6	5
		C_n	0.02	0.03	0.08	0.11	0.76
	43.3	E_n	0.00	0.05	0.20	0.43	0.32
		Q_n	0.10	0.15	0.16	0.12	0.06
		G_n	0.74	0.67	0.64	0.83	1.51
		V_n	0	8	6	5	5
		C_n	0.00	0.01	0.04	0.11	0.84
Squalus acanthias L. (EN)	19.0	E_n	0.00	0.12	0.17	0.25	0.46
		Q_n	0.07	0.09	0.06	0.03	0.02
		G_n	1.43	1.00	0.85	0.75	0.75
		C_n	0.01	0.06	0.05	0.08	0.80
	112.0	E_n	0.00	0.06	0.13	0.23	0.58
		Q_n	0.07	0.10	0.06	0.04	0.02
		G_n	1.66	0.91	0.89	1.00	1.00
		C_n	0.00	0.02	0.03	0.06	0.89
Lebistes reticulatus Peters (EN)	0.66	E_n	0.00	0.02	0.26	0.32	0.40
		Q_n	0.16	0.20	0.14	0.08	0.06
		G_n	1.34	1.05	0.57	0.50	0.33
		V_n	0	6	8	8	8
		C_n	0.01	0.02	0.08	0.14	0.75
	1.65	E_n	0.00	0.05	0.12	0.37	0.46
		Q_n	0.14	0.21	0.19	0.13	0.06
		G_n	1.43	0.88	0.75	0.50	0.30
		V_n	0	7	8	7	8
		C_n	0.00	0.00	0.03	0.03	0.94

(camouflage, protection, terrestrial locomotion, etc.) as a rule reduces the body's capacity to act as an axial undulatory propulsor and as a result the tail becomes its main or only element. In Ophidia, during both aquatic and terrestrial life, the entire body serves as a propulsor.

The ontogenetic adaptive changes in the functional characteristics of the axial undulatory propulsor are greater the broader the variation of the Reynolds number with age: they are greatest in spawning fishes and least pronounced in small nektonic reptiles (Hydrophidae), sirenians and cetaceans (Tables 8, 11). In the ontogeny of fishes the propulsive mechanism changes in most cases from the anguilliform towards the scombroid type as a result of adaptation to locomotion at ever increasing speed (Aleyev 1963a): the body becomes relatively thicker and less flexible, and the single fin running around its surface gets divided into separate fins with different functions, with the propulsive action transferred more and more to the tail fin (Tables 8, 11, Figure 47). All spawning fishes initially have an anguilliform type of propulsor (Aleyev 1963a). In viviparous fishes and cetaceans, in the range $L_a < 450$ cm, the changes occurring in the structure of the propulsive mechanism with age are restricted however to changes in the shape of the tail fin towards stronger concavity (Table 8, *Squalus*, *Delphinus*).

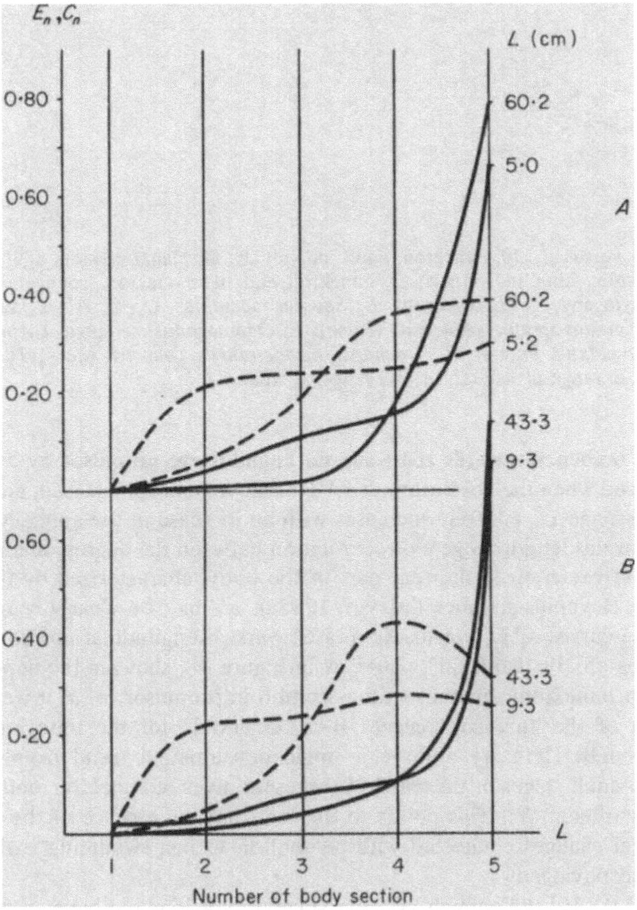

Figure 47. Variation of body flexibility E_n and distribution C_n of propulsive action along longitudinal body axis in nektonic animals with age. *A*, *Pomatomus saltatrix* (L.), *B*, *Trachurus mediterraneus ponticus* Aleev.

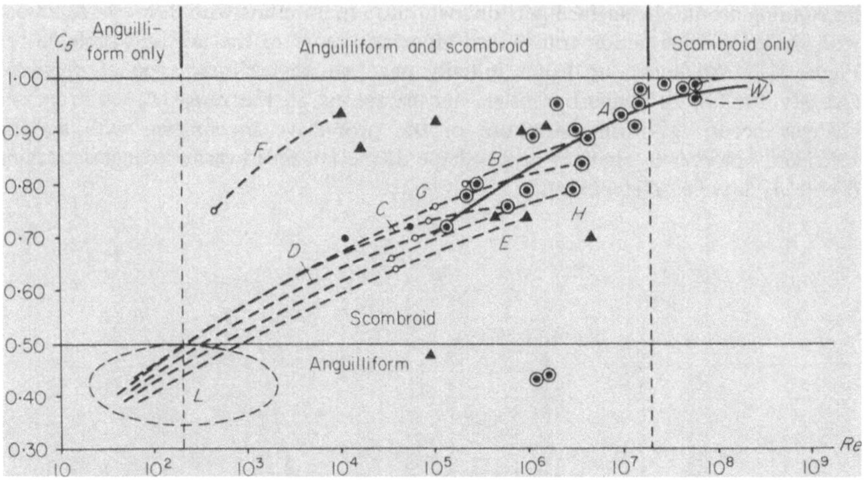

Figure 48. C_5 versus *Re*. ⊙, eunekton, adult individuals; ●, planktonekton, adult individuals; ▲, benthonekton, adult individuals; ○, eunekton and benthonekton, young individuals. *A*, eunekton, phylogeny; *B-H*, ontogeny. *B. Squalus acanthias* L., *C, Alosa kessleri pontica* (Eichw.); *D, Sprattus sprattus phalericus* (Risso); *E, Odontogadus merlangus euxinus* (Nordm); *F, Lebistes reticulatus* Peters; *G, Trachurus mediterraneus ponticus* Aleev; *H, Pomatomus saltatrix* (L.). *L*, range of fish larvae; *W*, range of whales.

Since **the tendency towards replacing an anguilliform propulsor by a scombroid** one is observed when the absolute speed V_m of movement is increased, and since this speed in the range $L_a < 450$ cm increases with an increase in the animal's length L_a, we observe in this length range a direct relation between the degree of concentration of the propulsive action in the rear part of the body, characterized by the value of C_5, and the Reynolds number (Aleyev 1963a), as may be clearly seen from the nature of the curves of C_5 versus *Re* and C_n versus longitudinal position along the body (Figures 45, 48, Table 11). Curve A in Figure 48, showing the dependence of C_5 on *Re* for eunektonic animals with a scombroid propulsor, is, as it were, a direct continuation of the analogous curves B–E, G and H for the ontogeny of these nektonic animals. Here we observe a single development trend from fish larvae (region L) to adult stages of the smallest representatives of eunekton, and further, to adult whales (region W). This points to the fact that the essence of the functional-morphological changes connected with adaptation to fast swimming is the same in ontogeny and phylogeny.

For $Re < 2.0 \times 10^2$ the animals at every stage of development have an anguilliform propulsor only ($C_5 \leq 0.50$) and for $Re > 2.0 \times 10^7$ the scombroid type only ($C_5 > 0.50$). However, in the vast range $2.0 \times 10^2 \leq Re \leq 2.0 \times 10^7$ there occur both anguilliform and scombroid propulsors, depending in each individual case on the particular ecological features of the species (Figure 48). Thus, greater concentration of the propulsive action in the rearmost part of the body, aimed at limiting the energy expended on movement and reflected in higher C_5 values, takes place both at higher absolute swimming speeds V_m and higher Reynolds numbers and at an increased relative swimming speed V_r with $Re \approx$ constant.

A typical anguilliform propulsive mechanism comprises several simultaneously functioning elementary sections each of which creates during a single period of the body's propulsive undulations an elementary propulsive force p changing according to a sine law from 0 to 1. However the addition of several simultaneous elementary forces results in the propulsive mechanism as a whole creating at any given moment a total propulsive force P whose magnitude remains practically unaltered throughout a single propulsive wave period.

In the case of a typical scombroid propulsor the tail fin has been transformed into a single, only slightly elastic plane, and according to its functional characteristics can be regarded to a sufficient approximation as an indivisible elementary section, similarly creating an elementary propulsive force p changing according to a sine law. However, superimposed on this variation of the propulsive force during each period of the propulsive body undulations is the inertia of the propulsive process, so that the forward movement in this case too remains essentially uniform. Experimental results obtained (Komarov 1971) by means of the 'Nekton' ultrasonic information system by sounding swimming fishes and represented in the form of oscillograms of instantaneous swimming speeds (with about 100 information pulses per second) and synchronous cine-records showed that during the propulsive cycle the speed of progression is independent of the tail fin's position relative to the longitudinal axis of the body.

In all nektonic animals the actual swimming speed as a rule constantly changes owing to changes in the mode of the propulsor's performance: all nektonic animals alternately speed up and brake. A scombroid propulsor, all other conditions being unchanged, is capable of ensuring higher speeds than an anguilliform propulsor, and accordingly is better for providing locomotion characterized by the alternation of short bursts of swimming with long stretches of deceleration, which (Shebalov 1969) affords the greatest advantages as regards diminishing friction drag, i.e. the basic kind of hydrodynamic resistance encountered by nektonic animals (see chapter V).

2. The Pseudo-axial Undulatory Propulsors

A variant of the axial undulatory propulsive mechanism is encountered in xeronektonic pinnipeds belonging to the families Phocidae and Odobenidae, and also in the nektoxeric *Enhydra lutris.* The propulsor of these animals may be described as pseudo-axial (Aleyev 1973b): functionally it is similar to the tails of fishes (Phocidae and Odobenidae) or cetaceans (*Enhydra*), however, in distinction from such tails, it is formed not by the axial skeleton, but by the posterior limbs (Figures 49, 50). We studied the pseudo-axial undulatory propulsive mechanism (Aleyev 1973a) using the

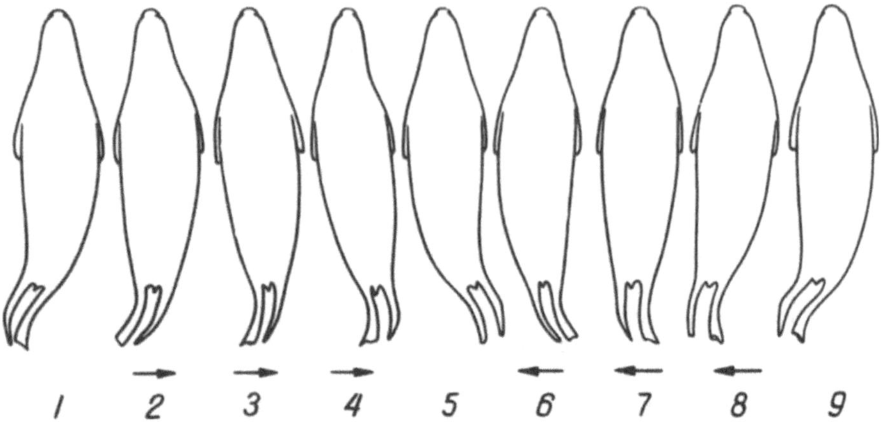

Figure 49. Propulsive movements of *Pagophoca groenlandica* (Erxl.). Arrows indicate direction of transverse movement of hind flippers. Filmed from helicopter.

examples of Phocidae (*Phoca* and *Pagophoca*), Odobenidae (*Odobenus*) and partly *Enhydra*.

In Phocidae and Odobenidae, during rapid swimming in a straight line the fore-flippers are, as a rule, pressed to the body, while the posterior ones are extended backwards with their soles together, situated vertically like the tail fin of fishes, and perform lateral movements together with the rear part of the body, being deflected together now to the right, now to the left (Allen 1880; Abel 1912; Howell 1929; Smirnov 1929; Ognev 1935; Pikharev 1940; Frechkop 1955; Kenyon & Rice 1959; Gambarjan & Karapetjan 1961; Mansfield 1963; Ray 1963; King 1962, 1964; Mordvinov 1968; A. S. Sokolov 1969a; Aleyev 1973b). When moving towards the body each of the flippers is spread to the utmost, providing the maximum working surface, whereas when moving away from the body, the toes are brought together so that the flipper's working surface is reduced (Figures 49, 50). Since during the movement of one flipper towards the body the other moves away from it, the total propulsive force created by both flippers is absolutely constant during their deflexion either to the right or to the left, i.e. the propulsor as a whole is bilateral, which, as has been mentioned above, is one of the crucial characteristics of an undulatory propulsive mechanism. Both Phocidae and Odobenidae are capable of swimming by using their fore-flippers as oars (see below), but this is not their habitual mode of locomotion.

The body of *Enhydra* (NX) performs wavelike propulsive movements in the vertical direction, and the posterior limbs, transformed into flippers, which perform the basic propulsive action, are situated during swimming accordingly, i.e. horizontally, forming together with the tail (compressed between them) a single horizontal plane (Barabash-Nikiforov 1933, 1947; Barabash 1937; Grasse 1955; Gambarjan & Karapetjan 1961; A. S. Sokolov 1970).

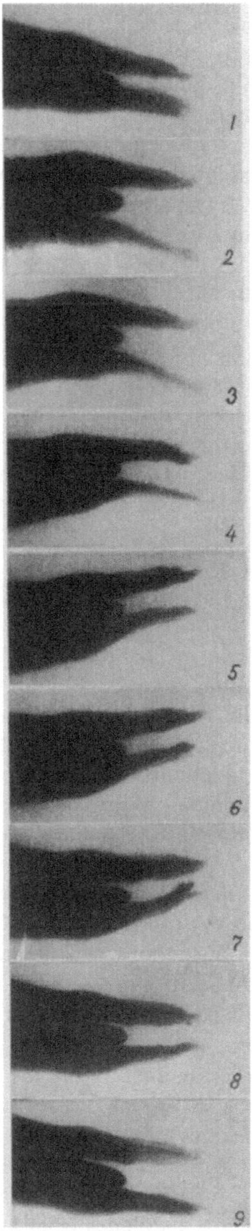

Figure 50. Propulsive movements of posterior flippers of *Pagophoca groenlandica* (Erxl.). Viewed from above. Filmed in biohydrodynamic channel.

The values of C_n which we obtained for *Pagophoca groenlandica* (Table 10) characterize its pseudo-axial propulsive mechanism as a scombroid one ($C_5 = 0.95$), which doubtless it is. This is also the case in all the other representatives of Phocidae and Odobenidae, as well as in *Enhydra*.

In the ontogeny of Phocidae, Odobenidae and *Enhydra* the pseudo-axial propulsor does not undergo any essential functional-morphological changes, which can probably be explained by the very small variation of *Re* in these animals with age.

The phylogeny of the pseudo-axial undulatory propulsor followed the path of improving the posterior limbs as the basic propulsive elements and gradual atrophy of the tail, which has eventually led to a terminal position of the posterior limbs and maximum concentration of the propulsive action at the rear end of the body. The proximal parts of the limbs grew shorter and became covered by the skin sac of the body; the foot lengthened, became flatter and turned into a more-or-less perfect flipper. We can observe an earlier stage of all these transformations in nektoxeric forms, the extant *Enhydra* and the Myocene *Semantor*, and a later stage in xeronektonic Phocidae and Odobenidae. As is clear from the above, the pseudo-axial propulsor is analogous to the axial one, though not homologous to it.

3. Peripheral Undulatory Propulsors

We call an undulatory propulsive mechanism based on undulations of peripherally placed flippers and not including the body's longitudinal axis a peripheral one (Aleyev 1973b). Such a propulsor occurs only in Cephalopoda, Chondrichthyes and Osteichthyes; it never arises in secondary aquatic nektonic organisms since they do not have fins furnished with adequate musculature. We studied (Aleyev 1958a, 1963a, 1973b) the peripheral undulatory propulsor using specimens of Cephalopoda (*Todarodes*), Chondrichthyes (Rajidae, Mobulidae and Dasyatidae) and Osteichthyes (*Zeus*, Syngnathidae, etc.).

Locomotion by means of undulatory fin movements is based on the wavelike motions of the fin's blade due to strictly rhythmic contractions of special muscles which directly change the form of the fin when situated within the blade itself, as in the case of cephalopods (Tompsett 1939), and facilitate consecutive transverse deflexions of the fin's supporting rays when situated outside the blade, as in fishes (Aleyev 1963a). The peripheral undulatory propulsor in nektonic cephalopods is based on the mantle fins. In fishes, undulations in the various cases are performed by the dorsal (Gymnarchidae), anal (Gymnotidae) or pectoral fins (Rajiformes, Dasyatiformes and Torpediniformes), of some combination of these fins operating simultaneously (Zeiformes and Tetrodontiformes). Depending on the relative length of the base of an undulating fin (Figure 51), a different number of propulsive wavelengths can be fitted in across its blade: from a small fraction of a wavelength (as in Molidae and most other Tetrodontiformes) to several wavelengths (*Gymnarchus*, etc.; see Figures 52, 63A). As it sets in motion, as a rule, only relatively small

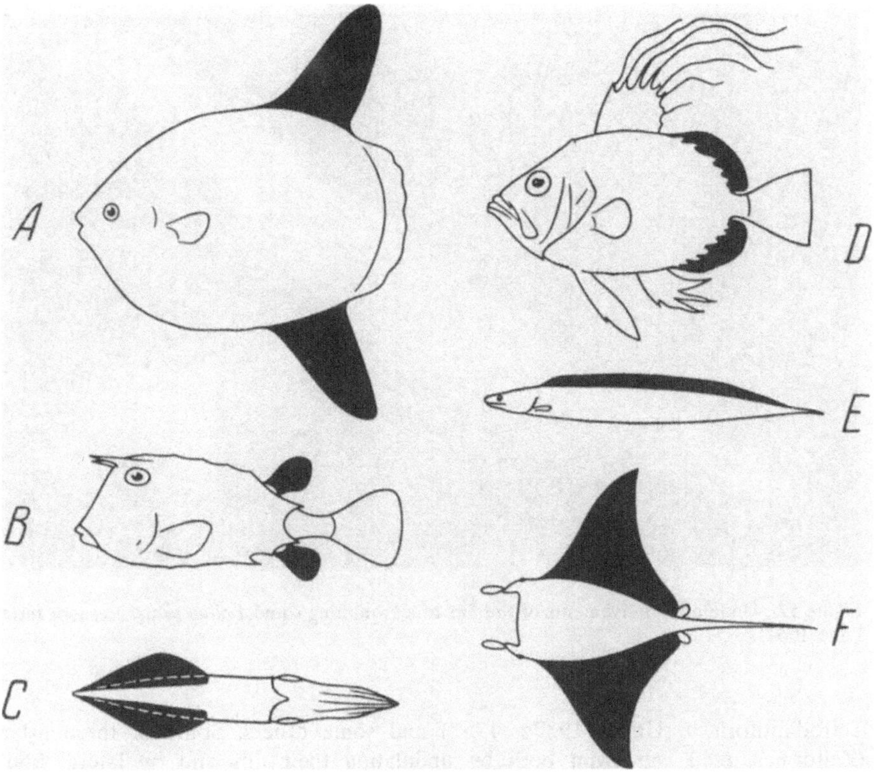

Figure 51. Different shapes of undulating fins (shaded) in nektonic animals. *A, Mola mola* L.; *B, Acanthostraction tricornis* (L.); *C, Loligo vulgaris* Lam.; *D, Zeus faber pungio* Val.; *E, Gymnarchus niloticus* Cuv.; *F, Manta birostris* (Walb.).

masses of water, a peripheral undulatory propulsor is in the majority of cases capable of producing comparatively low swimming speeds only.

The mantle fins of cephalopods are a pair of leaflike muscular formations with a cartilaginous base and streaked with muscles, which stretch in three mutually perpendicular directions: along the fin, across the fin and normal to it (Tompsett 1939). Cephalopods use undulation of these fins only when swimming slowly (Figures 52, 63A); for rapid movement they always switch on their hydrojet propulsor.

Among nektonic and nektobenthic fishes, swimming by undulation of the fins is characteristic of Rajiformes, Dasyatiformes, Torpediniformes (Schlesinger 1911b; Andriashev 1946; Aleyev 1963a), Notopteridae (Schlesinger 1910b, 1911b), Gymnarchoidei (Schlesinger 1911b; Gray 1953a; Lissmann 1961), Gymnotoidei (Sachs 1881; Schlesinger 1910a), Zeiformes (Andriashev 1946; Aleyev 1958a, 1973b),

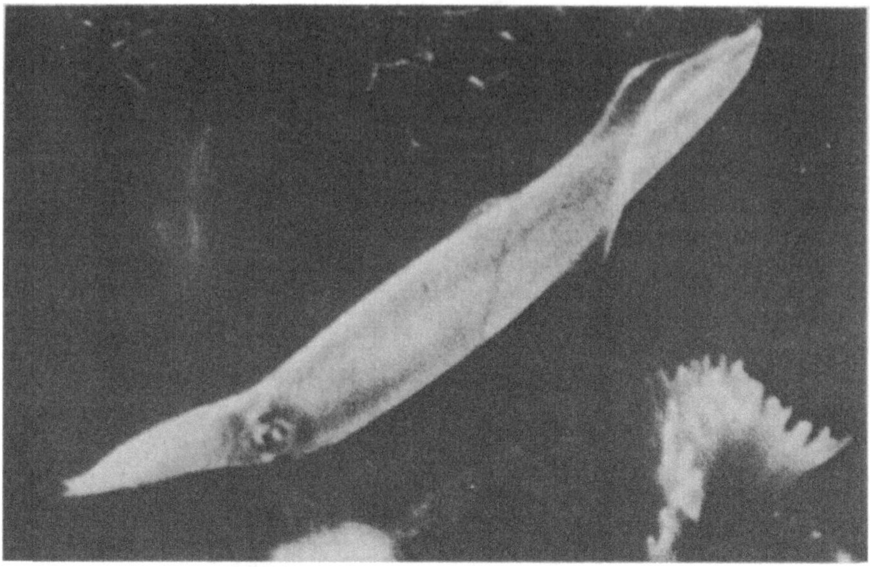

Figure 52. Undulating movements of the fins of a swimming squid *Loligo pealei* Lesueur (after Lane 1957).

Tetrodontiformes (Harris 1937a, 1953) and some others. Some of these fishes (Zeiformes, etc.) can swim both by undulating their fins and by lateral body undulations. For others (Rajiformes, Dasyatiformes and Torpediniformes) fin undulation is the only means of locomotion. In many cases, in addition to forward movement, fishes can use a peripheral undulatory propulsor also for swimming backwards, as is observed, for example, in *Gymnarchus* (Lissmann 1961).

The peripheral undulatory propulsor occurs as the only propulsive mechanism almost exclusively in benthic (Rajiformes) and planktonic (Syngnathiformes) species, but even here it is frequently combined with an axial undulatory propulsor (Pleuronectiformes). In benthonektonic fishes the peripheral undulatory propulsor in the great majority of cases is not the only one, being as a rule combined with an axial undulatory propulsor (Gymnarchoidei, Notopteridae, Zeiformes, Tetrodontiformes ex p.); here the basic propulsive action can be performed by the peripheral propulsor, whereas the axial one is of very minor importance and is used only rarely (*Zeus*, etc.). Among eunektonic fishes we find a peripheral undulatory propulsor in Mobulidae and Molidae only, however in both groups it is formed in the benthic stage. Thus the incidence of the peripheral undulatory propulsor and the amount of work it carries out increase with transition from eunektonic to benthic and planktonic forms.

In ontogeny the peripheral undulatory propulsor in the overwhelming majority of cases either develops in parallel with some other propulsor, as in all cephalopods, which from the very beginning of the post-embryonic period always have a functioning hydrojet propulsor, or appears to replace some other propulsive mechanism, as in all fishes, in which the peripheral undulatory propulsor replaces an axial undulatory one. Only in a few cases do the young emerge from the egg membranes already having a peripheral undulatory propulsor only, a characteristic of adult individuals (Rajiformes, Dasyatiformes and Torpediniformes).

In phylogeny the peripheral undulatory propulsor also appears always in the presence of either a hydrojet (cephalopods) or an axial undulatory (fishes) propulsor during the development of the initially secondary functions of the fins.

3. Swimming. Paddling propulsors

If the undulatory propulsor is the most common nektonic aquatic propulsive mechanism, coming next in frequency among nektonic animals from different systematic groups is a propulsive mechanism operating on the principle of oars, which we call a paddling propulsor (Aleyev 1973b). This is found to a different extent in many fishes, aquatic Testudinata, Placodontia, Sauropterygia, Hesperornithes, Sphenisciformes, Pinnipedia and some others. The paddling mode of swimming is also characteristic of the majority of nektoxeric animals (adult Anura, waterfowl and most of the semiaquatic mammals). We studied the paddling propulsor (Aleyev 1963a, 1973b) using specimen of different representatives of the Osteichthyes (Labridae, Pomacentridae, Gasterosteidae, etc.), Chelonioidea (*Chelonia, Caretta* and *Eretmochelys*), Sphenisciformes (*Eudyptes*) and Pinnipedia (*Phoca, Pagophoca, Odobenus* and *Arctocephalus*). Propulsive paddling in nektonic animals is always carried out by the limbs, as a rule predominantly or exclusively by the anterior limbs (fishes, Testudinata, Placodontia, Sauropterygia, Sphenisciformes and Otariidae), and more seldom by the posterior limbs (*Hesperornis*). The limbs performing the function of a paddling propulsor are always transformed in some measure, depending on the degree of the animal's adaptation to the aquatic mode of life; into fins or flippers which have a well-streamlined cross-section (fishes, Testudinata, Placodontia, Sauropterygia, Sphenisciformes, Otariidae, etc.; Figures 53A–D,F) or are well streamlined when folded (*Hesperornis*; Figure 53E). They are distinguished by a marked ability to deflect themselves at considerable angles in all directions and to rotate (the entire limb or its distal part) about their longitudinal axis; the latter is particularly characteristic of non-folding flippers and fins and enables considerably changes in the fin's angle of attack. Besides, characteristic of reptiles, birds and mammals are elongation of the toes, a reduction of their autonomy and the conversion of the palm (or foot) into a single, outwardly more-or-less monolithic plane, as well as shortening of the limb's two proximal elements (i.e. the shoulder and antibrachium or thigh and shin), which leads to a reduction in the paddling

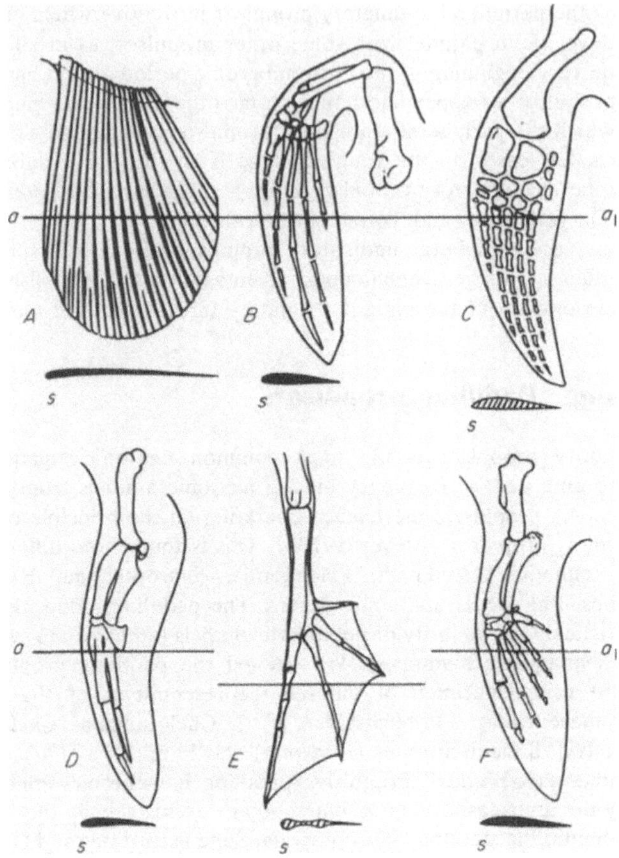

Figure 53. Fins functioning as paddles. *A, Crenilabrus tinca* (L.); *B, Chelonia mydas* (L.); *C, Thaumatosaurus victor* Fraas (after Fraas 1910, with modifications); *D, Aptenodytes patagonicus* Miller (after Abel 1912); *E, Hesperornis regalis* Marsh (after Heilmann 1916; see Abel 1922); *F, Otaria byronia* (Blainv.) (after Frechkop 1955, with modifications). *s*, cross-section along *aa*.

moment of the 'arm' of the limb and increases its relative power (Figures 53B–F). The most general functional characteristic of the paddling propulsor, constituting its fundamental distinction from a propulsor of the undulatory type, is the presence of a working surface on only one side of the working elements.

The working cycle of a paddling propulsor, studied in detail in specimens of various Actinopterygii (Labridae, etc.) Chelonioidea and Pinnipedia, consists of two phases (Figures 54–56, 58, 61).

Phase I: idle stroke. This consists of drawing the working element (fin or flipper) away from the body to its forward position, during which it is situated 'edgewise' relative to the flow (fishes, Testudinata, Placodontia, Sauropterygia, Sphenisciformes and Otariidae) or is folded (*Hesperornis*); in both cases the frontal area of the working element and, consequently, the drag force $-p$ are minimal (Figure 54A, positions 1–6).

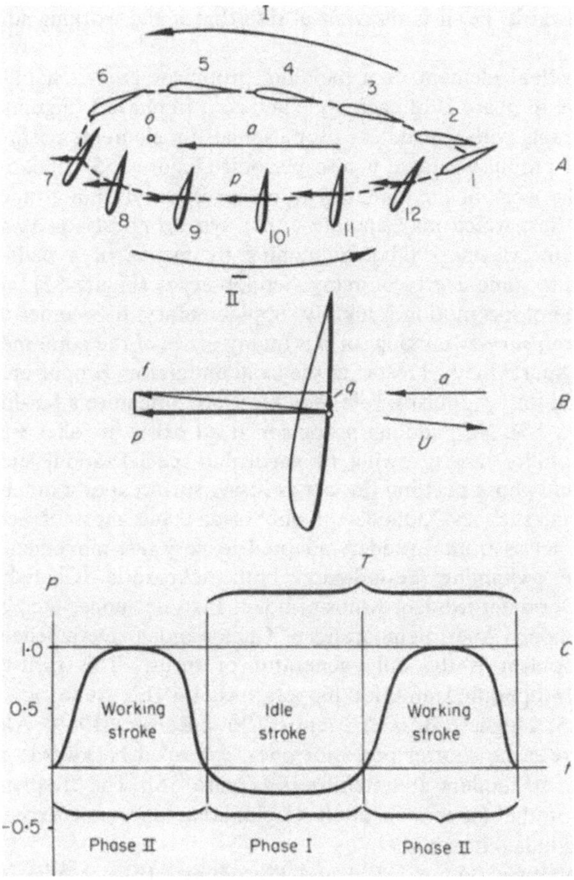

Figure 54. Schematic diagrams of performance of a paddling propulsor. *A*, motion of working element. *I*, idle stroke; *II*, working stroke; *a*, direction of animal's movement; *O*, point of attachment of working element to trunk; *p*, propulsive force; Arabic numerals indicate successive positions of working element. *B*, forces acting on working element. *f*, total drag force; *p*, *q*, its components; *U*, direction of movement of working element; *a*, direction of animal's movement. *C*, total propulsive force *P* created by propulsor as a function of time. See text for further explanations.

Phase II: working stroke. This consists of drawing the working element towards the body into the rearmost position, during which it is perpendicular to the direction of movement and is extended to the utmost, ensuring the maximum frontal area (Figure 54A, positions 7–12), maximum frontal drag and accordingly, the maximum propulsive force $+p$ (Figures 54A,B). During this phase, in Actinopterygii the fin is situated with its leading edge upwards, i.e. its dorsal side is the working surface (Figure 55), whereas in Reptilia, Sphenisciformes and Mammalia its leading edge is turned downwards, i.e. it is the ventral side that is the working surface (Figures 56, 58, 61).

Each individual element of a paddling propulsor creates a pulsating propulsive force, positive in phase II of each cycle and zero in phase I (Figures 54A,C). In most nektonic animals with a paddling propulsor all its elements perform synchronously and the total propulsive force is also periodic (Figures 55A, 56, 58, 61).

In fishes the work of the paddling propulsor is based on non-undulatory action of the pectoral fins, which may operate either synchronously or asynchronously (Andriashev 1946; Aleyev 1963a). Swimming by means of a paddling propulsor is characteristic to some degree of many Actinopterygii (Figure 55) but in a majority of cases this type of locomotion is merely supplementary. It becomes the main one only in some littoral, slow-swimming species (many fishes of the coral reefs, etc.), however during quick spurts they all resort to the axial undulatory propulsor. With an increase in the pectoral fins' propulsive role they most often acquire a fan-like, rounded form (Figures 53A, 55). No paddling propulsor at all exists in fishes whose pectoral fins have lost mobility largely owing to particular specialization, such as sharks and Acipenseridae (whose pectoral fins act as lifting surfaces) or a number of particularly fast swimmers, such as Xiphiidae, Istiophoridae and most of Scombridae (whose pectoral fins act as frontal rudders adapted to very fast movement).

In all the swimming **Testudinata**, both nektoxeric (Chelydridae, Chelyidae, Emydinae, Dermatemydidae, Kinosternidae, Platysternidae, etc.) and xeronektonic (Thalassemydidae, Apertotemporalidae, Chelonioidea, Dermochelyoidea, etc.), the paddling propulsor is the only generator of thrust. This thrust is produced by synchronously operating anterior flippers (Abel 1912; Sukhanov 1964; Schubert-Soldern 1966; Hughes, Bass & Mentis 1967; Bellairs 1969; Aleyev 1973b, and others), whereas the shorter posterior ones, extended backwards during swimming, usually serve as rudders and stabilizers (Figure 56). The frequency of propulsive movements of the flippers in adult Cheloniidae does not exceed 0.5 c/s, but in younger individuals it is higher.

In Sauropterygia (Figure 57A) and **Placodontia** (Figure 57B) the posterior and anterior limbs had, as a rule, a more-or-less similar structure (Abel 1912; Novozhilov 1964a–d; Tatarinov 1964h–j; Newmann & Tarlo 1967) and resembled the flippers of extant sea turtles (Figures 53C, 57), the basic propulsive action probably being carried out, just as in turtles, by the anterior limbs. The transformation of the limbs into flippers was accompanied by shortening of the shoulder and lower leg, and in Sauropterygia also by strongly pronounced hyperphalangism

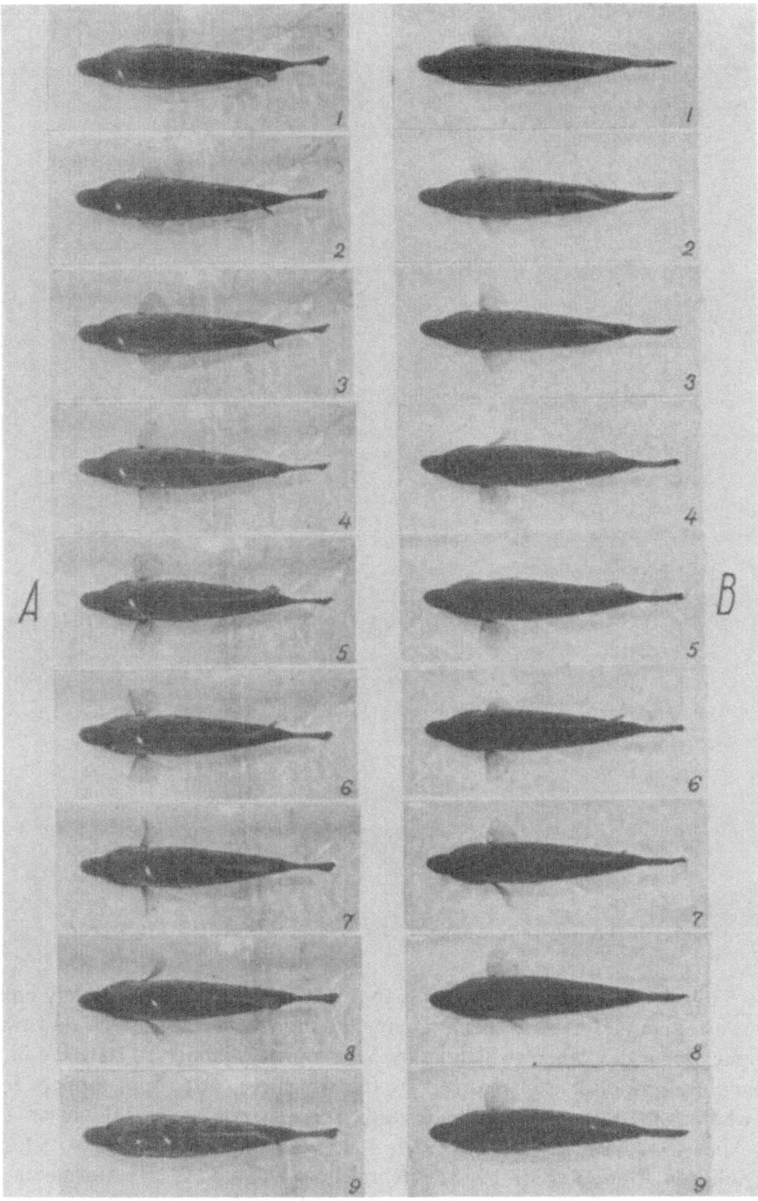

Figure 55. Paddling propulsor of *Labrus viridis* L. during (*A*) synchronous and (*B*) asynchronous operation of pectoral fins. Cine-record.

Figure 56. Propulsive movements of *Chelonia mydas* (L.). Cine-record.

(Figure 53C): in the girdles of the limbs the ventral elements were greatly enlarged with powerful muscles fixed to them, moving the flippers backwards and downwards.

In Sphenisciformes, when swimming in the completely immersed state the function of a paddling propulsor is performed by the wings converted into flippers (Figure 53D), whereas the legs, extended backwards, serve as rudders and stabilizers (Chun 1900; Abel 1912; Murray-Levick 1914, 1915; Murphy 1915; Brooks 1917; Neu 1931; Simpson 1946; Berlioz 1950a; Oehmichen 1950a,b; W. B. Alexander 1955; see also Figure 58). When swimming at the surface of the water the wings again constitute the main propulsor, the legs fulfilling mainly the functions of rudder and stabilizer and only rarely creating additional thrust by operating as paddles (Neu

1931). Thanks to the lateral expansion and dorso-ventral compression of the bones of the shoulder and forearm, the wing of the penguin is very thin (Figure 53D), on account of which its total drag is noticeably reduced at $\alpha = 0$, during the idle stroke. Also, small scale-like feathers cover the wing's entire surface, making it smooth (Figure 59). A characteristic of the paddling propulsor of penguins is the comparatively high frequency of propulsive movements of the wings. Thus, in *Aptenodytes patachonica* it reaches up to 120 strokes per minute (Neu 1931), and in *Pygoscelis papula,* at a swimming speed of up to 10 m/s (Murphy 1915), up to 200 strokes per minute (Brooks 1917).

The now extinct non-flying *Pinguinus impennis* (Alciformes), which was probably xeronektonic, also swam with the aid of its wings.

In the Cretacian xeronektonic **Hesperornithes,** which had no wings at all, it wings being represented morphologically by the rudimentary shoulder bone concealed under the skin, the only generators of propulsive forces during swimming were the legs (Abel 1912, 1922; Dabelow 1925; Piveteau 1950; Dementyev 1964), operating

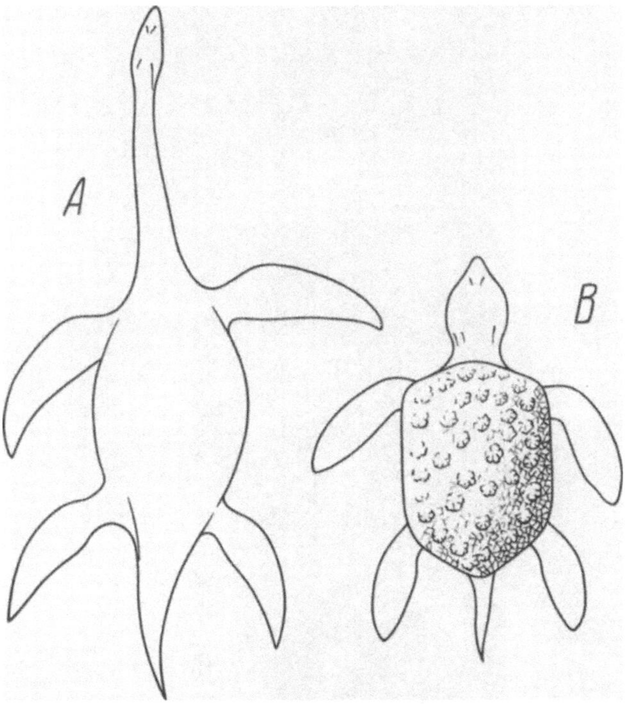

Figure 57. Sauropterygia (*A*) and Placodontia (*B*) with a paddling propulsor. *A, Cryptocleidus oxoniensis* Phillips, author's reconstruction from material in Andrews (1910) and Abel (1912); *B, Placochelys placodonta* Jaeckel, author's reconstruction from material in Jaeckel (1907).

Figure 58. Swimming *Pygoscelis papua* (Forster) (after Neu 1931).

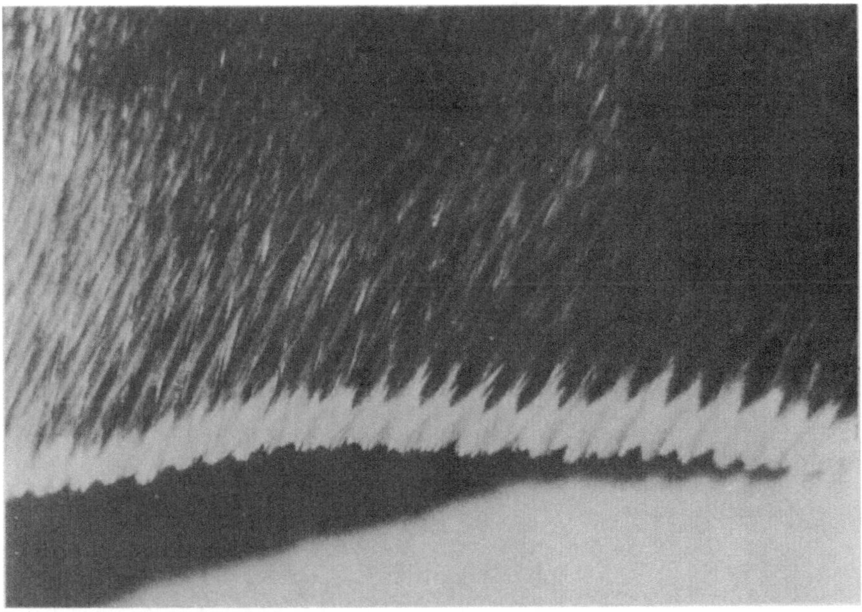

Figure 59. Feathers covering the wing of *Eudyptes chrysolophus* Brandt. Photograph.

as a paddling propulsor. The legs of *Hesperornis* resembled in structure those of extant *Phalacrocarax*, i.e. the swimming membrane linked all four toes (Figures 53E, 60A). The propulsive movements of the legs of *Hesperornis* were by every indication analogous to the leg movements of extant Gaviidae and Podicipedidae, i.e. its legs moved practically in a horizontal plane (Dabelow 1925), without creating any detrimental pitching moments (Figure 60B,D), which is of great importance when swimming fully immersed; with leg movements in a vertical or nearly vertical plane, as in Anatidae, this moment inevitably becomes quite considerable (Figure 60C).

The paddling propulsor of the **pinnipeds** is based on anterior limbs converted into flippers. In Phocidae it is a secondary propulsor, the basic propulsive action being carried out by a pseudo-axial type of propulsor; in Odobenidae the paddling propulsor is used more often and is of approximately the same importance as the pseudo-axial one; in Otariidae the paddling propulsor is practically the only generator of propulsive thrust in water and attains the highest perfection, while the posterior flippers fulfil the role of rudders and stabilizers (Abel 1912; Frechkop 1955; Backhouse 1961; Gambarjan & Karapetjan 1961; Mansfield 1963; Ray 1963; King 1964; Mordvinov 1968; see also Figure 61). Accordingly the hydrodynamic properties of the anterior flippers in pinnipeds improve through the series Phocidae–Odobenidae–Otariidae (Murie 1871, 1872, 1874; Mori 1958; Scheffer 1958; King

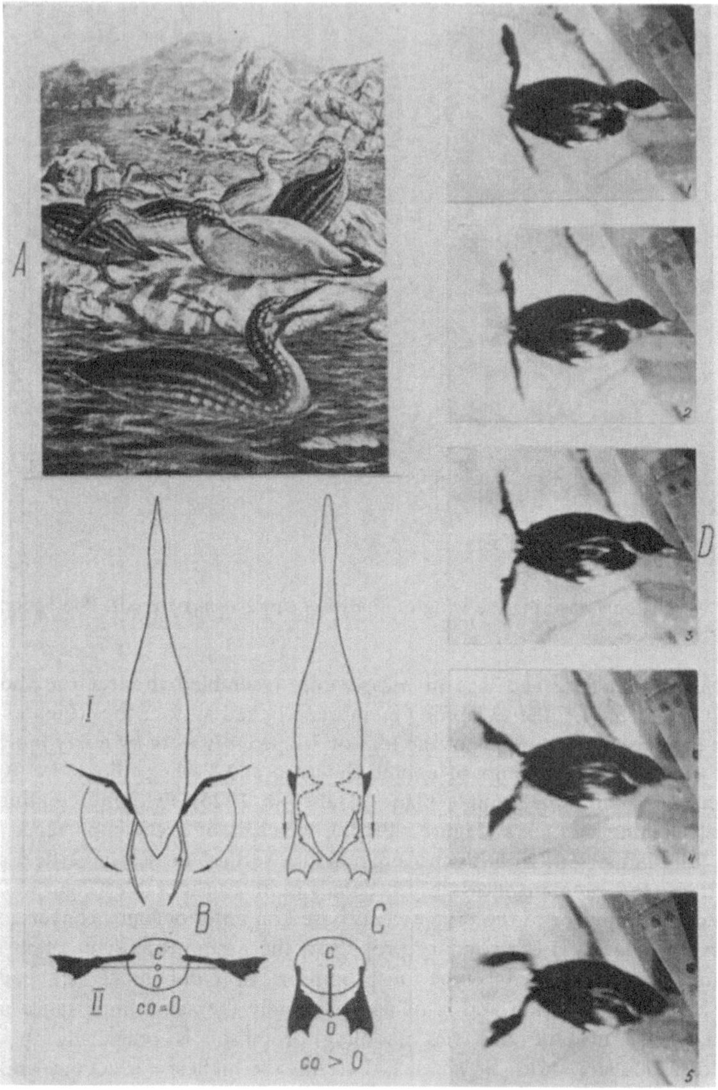

Figure 60. A, *Hesperornis regalis* March, reconstruction (after Heilmann 1916; see Abel 1922); *B*, *C*, schematic movements of bird's legs during swimming in Gaviidae (*B*) and Anatidae (*C*); *I*, top view, *II*, projection onto transversal plane; *c*, centre of gravity; *o*, point of application of propulsive force, *co*, arm of detrimental pitching moment. *D*, swimming of *Podiceps ruficollis* (filmed by Yu. E. Mordvinov).

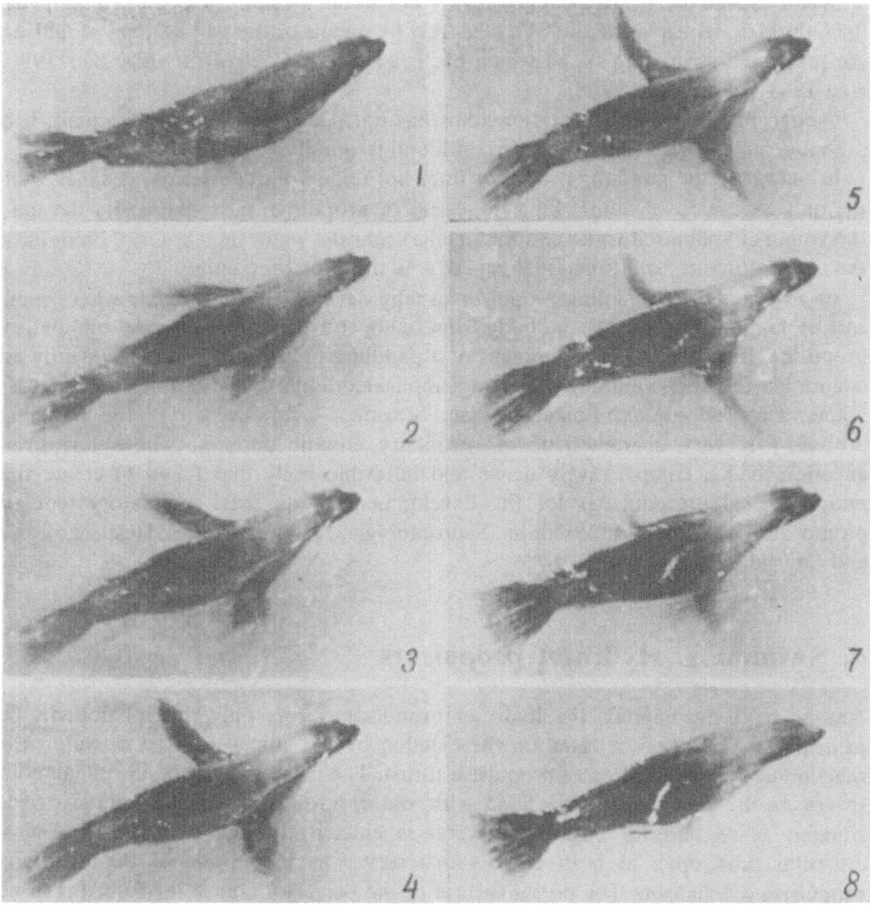

Figure 61. Propulsive movements of *Arctocephalus pusillus* (Schreb.) (after Morvinov 1968).

1964; A. S. Sokolov 1966): the flippers grow longer and flatter and assume the shape of a wing with the claws reducing more and more. The area of the anterior flippers is about 30% of the body's cross-sectional area in Phocidae, 50% in Odobenidae and 100% in *Eumetopias*; the musculature of the anterior limbs is only 2–3 times heavier than the muscles of the posterior limbs in Phocidae, 4 times in Odobenidae and 8 times in *Eumetopias.*

The frequency of propulsive movements of the anterior flippers in pinnipeds is comparatively low: in *Arctocephalus pusillus*, for example, according to our data, it does not exceed 2 c/s. Also comparatively low is the swimming speed of pinnipeds: for Otariidae, which swim almost exclusively by means of the paddle-type propulsor, the maximum swimming speed which has been recorded is 9.0 m/s (Scheffer 1950; Ray 1963; Mordvinov 1968).

Though paddling by the anterior limbs has not been described for sirenians, it is probable, since these limbs possess considerable mobility.

In ontogeny the paddling propulsor does not undergo any essential changes, with the only exception of diminishing frequency of propulsive movements. By the time the young of Sphenisciformes and Pinnipedia enter the water their anterior limbs have the same structure and functional capacity as those of the adults.

In phylogeny the paddling propulsor usually develops in those cases where some sort of adaptation prevents the body from being converted into an axial undulatory propulsor. In fishes, the development of a paddling propulsor is most frequently an adaptation to slow swimming among a complicated labyrinth of obstacles (Gasterosteidae, Chaetodontidae, Pomacentridae, Scaridae, Labridae, etc.). The paddling propulsor is very characteristic of secondary aquatic animals, whose terrestrial ancestors had a comparatively dense and inflexible body that failed to create the morphological prerequisites for the development of an axial undulatory type of propulsor (Testudinata, Placodontia, Sauropterygia, Sphenisciformes, Hesperornithes and Otariidae).

4. Swimming. Hydrojet propulsors

Among nektonic animals the hydrojet propulsor occurs only in cephalopods. Its principle of operation is based on the ejection by the animal of a jet of water, the reaction against which creates propulsive thrust. The mantle cavity of all cephalopods serves as the reservoir, being filled with water through a slit-like mantle orifice situated at its anterior edge. The water is ejected through a funnel, a conical muscular tube open at both ends, which serves as the nozzle of the hydroject propulsive mechanism. The dorsal surface of the posterior end of the funnel is fused with the ventral surface of the anterior part of the mollusc's body and opens into the mantle cavity; the anterior end, which is narrower, opens into the environment (Figure 62). The anterior end of the funnel is capable of freely changing its direction by bending downwards or sideways, so that the mollusc can instantly change the direction of the jet's thrust, and accordingly that of its movement.

The hydrojet propulsor of the cephalopods has reached the greatest perfection in nektonic Teuthoidea (Figure 63). Unencumbered by an outer shell, the mantle of a Teuthoidea constitutes a distensible sac with annular and longitudinal muscles in its walls; the former compress the mantle, the latter bend it (Kondakov 1940). The anterior edge of Teuthoidea's mantle is, as a rule, free all along the body's perimeter and during relaxation of the annular muscles becomes slightly detached from the head all along its length, forming a mantle slit. The opening of this slit is also

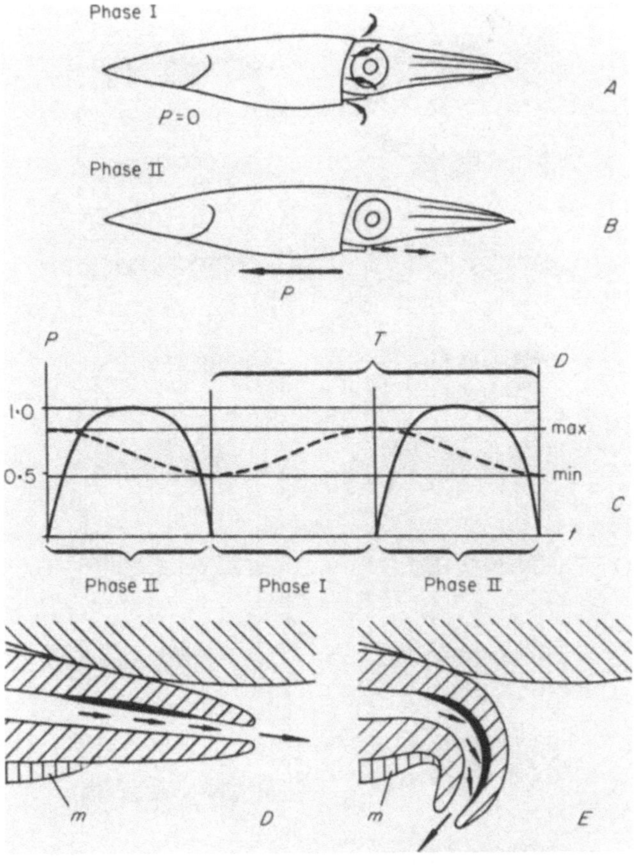

Figure 62. Diagram of operation of hydrojet propulsor of Teuthoidea. *A*, intake of water into mantle cavity; *B*, ejection of water from mantle cavity. Short arrows indicate direction of movement of water; *P*, propulsive force created by propulsor. *C*, propulsive force *P* (continuous line) and maximum transverse diameter *D* of body (dashed line) as functions of time *t*. *T*, working stroke of hydrojet propulsor. *D*, *E*, median section through funnel at moment of water ejection when swimming with (*D*) mantle end and (*E*) head forward. ▬, funnel valve; *m*, mantle. See also text.

facilitated by the relaxation of the *m. m. retractores capitis*, which pulls in the head, as a result of which the head protrudes forward somewhat. During contraction of the annular muscles of the mantle's edge it is closely pressed to the head all along its length; simultaneously the *m. m. retractores capitis* contracts, owing to which the head is pressed still closer to the edge of the mantle. Besides, when the mantle slit shuts, the anterior edge of the mantle is clasped by special cartilages acting on the

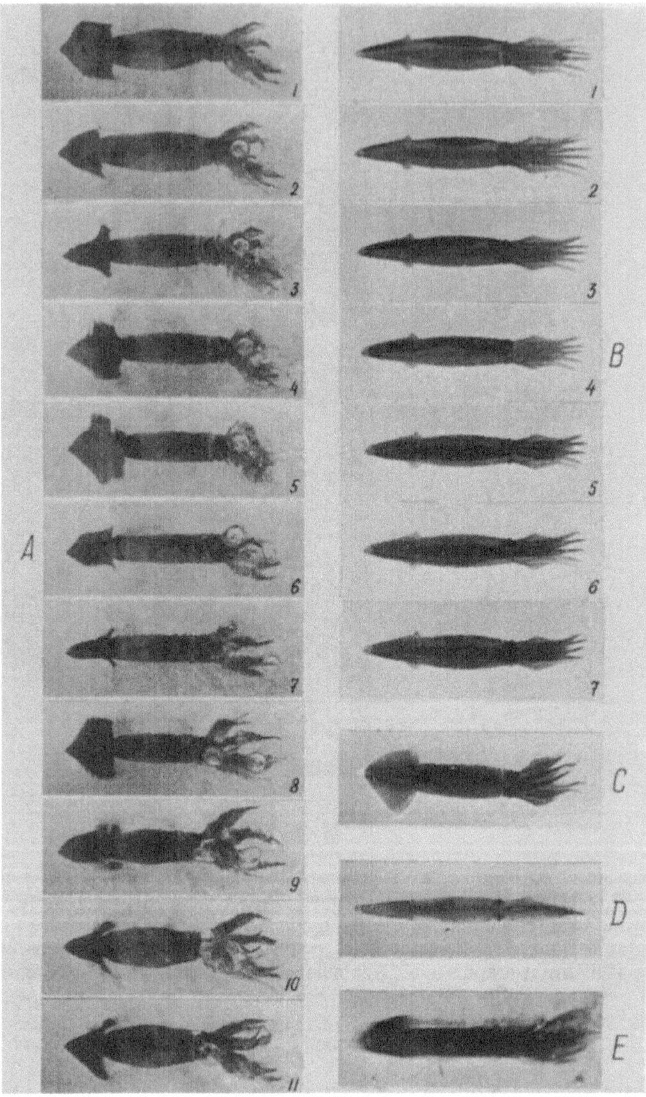

Figure 63. Todarodes pacificus (Steenst.). *A*, slow movement hydrojet and peripheral undulatory propulsors working simultaneously; *B*, fast movement, hydrojet propulsor operating alone, mantle fins folded; *C*, slow movement; *D*, very fast movement, mantle fins completely folded; *E*, attempt to visualize flowing stream by Aleyev–Ovcharov method (1969), azure paste applied to mantle fins. Filmed by B. V. Kurbatov.

principle of cuff-links, which enter corresponding depressions in the inner surface of the mantle's edge. In some cases, for example in *Symplectoteuthis*, the mantle and funnel cartilages grow close together at the point of their connection (Akimushkin 1963; Zuyev 1966).

In the walls of the funnel in Teuthoidea, just as in other Decapoda (Sepioidea), there are annular muscles whose contraction causes the lumen of the funnel to diminish; by changing the diameter of its funnel's orifice the mollusc can control the speed of the water jet leaving the funnel and accordingly the speed of its movement. There is a special lobe-like outgrowth on the inner dorsal surface of the funnel, its apex being directed towards the funnel's narrow end: this is the so-called funnel valve (Jatta 1893; Kondakov 1940; Akimushkin 1963). The degree of funnel-valve development in various cephalopods corresponds to the degree of their mobility: it is best developed in nektonic Teuthoidea, less developed in planktonic Teuthoidea and nektobenthic Sepioidea, and is absolutely lacking in benthic Octopoda. The function of the funnel valve is to ensure the mechanical strength of the funnel's dorsal wall necessary when the mollusc swims with its head forward and the external end of the funnel is bent towards the mantle end of the body. In this case the water jet from the funnel hits the funnel valve, which covers the dorsal wall of the funnel (Figure 62E). During normal swimming with the mantle end forwards, when the funnel is extended the funnel valve is pressed to its dorsal wall (Figure 62D). At the moment when water is being drawn into the mantle cavity the funnel's orifice is closed by bringing its dorsal and ventral walls together, a process in which the funnel valve takes no part, remaining pressed to the funnel's dorsal wall (Zuyev 1966).

The working cycle of the hydrojet propulsor in Teuthoidea, which we studied in specimens of *Todarodes pacificus*, consists of two phases clearly separated in time and having about the same duration (Figures 63A,B).

Phase I: Here water is drawn into the mantle cavity through the wide-open mantle slit, possibly not only because of relaxation of the annular muscles of the mantle and other muscles, but also because of active work by certain muscles. The lumen of the funnel is closed. Throughout this phase of the cycle no propulsive force is created (Figure 62A). When the mantle cavity is sufficiently full of water, the mantle slit closes and phase II of the cycle begins.

Phase II: This is the process of ejecting water from the mantle cavity through the funnel. The mantle slit is closed (Figure 62B). Propulsive thrust is continuously created throughout this phase of the cycle. After a certain proportion of the water in the mantle cavity has been ejected, the mantle slit opens, while the lumen of the funnel closes, i.e. a new cycle begins.

Since at the swimming speeds of which Teuthoidea are capable, reaching 15–16 m/s (Gronningsaeter 1946; Akimushkin 1963; Zuyev 1964c), the frequency of operation of the hydrojet propulsor is high–in the squids *Todarodes sagittatus* and *Symplectoteuthic oualaniensis*, for example, it reaches 5 c/s (Zuyev 1966) and possibly more–the pulsating character of the jet thrust does not in practice make the forward movement of the mollusc non-uniform. The movement becomes irregular only as a result of changes in the propulsor's mode of operation, which for cephalopods are as common as for all other nektonic animals. The rapidly alternating working phases of the hydrojet propulsor's cycle require rather perfect control of the muscular apparatus by the nervous system. This is probably why the nerve fibres of fast-swimming nektonic Teuthoidea are extremely thick (Young 1938, 1944; Pumphrey & Young 1938; Lane 1957; Akimushkin 1963).

The volume of water ejected by nektonic Teuthoidea from the mantle cavity during a single working cycle of the hydrojet propulsor is equal to about half the total volume of the mollusc (Zuyev 1966). This inevitably leads to pronounced changes in the body's maximum cross-section during the cycle. According to data we obtained as a result of analysing ciné records of the movement of *Todarodes pacificus* (Figure 63A,B), during slow swimming the maximum horizontal diameter of this squid's mantle at the end of phase I of the working cycle comes to about 30% of its length, whereas at the end of phase II it constitutes only about 21% (Figure 63A); during rapid swimming these diameters are equal to 23% and 21% respectively (Figure 63B), i.e. in this case changes in body shape are less significant, probably because of the need to reduce the maximum cross-section of the body, as a greater cross-section inevitably leads to higher total drag.

Thus we observe a certain functional analogy between a hydrojet and an axial undulatory propulsor: in both cases higher swimming speeds are attained by increasing the propulsor's working frequency, while reducing the amplitude of changes in body shape and improving streamlining. Here a feature common to all nektonic animals is manifested, stemming from the growing need to limit the energy expended on swimming when building up speed.

In Sepioidea, which in most cases are not nektonic, the hydrojet propulsor is not as perfect as in Teuthoidea, which is manifested in a reduction of the relative capacity of the mantle cavity, a shortening of the mantle slit, weak development of the funnel valve and some other specific features.

Judging by everything, the hydrojet propulsor of nektonic Belemnoidea was similar to that of extant Teuthoidea and Sepioidea in all its basic features.

In the ontogeny of Teuthoidea the hydrojet propulsor is the main one at every stage of the post-embryonic period, and in its very early stages, when the mantle fins are as yet incapable of undulating movements, the only source of propulsive thrust. Since in the course of ontogeny the relative thickness of the body diminishes in Teuthoidea (Kondakov 1940; Zuyev 1966), all the organs situated along the perimeter, including the hydrojet propulsor's funnel, become relatively closer to the body's longitudinal axis, which helps to reduce the pitching moment M_p created by the propulsor.

Obviously, **in the phylogeny** of cephalopods the exchange of water in the mantle cavity was originally necessary, just as in other molluscs, for maintaining the respiratory process only. Later, the reaction against the stream ejected from the mantle cavity came to be used as one of the means of creating propulsive thrust, at first rather low, as in some Bivalvia (*Pecten*). Since there was no morphological basis for the development of an axial undulatory propulsor because the mollusc was clad in an outer shell, the progressive evolution of the hydrojet method of locomotion has led to the combination of the respiratory act with the act of creating propulsive thrust, which entailed profound changes in the animal's entire organization.

In Nautiloidea and particularly, one may presume, in Ammonoidea, the hydrojet propulsor attained the maximum perfection possible within the framework of the former structure of the mollusc. Further progress in propulsor development followed the path of converting the entire mantle cavity into the propulsor's working chamber. In this connection the process of shell reduction began, ending in its complete elimination: remnants of the shell found themselves 'buried' inside the soft body, where they either preserved the functions of supporting skeletal formations (Teuthoidea) or combined supporting and hydrostatic functions (Belemnoidea and Sepioidea), or eventually turned into absolutely negligible rudiments without any essential function (Octopoda), or even completely disappeared (some Sepiolidae). Rid of an outer shell, the mantle was now able to stretch and its muscular walls turned into an apparatus ejecting water from its cavity. The volume of the propulsor's working chamber steeply increased, as did the strength of the muscles reducing this chamber's volume, with the result that the propulsor's relative capacity increased steeply, along, consequently, with the speed of locomotion. As a result of this transformation the mollusc turned from a comparatively slow-moving creature living in a shell into a true nektonic animal, with a well-streamlined body shape.

A body shape close to axisymmetric, as in Belemnoidea and Teuthoidea, is most advantageous for jet propulsion, as given such a body shape, the intake of water through a peripheral mantle slit also constitutes an axisymmetric process and creates almost no moment about the body's longitudinal axis. The not quite axisymmetric arrangement of the hydrojet propulsor's nozzle in Teuthoidea creates a small pitching moment which, however, is easily neutralized by a reverse moment created by the action of very flexible tentacles.

5. Flight. Aerial propulsors

The capacity to fly is found among nektonic animals in representatives of two classes only: the Cephalopoda and Osteichthyes. Such among the cephalopods are some Teuthoidea, and of the Osteichthyes some Hemirhamphidae (*Oxyporhamphus micropterus and Euleptorhamphus viridis*) and all Exocoetidae, Gasteropelecidae and Pantodontidae. Swimming in the body of the water is the main mode of locomotion for all these animals, flight being only secondary. The established opinion that Dactylopteridae are capable of flying has not been confirmed by documented evidence, and according to all indications is a misunderstanding (Hubbs 1933); Bertin 1958a; Herald 1962; Parin 1971b). Two kinds of flight are distinguished: passive, i.e. gliding, and active, i.e. flapping.

1. Passive Flight

The majority of flying nektonic animals are capable of passive flight only. This is true of all flying Teuthoidea, and among the fishes, of Hemirhamphidae, Exocoetidae and Pantodontidae. Since passive flight consists of a jump out of the water followed by gliding, it does not presuppose the creation of propulsive thrust in the air. Acceleration in the water preceding flight always takes place on account of the operation of the usual aquatic propulsor: in Teuthoidea a hydrojet, and in fishes an axial undulatory propulsor.

Capable of flying among the **Teuthoidea** are only a few species, and mostly small individuals up to 30 cm long at that, particularly representatives of the genera *Ommastrephes, Symplectoteuthis, Todarodes* and *Onychoteuthis*, distinguished by having the highest speeds (Verrill 1879–1882; Rush 1892; Abel 1916; Rees 1949; Arata 1954; Lane 1957; Heyerdal 1962; Akimushkin 1963; Clarke 1966).

Either when escaping from enemies or when pursuing their prey, squids are capable of jumping out of the water to heights of up to 7 m and of flying, by gliding, over distances of up to 50 m or more, depending to a considerable extent on the velocity and direction of the wind (Rush 1892; Rees 1949; Arata 1954; Akimushkin 1963, and others). Before jumping out of the water a squid must develop a speed of at least 14–16 m/s. During gliding through the air the mantle fins are widely extended and fulfil the function of lifting surfaces. A certain elongation of these fins in flying forms has been noted (Zuyev 1966) and perhaps shows an analogy with flying fishes. Unquestionably, in the phylogeny of the Teuthoidea, in the process of adapting to flight the development of the supporting function of the mantle fins was restricted by their terminal situation: they could increase in size only so much that the lift force they created still did not produce any noticeable nose-up moment. Yet moving the mantle fins closer to the centre of gravity, i.e. to the head, would complicate their functions as rudders (see chapter VI) and as apparatus actively creating lift in certain swimming modes (see chapter III) and, besides, would conflict with the function of the mantle as an extensible sac markedly changing its form during the operation of the hydrojet propulsor (Figure 63).

In fishes adapted to passive flight – Hemirhamphidae, Exocoetidae and Pantodontidae – the pectoral fins are not situated so terminally as the mantle fins of Teuthoidea, and in Exocoetidae are situated in the immediate proximity of the centre of gravity (Figures 64B–D), i.e. in this sense nothing prevents the development of the pectoral fins as lifting surfaces. Moreover, in some Exocoetidae (*Hirundichthys, Cypselurus* and *Cheilopogon*) the ventral fins also play an essential role in creating a lift force, owing to which adjustment of the point of application of the total lift force to the location of the centre of gravity is simplified still more. By changing the degree of extension of the pectoral fins, i.e. by drawing them away from or towards the body, a gliding fish is capable of considerably shifting the point of application of the lift force they create, and, given the same degree of pectoral-fin deflexion from the body, the longitudinal shift in the point of application of the lift

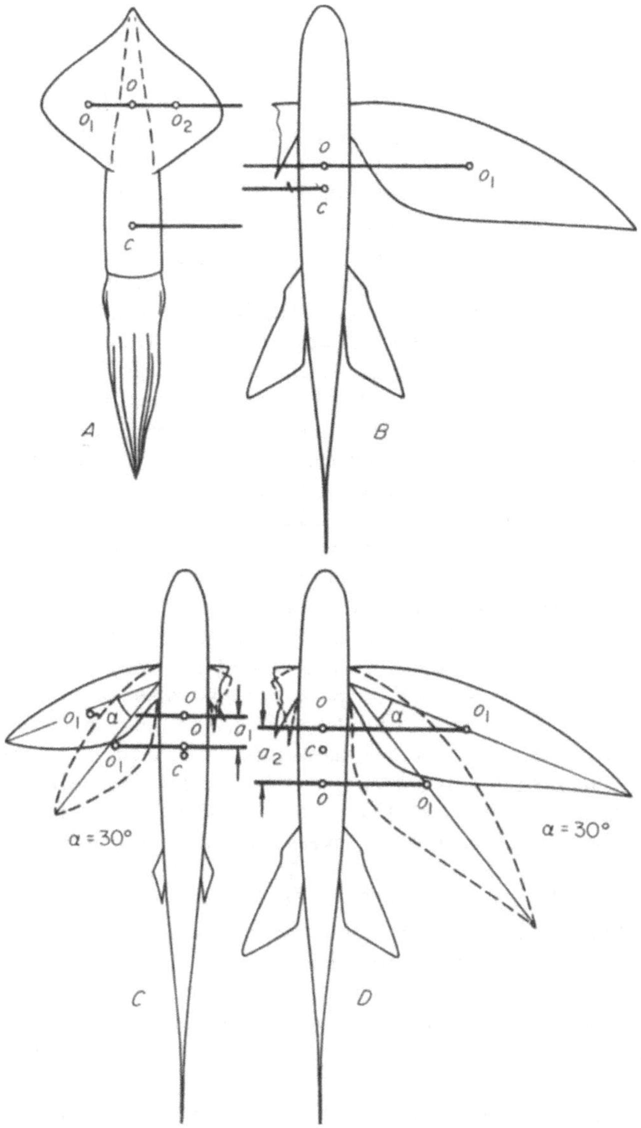

Figure 64. Position of fins creating lift force during gliding in (*A*) Teuthoidea and (*B*-*D*) Exocoetidae. *o*, point of action of lift force created by fins; *c*, centre of gravity; o_1, centre of blade of pectoral fin; a_1, a_2, magnitude of shift in *o* with short and long pectoral fins respectively. *A*, *Onychoteuthis banksii* (Leach); *B*, *D*, *Hirundichthys rondeletii* (Cuv. et Val.); *C*, *Oxyporhamphus micropterus* (Cuv. et Val.).

force created by the fins is all the more considerable the longer they are (Figures 64C,D: $a_1 < a_2$). It should also be noted that the thin elastic fins gently reinforced with rays of Hemirhamphidae, Exocoetidae and Pantodontidae have a relatively higher vertical breaking strength than the thick gelatinous fins of squids. All this furnishes the possibility of a much deeper adaptation of members of the above-mentioned fish families, Exocoetidae above all, to gliding flight than proved possible on the basis of the morphological organization of the Teuthoidea.

Numerous investigations have been devoted to some degree or another to describing the structure and gliding flight of the marine flying fishes Hemirhamphidae and Exocoetidae (Möbius 1878; Bois-Reymond 1891; Seitz 1891; Dahl 1892a,b; Gill 1905; Hankin 1920; Nichols & Breder 1928; Shuleikin 1928, 1929; Breder 1929, 1930, 1934, 1937; Garnett 1929; Calderwood 1931; Hubbs 1933, 1935, 1936; Bruun 1935; E. L. Gill 1935; Tewes 1936; Edgerton & Breder 1941; G. S. Carter 1945; Tortonese 1950; Mohr 1954; Bertin 1958a; Klausewitz 1959; Parin 1961a,b, 1971a; Aleyev 1963a; Stephens 1964 and others). For Exocoetidae it is known (Möbius 1878; Shuleikin 1928; Breder 1929, 1930, 1934; Calderwood 1931; Hubbs 1933, 1935, 1936; Edgerton & Breder 1941, and others) that before flight the fish develops a speed of about 10 m/s in the water, with its paired fins closely pressed to the body. Then the fish emerges onto the surface of the water and when the body is already in the air the pectoral fins open up, followed by the pelvic ones (in cases where they take part in creating a lift force). The setting of the pectoral fins at a small positive angle of attack ensures creation of the lift force necessary to support the fish in the air. The flattened ventral side of the body plays the role of an additional lifting surface. For a short period of time when the trunk of the fish is already wholly in the air (measured in fractions of a second), the lower blade of the tail fin is still in the water and continues to do work, creating a propulsive force (Figure 65). The propulsor is operating at this moment with a frequency of up to 50–70 c/s, which is probably a kind of 'record' for fishes. It is in this brief period that the main 'speed-up' before take-off takes place: the speed increases from 10 to 16–20 m/s, after which the fish is in flight. The more strongly developed lower lobe of the tail fin in Exocoetidae and flying Hemirhamphidae (*Euleptorhamphus* and *Oxyporhamphus*; see Figures 66A,B) is connected precisely with the fact that during take-off this lower lobe acts as the main propulsor. The detachment of the fish's trunk from the surface of the water at the moment of take-off is made easier, probably, by a special projection on the lower surface of the fish's head (Figure 66B), which plays the role of a 'redan' (Shuleikin 1928); this however, is not generally recognized (Breder 1930; Parin 1961b). The fish glides with immobile, widely spread fins until its speed falls to a certain limit below which the flight can continue no longer. The duration of a 'non-stop' flight does not usually exceed 2 s, over a distance of up to 50 m at a height of up to 6–7 m above the surface. By touching the wave crests with its tail the fish may create new propulsive impulses and prolong its total flight time to more than 15 s, covering up to 400 m or more, depending on the direction and velocity of the wind.

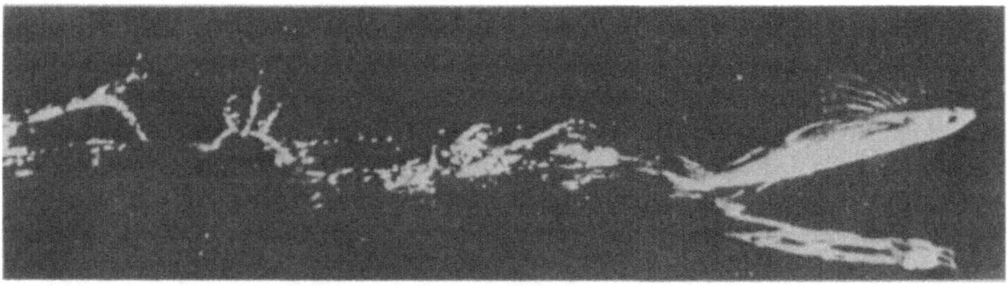

Figure 65. Flying fish (Exocoetidae) before take-off. Lower lobe of tail fin is still in water and continues to work. After Edgerton & Breder (1941).

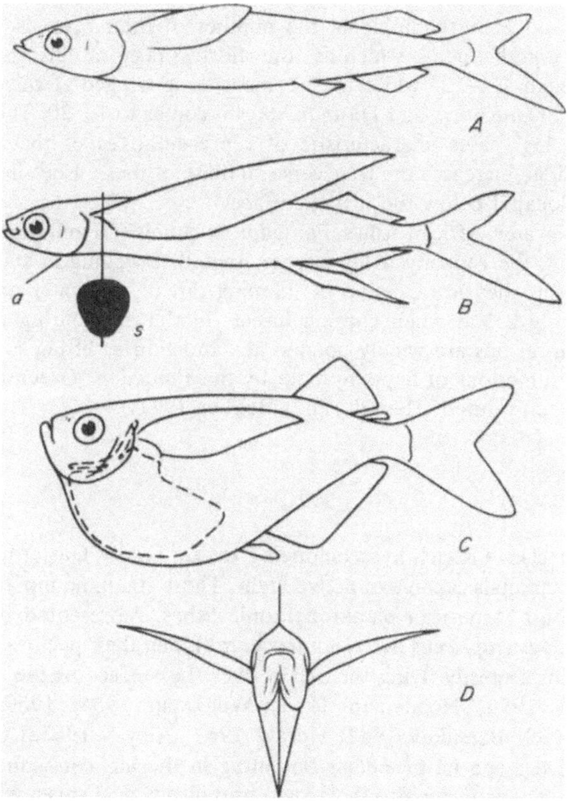

Figure 66. A, *Oxyporhamphus micropterus* (Cuv. et Val.); *B, Hirundichthys rondeletii* (Cuv. et Val.); *C, D, Gasteropelecus levis* (Eigenmann) (after Ostheimer 1966; Bertin 1958a, with modifications). *a*, protuberance possibly playing the role of a redan; *s*, body cross-section; – – –, coracoideum.

In flying Hemirhamphidae (*Oxyporhamphus micropterus* and *Euleptorhamphus viridis*) the morphological adaptation associated with flight are much less developed than in Exocoetidae, which is manifested primarily by the relatively shorter pectoral fins (Figures 64C, 65A). The capacity to fly in Hemirhamphidae is, accordingly, lower: their flight distance (Parin 1971a) does not exceed 50–60 m.

Among the adaptations in Exocoetidae and Hemirhamphidae associated with flight we should note first of all the extremely strong elongation of the paired fins, particularly the pectoral ones (Figure 66B): the length of the pectoral fins in *Exocoetus*, for example, reaches up to 80% of the body's length to the end of the vertebral column. No less characteristic is dorsoventral elongation of the cross-sections of the pectoral fins' rays, increasing their dorsoventral rigidity (Breder 1930). With the progressive development of the pectoral fins as lifting surfaces in Exocoetidae and Hemirhamphidae, the number of their rays also increases (Parin 1961b): in *Oxyporhamphus*, which has the shortest pectoral fins, the number of rays per fin comes to 11–13, while in the most 'long-winged' *Exocoetus*, *Cypselurus*, *Cheilopogon*, *Prognichthys* and *Hirundichthys* it comes to 13–20. The high location of the pectoral fins' bases characteristic of representatives of the Exocoetidae and Hemirhamphidae increases the transverse stability of these fishes in flight (the centre of gravity is located below the lifting surfaces).

In the freshwater African fishes *Pantodon buchholtzi* (Pantodontidae), which are 10–12 cm long, the capacity to fly is very limited. Dwelling in stagnant overgrown bodies of water, they are capable of jumping out of the water and flying over its surface for only 2–3 m when chasing insects in the air. During such 'flight' their enlarged pectoral fins are widely opened and function as lifting surfaces, remaining immobile. Assumptions of flapping flight by the *Pantodon* (Greenwood & Thomson 1960) have been refuted (Herald 1962; Bekker 1971).

2. Active Flight

Fishes of the class Osteichthyes belonging to the family Gasteropelecidae are the only nektonic animals capable of active flight. These small, no more than 10 cm long, freshwater South American planktonektonic fishes, represented by several genera (*Carnegiella*, *Gasteropelecus* and *Thoracocharax*), flap their pectoral fins during flight like birds' wings, noisily flying for 3–5 m over the surface of the water (Ridewood 1913; Myers 1950; Hoedeman 1952; Weitzman 1954, 1960; Bertin 1958a; Klausewitz 1959; Barsukov 1962; Herald 1962; Aleyev 1963a; Ostheimer 1966). Thus they have a special propulsor operating in the air, consisting of the pectoral fins, which, by the way, are also their main propulsive mechanism in the water, where they function as paddles.

In connection with adaptation to active flight, the pectoral fins of flying Gasteropelecidae are noticeably enlarged, though not to the degree we find in the most 'long-winged' Exocoetidae (Figures 66C,D). Extremely strongly developed in the brachial girdle is the coracoideum, to which the muscles that bring down the pectoral fins are fastened (Figure 66C). The weight of the brachial-girdle muscles comprises about 25% of the total weight of the fish (Ridewood 1913; Klausewitz 1959).

Apparently, the lift force necessary for the flight of Gasteropelecidae is created by the pectoral fins when the vertical component F_1 of the force transverse to these fins emerging during their downward stroke is higher than the analogous force F_2 emerging during their upward movement. On the basis of the results of modelling the flapping flight of various animals (Houghton 1964) and drawing an analogy between the pectoral fins of Gasteropelecidae and the wings of bees and horse-flies, one may take it that the condition $F_1 > F_2$ is achieved in Gasteropelecidae on account of helical bending of the fin's blade along its longitudinal axis.

3. Phylogeny of Adaptations Associated with Flight

The capacity to fly in nektonic animals, both cephalopods and fishes, emerged as an adaptation affording escape from pursuing predators (Nikolsky 1963; Parin 1961b; Akimushkin 1963; Aleyev 1963a, and others).

The initial prerequisite for developing a capacity to fly in phylogeny was simply leaping out of the water to escape from predators, which is characteristic of, among others, many small pelagic fishes (*Alburnus*, *Rutilus*, etc.). The gradual emergence of the above-described special adaptations associated with prolonging the stay in the air, primarily the development of lifting surfaces based on the pectoral or other fins, can be seen in the example of marine flying fishes in the Hemirhamphidae-Exocoetidae series (Figures 66A,B).

6. Locomotion on the surface of solid substrata. Tactile propulsors

Locomotion upon the surfaces of solid substrata is not on the whole characteristic of nektonic animals. It is an attribute, to a small extent, of individual representatives of benthonekton, and is regularly used by xeronektonic forms only.

One may note the burrowing in the ground and locomotion in the ground by the benthonektonic *Gymnammodytes*, carried out by means of movements characteristic of the axial undulatory propulsor. Catadromous eels *Anguilla* are capable in the benthonektonic stage of crawling over land from one body of water to another and also of burrowing in the ground. None of these modes of locomotion over the bottom of a body of water or on land, and equally, none of these methods of burrowing is associated with any special tactile propulsor.

Xeronektonic animals are all capable of locomotion on a solid substratum, land, floating ice, etc., and characteristically have some special tactile propulsive mechanism, its degree of perfection being in inverse proportion to the degree of adaptation to nektonic life.

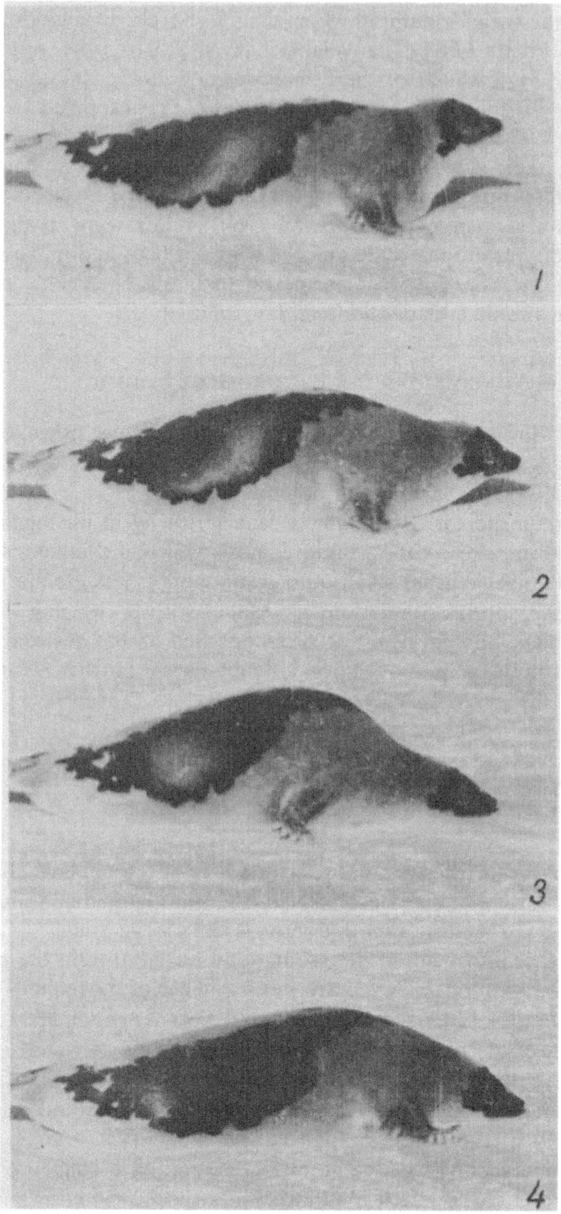

Figure 67. Caterpillar movement of *Pagophoca groenlandica* (Erxl.) on ice. Cine-record.

Without examining the tactile propulsive mechanisms of terrestrial animals, we merely note that they provide for one of four modes of locomotion, crawling, walking, running and jumping, or various combinations of them (Gray 1953a). In Ophidia, both the purely terrestrial forms and the eunektonic forms move with the aid of an axial undulatory propulsor. A propulsive apparatus of the walking type always regresses during transition from terrestrial to eunektonic forms. The initial stage of such regression and the first step towards the assertion of a nektonic propulsor can be seen in nektoxeric forms (Crocodylidae, *Castor*, *Enhydra*, Chelydridae, Emydinae, *Ornithorhynchus*, etc.).

In xeronektonic animals the walking apparatus regresses, as a rule, rather strongly and, along with this, their nektonic propulsors are nearly 'perfect', so that swimming is always the basic mode of locomotion. In connection with the far-reaching regressive changes in the walking apparatus of xeronektonic animals, we observe in some cases the emergence of new tactile propulsors specific to xeronekton. Thus the phocidae move on a solid substratum by using the anterior flippers and the body itself, i.e. here is a case of an absolutely individual tactile propulsive mechanism not characteristic of terrestrial mammals. It is well known (Tikhomirov 1966; Doux-champs 1969, and others) that in different cases the anterior flippers of Phocidae may be thrown forward either alternately, i.e. we observe lateral bending of the body, or simultaneously; the latter mode of locomotion may be described as 'caterpillar-like' since it resembles in principle the movement of a caterpillar (Figure 67).

We find an individual tactile propulsive mechanism also in *Trichechus*. The mode of movement of *Trichechus* on land, in itself a rare and brief occurrence, is actually crawling by means of anterior flippers and lateral bending of the body (Petit 1955). We see in this example one of the latest stages in the regressive development of the walking apparatus of xeronektonic animals.

7. Other modes of movement

Among the other modes of movement of nektonic animals, which are rare exceptions only, we should examine **movement by attachment to swimming objects**, characteristic exclusively of fishes and fish-like organisms. This includes certain cases of parasitism, commensalism and symbiosis.

The cases of parasitism among nektonic fishes are limited, apparently, to parasitism of the males, which is known for many Ceratioidei (Regan 1925, 1926; Regan & Trewavas 1932). The parasitic males fuse with a female and cannot move independently; their dimensions diminish and their fins and eyes remain underdeveloped.

Of considerable interest are cases of attachment to the surface of swimming objects not associated with parasitism, a classical example of which is afforded by the Echeneidae, whose first dorsal fin has turned into a powerful sucker (Figure 68) used by these fishes to attach themselves to sharks, ships and other swimming objects. In a

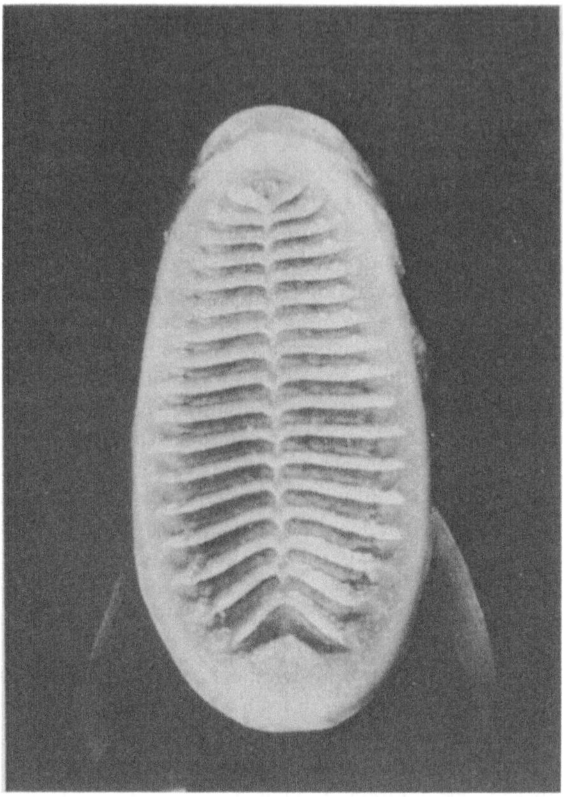

Figure 68. Remora remora (L.). Sucker formed by first dorsal fin. Photograph.

number of cases attachment to a swimming object can take place by means of hydrodynamic forces, as is observed in the *Naucrates*, which accompany sharks, swimming in direct proximity to them, and, despite their very small size compared with the shark, never lag behind it. This is explained (Shuleikin 1958a,b) by the fact that the *Naucrates* is within the boundary layer surrounding the shark's body, the thickness of this layer being up to 20 cm or more. If the pilot fish during the shark's movement were to stray by chance beyond the boundary layer, it would be immediately 'pulled back' by a hydrodynamic force that emerges during movement of the two fishes on parallel courses in the potential stream. **Piloting** is characteristic of some other Carangidae as well and has been reported (Aleyev 1963a) also for *Trachurus mediterraneus ponticus* in *Acipenser güldenstädti colchicus*, *Huso huso* and *Chelonia mydas* (Figure 69).

In the development of adaptations to movement by attachment to moving objects, we may distinguish three stages (Aleyev 1963a).

During the first stage, which may be described as the stage of optional piloting, though not yet an ecological necessity, piloting may take place under certain conditions. There are no features associated with piloting noticeable in the morphology of the piloting fish at this stage. A case in point is the *Trachurus mediterraneus*

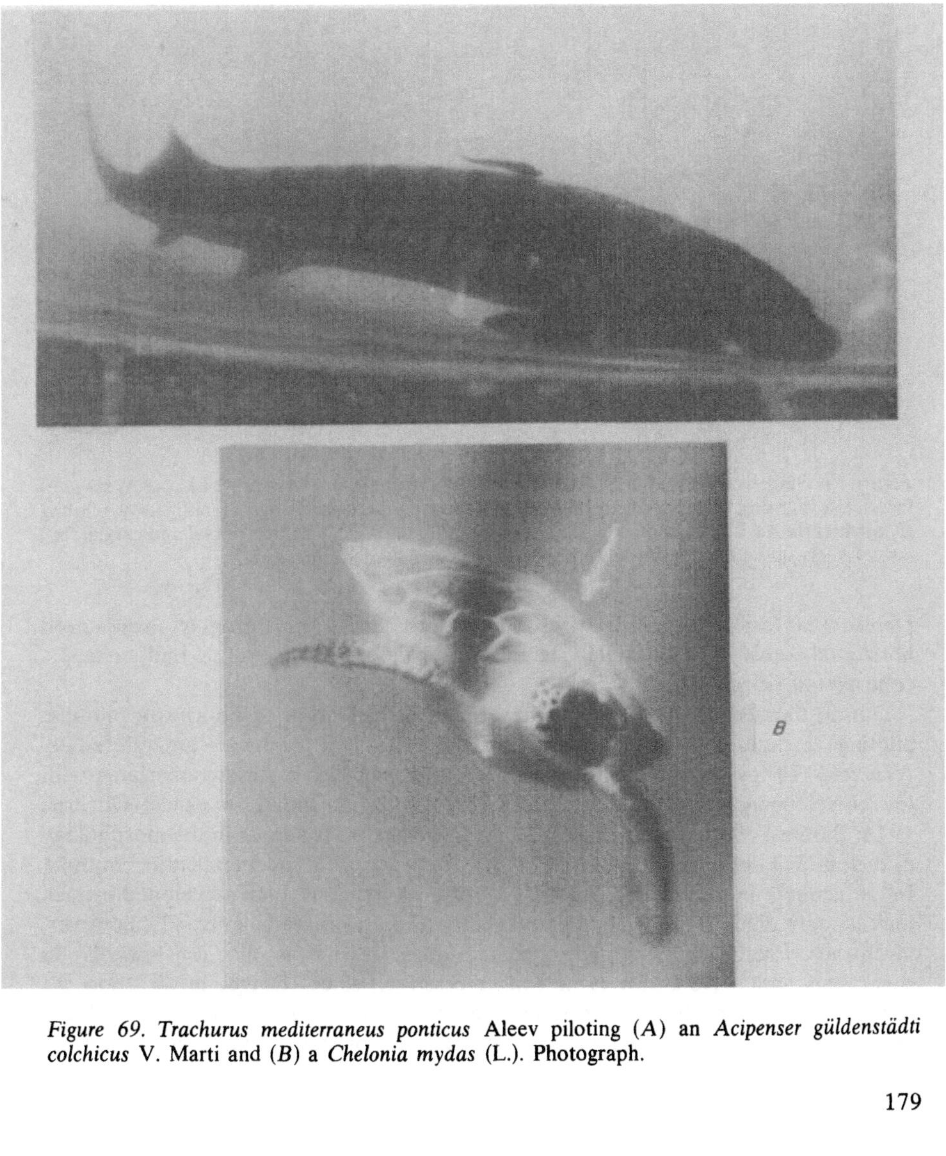

Figure 69. Trachurus mediterraneus ponticus Aleev piloting (*A*) an *Acipenser güldenstädti colchicus* V. Marti and (*B*) a *Chelonia mydas* (L.). Photograph.

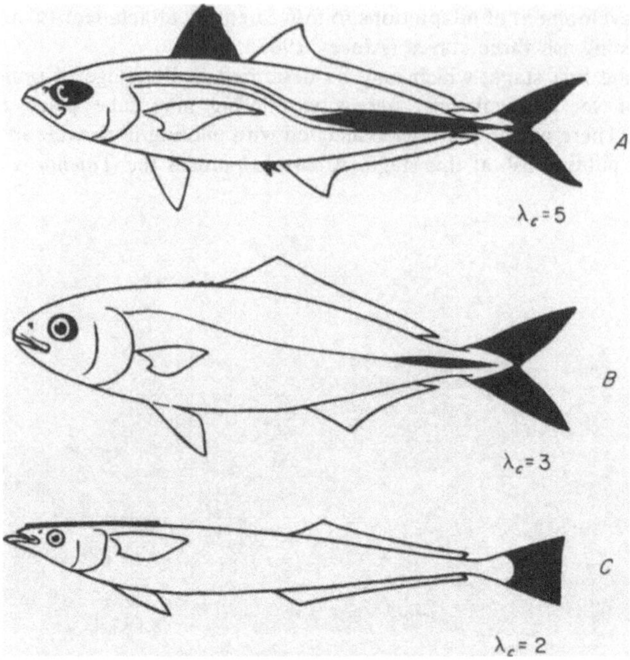

Figure 70. Stages of adaptation to movement by attachment to swimming objects. *A*, stage of facultative piloting (*Trachurus mediterraneus ponticus* Aleev); *B*, stage of obligatory piloting (*Naucrates ductor* L.); *C*, contact stage (*Echeneis naucrates* L.). The first dorsal and caudal fins, lateral keels on caudal peduncle and fatty eyelids are shaded. See text.

ponticus; a fast pelagic fish with a large first dorsal fin, strongly pronounced horizontal keels on the caudal stem, well-developed fatty eyelids and a deeply concave tail (Figure 70A).

During the second stage, which may be called the stage of obligatory piloting, piloting becomes a necessary feature of the fish's life. Examples are offered by *Naucrates.* In this case, the ecology of the piloting species is closely associated with the object being piloted. In *Naucrates* this association greatly expands (Gilchrist 1918; Barnard 1926; Shuleikin 1958a). At this stage traits appear in the morphology of the piloting fish which indicate a diminishing capacity for independent locomotion. In *Naucrates,* in particular, the first dorsal fin remains underdeveloped, which undoubtedly diminishes the manoeuverability of the fish (see chapter VI); however, given the constant action of hydrodynamic forces pulling the pilot fish towards the shark, this manoeuverability is no longer essential, since changes in direction are determined by the shark. The lateral horizontal keels on the tail stem of the *Naucrates* are now not pronounced, the concavity of the tail fin is comparatively shallow and the fatty eyelids, which act as fairings for the eyes (see chapter V), are rudimentary (Figure 70B). All this conforms to the pilot fish's movement in the boundary layer at relatively low speeds.

During the third stage, which may be called the contact stage, in addition to the piloting itself, contact with the object being piloted (in particular, a shark) now becomes ecologically necessary, since there appear morphological traits of deep specialization associated both with attachment and, at the same time, with a sharply deteriorating capacity for independent movement. An example of this is the Echeneidae. Here the first dorsal fin does not just lose its usual function as in *Naucrates*, but acquires a new function associated with attachment. Lateral keels on the tail stem, concavity in the tail fin and fatty eyelids, all attributes of fast swimmers, are absent (Figure 70C).

One of the methods of 'motorless' movement by nektonic animals is the widely known **swimming of dolphins on the crests of waves** (Matthews 1948; Woodcock 1948; Woodcock & McBride 1951; Hayes 1953; Scholander 1959; Fejer & Backus 1960; Gordon 1961; Perry, Acosta & Kiceniuk 1961; Yuen 1961; Korotkin 1973, and others). During such swimming, the dolphin takes up a position with its head pointing in the direction of the wave's movement, on its frontal slope, and makes no swimming movements. The propulsive force appears on account of the moving masses of water in the propagating wave: the wave, as it were, pushes the dolphin, imparting to the animal a certain speed in the direction of its longitudinal axis.

8. Phylogeny of adaptations associated with creating propulsive force

In the course of the phylogeny of the overwhelming majority of nektonic animals an undulatory, most often an axial undulatory, type of propulsor is formed. This type of propulsor is the rule for nekton, whereas all the other types are nothing more than exceptions. Table 12 shows that out of the twelve classes of animals containing nekton the undulatory propulsor occurs as the main propulsor in eleven of them, and only in xeronektonic Aves is replaced by the paddling type. Of these eleven classes in which the undulatory propulsor occurs as the main one, it is represented exclusively by the axial type in nine and only in cephalopods exclusively by the peripheral type and only in Osteichthyes by both the axial and the peripheral type. **The advantages of the axial undulatory propulsor** are due, above all, to three circumstances.

First, to the fact that the axial undulatory propulsor allows as does none other the maximum variety in the distribution of the propulsive action along the body's longitudinal axis, thereby creating the broadest opportunities for adaptation to the most diverse regimes of movement.

Second, to the fact that, being based on the animal's most powerful musculature, the longitudinal trunk muscles, the axial undulatory propulsor has a very high

Table 12. Basic propulsors of nektonic animals (marked by symbol +): undulatory (UP), paddling (PP) and hydrojet (HJP).

Classes of animals	UP	PP	HJP	Classes of animals	UP	PP	HJP
Benthonekton				*Xeronekton*			
Cephalopoda	+	–	+	Amphibia	+	–	–
Diplorhina	+	–	–	Reptilia	+	+	–
Monorhina	+	–	–	Aves	–	+	–
Placodermi	+	–	–	Mammalia	+	+	–
Acanthodei	+	–	–				
Chondrichthyes	+	–	–	*All nekton*			
Osteichthyes	+	+	–	Cephalopoda	+	–	+
Reptilia	+	+	–	Sagittoidea	+	–	–
Mammalia	+	+	–	Diplorhina	+	–	–
				Monorhina	+	–	–
Planktonekton				Placodermi	+	–	–
Cephalopoda	+	–	+	Acanthodei	+	–	–
Sagittoidea	+	–	–	Chondrichthyes	+	–	–
Osteichthyes	+	–	–	Osteichthyes	+	+	–
				Amphibia	+	–	–
Eunekton				Reptilia	+	+	–
Cephalopoda	+	–	+	Aves	–	+	–
Placodermi	+	–	–	Mammalia	+	+	–
Acanthodei	+	–	–				
Chondrichthyes	+	–	–				
Osteichthyes	+	–	–				
Reptilia	+	?	–				
Mammalia	+	–	–				

relative power, which, all other conditions being unchanged, allows absolute swimming speeds to be increased to a maximum.

Third, to the fact that the axial undulatory propulsor in its scombroid version allows reduction to a minimum of the irrational waste of energy associated with movement in a viscous fluid (water) by using again the boundary layer's kinetic energy.

The paddling propulsor occurs as the main one in representatives of four classes. In three of them (not Aves), in different cases both the paddling and the undulatory propulsor may act as the main one. In nektonic Osteichthyes, in the overwhelming majority of cases the paddling propulsor is combined with the axial undulatory type used in the 'boosting' regime.

Lastly, the hydrojet propulsor occurs in representatives of only one of the twelve nektonic classes.

The development of paddling or hydrojet propulsors in the phylogeny of certain groups of nektonic animals was usually due to some factor hindering the conversion of the nektonic animal's body into an axial undulatory propulsive mechanism. Such a factor in the overwhelming majority of cases was the loss of the capacity for vermicular undulations of the body, which is preserved to one degree or another in nearly all animals and disappears only in case of some rather deep morphological change connected with some far-reaching unilateral adaptation. It is this sort of adaptation that is characteristic of molluscs, birds and some reptiles (Testudinata, etc.). **The predominance of the axial undulatory propulsor** is undoubtedly the most characteristic feature of the phylogenetic development of the modes of locomotion of nektonic animals.

Another characteristic of the phylogenetic development of nektonic propulsors is **the tendency to free the animal's forward movement from the effects of the intermittent character of the propulsive force generated**. This is achieved through multiplying the number of simultaneously and asynchronously operating elementary propulsors, as for example, in the anguilliform variant of the axial undulatory propulsor, or by increasing the propulsor's working frequency, which is characteristic, for example, of the scombroid variant of the axial undulatory propulsor and the hydrojet propulsor. In both cases the inertia of the hydromechanical processes characterizing the movement overcomes the influence of the intermittent character of the propulsive force, i.e. the animal's movement during steady operation of the propulsor turns out to be uniform.

The non-uniformity of locomotion characteristic of nektonic animals is due in general not to the intermittent character of the propulsive force, but to the continual, mostly aperiodic changes in the propulsor's operational mode. Only in the cases of comparatively primitive low-frequency propulsors, such, for example, as the paddling propulsors of the Testudinata and Otariidae, do we actually observe the influence of the intermittent character of the propulsive force on the mode of progression: in such cases the non-uniformity of locomotion can be due both to the intermittent character of the propulsor's performance and to changes in its operational mode. This, however, is not characteristic of nektonic animals.

Lastly, the third fundamental feature of the phylogenetic development of nektonic propulsive mechanisms is **the progressive tendency to utilize the kinetic energy of the boundary layer**, which is achieved by shifting the propulsor to the posterior end of the animal's body, i.e. to the zone where the boundary layer is thickest and where it carries the maximum kinetic energy. We find the best adaptation for utilizing the kinetic energy of the boundary layer in the scombroid variant of the axial undulatory propulsor, since its working element is attached to the rearmost pont of the body on the axis of the wake flow. This is achieved to a lesser degree by the hydrojet propulsor of cephalopods, its nozzle being situated not at the rearmost (relative to the direction of movement) point of the body nor coaxial with the wake flow. The peripheral undulatory and paddling propulsors can hardly utilize the kinetic energy of the boundary layer at all, since their working elements are situated entirely outside the body's boundary layer and the wake flow.

Thus the tendency to utilize the kinetic energy of the boundary layer also creates prerequisites for the predominant development in nektonic animals of the scombroid variant of the axial undulatory propulsor. It is this that determines the fact that typical scombroid propulsors coexist with typical anguilliform ones within a rather wide range of Reynolds numbers: from 2.0×10^2 to 2.0×10^7.

V. REDUCING RESISTANCE TO MOVEMENT

1. General

The development of adaptations aimed at reducing hydrodynamic resistance is one of the most important aspects of adaptation to the nektonic mode of life and constitutes one of the most general specific features of nektonic animals, since the total hydrodynamic resistance encountered by a nektonic animal determines the overall expenditure of energy required by the given ecology. The task of reducing the hydrodynamic resistance encountered by the animal boils down, as we have seen, to reducing form drag, friction drag and induced drag.

The methods of reducing form drag are aimed, in effect, at reducing vortex formation and consist of preventing boundary-layer separation. We know from hydrodynamics (Prandtl 1951; Schlichting 1956; Martynov 1958, and others) that there are several ways in which boundary-layer separation may be prevented. In nektonic animals we encounter mainly three methods, enumerated below in order of diminishing incidence.

1. Laminarization of body shape is attained through smoothness of its general contours and location of its maximum cross-section close to the middle of the longitudinal axis. In certain regimes the stream flows over bodies of laminarized shape without, or nearly without boundary-layer separation, owing to which vortex formation is minimal. Flow without separation is assisted in this case also by smoothness of the body surface, i.e. the absence on it of any roughness, which, particularly on the contraction area, easily cause separation.
2. Transition to turbulence of the boundary layer at the site of laminar separation or somewhat upstream, makes the boundary layer more stable, thanks to which the site of separation shifts far backwards, to the rear part of the body, and the dimensions of the vortex system appearing are considerably diminished.
3. Sucking away the boundary layer draws off the retarded fluid. This tends to increase the speed of movement of the boundary layer, which also prevents the appearance in it of reverse flows.

Methods of reducing friction drag consist primarily of preserving laminarity of the boundary layer, which is achieved by increasing the speeds in it and improving the smoothness of the surface in the stream.

The speeds in the boundary layer increase, in particular, over the contraction area of a streamlined body. An increase in the speeds in a given boundary layer can be achieved, therefore, by extending this area to over a considerable part of the streamlined body. However, excessive elongation of the contraction area leads to an excessive shortening of the diffusion area, which entails an increase in the angle turned through, facilitating boundary-layer separation on the diffusion area, i.e. leads to greater vortex formation, which is a disadvantage. Hence the ration of the sizes of the contraction and diffusion areas of well-streamlined bodies should be such as to facilitate the maximum elongation of the laminar region of the stream, yet not lead to marked intensification of vortex formation over the diffusion area. Such, precisely, are the above-mentioned laminarized profiles for which the distance from the frontal point to the point of maximum thickness is 40–50% or even more of the length of the profile's chord (Martynov 1958). The shape of such profiles strikingly resembles that of the longitudinal projections of the trunks of some fast-swimming nektonic animals, representatives of Scombridae in particular (Figure 71A). In the case of a laminarized profile, the point of minimum dynamic pressure is shifted markedly backwards, which directly determines the increasingly posterior position of laminar-to-turbulent boundary-layer transition compared with a profile with a more anterior location of the zone of maximum thickness (Figure 71B).

Greater smoothness of the part of the body surface in the stream, apart from the absence of any large projections or roughness, is determined by its degree of aerodynamic smoothness, which is characterized (Martynov 1958, and others) by the magnitude of its relative roughness, i.e. by the ratio of the average height k of the roughness elements to a characteristic linear dimension of the body concerned, in this case the effective length L_c of the animal's body. Regarded here as roughness is the presence on the body surface of closely situated projections or depressions whose separation is of the same order as their height. Different degrees of roughness have different effects on friction drag. Here three cases may be distinguished.

When the roughness elements are so small as to cause no increase in friction drag compared with that on a smooth plate, the surface is called aerodynamically smooth, and its roughness admissible; in a turbulent boundary layer such roughness elements are wholly within the laminar sublayer. The height k_{adm} of the largest admissible roughness is determined by the formula (Schlichting 1956)

$$k_{\text{adm}} \leqslant 100 \frac{l}{Re_l}, \tag{26}$$

where l is the length of the body in the stream, in our case L_c. When the roughness is within the largest admissible size, then the relative roughness does not affect flow in the boundary layer and the friction drag depends on the Reynolds number only.

When the height of the roughness elements is of the same order as the thickness of the laminar sublayer, flow in the boundary layer is determined by both the Reynolds number and the relative roughness.

When the height of the roughness elements notably exceeds the thickness of the laminar sublayer, so that flow over the elements proceeds with separation, the flow characteristics no longer depend on the Reynolds number, but are determined by the magnitude of the relative roughness only (Schlichting 1956; Martynov 1958).

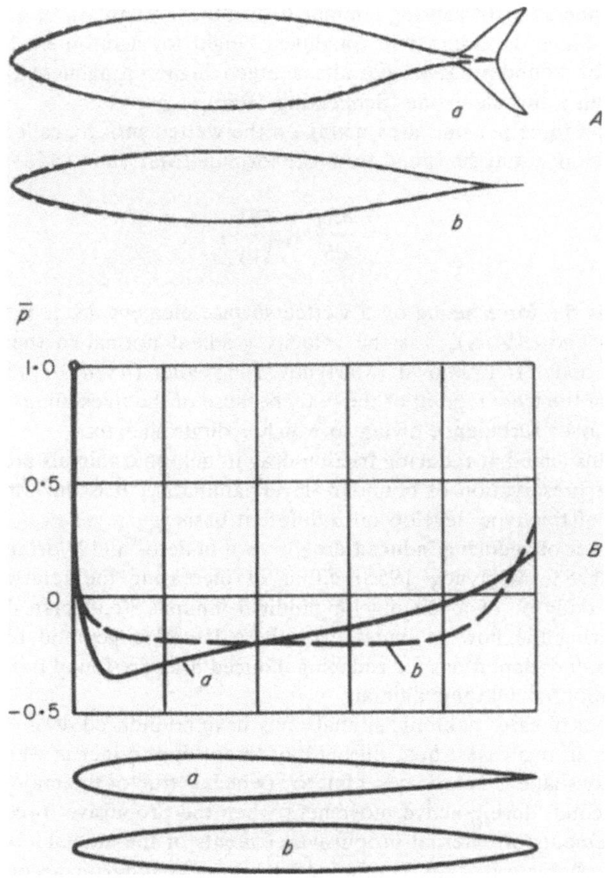

Figure 71. Laminarized body shapes. *A*, laminarized profile used in aerodynamics (dashed line, after Martynov 1958) and longitudinal projections of the trunk of *Scomber scombrus* L. (continuous line). *a*, vertical; *b*, horizontal; in both cases the laminarized profile has been reduced to a relative thickness equal to the maximum transverse diameter of the fish. *B*, distribution of dynamic pressure $\bar{p}$ over the surface of (*a*) conventional and (*b*) laminarized profiles.

An increase in the roughness height in a laminar boundary layer to the so-called critical height k_{cr} causes transition of the laminar flow to turbulence owing to which the point of boundary-layer transition moves forwards, i.e. upstream. Here the total drag increases when friction drag predominates, as in the case of well-streamlined bodies, such, for example, as the bodies of nektonic animals, but diminishes when form drag is predominant, which is characteristic of poorly streamlined bodies. The critical roughness height causing laminar-to-turbulent transition of a boundary layer is about 10 times the admissible roughness height for a turbulent boundary layer, i.e. a laminar boundary layer permits a much higher roughness before the drag increases than a turbulent one (Schlichting 1956).

The friction force per unit area acting on the wetted surface, called the tangential stress of friction τ, may be found from the formula (Martynov 1958)

$$\tau = \frac{dXf}{dS} = \mu\left(\frac{dV}{dy}\right)_{y=0}, \tag{27}$$

where dXf is the force acting on a wetted surface element dS, μ is the viscosity of the medium and $(dV/dy)_{y=0}$ is the velocity gradient normal to the surface of the streamlined body. It is known (Martynov 1958) that $(dV/dy)_{y=0}$ diminishes with distance from the frontal point of the body because of the thickening of the boundary layer (laminar or turbulent), owing to which τ diminishes too.

Adaptations aimed at reducing friction drag in nektonic animals are based in most cases on the preservation of boundary-layer laminarity. It is only sporadically that adaptations of this type develop on a different basis.

The methods of reducing induced drag known in aero- and hydrodynamics consist (Patrashev 1953; Martynov 1958) either of increasing the relative span of the streamlined body or of arranging longitudinal laminae ('end plates') on its lateral edges hindering the flow of liquid over these lateral edges and reducing trailing vorticity. Similar adaptations for reducing induced drag are found in the morphological organization of nektonic animals.

In a number of cases nektonic animals may have no induced drag. This may occur, in particular, in two cases: first, during movement due to inertia when the nektonic animal's body shape is nearly axisymmetric (which is true of the majority of nektonic animals); second, during active movement when the propulsive force is not created by the movements of external propulsive elements of the animal's body, which, for example, is characteristic ot Teuthoidea when swimming on account of hydrojet thrust with the mantle fins pressed to the body. During the operation of axial undulatory or paddling propulsors there is always some induced drag, of greater or lesser importance, owing to the flow of water across the lateral edges of the propulsive elements.

Most of the adaptations found in nektonic animals associated with reducing resistance to movement facilitate reduction of form drag and friction drag simultaneously, thus providing a manifestation of the multiple functions of the body's structure. At the same time there are also adaptations with a narrower functional spectrum aimed at reducing either form drag, friction drag or induced drag alone. All these adaptations are intimately interrelated and it is virtually impossible to examine each of them in complete isolation. However, two large groups of adaptations stand apart, one aimed predominantly at preventing boundary-layer separation, i.e. at the reduction or elimination of vortex formation, and the other predominantly at the preservation of boundary-layer laminarity. And it is with these two closely interrelated groups of adaptations that the subsequent material is concerned.

The problem of reducing hydrodynamic drag is one with many aspects, including various morphological, physiological, biochemical and bionic ones. Below we mainly look into those aspects of the problem which are of the greatest interest for characterizing nekton as a while, all the others being touched upon merely as required to outline the basic material.

2. Features of flow over body

The data now available concerning the flow over nektonic animals' bodies are comparatively few (Hill 1950; Steven 1950; Gero 1952; Rosen 1959, 1961; Hertel 1963, 1966, 1967a–d; Liebe 1963; Aleyev 1970a,b, 1973b; Aleyev & Ovcharov 1969, 1971, 1973a,b; Ovcharov 1971, 1974; Shakalo & Buryanova 1973; Romanenko & Yanov 1973, and others), however they allow assessment with sufficient reliability of the basic, most common features of the flow over the bodies of nektonic animals.

1. Vortex Formation

In the majority of cases, the locomotion of nektonic animals is accompanied by the formation of vortices, as is testified by direct experimental data (Rosen 1959, 1961; Hertel 1963, 1966, 1967a–d; Liebe 1963; Aleyev & Ovcharov 1969, 1971, 1973a,b; Ovcharov 1971, 1974, and others). However, vortices and other visible disturbances in the flow around the body observed during the locomotion of nektonic animals sometimes only arise as a result of the specific conditions of the experiment, and do not occur when the animal moves about in its natural environment.

Thus Shuleikin (1934), when analysing ciné records of a swimming *Scomber* not intended for studying the flow pattern (Gray 1933c), drew attention to eddy lines on the lateral surface of the fish's body and concluded on this basis that quite intensive vortex formation takes place during movement even around such a well-streamlined fish as the *Scomber.* But these ciné records were obtained by filming the fishes in a shallow bath close to the surface of the water, and the shadows Shuleikin noted on the frames are nothing but surface water waves engendered by the propulsive undulations of the fish's body (Aleyev & Ovcharov 1969).

Similarly, vortices obtained in tests with *Brachydanio albolineatus* (Rosen 1959, 1961) also represent a flow picture distorted by the conditions of the experiment – the closeness of the swimming fish to the surface of the water – as has been proved experimentally (Aleyev & Ovcharov 1971). Clearly, these tests (Rosen 1961) give no grounds for speaking about any 'useful' role played by vortices: vortices can be nothing but detrimental since they always involve the wasting of at least some of the energy spent by the organism on locomotion. The view (Rosen 1961) that a system of vertical vortices should be assumed as the basic hypothesis of fish locomotion has not been supported by experimental results (Aleyev & Ovcharov 1969, 1971, 1973a,b and others). Similar methodological errors occurred in tests carried out by Hertel (1967a). The vortices observed in these tests in the hydrodynamic wakes behind a trout and an eel provide no grounds for concluding that vortex formation takes place actually on the fish's body. As has already been shown (Aleyev & Ovcharov 1969) by using more objective methods of visualizing the flow, also with a trout swimming in the water sufficiently far away from the surface, there are no vortices near the fish's body, whereas vortex formation in the slip-stream can be traced in every detail.

In Rosen's tests (1961) with the dolphin *Lagenorhynchus obliquidens* leaping out of the water and swimming through a dust screen formed by slowly falling particles of a plastic meterial, the picture of the flow is distorted by the proximity of the water's surface and there is no way of ascertaining the presence of vortices on the animal's body. Photographs taken by Rosen during other tests (1961) with the forebody of a swimming dolphin painted with fluorescent paint point to the absence of any vortices on the body: the water stained with the fluorescent substance flows in a clear-cut ribbon all along the dolphin's body with no indication of vortices and only in front of the tail flipper slightly and very smoothly parts from the body, without any vortex movements, which is characteristic of the usual picture of boundary-layer separation. Observations of a swimming dolphin in condition of sea fluorescence (Hill 1950; Steven 1950) show that the animal leaves behind nothing but a narrow bright strip without any trace of vortices.

When speaking of vortex formation during the swimming of nektonic animals it is necessary to distinguish clearly two cases: 1. the appearance of vortices in the hydrodynamic wake of the swimming animal and 2. the emergence of breakaway vortices actually on the body of the swimming animal.

The presence of vortices in the hydrodynamic wake behind a swimming animal with an axial undulatory propulsor is a firmly established fact and is due to the very essence of the propulsive action of this type of propulsive mechanism (Aleyev 1959b, 1963a, 1973b, 1974; Aleyev & Ovcharov 1969, 1971, 1973a,b; Hertel 1967b, and others). As for **the separation vortices that could appear right on the body** of the swimming animal, nobody has so far proved their presence during the swimming of nektonic animals in their natural habitat, and all reverences to the appearance of such vortices (Shuleikin 1934, 1953; Rosen 1959, 1961; Hertel 1967b, and others) are based on artefacts that can be specially modelled in experimental conditions (Aleyev & Ovcharov 1971). Both these facts, the presence of vortices in the hydrodynamic wake and the absence of separation vortices immediately at the animal's body, have been established and studied by us in detail (Aleyev & Ovcharov 1969, 1971, 1973a,b, and others) during an experimental investigation of flow over the fishes *Acipenser, Salmo, Anguilla, Chalcalburnus, Carassius, Mugil, Diplodus, Spicara, Pomatomus, Trachurus,* etc. The tests were conducted in two water channels: a small and a large one. The length, width and height of the small channel were

2.00, 0.24 and 0.22 m respectively and the effective depth of the water layer 0.20 m, while the dimensions of the large channel were 4.00, 0.40 and 0.40 m and the effective water depth 0.30 m. The lengths of the fishes studied varied from 10 to 60 cm with body heights from 2 to 7 cm. Such a relationship between the sizes of the fishes and the channels ensured that the influence of the water surface and channel walls on the propulsive process was negligible. The flowing stream was visualized by injecting a special staining paste into the oral cavity of the fish using a syringe, the paste being based on the histological staining agent Azure-2. During the experiments the water in the channel – sea or fresh water, according to the ecology of the fishes concerned – was immobile. The movement of the fish and the trail behind were recorded on 35 mm film by means of filming from above with a Konvas camera, as well as by synchronous two-view filming, from above and laterally, with two Konvas cameras, which allowed presentation of a stereometric picture of the flow. The speeds of the fishes, determined frame by frame, in different cases varied from 0.5 to 2.5 m/s at *Re* values from 5.0×10^4 to 5.0×10^5. During the movement of the fish water stained with Azure issued from the gill slits in contrasting blue streams, washing practically the entire body and creating a flow picture optimal for visual observation and filming (Figures 72–76). All in all, more than 3000 tests were staged.

Both during active swimming and during movement through inertia, the individual volumes of water ejected from the gill slits, corresponding to individual exhalations, remained visible right down to speeds of locomotion of 2 m/s (Figures 72–76). This shows that **the action of the gill apparatus,** along with the propulsive undulations of the body, is one of the most important causes of the unsteady character of flow over fishes, which agrees with an earlier theory (Breder 1924, 1965; Burdak 1968) about the influence of the respiratory process on fish locomotion. In the case of **the passive type of respiration** which is observed in fishes with higher swimming speeds (Hughes 1960a,b; Walters 1962; von Wahlert 1964, and others), the water from the gill slits enters the boundary layer more or less uniformly, so that the gill apparatus loses its importance, partially or completely, as a cause of unsteady flow. During our tests with *Spicara* and *Trachurus* with V_r between 3 and 5 L_a/s, during active respiration the laminar portion of the boundary layer made up 30–40% of the fish's absolute length L_a (Figure 73), whereas during passive respiration this portion notably increased, to 45–60% of L_a (Figure 74). This confirms the view (Burdak 1969a) that a functioning gill apparatus in fishes has a turbulizing effect on the boundary layer. The turbulent character of the stream of the water ejected from the gill slits during active respiration is clearly visible on the cine records we obtained (Figures 72B, 73 and others).

The results of our tests allow us to conclude that both during active swimming and during movement by inertia there are no large vortices associated with general boundary-layer separation in the flow over the fish's body. In all cases, without exception, we observed only **flow without separation over the fish's body** throughout its length, including the entire tail fin (Figures 72–76).

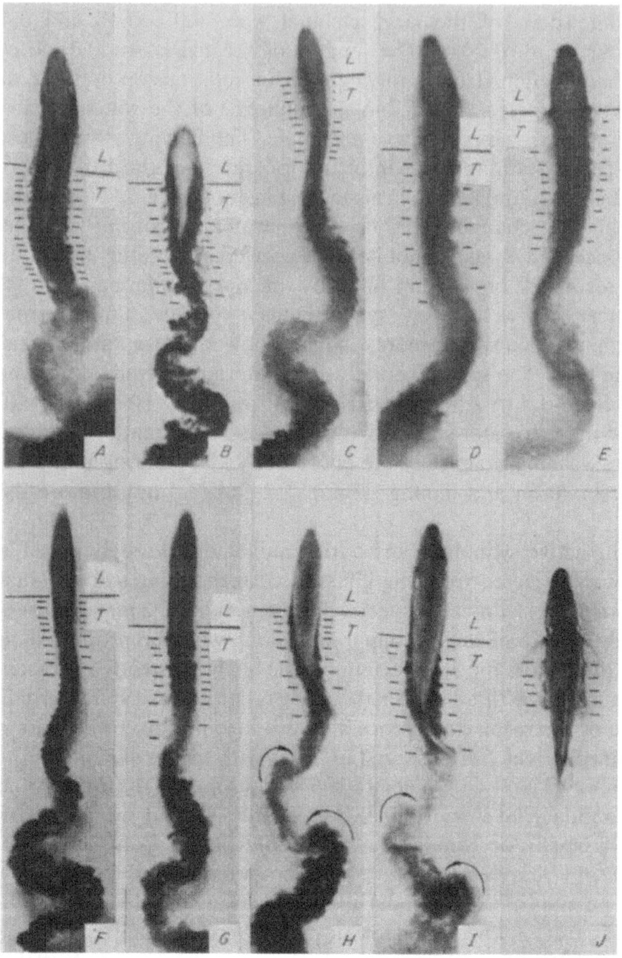

Figure 72. Visualization of fish movement. *A*, *Salmo gairdneri irideus* Gibb., $L_a = 16.3$ cm, $V = 0.75$ m/s, $Re = 1.2 \times 10^5$; *B*, *Carassius auratus* (L.), $L_a = 12.2$ cm, $V = 0.29$ m/s, $Re = 3.5 \times 10^4$; *C*, *Mugil auratus* Risso, $L_a = 12.0$ cm, $V = 0.95$ m/s, $Re = 1.1 \times 10^5$; *D*, *Mugil saliens* Risso, $L_a = 19.0$ cm, $V = 0.76$ m/s, $Re = 1.4 \times 10^5$; *E*, *Mugil auratus* Risso $L_a = 18.0$ cm, $V = 0.48$ m/s, $Re = 8.7 \times 10^4$; *F*, *Chalcalburnus chalcoides* (Gül.) $L_a = 13.5$ cm, $V = 0.84$ m/s, $Re = 1.1 \times 10^5$; *G*, *Spicara smaris* (L.), $L_a = 15.9$ cm, $V = 0.46$ m/s, $Re = 7.4 \times 10^4$; *H*, *Diplodus annularis* (L.), $L_a = 16.4$ cm, $V = 2.00$ m/s, $Re = 3.3 \times 10^5$; *I*, *Diplodus annularis* (L.), $L_a = 16.4$ cm, $V = 0.76$ m/s, $Re = 1.2 \times 10^5$; *J*, *Diplodus annularis* (L.), $L_a = 16.4$ cm, $V = 0.36$ m/s, $Re = 4.5 \times 10^4$. Horizontal line indicates point of laminar (*L*) to turbulent (*T*) boundary-layer transition; dashes mark individual microvortices in the turbulent boundary layer; bent arrows indicate sense of vorticity in the wake.

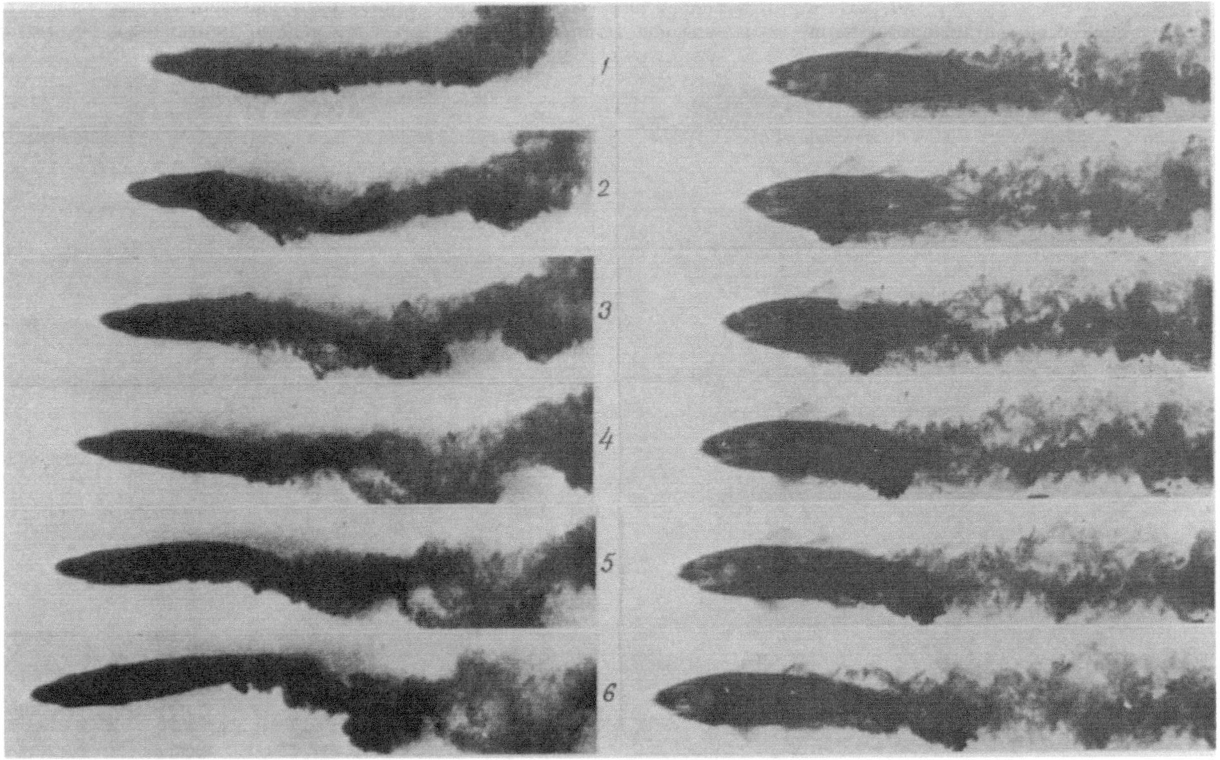

Figure 73. Visualization of flow over *Trachurus mediterraneus ponticus* Aleev during active movement and active breathing. Synchronous film frames (plan and side views) from filming with two cameras. $L_a = 13.6$ cm, $V = 0.40$ m/s, $Re = 5.4 \times 10^4$. See text.

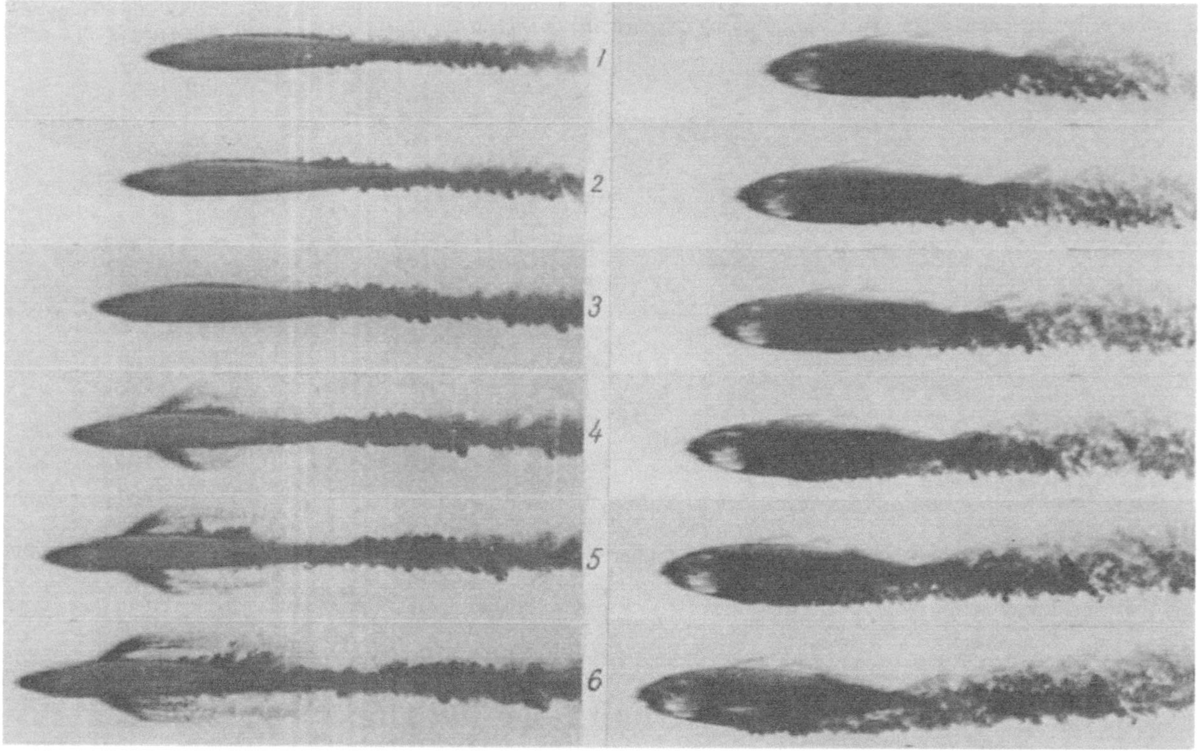

Figure 74. Visualization of flow over *Trachurus mediterraneus ponticus* Aleev during movement by inertia and passive breathing. Synchronous film frames (plan and side views) from filming with two cameras. $L_a = 13.6$ cm, $V = 0.52$ m/s, $Re = 7.1 \times 10^4$. See text.

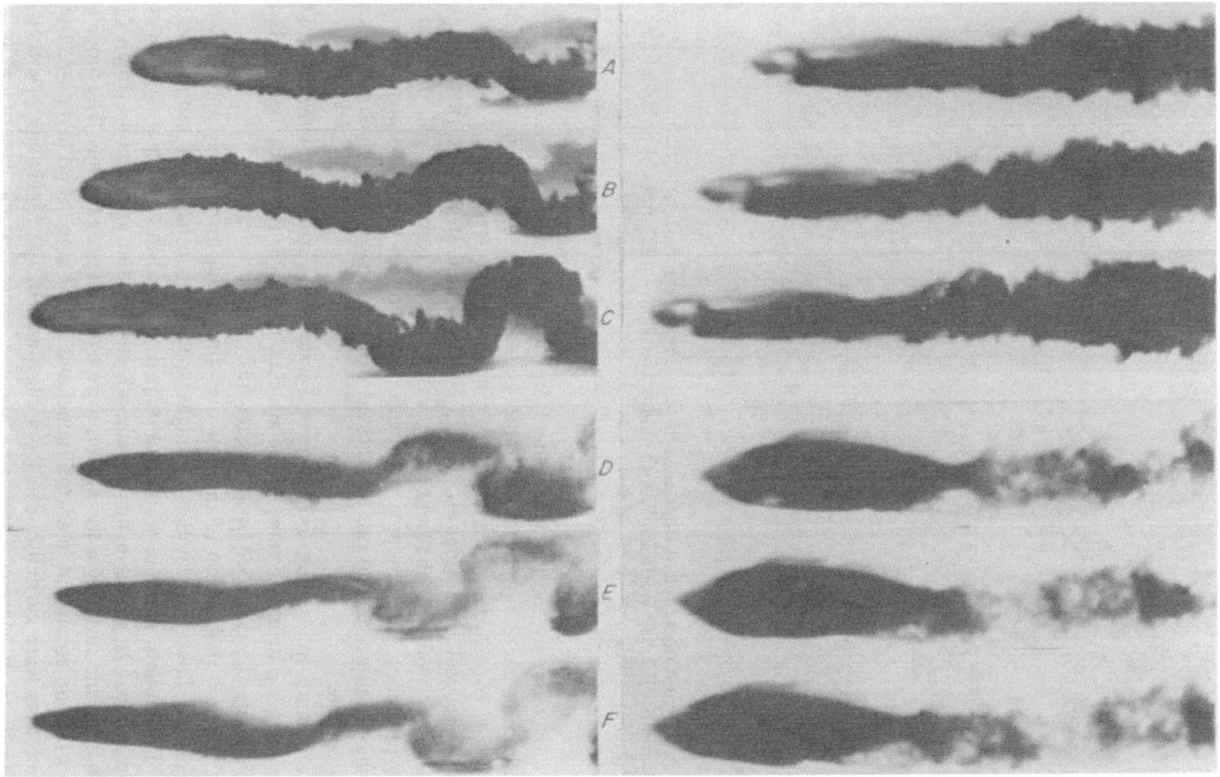

Figure 75. Visualization of flow over (*A-C*) *Mugil aurarus* Risso and (*D-F*) *Spicara smaris* (L.). Synchronous frames (plan and lateral views) from filming with two cameras. For *Mugil* $L_a = 18.9$ cm, $V = 0.60$ m/s, $Re = 1.1 \times 10^5$; for *Spicara* $L_a = 17.5$ cm, $V = 0.49$ m/s, $Re = 8.6 \times 10^4$. See text.

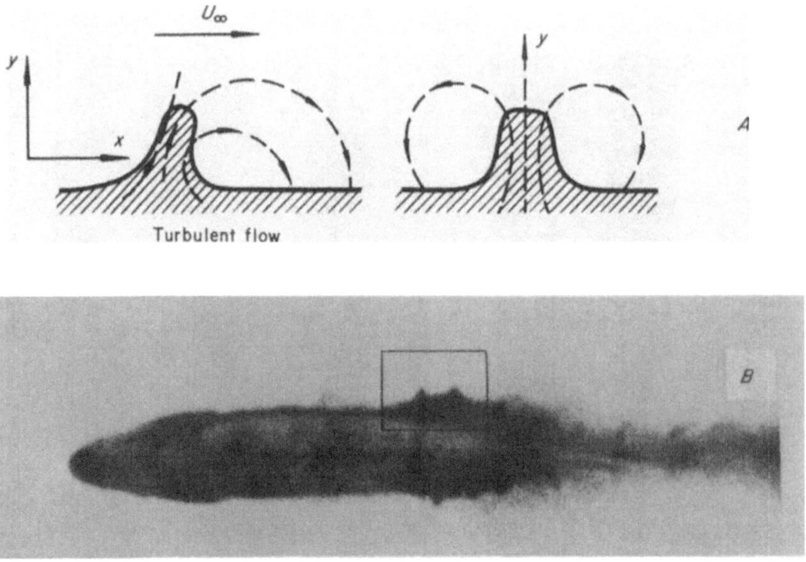

Figure 76. *A*, diagram of turbulent displacement of stream in external part of boundary layer (data of Townsend 1957; see Rotta 1967, with modifications). ———, border of turbulent fluid; ‐ ‐ ‐ ‐, streamlines relative to free stream; U_∞, direction of potential flow. *B*, turbulent boundary layer on body of *Diplodus*; frame clearly shows two wavy turbulent formations.

During movement by inertia, when there are no propulsive undulations of the body, neither are there any vortices in the hydrodynamic wake behind the fish (Figures 72, 74, 76). During active swimming, however, the wake always twists into vortices with vertical axes (Figures 72, 73), from which it follows that vortex formation in the wake is the immediate result of the operation of the axial undulatory propulsor. The intensity of vortex formation in the wake increases at higher Reynolds numbers, and can be assessed from the speed with which the wake twists into vortices.

The first attempts to visualize the flow over Teuthoidea, carried out with the squid *Todarodes pacificus*, bear testimony of the attached flow over the entire body of the mollusc, including the extended arms (Figure 63E).

In order to study the pattern of the flow over the bodies of nektonic animals we investigated also flow over models of eight species of fishes, sea reptiles and cetaceans. The models, made of wood and shaped to conform to inertial movement of the animals, had an aerodynamically smooth surface and an effective length of 1.00 m. Glued to the surfaces of the models were light indicator threads. The models were tested 1. in an air stream in a wind tunnel and 2. by being towed in a pool. The aerodynamic tests were carried out in the wind tunnel of the Mechanics Scientific-Research Institute of Moscow University, and the hydrodynamic experiment in a pool 50 m long and 25 m wide with a water depth of about 3 m. In the pool the models were towed by an electric winch at a speed of about 4 m/s. The arrangement of each model's longitudinal axis strictly in the direction of towing was maintained by a swimmer who was towed along with the model (Figure 77E). The towing of the models was accompanied by continuous underwater filming with a Konvas camera.

Tested in the wind tunnel were the models of *Auxis*, *Xiphias* and *Istiophorus*, at $Re = 5.2 \times 10^6$ (Aleyev 1970b, and others). The models towed in the pool were those of *Thunnus*, *Phocoena*, *Delphinus*, *Chelonia* and *Crytocleidus*, at $4.0 \times 10^6 \leqslant Re \leqslant 4.4 \times 10^6$. At these *Re* values the fluid flowed over all the models without separation of the boundary layer, as is testified by the strictly longitudinal position of the indicator threads over the entire surface of the models (Figures 77–79). Some vibration of the threads at the rear ends of the models points to the turbulent state of the boundary layer there.

Thus, investigation into the pattern of the flow over living nektonic animals (Rosen 1961, Aleyev & Ovcharov 1969, 1971, 1973a,b, etc.) and also studies of the flow over models of nektonic animals (Aleyev 1970b, 1974, etc.) led us to the conclusion that flow without separation is the general rule for nektonic animals, and constitutes and essential feature of their movement directly aimed at reducing form drag.

2. Boundary Layer

The boundary layer on a nektonic animal, depending on the body shape, regime of locomotion and *Re* value, may be laminar as well as turbulent (Hill 1950; Steven 1950; Gero 1952; Rosen 1961; Aleyev & Ovcharov 1969, 1971, 1973a,b; Aleyev 1970b, 1973b, 1974, and others). Our experiments with fishes (Aleyev & Ovcharov 1969, 1971, 1973a,b) show that at $Re > 10^5$ the boundary layer is mixed in all the species studied, consisting of laminar and turbulent portions, and that the **boundary layer's point of laminar-to-turbulent transition** approximately corresponds to the border between the contraction and diffusion areas of the body (Figures 72–76). Many of our ciné records distinctly show the crests of wavy turbulent formations, which represent streams of turbulent fluid emerging from the inner part of the boundary layer into the relatively non-turbulent flow outside the layer (Figures 72H,I, 76). Our experiments confirm the view (Burdak 1968, 1969a) that the fins of fishes constitute a turbulizing factor: deflexion of the fins away from the body is in all cases conducive to shortening of the laminar section of the flow (Figures 72D,E,H–J, 74).

At $Re < 10^5$ the boundary layer may remain laminar also in the upstream part of the diffuser. In small Scombridae with a very well streamlined, laminarized body shape the laminar boundary layer in the upstream part of the diffuser is apparently preserved also at $Re > 10^5$, according to the results of experiments with Scombridae models, described below (Figures 77A, 85A).

A special case of flow over Xiphioidae will also be described.

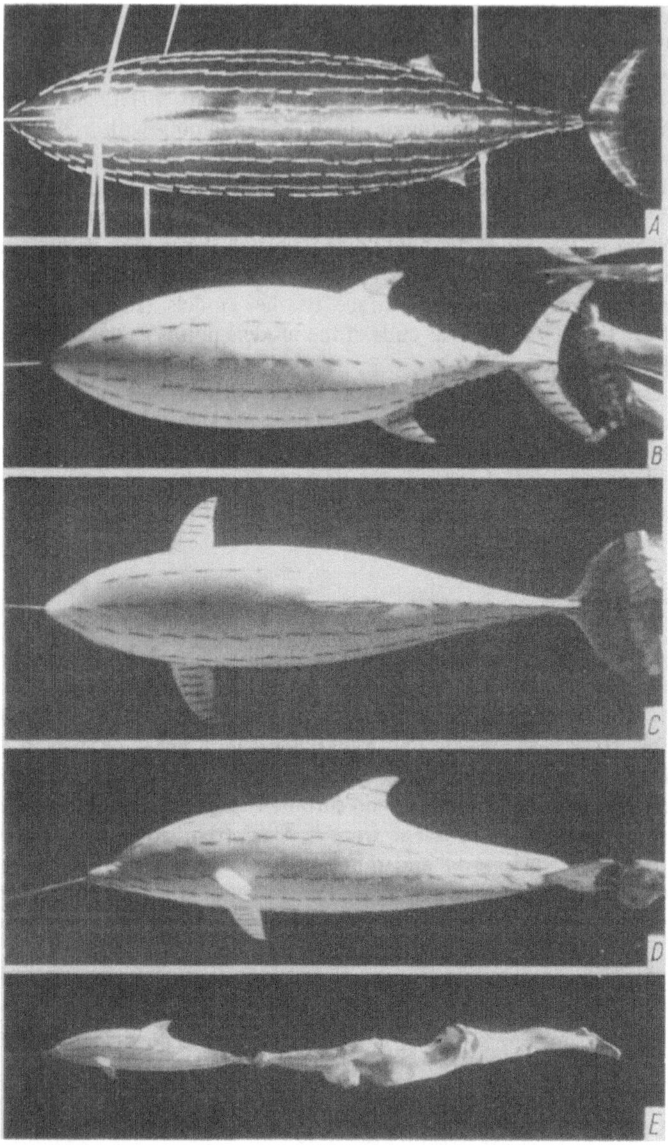

Figure 77. Flow over models of nektonic animals. *A*, model of *Auxis thazard* (Lac.) in aerodynamic tunnel; *B-E*, models towed in pool. *B*, *Thunnus alalunga* (Bonnat.); *C*, *Phocoena phocoena* (L.); *D*, *Delphinus delphis ponticus* Barab.; *E*, general view of model with swimmer operator during towing. See text.

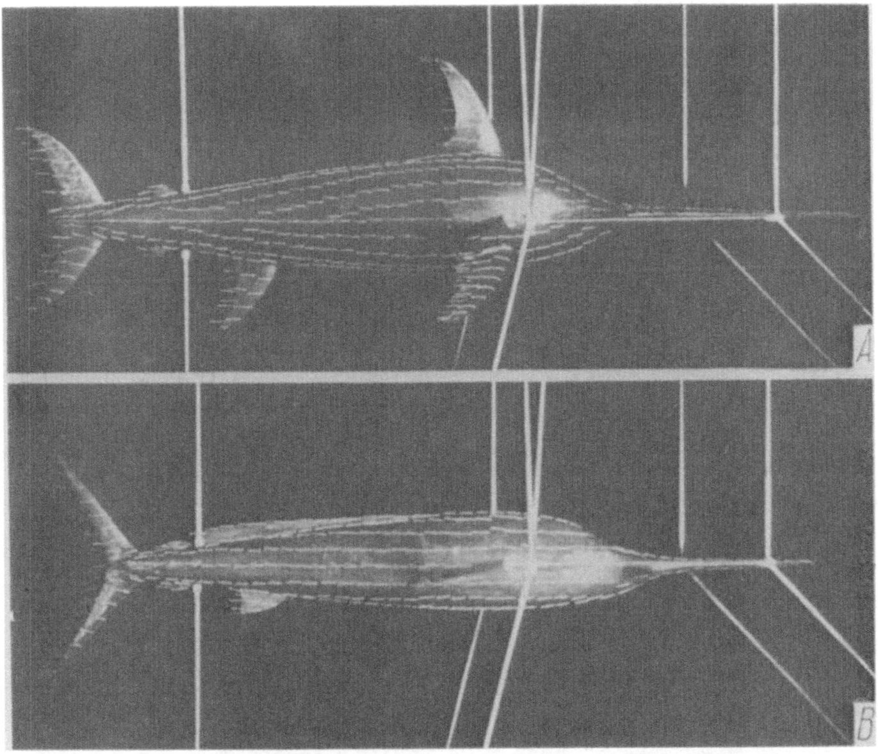

Figure 78. Models of (*A*) *Xiphias gladius* L. and (*B*) *Istiophorus platypterus* (Show and Nodder) in aerodynamic tunnel. See text.

The first fully objective investigation into the boundary layer on cetaceans consisted (Rosen 1961) of tests on dolphins with fluorescent paint for visualizing the flow. It was shown that the boundary layer is laminar only on the contraction area, up to the dorsal fin, which corresponds to a local *Re* value in the transition area of about 3.0×10^6. This has been confirmed also (Romanenko & Yanov 1973) by the results of instrumental measurements of turbulent pulsations in the boundary layer on a live dolphin.

More evidence concerning the point of boundary-layer transition is provided by the distribution of dynamic pressure over the surface of the animal's body, which we investigated on models of 19 species of various nektonic animals: cephalopods, fishes, ichthyosauras, penguins, cetaceans and pinnipeds. Pressure measurements were taken in the *Re* range 10^6–10^7, within which the transition point approximately coincides (Schlichting 1956) with the point of minimum dynamic pressure.

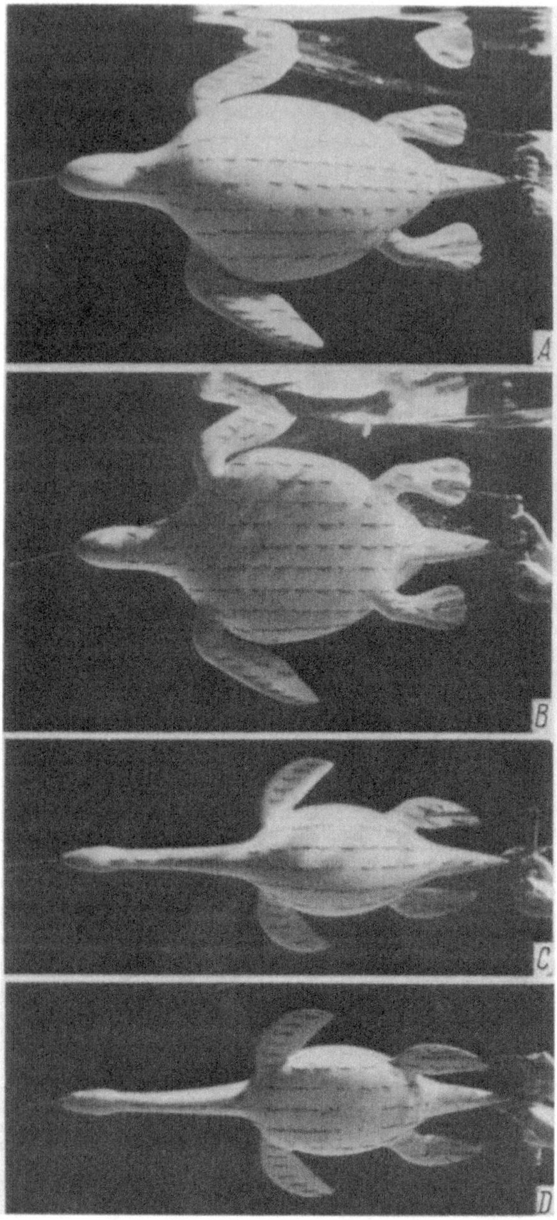

Figure 79. Flow over models towed in pool. *A, B, Chelonia mydas* L.; *C, D, Cryptocleidus oxoniensis* Phill. See text.

The models of 16 species of nektonic animals were based on our direct measurements of the animals. The model of *Stenopterygius quadriscissus*, however, was based on the author's reconstruction from paleontological material (Abel 1912; Huene 1956; Romer 1956; Müller 1968, and others) and the models of *Balaenoptera physalus* and *Physeter catodon* on data from the literature (Sleptsov 1952; Bourdelle & Grassé 1955; Tomilin 1957, 1962a; Chepurnov 1968, and others). Sexually mature individuals were always taken as the prototypes of the models, which were made of wood and had an aerodynamically smooth surface and an effective length of about 100 cm; their shape conformed to that of the animals during inertial movement. About 20 copper draining tubes with an inner diameter of 2.5 mm were inserted into the body of each model. These tubes emerged at one end in three rows on the surface of the model (Figures 80–90), while the other ends emerged in a single bunch at the rear of the model and were connected to a multitube pressure gauge, each of its tubes corresponding to a certain point on the surface of the model. The models were tested in a horizontal water channel with an open working section and an open-return operational circuit. The degree of stream non-uniformity in the trough was about 3% and the degree of stream turbulence in the working section about 2%. The working section of the trough was 50.0 cm wide, 60.0 cm high and 150.0 cm long and was made of transparent plastic material, which allowed observation of the model during the tests (Figure 80). Pressure measurements were taken at

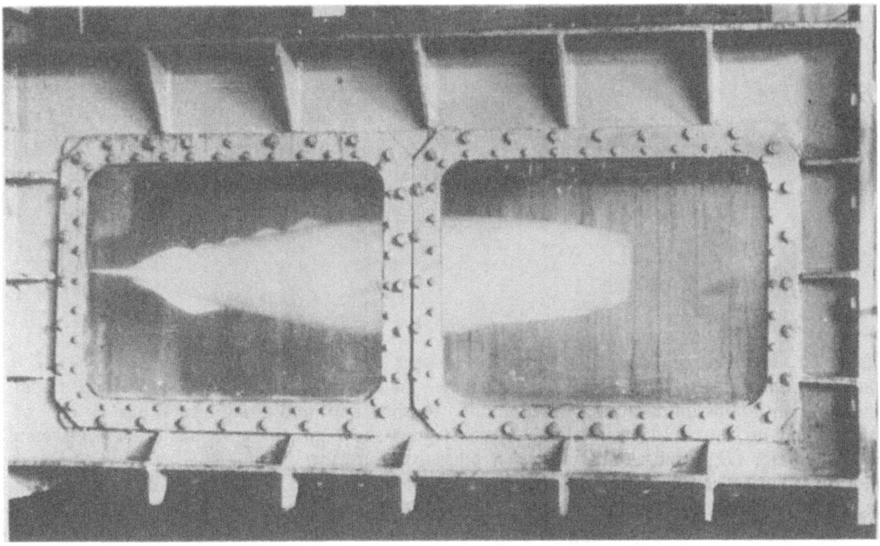

Figure 80. Working section of trough at Marine Test Station with model of *Physeter catodon* L. in it.

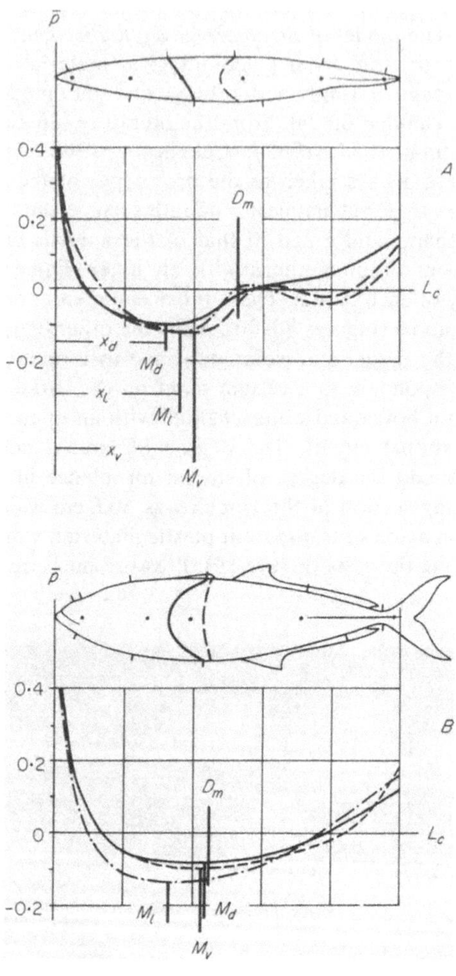

Figure 81. Distribution of dynamic pressure $\bar{p}$ over the surface of models of (*A*) *Symplectoteuthis oualaniensis* (Less.) and (*B*) *Trachurus mediterraneus ponticus* Aleev along dorsal midline (continuous line), along ventral midline (dashed line) and along lateral line at level of longitudinal body axis (dash-dot line). Vertical lines show locations of maximum transverse body diameter (D_m) and points of lowest pressure along the dorsal (M_d), ventral (M_v) and lateral (M_l) measurement lines. *A*, effective length of model $L_c = 100.0$ cm, flow velocity in trough $V_m = 4.0$ m/s, Reynolds number for model $Re_m = 4.0 \times 10^6$. *B*, $L_c = 100.0$ cm, $V_m = 3.0$ m/s, $Re_m = 3.5 \times 10^6$. The diagrams of the models show the arrangement of the outlets of the draining pipes in the dorsal (tick marks), ventral (tick marks) and lateral (points) rows, as well as the border between the contractor and diffuser areas (dashed line) and the location of the lowest dynamic pressure (continuous line).

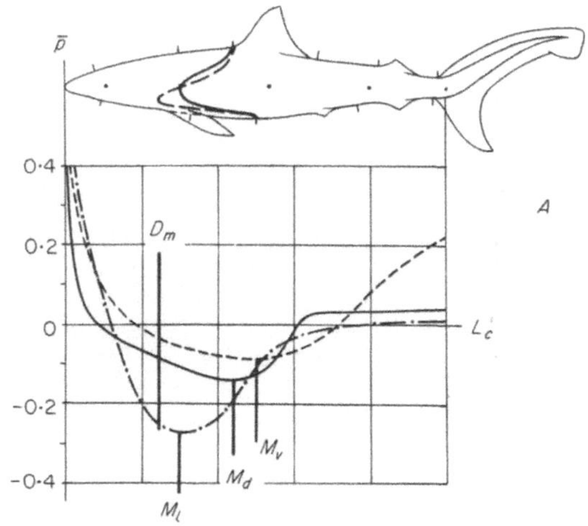

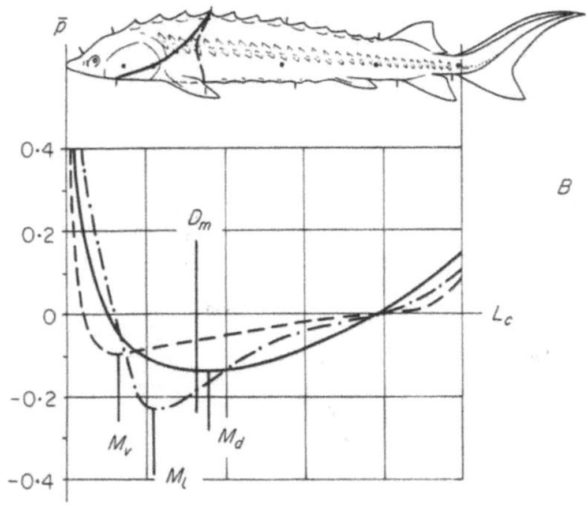

Figure 82. Distribution of dynamic pressure $\bar{p}$ over the surface of models of (A) *Prionace glauca* (L.) and (B) *Acipenser güldenstädti colchicus* V. Marti. A, $L_c = 74.0$ cm, $V_m = 6.0$ m/s, $Re_m = 6.0 \times 10^6$; B, $L_c = 76.0$ cm, $V_m = 4.0$ m/s, $Re_m = 4.0 \times 10^6$. Other notation as in figure 81.

values of Re characteristic of the modelled animals, from 3.5×10^6 to 9.5×10^6, which corresponded to flow velocities in the trough from 3.0 to 8.0 m/s.

The magnitude of the pressure coefficient $\bar{p}$ was found according to the generally accepted formula (Martynov 1958)

$$\bar{p} = \frac{2(p - p_o)}{\rho V^2}, \tag{28}$$

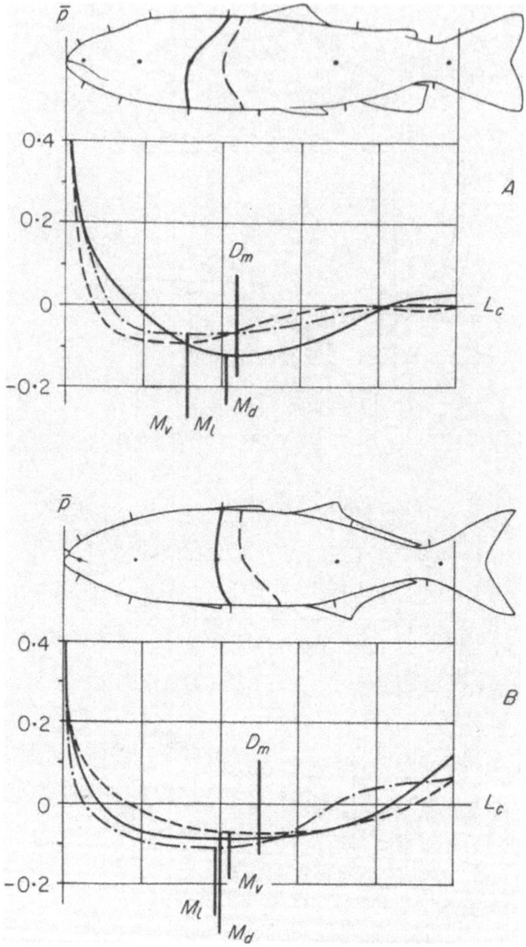

Figure 83. Distribution of dynamic pressure $\bar{p}$ over the surface of models of (*A*) *Salmo trutta labrax* Pall. and (*B*) *Pomatomus saltatrix* (L.). *A*, $L_c = 100.0$ cm, $V_m = 6.0$ m/s, $Re_m = 6.6 \times 10^6$; *B*, $L_c = 100.0$ cm, $V_m = 6.0$ m/s, $Re_m = 6.9 \times 10^6$. Other notation as in figure 81.

where p is the pressure at the point under investigation, found from the height of the water column in the corresponding tube of the multitube gauge, p_0 is the static pressure in the free stream, ρ is the water density and V is the velocity of the free stream.

The inertial-movement values of $\bar{p}$ we obtained for the 19 species of nektonic animals are shown in Table 13, where a definite value of $\bar{x}$ corresponds to each $\bar{p}$

value, the $\bar{x}$ value indicating the distance from the frontal point of the body's effective length L_c to the point under investigation, expressed as a fraction of L_c.

Table 13 and Figures 81–90 show that the line of minimum dynamic pressure is situated in every case either in direct proximity to the border between the contractor and the diffuser, indicating a boundary layer laminar on the contraction area and turbulent on the diffusion area, or markedly ahead of this border, indicating turbulent flow not only over the diffuser, but also over the downstream part of the

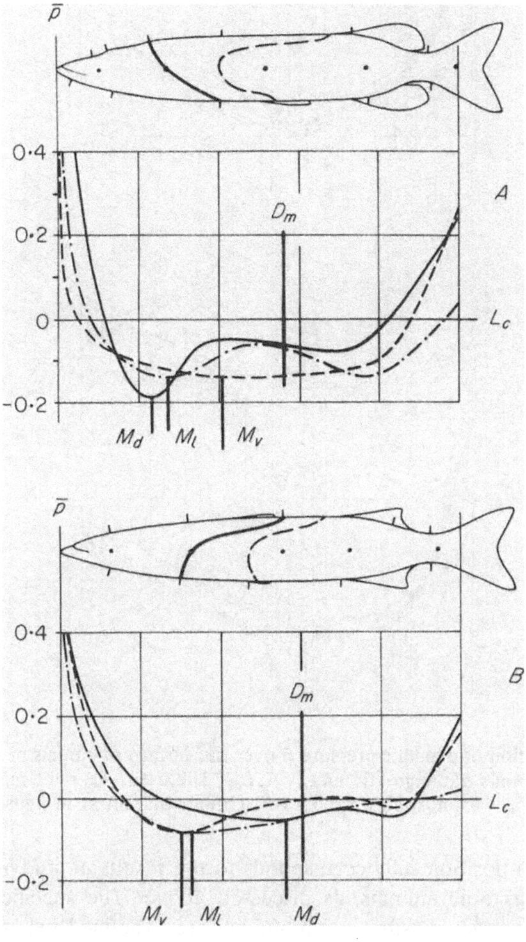

Figure 84. Distribution of dynamic pressure $\bar{p}$ over the surface of models of (*A*) *Esox lucius* L. and (*B*) *Sphyraena barracuda* (Walb.). *A*, $L_c = 100.0$ cm, $V_m = 6.0$ m/s, $Re_m = 7.0 \times 10^6$; *B*, $L_c = 100.0$ cm, $V_m = 6.0$ m/s, $Re_m = 6.9 \times 10^6$. Other notation as in figure 81.

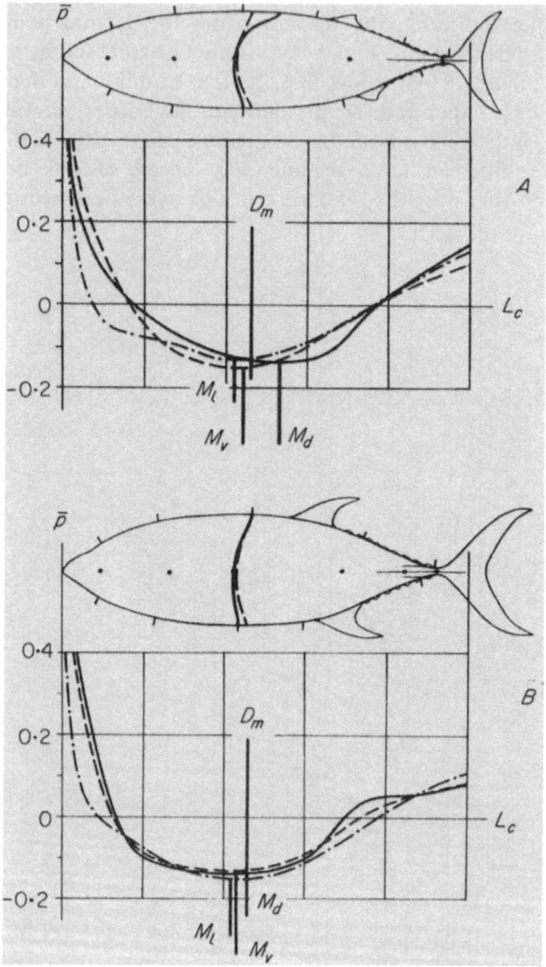

Figure 85. Distribution of dynamic pressure $\bar{p}$ over the surface of models of (*A*) *Auxis thazard* (Lac.) and (*B*) *Thunnus alalunga* (Bonnat.). *A*, $L_c = 100.0$ cm, $V_m = 4.0$ m/s, $Re_m = 4.3 \times 10^6$; *B*, $L_c = 100.0$ cm, $V_m = 6.0$ m/s, $Re_m = 6.9 \times 10^6$. Other notation as in figure 81.

contractor. Such a position fully corresponds to the results of studying the boundary layer on living nektonic animals, as discussed above. The specific features of the distribution of dynamic pressure in Xiphioidae are examined below.

The comparatively close results of the studies of the boundary layers on living animals and their models point to the fact that the ratio of the laminar and turbulent sections of flow over nektonic animals is mainly determined in a typical case by the specific features of the overall body shape and to a lesser degree by the special properties of the skin. Physiological data corroborate this conclusion. For example, calculations (Gray 1936b) show that during movement in the turbulent regime a dolphin should possess a power seven times higher than its muscles can provide. However an experimental check on this phenomenon (Lang 1963; Lang & Norris 1966; Focke 1965; Norris 1965; Semenov 1969; Karandeyeva, Protasov & Semenov 1970, and others) showed that the difference between the theoretical and actual

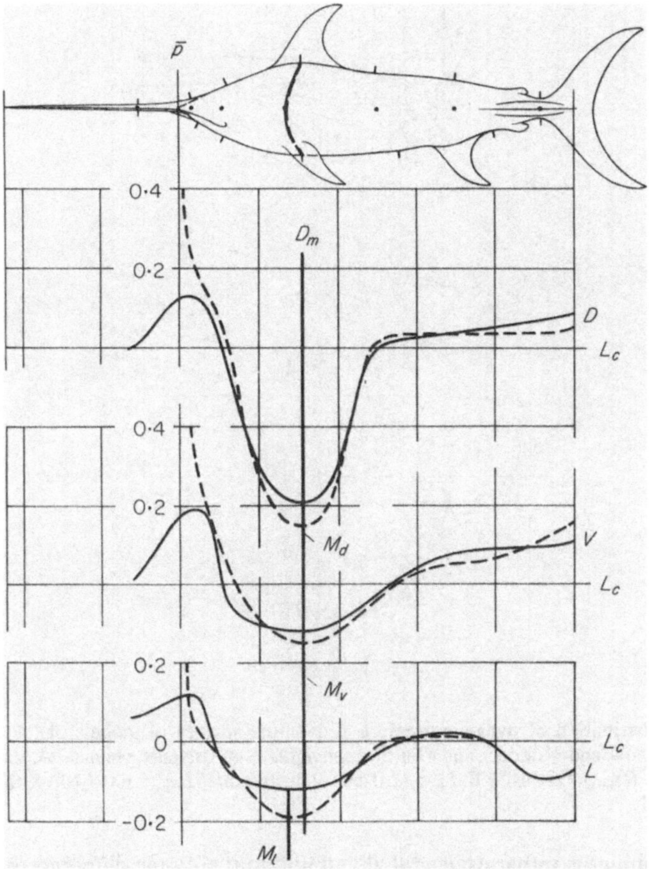

Figure 86. Distribution of dynamic pressure $\bar{p}$ over the surface of a model of *Xiphias gladius* L. along the dorsal (*D*), ventral (*V*) and lateral (*L*) midlines (continuous lines). ---, same model with conical fairing instead of rostrum. For model with rostrum, $L_c = 65.0$ cm, $V_m = 6.0$ m/s, $Re_m = 6.7 \times 10^6$; for model with conical fairing, $L_c = 65.0$ cm, $V_m = 6.0$ m/s, $Re_m = 3.9 \times 10^5$. Other notation as in figure 81.

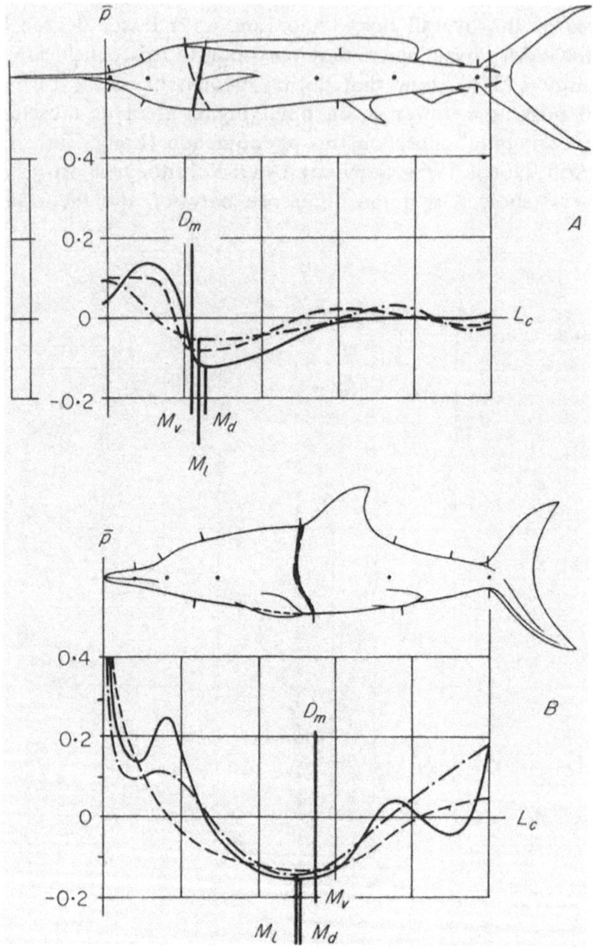

Figure 87. Distribution of dynamic pressure $\bar{p}$ over the surface of models of (*A*) *Istiophorus platypterus* (Show and Nodder) and (*B*) *Stenopterigius quadriscissus* Quenst. *A*, $L_c = 80.0$ cm, $V_m = 6.0$ m/s, $Re_m = 7.6 \times 10^6$; *B*, $L_c = 82.0$ cm, $V_m = 6.0$ m/s, $Re_m = 6.0 \times 10^6$. Other notation as in figure 81.

power of dolphins is rather exaggerated and that in reality the difference is not more than a factor of two.

The comparatively very smooth variation of the dynamic pressure along the body's longitudinal axis and its noticeable rise to positive values in the rear area in all the species investigated (Figure 81–90) correspond to the good streamlining of the animals' bodies.

All the above leads to the conclusion that at $Re > 10^5$ the boundary layer over the body's diffusion area is turbulent in all nektonic animals. The preservation of boundary-layer laminarity throughout the length of the animal's body, judging by everything, is possible only in comparatively small nektonic animals at $Re < 10^5$ (Sagittoidea, small cephalopods and fishes). The assumption of the possibility of preserving boundary-layer laminarity right up to Re values of the order of 4.0×10^7 (M. O. Kramer 1960a–c, 1962; Lang 1963, and others) can be accepted only with

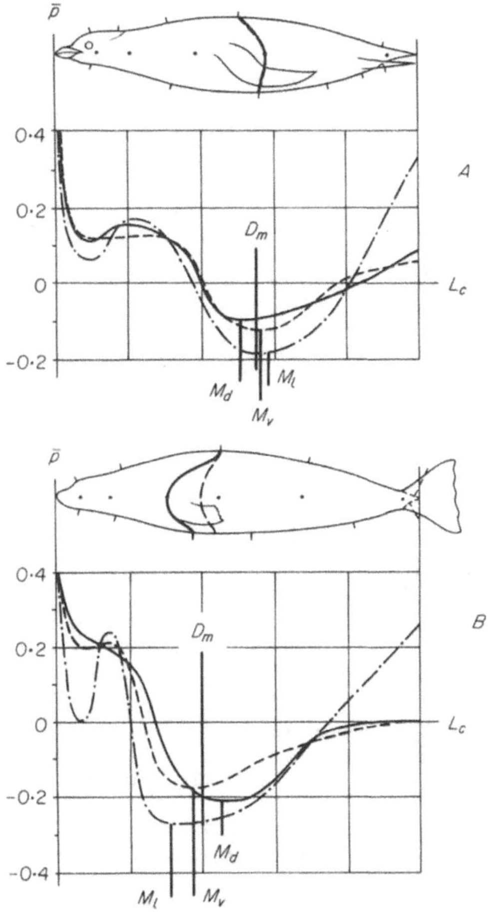

Figure 88. Distribution of dynamic pressure $\bar{p}$ over the surface of models of (*A*) *Eudyptes chrysolophus* Brandt and (*B*) *Pagophoca groenlandica* (Erxl.). *A*, $L_c = 90.0$ cm, $V_m = 4.0$ m/s, $Re_m = 4.0 \times 10^6$; *B*, $L_c = 113.0$ cm, $V_m = 6.0$ m/s, $Re_m = 7.0 \times 10^6$. Other notation as in figure 81.

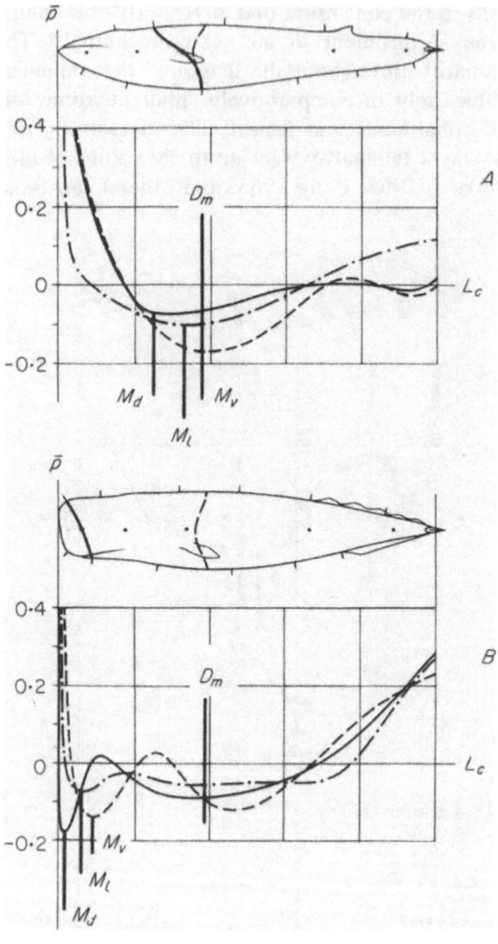

Figure 89. Distribution of dynamic pressure $\bar{p}$ over the surface of models of (*A*) *Balaenoptera physalus* (L.) and (*B*) *Physeter catodon* L. *A*, $L_c = 100.0$ cm, $V_m = 8.0$ m/s, $Re_m = 8.0 \times 10^6$; *B*, $L_c = 100.0$ cm, $V_m = 8.0$ m/s, $Re_m = 8.0 \times 10^6$. Other notation as in figure 81.

respect to laminarity over the contraction area, which, in some cases, may actually be preserved up to local *Re* values in the transition area of the order of 10^7.

3. Unsteadiness of Flow Over Body

A crucial feature of the flow over nektonic animals is its **unsteadiness**, caused by periodic changes in body shape connected with locomotion, manoeuvering, breathing, etc. (Breder 1924; Gray 1936b; Richardson 1936; M. F. Osborn 1961;

Walters 1962; Aleyev 1963a, 1973b, 1974; Pershin 1965a,b, 1967; Zuyev 1966; Kleinenberg & Kokshaiski 1967; Burdak 1968; Lyapin 1971; Shakalo & Buryanova 1973, and others). In cephalopods it is due to periodic changes in relative body thickness (Figure 63) and the periodic sucking off of the boundary layer through the mantle slit; in all nektonic animals with an undulatory propulsor it is due to the undulations of the body or fins (Figures 25–29), and in fishes also to respiratory

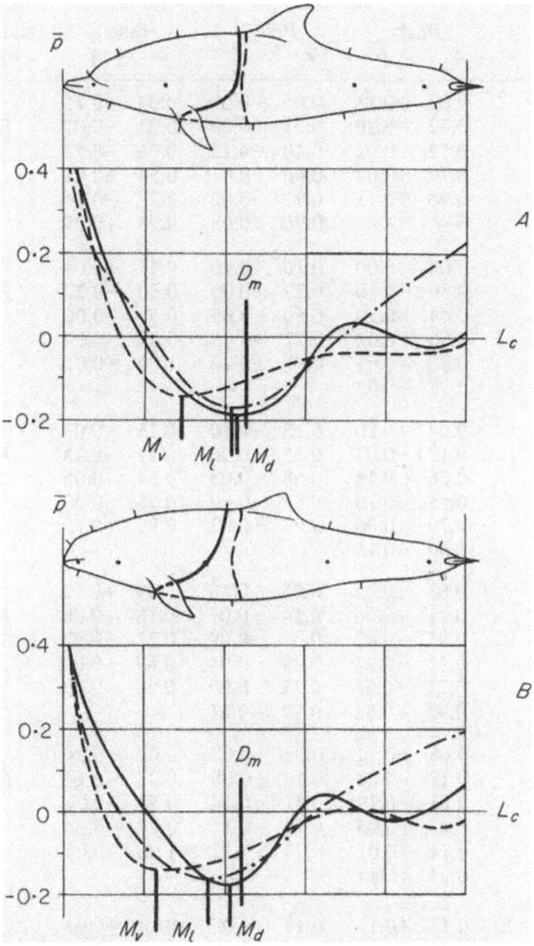

Figure 90. Distribution of dynamic pressure $\bar{p}$ over the surface of models of (*A*) *Delphinus delphis ponticus* Barab. and (*B*) *Phocoena phocoena* (L.) *A*, $L_c = 100.0$ cm; $V_m = 6.0$ m/s, $Re_m = 6.3 \times 10^6$; *B*, $L_c = 100.0$ cm, $V_m = 6.0$ m/s, $Re_m = 6.3 \times 10^6$. Other notation as in figure 81.

Table 13. Dynamic pressure $\bar{p}$ at various longitudinal distances $\bar{x} = xL_c^{-1}$ from frontal point (relative to direction of movement) of effective body length L_c for nektonic animals according to model trials. (This definition of $\bar{x}$ makes $\bar{x}$ values for *Xiphias* rostrum negative). Figure for points of minimum pressure are indicated in bold type.

Species, nektonic type, absolute length L_a of modelled specimens (cm) and body shape index Y	Back $\bar{x}$	Back $\bar{p}$	Belly $\bar{x}$	Belly $\bar{p}$	Side $\bar{x}$	Side $\bar{p}$	Maximum Reynolds number for live specimen (Re) and for model (Re_m) and flow velocity V_m over model (m/s)
Symplectoteuthis	0.03	+0.13	0.05	+0.07	0.04	+0.15	$Re = 4.0 \times 10^6$
oualaniensis	0.32	**−0.10**	0.21	−0.09	0.21	−0.07	$Re_m = 4.0 \times 10^6$
(Less.) (EN)	0.52	−0.02	0.40	**−0.12**	0.36	**−0.12**	$V_m = 4.0$
$L_a = 45.2$	0.79	+0.04	0.60	+0.01	0.54	+0.01	
$Y = 0.53$	0.96	+0.13	0.77	−0.02	0.77	−0.04	
	—	—	0.96	+0.08	0.96	+0.05	
Prionace glauca	0.08	0.00	0.10	+0.10	0.11	+0.04	$Re = 4.0 \times 10^7$
(L.) (EN)	0.30	−0.10	0.33	−0.06	0.30	**−0.27**	$Re_m = 6.0 \times 10^6$
$L_a = 403.0$	0.44	**−0.14**	0.50	**−0.09**	0.54	−0.06	$V_m = 6.0$
$Y = 0.25$	0.66	+0.02	0.77	+0.05	0.80	0.00	
	0.82	+0.03	0.97	+0.20	1.00	+0.01	
	0.98	+0.03	—	—	—	—	
Acipenser	0.04	+0.20	0.13	**−0.10**	0.15	−0.07	$Re = 4.5 \times 10^6$
güldenstädti	0.12	−0.03	0.35	−0.06	0.22	**−0.23**	$Re_m = 4.0 \times 10^6$
colchicus V.	0.36	**−0.14**	0.58	−0.03	0.54	−0.06	$V_m = 4.0$
Marti (BN)	0.55	−0.10	0.84	0.00	0.78	0.00	
$L_a = 150.0$	0.79	0.00	0.98	+0.07	0.99	+0.10	
$Y = 0.33$	1.00	+0.15	—	—	—	—	
Salmo trutta	0.03	+0.25	0.05	+0.10	0.04	+0.12	$Re = 1.2 \times 10^7$
labrax Pall.	0.11	+0.06	0.14	−0.07	0.18	−0.06	$Re_m = 6.6 \times 10^6$
(BN)	0.21	−0.03	0.31	**−0.09**	0.31	**−0.07**	$V_m = 6.0$
$L_a = 110.0$	0.41	**−0.12**	0.49	−0.04	0.69	−0.03	
$Y = 0.44$	0.74	−0.04	0.71	0.00	0.98	0.00	
	0.92	+0.02	0.92	−0.01	—	—	
Esox lucius	0.08	+0.14	0.04	+0.02	0.09	+0.04	$Re = 10^7$
L. (BN)	0.12	−0.03	0.14	−0.08	0.27	**−0.14**	$Re_m = 7.0 \times 10^6$
$L_a = 103.4$	0.23	**−0.19**	0.41	**−0.14**	0.52	−0.06	$V_m = 6.0$
$Y = 0.54$	0.41	−0.05	0.78	−0.11	0.77	−0.14	
	0.74	−0.07	0.94	+0.11	1.00	+0.04	
	0.93	+0.14	—	—	—	—	
Sphyraena	0.13	+0.12	0.12	+0.05	0.13	+0.02	$Re = 2.0 \times 10^7$
barracuda	0.32	−0.02	0.30	**−0.08**	0.33	**−0.08**	$Re_m = 6.9 \times 10^6$
(Walb.) (EN)	0.56	**−0.05**	0.47	−0.02	0.56	−0.05	$V_m = 6.0$
$L_a = 130.0$	0.69	−0.02	0.70	0.00	0.73	−0.02	
$Y = 0.60$	0.83	−0.04	0.93	+0.06	0.94	0.00	
	0.93	+0.07	—	—	—	—	

Table 13. (*contd.*)

Species, nektonic type, absolute length L_a of modelled specimens (cm) and body shape index Y	Back		Belly		Side		Maximum Reynolds number for live specimen (Re) and for model (Re_m) and flow velocity V_m over model (m/s)
	$\bar{x}$	$\bar{p}$	$\bar{x}$	$\bar{p}$	$\bar{x}$	$\bar{p}$	
Pomatomus	0.01	+0.25	0.04	+0.12	0.04	+0.02	$Re = 1.2 \times 10^7$
saltatrix	0.04	+0.10	0.17	0.00	0.18	−0.09	$Re_m = 6.9 \times 10^6$
(L.) (EN)	0.17	−0.06	0.40	**−0.07**	0.39	**−0.11**	$V_m = 6.0$
$L_a = 99.0$	0.40	**−0.09**	0.69	−0.06	0.70	0.00	
$Y = 0.50$	0.73	−0.05	0.94	+0.02	0.98	+0.06	
	0.94	+0.09	—	—	—	—	
Trachurus	0.05	+0.19	0.06	+0.12	0.08	+0.06	$Re = 3.5 \times 10^6$
mediterraneus	0.08	+0.10	0.22	−0.06	0.27	**−0.12**	$Re_m = 3.5 \times 10^6$
ponticus	0.22	−0.06	0.43	**−0.10**	0.39	**−0.12**	$V_m = 3.0$
Aleev (EN)	0.44	**−0.08**	0.85	0.00	0.71	−0.03	
$L_a = 56.0$	0.73	−0.03	0.95	+0.11	0.99	+0.15	
$Y = 0.45$	0.95	+0.09	—	—	—	—	
Auxis thazard	0.12	+0.05	0.12	+0.08	0.11	−0.05	$Re = 5.0 \times 10^6$
(Lac.) (EN)	0.27	−0.07	0.27	−0.11	0.27	−0.09	$Re_m = 4.3 \times 10^6$
$L_a = 56.0$	0.44	**−0.13**	0.44	**−0.15**	0.42	**−0.13**	$V_m = 4.0$
$Y = 0.46$	0.62	**−0.13**	0.68	−0.04	0.70	−0.03	
	0.83	+0.04	0.88	+0.05	0.92	+0.09	
	0.93	+0.12	—	—	—	—	
Thunnus alalunga	0.09	+0.13	0.09	+0.08	0.09	0.00	$Re = 1.6 \times 10^7$
(Bonnat.) (EN)	0.23	−0.12	0.23	−0.10	0.26	−0.11	$Re_m = 6.9 \times 10^6$
$L_a = 89.0$	0.46	**−0.14**	0.43	**−0.13**	0.42	**−0.15**	$V_m = 6.0$
$Y = 0.46$	0.76	+0.03	0.59	−0.09	0.69	−0.06	
	0.92	+0.06	0.89	+0.06	0.84	+0.04	
	—	—	—	—	0.92	+0.08	
Xiphias gladius	−0.10	+0.01	−0.10	+0.03	−0.10	+0.07	$Re = 1.1 \times 10^8$
L. (EN)	0.03	+0.13	0.05	+0.19	0.03	+0.12	$Re_m = 6.7 \times 10^6$
$L_a = 356.0$	0.12	−0.01	0.12	−0.04	0.12	−0.08	$V_m = 6.0$
$Y = 0.31$	0.31	**−0.39**	0.31	**−0.12**	0.27	**−0.12**	
	0.50	0.00	0.55	+0.01	0.50	−0.03	
	0.70	+0.04	0.80	+0.09	0.70	+0.02	
	0.91	+0.07	0.92	+0.09	0.92	−0.09	
Same model	0.03	+0.21	0.05	+0.27	0.03	+0.05	$Re_m = 3.9 \times 10^6$
with conical	0.12	+0.02	0.12	+0.05	0.12	−0.06	$V_m = 6.0$
fairing instead	0.31	**−0.45**	0.31	**−0.15**	0.27	**−0.19**	
of rostrum	0.50	+0.01	0.55	0.00	0.50	−0.01	
$Y = 0.31$	0.70	+0.03	0.80	+0.05	0.70	+0.01	
	0.91	+0.03	0.92	−0.11	0.92	−0.09	

Table 13. (*contd.*)

Species, nektonic type, absolute length L_a of modelled specimens (cm) and body shape index Y	Back		Belly		Side		Maximum Reynolds number for live specimen (Re) and for model (Re_m) and flow velocity V_m over model (m/s)
	$\bar{x}$	$\bar{p}$	$\bar{x}$	$\bar{p}$	$\bar{x}$	$\bar{p}$	
Istiophorus	0.01	+0.06	0.01	+0.10	0.01	+0.09	$Re = 6.1 \times 10^7$
platypterus	0.10	+0.14	0.10	+0.10	0.11	+0.01	$Re_m = 7.6 \times 10^6$
(Show & Nodder)	0.20	0.00	0.22	**−0.08**	0.24	**−0.05**	$V_m = 6.0$
(EN)	0.26	**−0.12**	0.60	+0.02	0.55	−0.02	
$L_a = 183.0$	0.61	−0.01	0.96	−0.02	0.75	+0.03	
$Y = 0.22$	0.96	0.00	—	—	0.96	−0.03	
Stenopterygius	0.09	+0.14	0.24	−0.05	0.08	+0.09	$Re = 2.4 \times 10^7$
quadriscissus	0.16	+0.30	0.55	**−0.13**	0.16	+0.11	$Re_m = 6.0 \times 10^6$
Quenst. (EN)	0.25	+0.02	0.77	−0.03	0.29	−0.03	$V_m = 6.0$
$L_a = 210.0$	0.51	**−0.15**	1.00	+0.05	0.50	**−0.14**	
$Y = 0.55$	0.75	+0.04	—	—	0.74	0.00	
	0.90	−0.04	—	—	1.00	+0.19	
	1.00	+0.19	—	—	—	—	
Eudyptes	0.10	+0.10	0.10	+0.12	0.12	+0.06	$Re = 5.0 \times 10^6$
chrysolophus	0.18	+0.16	0.32	+0.12	0.21	+0.17	$Re_m = 4.0 \times 10^6$
Brandt (XN)	0.31	+0.13	0.56	**−0.12**	0.39	−0.04	$V_m = 4.0$
$L_a = 69.6$	0.51	**−0.10**	0.75	−0.02	0.58	**−0.18**	
$Y = 0.55$	0.91	+0.03	0.84	+0.02	0.92	+0.20	
	—	—	0.90	+0.04	—	—	
Pagophoca	0.06	+0.24	0.07	+0.20	0.07	0.00	$Re = 1.2 \times 10^7$
groenlandica	0.18	+0.17	0.16	+0.21	0.15	+0.24	$Re_m = 7.0 \times 10^6$
(Erxl.) (XN)	0.45	**−0.21**	0.37	**−0.18**	0.30	**−0.27**	$V_m = 6.0$
$L_a = 197.0$	0.69	−0.06	0.53	−0.12	0.44	−0.26	
$Y = 0.40$	0.99	0.00	0.82	−0.03	0.68	−0.08	
	—	—	0.98	0.00	0.96	+0.20	
Balaenoptera	0.15	+0.03	0.06	+0.30	0.08	0.00	$Re = 2.1 \times 10^8$
physalus	0.25	**−0.07**	0.17	0.00	0.20	−0.07	$Re_m = 8.0 \times 10^6$
(L.) (EN)	0.46	−0.05	0.38	**−0.17**	0.33	**−0.10**	$V_m = 8.0$
$L_a = 2500.0$	0.70	0.00	0.63	−0.06	0.73	+0.03	
$Y = 0.38$	0.92	−0.01	0.79	+0.01	0.91	+0.09	
	—	—	0.94	−0.02	—	—	
Physeter catodon	0.02	**−0.18**	0.02	+0.20	0.06	**−0.08**	$Re = 1.1 \times 10^8$
(L.) (EN)	0.11	+0.01	0.09	**−0.16**	0.17	−0.03	$Re_m = 8.0 \times 10^6$
$L_a = 1800.0$	0.32	−0.09	0.26	+0.02	0.34	−0.06	$V_m = 8.0$
$Y = 0.39$	0.48	−0.08	0.46	−0.12	0.67	−0.05	
	0.67	−0.01	0.63	−0.05	0.89	+0.12	
	0.86	+0.13	0.76	+0.07	—	—	

Table 13. (contd.)

Species, nektonic type, absolute length L_a of modelled specimens (cm) and body shape index Y	Back $\bar{x}$	Back $\bar{p}$	Belly $\bar{x}$	Belly $\bar{p}$	Side $\bar{x}$	Side $\bar{p}$	Maximum Reynolds number for live specimen (Re) and for model (Re_m) and flow velocity V_m over model (m/s)
Delphinus	0.07	+0.27	0.12	+0.03	0.07	+0.21	$Re = 2.7 \times 10^7$
delphis ponticus	0.14	+0.10	0.29	**−0.14**	0.25	−0.04	$Re_m = 6.3 \times 10^6$
Barab. (EN)	0.28	−0.12	0.47	−0.08	0.42	**−0.17**	$V_m = 6.0$
$L_a = 178.5$	0.43	**−0.19**	0.70	−0.04	0.72	+0.02	
$Y = 0.45$	0.71	+0.03	0.93	−0.04	0.90	+0.15	
	0.93	−0.02	—	—	—	—	
Phocoena phocoena	0.02	+0.33	0.04	+0.17	0.04	+0.22	$Re_m = 1.4 \times 10^7$
(L.) (EN)	0.08	+0.20	0.23	**−0.14**	0.14	0.00	$Re_m = 6.3 \times 10^6$
$L_a = 128.0$	0.18	+0.04	0.42	−0.11	0.36	**−0.16**	$V_m = 6.0$
$Y = 0.44$	0.41	**−0.18**	0.75	0.00	0.66	+0.02	
	0.66	+0.03	0.95	−0.04	0.91	+0.15	
	0.82	−0.02	—	—	—	—	
	0.94	+0.01	—	—	—	—	

movements accompanied by the expulsion of turbulent water into the boundary layer; while in animals with a paddling propulsor it is due to the periodic flapping of paired limbs (Figures 55, 56, 58, 61), etc. Thus, for example, during the propulsive undulations of the bodies of fishes the boundary layer on the leading (lateral) surface of the body, which is being drawn towards the stream, is always thinner, has higher flow speeds in it and has a longer laminar section than that on the trailing side, which at the given moment is being drawn away from the stream (Figures 72, 73). The attached flow over nektonic animals is due not only to their well-streamlined body shape (Aleyev 1955, 1957c, 1962, 1963a,c, 1964a, 1965b, 1969a, 1970b; Lang 1966a,b, and others) but also to its unsteadiness during movement (Gray 1936b; Lyapin 1971; Aleyev 1974). Changes in the shape of the nektonic animal's body during swimming facilitate the elimination of possible boundary-layer separation zones, and are particularly characteristic of axial undulatory and hydrojet propulsors (Aleyev 1973b, 1974). The boundary layer on cephalopods and fishes is strongly influenced by respiratory movements since these also alter the body's external shape. Important also are changes in swimming speed. The most advantageous locomotion regime for reducing friction drag is when short bursts alternate with long stretches of deceleration (Shebalov 1969). As the author's many years of observation and the results of special filming both in the natural environment and in biohydrodynamic channels show, this regime of locomotion is indeed characteristic of cephalopods, fishes, Chelonioidea, cetaceans and pinnipeds (Aleyev 1974).

3. Preventing boundary-layer separation

1. Slenderness Ratio and Laminarization of Body Shape

A. Slenderness ratio of body

One of the adaptations aimed at preventing boundary-layer separation common to nektonic animals is **the body's comparatively high slenderness**, i.e. the body's thickness ratio U does not exceed 0.40, which ensures smoother variation of the dynamic pressure along the longitudinal axis and a calmer confluence of the stream behind the body than would occur under the condition $U > 0.40$. It is this that facilitates a reduction in form drag for elongated bodies (Aleyev 1972a).

The degree of the animal's slenderness can be characterized in some measure also by the body's relative maximum diameter D_m:

$$D_m = d_m L_c^{-1}, \tag{29}$$

where d_m is the maximum diameter of the body, i.e. the body's maximum height H (in most Actinopterygii, in Hydrophidae, in many Cetacea, etc.) or its maximum width I (in Sagittoidea, Chelonioidea, Pinnipedia, in most sharks and Teuthoidea, etc.). The locations of the body's maximum height and maximum width on its longitudinal axis do not usually coincide in nektonic organisms (Houssay 1912; Magnan & Lariboisière 1914; Aleyev 1962, 1963a; Zuyev 1966, and others). Individual protuberances on the body, such as the rostral plane in *Sphyrna*, occur exceedingly seldom in nektonic animals and should be neglected when determining d_m, since they do not characterize the body's general shape.

Of the greatest importance in regard to body-shape laminarization is **the ratio of the sizes of the contractor and the diffuser**, since this determines the general distribution of dynamic pressure along the body's longitudinal axis and decisively influences, therefore, the entire pattern of the flow (Aleyev 1962). One can practically always assess the contractor-diffuser ration from the location of the maximum transverse diameter of the body, which may be expressed by means of the index Y of body shape (Aleyev 1962):

$$Y = yL_c^{-1}, \tag{30}$$

where y is the distance from the body's maximum transverse diameter d_m to the foremost (in the direction of travel) point of the effective body length. The greater the value of Y, the greater is the contraction area, i.e. the higher the degree to which the body shape facilitates the preservation of flow laminarity. Optimal in this regard are Y values between 0.40 and 0.55. Below, by a laminarized body shape in nektonic animals, we shall mean a body shape characterized by $Y \leqslant 0.40$. The index Y is a rather universal criterion and has allowed assessment of the hydrodynamic properties of the bodies of the most diverse nektonic animals: fishes (Aleyev 1962, 1963a,c, 1964a, 1965b; Aleyev & Vodyanitsky 1966; Ovchinnikov 1966a,b, 1967a), cephalopods (Zuyev 1964b,d, 1966), cetaceans (Chepurnov 1968), pinnipeds (Alexeyev 1966) and other animals.

The set of parameters U, Y, D_m and S_0 permits an all-round characterization of the main elements of the general shape of nektonic bodies which influence the magnitude of the total drag. The values of U, Y, D_m and S_0 which we obtained from formulae (4), (30), (29) and (5) for adult individuals of 110 nektonic and 7 planktonic species are shown in Table 14. For *Serratosagitta*, for five species of nektonic Teuthoidea (*Symplectoteuthis*, *Illex*, two species of *Todarodes* and *Loligo forbesi*) and for seven species of large cetaceans (*Globicephalus*, *Balaena*, *Physeter*, *Orcinus* and three species of *Balaenoptera*), we calculated the values of U, D_m and Y from the material available in the literature (Burfield 1927; Sleptsov 1952; Bourdelle & Grassé 1955; Rose 1957; Tomilin 1957, 1962a; Zuyev 1966; Chepurnov 1968). For the other 95 species we obtained all the indices by direct measurement of the animals.

The maximum swimming speeds necessary for calculating the Reynolds numbers shown in Table 14 were found mainly as a result of the author's own determinations by methods described in Chapter IV, but also from the literature (Murphy 1915; Scheffer 1950; Tomilin 1957; Barsukov 1960a; Herald 1962; Walters 1962; Nikolsky 1963; Zuyev 1964a, 1966; King 1964; Sukhanov 1964; Mordvinov 1968, and also the studies quoted in Table 7). We determined the swimming speeds of *Ostracion* and *Chaetodon* from ciné records. The swimming speed of the ichthyosaurus *Stenopterygius* was calculated tentatively on the basis of morphological data. The speed was determined tentatively also for 7 more of the 118 species under investigation, on the basis of indirect data for related forms. Thus the speed of locomotion of *Tetrapturus* is assumed to equal that of *Istiophorus* of the same size (Table 7), the speed of *Cyanea* to equal that of *Rhyzostoma* of the same size and the speed of *Serratosagitta serratodentata* to equal that of *Sagitta setosa* of the same size. The speed of movement of *Mola* is taken to be 100 cm/s, which approximately corresponds to the relative swimming speed of some Tetrodontoidei with an analogous type of propulsor, determined from ciné records. The speeds of *Trichiurus*, *Trachypterus* and *Regalecus* were determined on the basis of the author's observations of the movement of various anguilliform and ribbon-like fishes, and are assumed to equal two lengths per second for *Trichiurus*, one length per second for *Trachypterus* and half a length per second for *Regalecus*, which should be close to the maximum speeds of these species.

Figure 1 and Table 14, where the species are arranged in order of increasing Re, show that U and D_m in plankton vary over a rather wide range, whereas in nektonic animals they vary over a much narrower range determined by the condition $U \leqslant 0.40$. The optimal, i.e. lowest, C_D values correspond to some mean range of U for which the sum $C_{Dp} + C_{Df}$ turns out to be minimal (Figure 1). The mean value of U for eunekton (Figure 1, curve A) increases from 0.15 at $Re = 10^5$ to 0.19 at $Re = 2.0 \times 10^7$, after which the curve $U = f(Re)$ for eunekton proceeds practically horizontally until extreme values of Re for nektonic animals of the order of 5.0×10^8.

Table 14. Relative body thickness U, body shape index Y, maximum body diameter D_m, reduced specific surface S_0 and Reynolds numbers Re for planktonic and nektonic animals.

Species and nektonic type	Absolute length L_a of animal (cm)	U	Y	D_m	S_0	Re
Plankton						
Mesopodopsis slabberi (V. Bened.) (EP)	1.5	0.10	0.40	0.10	—	1.3×10^3
Serratosagitta serrato-dentata (Krohn) (EP)	1.0	0.06	0.53	0.07	—	4.5×10^3
Aurelia aurita L. (EP)	10.0	1.00	0.50	1.00	—	10^4
Rhyzostoma pulmo Macri (EP)	100.0	0.60	0.33	0.60	—	10^5
Cyanea capillata L. (EP)	200.0	1.00	0.40	1.00	—	2.0×10^5
Trachypterus taenia Bl. et Schn. (NP)	65.0	0.07	0.20	0.14	4.80	3.3×10^5
Regalecus glesne (Ascan.) (NP)	600.0	0.03	0.28	0.08	4.70	2.0×10^7
Nekton						
Sagitta setosa Müll. (PN)	2.2	0.06	0.50	0.06	3.77	10^4
Lebistes reticulatus Peters (BN)	2.5	0.23	0.46	0.25	—	1.5×10^4
Gambusia affinis holbrooki Girard (BN)	2.0	0.20	0.45	0.25	—	1.5×10^4
Xiphophorus maculatus Günth. (BN)	4.0	0.28	0.53	0.41	—	2.8×10^4
Ostracion tuberculatus (L.) (BN)	12.0	0.34	0.41	0.35	2.72	3.0×10^4
Mollienisia velifera Reg. (BN)	4.4	0.29	0.54	0.34	—	3.3×10^4
Leucaspius delineatus (Heck.) (PN)	6.0	0.17	0.44	0.22	—	4.2×10^4
Clupeonella delicatula delicatula (Nordm.) (PN/EN)	7.8	0.16	0.43	0.28	—	5.0×10^4
Gasterosteus aculeatus L. (BN)	9.0	0.20	0.49	0.28	—	6.3×10^4
Xiphophorus helleri Heck. (BN)	8.8	0.22	0.50	0.30	—	7.0×10^4
Gymnammodytes cicerellus (Raf.) (BN)	10.3	0.07	0.33	0.09	3.51	9.0×10^4
Atherina bonapartei Boulenger (PN/EN)	9.1	0.14	0.37	0.18	—	9.0×10^4
Chaetodon striatus L. (BN)	12.0	0.27	0.56	0.66	3.19	1.1×10^5
Atherina mochon pontica Eichw. (EN)	12.2	0.14	0.40	0.18	—	1.2×10^5
Sprattus sprattus phalericus (Risso) (PN/EN)	12.0	0.12	0.41	0.21	—	1.2×10^5
Atherina hepsetus L. (EN)	13.5	0.13	0.39	0.15	—	1.3×10^5
Carassius auratus (L.) (BN)	12.5	0.28	0.49	0.45	—	2.0×10^5

Table 14. (*contd.*)

Species and nektonic type	Absolute length L_a of animal (cm)	U	Y	D_m	S_0	Re
Spicara smaris (L.) (EN)	18.0	0.22	0.40	0.29	2.96	2.0×10^5
Carassius carassius (L.) (BN)	20.0	0.28	0.50	0.44	—	2.2×10^5
Alosa caspia nordmanni Antipa (EN)	16.5	0.16	0.49	0.26	—	2.2×10^5
Engraulis encrasicholus ponticus Alex. (EN)	14.0	0.16	0.45	0.16	2.92	2.5×10^5
Zeus faber pungio Val. (BN)	27.3	0.27	0.53	0.52	—	2.7×10^5
Scardinius erythrophthalmus (L.) (BN)	22.0	0.22	0.50	0.37	—	2.9×10^5
Sardina pilchardus (Walb.) (EN)	14.5	0.17	0.48	0.22	—	3.0×10^5
Labrus viridis L. (BN)	32.1	0.21	0.48	0.30	—	3.7×10^5
Perca fluviatilis L. (BN)	24.0	0.22	0.41	0.34	—	4.0×10^5
Sciaena umbra L. (BN)	28.0	0.22	0.40	0.34	—	4.2×10^5
Serranus scriba (L.) (BN)	22.0	0.20	0.46	0.32	2.97	4.4×10^5
Rutilus rutilus rutilus (L.) (BN)	24.0	0.20	0.48	0.30	—	4.8×10^5
Alosa kessleri pontica (Eichw.) (EN)	27.9	0.16	0.49	0.23	2.93	5.0×10^5
Leuciscus leuciscus (L.) (BN)	20.2	0.18	0.45	0.24	—	5.4×10^5
Clupea harengus pallasi Val. (EN)	28.0	0.17	0.50	0.24	3.03	5.6×10^5
Odontogadus merlangus euxinus (Nordm.) (BN)	47.1	0.18	0.40	0.22	2.96	8.0×10^5
Abramis brama (L.) (BN)	60.0	0.23	0.54	0.41	2.99	9.0×10^5
Cyprinus carpio carpio L. (BN)	50.0	0.21	0.47	0.31	—	10^6
Belone belone euxini Günth. (EN)	50.4	0.05	0.56	0.07	3.95	10^6
Enhydrina schistosa (Daud.) (EN)	90.0	0.03	0.46	0.04	3.54	1.1×10^6
Umbrina cirrosa (L.) (BN)	42.0	0.22	0.43	0.31	—	1.2×10^6
Salmo trutta labrax m. fario L. (BN)	25.0	0.15	0.40	0.23	—	1.2×10^6
Anguilla anguilla (L.) (EN)	100.8	0.05	0.32	0.07	3.46	1.5×10^6
Mugil auratus Risso (BN)	41.4	0.19	0.47	0.22	2.80	1.5×10^6
Mugil saliens Risso (BN)	40.0	0.18	0.47	0.20	2.80	1.5×10^6
Gadus morhua macrocephalus Til. (BN)	72.0	0.18	0.36	0.20	—	1.8×10^6
Mola mola (L.) (EN)	200.0	0.30	0.50	0.55	3.06	2.0×10^6
Loligo vulgaris Lam. (EN)	26.5	0.14	0.50	0.14	2.68	2.0×10^6
Todarodes sagittatus (Lam.) (EN)	48.3	0.14	0.53	0.15	—	2.0×10^6
Symplectoteuthis oualaniensis (Less.) (EN)	45.2	0.14	0.53	0.14	2.68	2.0×10^6

Table 14.(contd.)

Species and nektonic type	Absolute length L_a of animal (cm)	U	Y	D_m	S_0	Re
Todarodes pacificus (Steenst.) (EN)	40.2	0.11	0.49	0.13	—	2.0×10^6
Loligo forbesi Steenst. (EN)	33.3	0.14	0.50	0.14	—	2.0×10^6
Illex coindeti (Verany) (EN)	26.1	0.12	0.56	0.12	—	2.0×10^6
Acipenser stellatus Pall. (BN)	140.0	0.12	0.46	0.13	3.43	2.0×10^6
Sphyraena sphyraena L. (EN)	42.1	0.10	0.55	0.12	—	2.0×10^6
Stizostedion lucioperca (L.) (BN)	60.0	0.17	0.43	0.22	—	2.4×10^6
Hirundichthys rondeletii (Cuv. et Val.) (EN)	24.0	0.14	0.51	0.16	3.00	2.5×10^6
Chelonia mydas (L.) (XN)	154.0	0.30	0.43	0.48	2.61	3.0×10^6
Eretmochelys imbricata (L.) (XN)	71.0	0.33	0.49	0.54	—	3.0×10^6
Caretta caretta (L.) (XN)	126.0	0.37	0.49	0.59	—	3.0×10^6
Salmo trutta L. (BN)	60.5	0.17	0.45	0.22	—	3.0×10^6
Scomber scombrus L. (EN)	50.0	0.18	0.50	0.20	2.69	3.0×10^6
Trachurus mediterraneus ponticus Aleev (EN)	56.0	0.18	0.45	0.20	2.80	3.4×10^6
Mugil cephalus L. (BN)	70.5	0.20	0.52	0.25	—	4.2×10^6
Squalus acanthias L. (EN)	120.0	0.14	0.25	0.14	3.90	4.2×10^6
Acipenser güldenstädti colchicus V. Marti (BN)	150.0	0.16	0.33	0.16	3.24	4.5×10^6
Trichiurus lepturus L. (EN)	150.0	0.03	0.15	0.05	4.35	4.5×10^6
Pygoscelis adeliae (Hombron and Jacquinot)(XN)	70.0	0.23	0.53	0.24	—	5.0×10^6
Eudyptes chrysolophus Brandt (XN)	69.6	0.24	0.55	0.25	2.64	5.0×10^6
Auxis thazard (Lac.) (EN)	52.1	0.20	0.46	0.23	2.62	5.0×10^6
Pomatomus saltatrix (L.) (EN)	99.0	0.17	0.50	0.22	2.96	6.0×10^6
Huso huso (L.) (BN)	200.0	0.17	0.36	0.18	—	8.0×10^6
Dermochelys coriacea L. (XN)	160.0	0.32	0.43	0.40	—	10^7
Esox lucius L. (BN)	103.4	0.14	0.57	0.17	2.91	10^7
Sarda sarda (Bl.) (EN)	73.2	0.20	0.49	0.24	2.77	10^7
Pagophoca groenlandica (Erxl.) (XN)	197.0	0.22	0.40	0.25	2.71	1.2×10^7
Salmo salar L. (BN)	111.5	0.19	0.46	0.21	—	1.2×10^7
Salmo trutta labrax Pall. (BN)	98.8	0.17	0.44	0.23	2.95	1.2×10^7
Phocoena phocoena (L.) (EN)	128.0	0.21	0.44	0.21	2.69	1.4×10^7
Arctocephalus pusillus (Schreb.) (XN)	200.0	0.19	0.50	0.23	2.76	1.6×10^7
Thunnus alalunga (Bonnat.) (EN)	89.0	0.22	0.46	0.27	2.65	1.6×10^7
Sphyraena barracuda (Walb.) (EN)	130.0	0.13	0.60	0.15	2.92	2.0×10^7

Table 14. (contd.)

Species and nektonic type	Absolute length L_a of animal (cm)	U	Y	D_m	S_0	Re
Sphyrna zygaena (L.) (EN)	199.0	0.16	0.43	0.16	—	2.0×10^7
Stenopterygius quadriscissus Quenst. (EN)	210.0	0.19	0.55	0.22	2.85	2.1×10^7
Delphinus delphis ponticus Barab. (EN)	178.5	0.21	0.45	0.21	2.70	2.7×10^7
Scomberomorus commersoni (Lac.) (EN)	120.0	0.15	0.55	0.20	2.89	3.0×10^7
Thunnus obesus (Lowe) (EN)	195.0	0.25	0.38	0.29	—	3.0×10^7
Thunnus albacora (Lowe) (EN)	150.0	0.21	0.40	0.26	—	3.1×10^7
Acanthocybium solandri (Cuv. et Val.) (EN)	150.0	0.12	0.43	0.15	—	3.2×10^7
Coryphaena hippurus L. (EN)	181.0	0.16	0.38	0.23	3.40	3.4×10^7
Prionace glauca (L.) (EN)	403.0	0.17	0.25	0.18	2.94	4.0×10^7
Globicephalus melas (Traill) (EN)	450.0	0.21	0.34	0.21	—	5.1×10^7
Tetrapturus belone Raf. (EN)	175.5	0.08	0.20	0.10	—	5.2×10^7
Tursiops truncatus (Montagu) (EN)	350.0	0.20	0.40	0.20	—	5.4×10^7
Istiophorus platypterus (Show & Nodder) (EN)	183.0	0.10	0.22	0.12	3.46	6.1×10^7
Thunnus thynnus (L.) (EN)	300.0	0.23	0.43	0.28	2.64	7.5×10^7
Balaena mysticetus L. (EN)	2000.0	0.27	0.47	0.30	2.56	10^8
Xiphias gladius L. (EN)	356.0	0.19	0.31	0.22	3.15	1.1×10^8
Physeter catodon L. (EN)	1800.0	0.19	0.39	0.20	2.70	1.1×10^8
Orcinus orca (L.) (EN)	800.0	0.26	0.40	0.26	—	1.2×10^8
Makaira indica (Cuv. et Val.) (EN)	450.0	0.18	0.25	0.23	2.96	1.6×10^8
Balaenoptera physalus L. (EN)	2500.0	0.16	0.38	0.17	2.71	2.1×10^8
Balaenoptera musculus L. (EN)	2750.0	0.17	0.40	0.18	—	2.8×10^8
Balaenoptera borealis Less. (EN)	2000.0	0.15	0.36	0.16	—	3.0×10^8

It is interesting to note that the slenderness ratios typical of eunekton (0.15–0.19) are close to the slenderness ratios ($U\approx0.15$) believed to be optimal (Gerasimov & Droblenkov 1962) for modern submarines (Figure 1a).

It should be noticed that in the range $Re>2.0\times10^7$ we see animals with the scombroid type of axial undulatory propulsor only, which fully corresponds to the higher relative power of this propulsor and its capacity to ensure high speeds. True, it has not been ruled out that the lower part of this range, approximately up to Re of the order of 5.0×10^7, includes also some Teuthoidea with a hydrojet propulsor.

Table 14 makes it possible to assess the magnitude of the reduced specific body surface area characteristic of nekton. In all the nektonic animals studied $S_0 < 4.50$; for eunekton and nekton as a whole the most characteristic S_0 range is from 2.60 to 3.50. In planktonic *Regalecus* and *Trachypterus* $S_0 > 4.50$; in *Trachypterus* $S_0 = 5.01$, which in the range $Re > 5.0 \times 10^5$ is probably close to the upper limit for swimming animals.

B. Laminarization of the body shape

All nektonic animals have a well-streamlined monolythic body with smooth flowing contours, devoid of any angular projections or depressions. At the front the nektonic body is in most cases bluntly rounded (Sagittoidea, most fishes, Hydrophidae, Mystacoceti, Physeteridae, Sirenia Pinnipedia, etc.) or has a somewhat pointed snout (squids, and among the fishes Sphyraenidae, most of the Scombridae, etc., most of the Ichthyosauria and Delphinidae, etc.), while it smoothly tapers at the rear into a more or less elongated cone (squids) or a pointed rib: the tail fin (Sagittoidea, all fishes and fish-like animals, Ichthyosauria, Sirenia and Cetacea). This is the shape characteristic of bodies over which fluid flows without boundary-layer separation. Many investigators have drawn attention to these specific features of the body shape of various nektonic animals (Parsons 1888; Abel 1912; Houssay 1912; Gregory 1928; Schmassmann 1928; Magnan & Sainte-Laguë 1928a,b, 1929a,b, 1930; Howell 1930; Le Danois 1939–1943; Aleyev 1955, 1957a–c, 1962, 1963a,c, 1964a, 1965b, 1969c, 1970b, 1972a,b, 1974; Tomilin 1957, 1962a–c, 1965; Greenway 1965; Zuyev 1966; Burdak 1968; Tweedie 1969; Matthews 1970, and others).

With increasing *Re* the shape of the body optimal for reducing resistance to movement changes essentially both in phylogeny and ontogeny, and the specific features of the nektonic body shape aimed at reducing the total drag change accordingly.

At $Re < 10^7$, one of the main devices for reducing hydrodynamic drag in nektonic animals is laminarization of the body shape which results in simultaneous reduction of form drag (on account of flow without separation) and friction drag (on account of preservation of boundary-layer laminarity over a greater or lesser part of the body surface). According to our investigations, laminarization of the body shape is ensured morphologically in nektonic animals mainly by two specific features: 1. by the location of the body's maximum cross-section close to the middle of its longitudinal axis and 2. by the absence of any notable roughness on the body.

The Y values for the animals tested vary from 0.15 to 0.60 (Table 14). The curve $Y = f(Re)$ for eunekton (Figure 91A) has distinct ascending and descending parts, with a maximum ($Y = 0.50$) at $Re = 5.0 \times 10^6$. In the interval $10^5 \leqslant Re \leqslant 6.0 \times 10^7$ this curve lies in the range $Y > 0.40$, which points to the predominance of laminarized body shapes within this interval. Generally speaking, laminarized body shapes occur among nektonic animals in a wider range of *Re*: approximately from 10^3 to 2.8×10^8 (Table 14, Figures 91, 92). It has been noted (Walters 1962) that among Scombridae

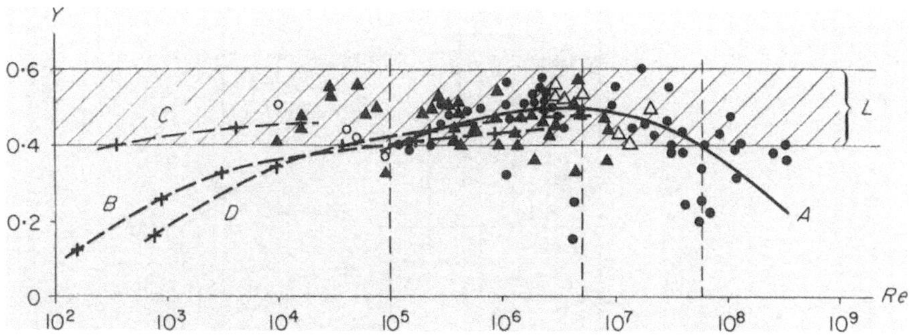

Figure 91. Y versus Re for nektonic animals. Each point corresponds to an individual species. *A*, eunekton, phylogeny; *B-D*, ontogeny. *B*, *Trachurus mediterraneus ponticus* Aleev; *C*, *Lebistes reticulatus* Peters; *D*. *Mugil saliens* Risso. *L*, zone of laminarized profiles. ●, eunekton; ○, planktonekton; ▲, benthonekton; △, xeronekton; +, eunekton and benthonekton, ontogeny. See text.

s.l. the smaller representatives have a laminarized body shape and apparently swim in the laminar regime, whereas large tuna fish (*Thunnus*, etc.) have a body shape which is closer to a non-laminarized one. However, in *Thunnus thynnus* $Y=0.43$, while in *Th. alalunga* $Y=0.46$ (Table 14), i.e. they have a laminarized body shape.

Among eunektonic forms the sharks are in a somewhat different position as regards the value of Y and usually have $Y<0.30$, since d_m represents the maximum width of the head. In sharks the wider frontal part of the body is connected with its functioning as a frontal horizontal rudder and also with the horizontal situation of the wide mouth slit.

With an increase in D_m the average Y values in nektonic animals somewhat increase (Figure 93), which may be directed at improving flow over the contraction area of the body.

Variations in U, D_m and Y, which we examined in 20 species of fishes and cetaceans, are given in Table 15. This table also gives data for four squid species from a paper (Zuyev 1966) which furnishes average body diameters, found as the mean of the body's maximum height H and maximum width I: having expressed these magnitudes in fractions of L_a we derived figures close to U (since the cross-sectional shape of a squid's body is nearly circular); just as in Table 14, the magnitudes of D_m are I values.

Since in the range $L_a<450$ cm the maximum swimming speed attainable by a nektonic animal is directly proportional to its length, within this range adaptation takes place in the ontogeny of all nektonic animals to movement at ever increasing speeds (Aleyev 1957b). This may be clearly traced in all eunektonic forms, particularly in fishes (Aleyev 1957b, 1962, 1963a, 1964a, and others) and Cephalopoda (Zuyev 1964d, 1966), but is not so pronounced in dolphins (Table 15, Figure 94),

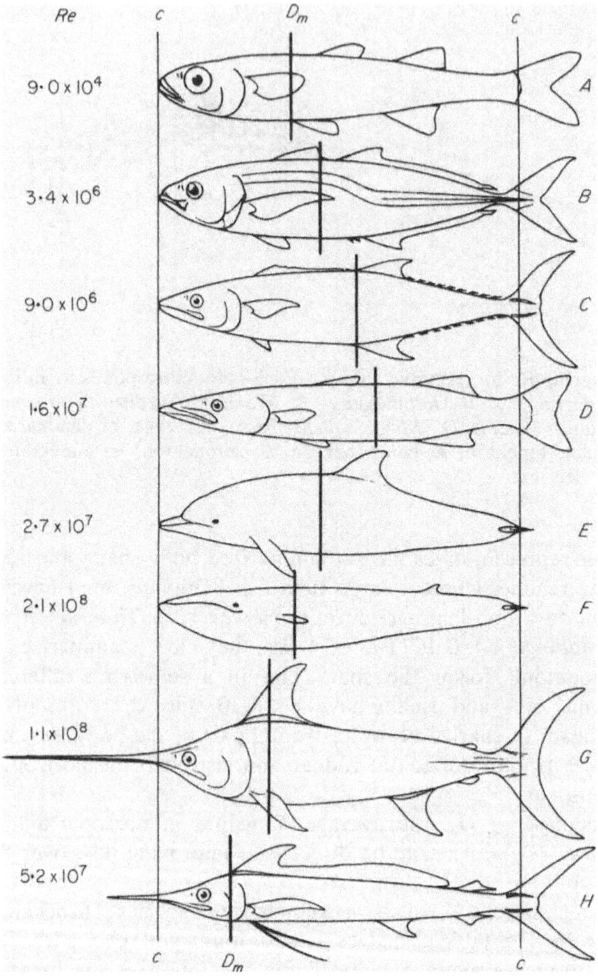

Figure 92. Position of maximum body diameter (D_m) in nektonic animals with different *Re* values. *c*, vertical lines through ends of effective body length L_c. *A*, *Atherina bonapartei* Boulenger, absolute length $L_a = 9.1$ cm; *B*, *Trachurus mediterraneus ponticus* Aleev, $L_a = 56.0$ cm; *C*, *Scomberomorus commersoni* (Lac.), $L_a = 68.7$ cm; *D*, *Sphyraena barracuda* (Walb.), $L_a = 130.0$ cm; *E*, *Delphinus delphis ponticus* Barab., $L_a = 178.5$ cm; *F*, *Balaenoptera physalus* (L.), $L_a = 2500$ cm; *G*, *Xiphias gladius* L., $L_a = 356.0$ cm; *H*, *Tetrapturus belone* Raf., $L_a = 175.5$ cm.

which is explained by their comparatively large linear dimensions and small variability of Re with age.

In the ontogeny of Teuthoidea U and D_m diminish, while Y increases (Table 15, Figure 94), which is directed at the formation of a more elongated, laminarized body shape and is conducive to higher swimming speeds.

The prelarvae of fishes with a yolk sac have high U and D_m values on account of the yolk sac being large (Table 15, Figures 94, 95). After the reabsorption of the yolk sac U, D_m and Y steeply decrease, which accords with the better functioning of the larva's anguilliform propulsor. Then, as the larva turns into a small fish, fishes with a scombroid propulsor show a more or less steep increase in Y, corresponding to the growing speeds of locomotion, and a slight increase in U and D_m, connected with the destruction of the parachute system and the anguilliform propulsor and the formation of a scombroid propulsor (Figure 94, *Sprattus*). In the case of early transition of the larva to fish eating we observe a considerable increase in the relative size of the jaw apparatus and a related temporary increase in D_m, which reaches a maximum by the time the larva turns into a small fish (Figure 94, *Trachurus*; Figure 95). Fishes which as adults have a large ventral keel as a rule show a zero or negligible decrease in D_m in ontogeny (Table 15, Figure 94, *Sprattus*). The reduced specific surface S_0 as a rule diminishes in the ontogeny of nektonic fishes, which

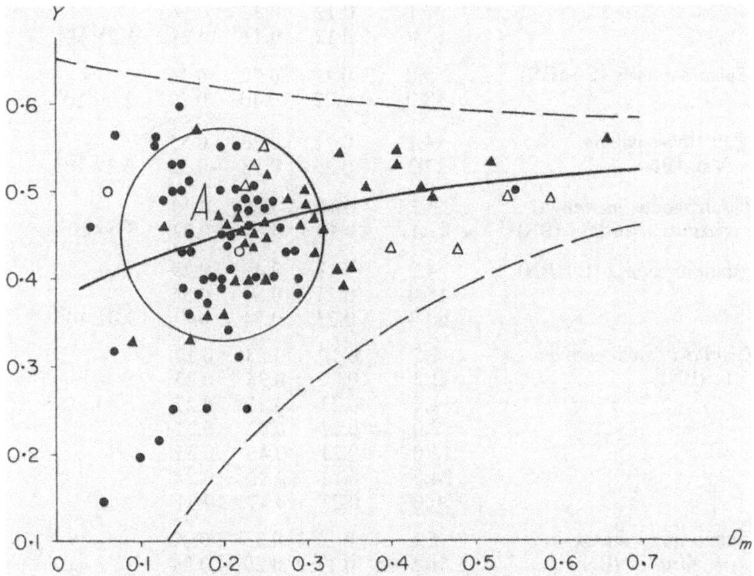

Figure 93. Y versus D_m for nektonic animals. Each point corresponds to an individual species. ●, eunekton; ○, planktonekton; ▲, benthonekton; △, xeronekton, Circle *A* corresponds to region, characteristic of eunekton.

Table 15. Variation of relative body thickness U, body shape index Y, maximum body diameter D_m and Reynolds number Re in nektonic animals with age.

Species and nektonic type	Absolute length L_a of animal (cm)	U	Y	D_m	Re
Lebistes reticulatus Peters (BN)	0.8	0.21	0.40	0.22	4.0×10^2
	1.6	0.23	0.45	0.24	4.0×10^3
	2.5	0.23	0.46	0.25	1.5×10^4
Clupeonella delicatula delicatula (Nordm.) (PN/EN)	3.0	0.13	0.35	0.20	—
	5.2	0.15	0.43	0.26	—
	7.8	0.16	0.43	0.28	5.0×10^4
Atherina mochon pontica Eichw. (EN)	1.9	0.14	0.33	0.15	1.6×10^3
	12.2	0.14	0.40	0.18	1.2×10^5
Sprattus sprattus phalericus (Risso) (PN/EN)	0.2	0.27	0.22	0.24	—
	0.45	0.15	0.18	0.17	—
	1.3	0.06	0.07	0.07	—
	2.6	0.10	0.13	0.10	—
	3.0	0.11	0.23	0.14	—
	4.7	0.11	0.34	0.17	—
	7.1	0.12	0.38	0.19	—
	12.0	0.12	0.41	0.21	1.2×10^5
Spicara smaris (L.) (EN)	5.0	0.19	0.38	0.25	—
	18.0	0.22	0.40	0.29	2.0×10^5
Zeus faber pungio Val. (BN)	14.1	0.27	0.50	0.55	—
	27.3	0.26	0.53	0.52	2.2×10^5
Odontogadus merlangus euxinus (Nordm.) (BN)	5.7	0.16	0.26	0.20	—
	47.1	0.18	0.40	0.22	8.0×10^5
Abramis brama (L.) (BN)	4.0	0.19	0.49	0.29	—
	15.0	0.21	0.54	0.36	—
	60.0	0.23	0.54	0.41	9.0×10^5
Cyprinus carpio carpio L. (BN)	1.1	0.19	0.28	0.17	—
	1.3	0.20	0.35	0.23	—
	1.7	0.21	0.37	0.27	—
	3.0	0.21	0.42	0.31	—
	12.0	0.21	0.46	0.35	—
	24.0	0.21	0.46	0.34	—
	50.0	0.21	0.47	0.31	10^6
Salmo trutta labrax m. fario L. (BN)	8.5	0.16	0.35	0.22	—
	16.5	0.17	0.39	0.24	—
	25.0	0.18	0.40	0.23	1.2×10^6
Mugil saliens Risso (BN)	0.8	0.19	0.17	0.23	8.0×10^2
	1.6	0.18	0.34	0.21	9.6×10^3

Table 15. (contd.)

Species and nektonic type	Absolute length L_a of animal (cm)	U	Y	D_m	Re
Mugil saliens Risso (BN)	3.5	0.17	0.40	0.20	4.2×10^4
	9.0	0.17	0.44	0.20	1.8×10^5
	40.0	0.18	0.47	0.20	1.5×10^6
Scomber scombrus L. (EN)	13.5	0.15	0.40	0.18	—
	16.8	0.15	0.43	0.18	—
	38.0	0.18	0.50	0.20	1.5×10^6
Loligo vulgaris Lam. (EN)	0.1–3.0	0.38	0.34	0.38	—
	3.1–6.0	0.19	0.38	0.19	—
	9.1–12.0	0.16	0.44	0.17	—
	24.1–27.0	0.14	0.50	0.15	2.0×10^6
	39.1–42.0	0.13	0.50	0.14	—
Symplectoteuthis oualaniensis (Less.) (EN)	0.1–3.0	0.23	0.38	0.23	—
	3.1–6.0	0.18	0.43	0.18	—
	6.1–9.0	0.15	0.45	0.15	—
	24.1–27.0	0.12	0.51	0.12	—
	42.1–45.0	0.14	0.53	0.14	2.0×10^6
Loligo forbesi Steenst. (EN)	0.1–3.0	0.23	0.36	0.24	—
	3.1–6.0	0.19	0.40	0.20	—
	9.1–12.0	0.16	0.46	0.17	—
	12.1–15.0	0.15	0.48	0.15	—
	33.1–36.0	0.14	0.50	0.15	2.0×10^6
Illex coindeti (Verany) (EN)	0.1–3.0	0.22	0.37	0.22	—
	3.1–6.0	0.17	0.40	0.18	—
	6.1–9.0	0.15	0.44	0.15	—
	9.1–12.0	0.14	0.49	0.15	—
	24.1–27.0	0.12	0.56	0.12	2.0×10^6
Acipenser stellatus Pall. (BN)	3.5	0.37	0.21	0.18	—
	24.0	0.12	0.47	0.13	—
	140.0	0.12	0.46	0.13	2.0×10^6
Trachurus mediterraneus ponticus Aleev (EN)	0.18	0.37	0.30	0.39	—
	0.23	0.18	0.13	0.20	1.5×10^2
	0.30	0.18	0.15	0.22	—
	0.52	0.18	0.21	0.29	—
	0.90	0.18	0.26	0.31	9.0×10^2
	1.6	0.17	0.32	0.28	4.4×10^3
	3.2	0.16	0.38	0.25	—
	10.5	0.16	0.41	0.23	1.4×10^5
	27.0	0.17	0.43	0.23	8.0×10^5
	56.0	0.18	0.45	0.23	3.4×10^6

Table 15. (*contd.*)

Species and nektonic type	Absolute length L_a of animal (cm)	U	Y	D_m	Re
Squalus acanthias L. (EN)	25.1	0.14	0.25	0.15	2.0×10^5
	47.9	0.13	0.23	0.14	9.1×10^5
	120.0	0.13	0.25	0.11	4.2×10^6
Acipenser güldenstädti colchicus V. Marti (BN)	3.7	0.18	0.26	0.21	—
	39.0	0.15	0.33	0.16	—
	98.0	0.15	0.33	0.16	—
	150.0	0.16	0.33	0.16	4.5×10^6
Sarda sarda (Bl.) (EN)	14.8	0.17	0.34	0.21	—
	49.0	0.18	0.49	0.21	—
	73.2	0.20	0.49	0.24	5.0×10^6
Pomatomus saltatrix L. (EN)	5.8	0.16	0.30	0.23	—
	12.1	0.18	0.46	0.27	—
	16.9	0.19	0.48	0.28	—
	24.7	0.19	0.49	0.28	—
	55.0	0.18	0.50	0.24	2.7×10^6
	94.1	0.18	0.50	0.24	8.0×10^6
Delphinus delphis ponticus Barab. (EN)	90.0	0.20	0.44	0.20	2.0×10^6
	120.0	0.21	0.45	0.21	—
	178.5	0.21	0.45	0.21	2.7×10^7
Xiphias gladius L. (EN)	57.4	0.12	0.30	0.15	—
	184.5	0.17	0.31	0.22	—
	356.0	0.19	0.31	0.22	1.1×10^8

is due to the elimination of planktonic adaptations and the progressive development of nektonic features of body shape. Thus, in *Mugil saliens*, at $L_a = 1.0$ cm and $Re = 8.0\times10^2$, $S_0 = 2.96$, whereas at $L_a = 40.0$ cm and $Re = 1.5\times10^6$, $S_0 = 2.80$; in *Trachurus mediterraneus ponticus*, at $L_a = 1.0$ cm and $Re = 9.0\times10^2$, $S_0 = 3.92$, and at $L_a = 56.0$ cm and $Re = 3.4\times10^6$, $S_0 = 2.80$. Apparently the only exceptions to this general rule are some nektonic Elasmobranchii, which have no planktonic stage of development.

In the range $L_a > 450$ cm, with the growth of the animals' absolute dimensions their maximum speed diminishes, owing to which adaptation to increasing swimming speeds no longer takes place in the ontogeny of nektonic animals in this range. A case in point is the ontogenetic development of whales, for which it has been shown (Chepurnov 1968) that with age they undergo no morphological changes towards improving the body's hydrodynamic properties.

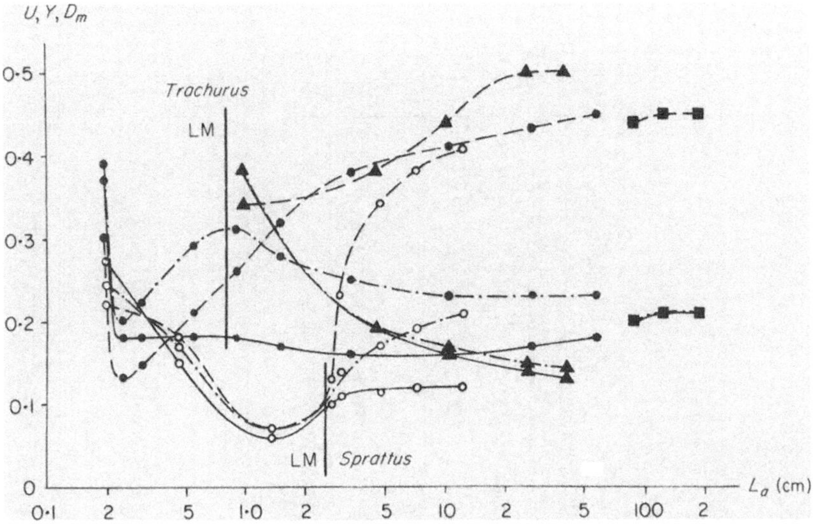

Figure 94. U versus L_a (continuous line), Y versus L_a (dashed line) and D_m versus L_a (dash-dot line) for the ontogeny of nektonic animals. ▲, *Loligo vulgaris* Lam.; ○, *Sprattus sprattus phalericus* Risso; ●, *Trachurus mediterraneus ponticus* Aleev; ■, *Delphinus delphis ponticus* Barab. Continuous vertical lines labelled LM correspond to the transition of larva into small fish.

C. Fairings, deflectors, filters

Nektonic animals have no bluff protuberances or depressions on the surface of the body. In species adapted to rapid movement all even slightly protruding structures (the mantle fins of Teuthoidea, the eyes of vertebrates, the fins or fishes, jaws, etc.) are usually covered by special **fairings**, or hidden in special recesses, grooves, etc., from which they extend only when in use. Thus, for example, Teuthoidea have a depression at the end of the mantle which receives the mantle fins when they are pressed to the body (Figure 63). The eyes of fast-swimming fishes are usually provided with special fairings, adipose eyelids, and the dorsal, pectoral and pelvic fins fold into special grooves or recesses behind special fairings made of transformed scales (Aleyev 1955; Burdak 1957); the caudal peduncle is covered by a special fairing made of skin or elongated scales: alae (Aleyev 1959d). Characteristically, adipose eyelids in ontogeny and phylogeny appear initially on the rearmost part of the eyes (*Caranx hippos*, *Trachurus*, etc.), i.e. on the diffusion part of the protruding structure (Figures 70A, 95) where vortices are more likely to appear (Aleyev 1955, 1957c). Along with their hydrodynamic function, adipose eyelids, being composed of parallel fibrils, polarize light and increase the accommodating capacity of the eye (Stewart 1962). The streamlining of the cross-section of the caudal peduncle in fast

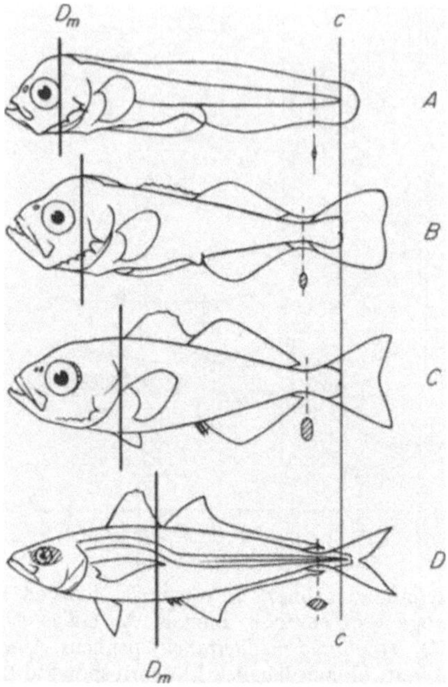

Figure 95. Variation in position of maximum cross-sectional body diameter D_m in the ontogeny of *Trachurus mediterraneus ponticus* Aleev. Fatty eyelids and cross-sections of caudal peduncle also shown. *c*, vertical line through end of effective body length. Length of fish to end of vertebral column (cm): *A*, 0.28; *B*, 0.68; *C*, 1.0; *D*, 43.3.

fishes and cetaceans is improved by means of tail keels (Walters 1962; Aleyev 1963a; see also Figure 46); in ontogeny these keels undergo progressive development (Figure 95), which corresponds to increasing speeds of movement.

In many fast-swimming fishes in whom the speed of transverse propulsive movements of the posterior part of the body is especially great (Scombridae, Xiphiidae, Istiophoridae, many Carangidae, etc.), small, separately set finlets appear behind the second dorsal and anal fins, forming completely or partially separate rays of these fins (Figures 92C,G,H, 96A,B,D,E, 97, 98A,B). Of similar form and relative size is the so-called adipose fin (Salmonoidei, Stomiatoidei, Scopeliformes, Characidae, etc.; see Figures 66C, 134A–C), as well as small second dorsal and anal fins found on the upper and lower edges of the caudal peduncle of some sharks (Figures 17A,B); their base becomes shorter and their rear edge longer, which makes the fin look like a little flag. The basic function of all these small finlets and separate rearmost rays of the dorsal and anal fins is to control the flow into the caudal-fin area (Walters 1962; Aleyev 1963a; Sheer 1964). When the caudal peduncle and fin are performing propulsive movements the stream flowing over them is mostly not strictly parallel,

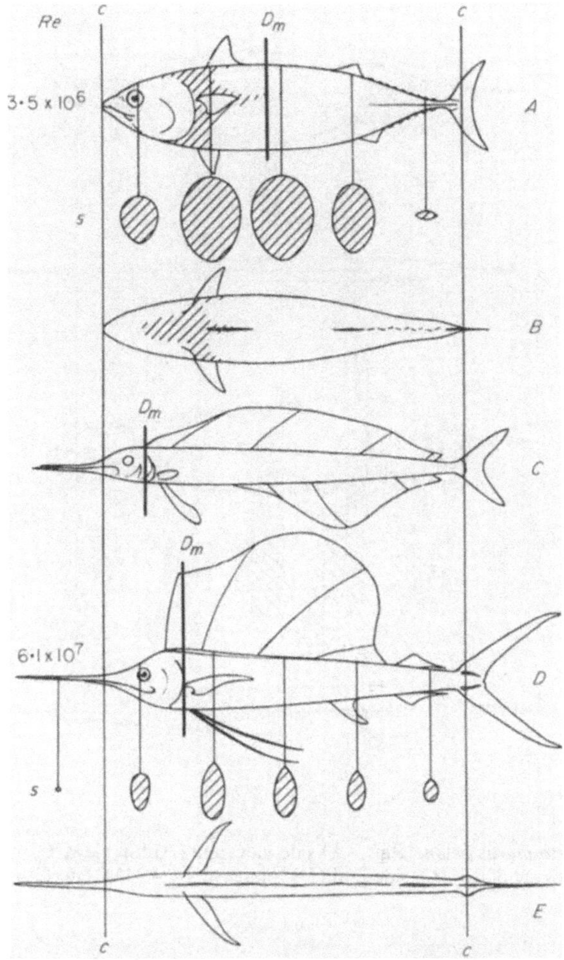

Figure 96. *A*, *B*, *Auxis thazard* (Lac.), (*A*) side view and (*B*) top view, $L_a = 42.1$ cm, surfaces covered by scales hatched; *C*, *Palaeorhynchus glarisianus* Blainville, $L_a = 60.0$ cm (after Woodward 1901, with modifications); *D*, *E*, *Istiophorus platypterus* (Show and Nodder), (*D*) side and (*E*) top view, $L_a = 183.0$ cm. *s*, body cross-section; *c*, vertical lines through ends of effective body length.

but is directed at some angle to the median plane of the fish, owing to which a more-or-less turbulent stream inevitably flows over the upper and lower edges of the caudal peduncle. The small finlets situated on these edges freely rotate to the right and left, forming together a screen capable of breaking up the vortices in the turbulent boundary layer, which helps to straighten the stream flowing on to the

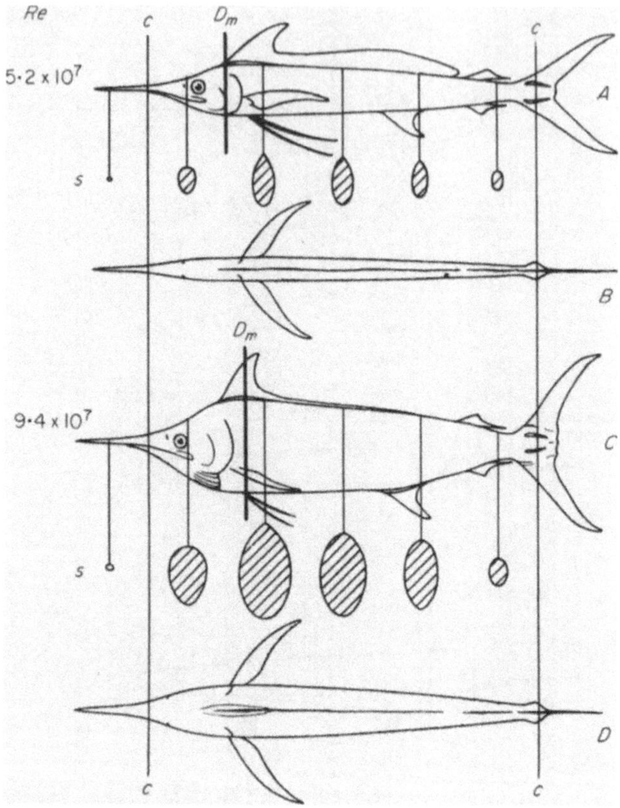

Figure 97. *A, B, Tetrapturus belone* Raf., (*A*) side view and (*B*) top view, $L_a = 175.5$ cm; *C, D, Makaira indica* (Cuv. et Val.), (*C*) side and (*D*) top view, $L_a = 245.0$ cm. Other notation as in figure 96.

caudal fin (Figure 99) and enhances, as a result, the effectiveness of the caudal fin as a propulsor. The above-mentioned finlets perform in this sense the same function as is carried out by special **damping screens and filters** in wind tunnels.

In many cases fast fishes (Scombridae, Istiophoridae and some sharks) have paired keels on each side of the caudal fin (Figure 100). These fulfil the role of **deflectors**, facilitating stricter rectilinear flow over the caudal fin and thus reducing induced drag (Aleyev 1963a). Since the rear ends of the keels are drawn together slightly, the stream passing between them is accelerated (Walters 1962), so that it is capable of travelling without separation for a greater distance than the stream flowing over the contractor, i.e. above and beneath it. This is why many fishes have a protuberance in the middle of the fin membrance opposite the end of the contractor (Figures 92, 93,

97, 100A): this is the spot where the potential vortex formation zone shifts backwards, and the zone of flow without separation over the caudal fin is therefore extended.

In a number of cases, also functioning as deflector/separators in nektonic animals are longitudinal crests or folds on the body, as a rule on its more convex side

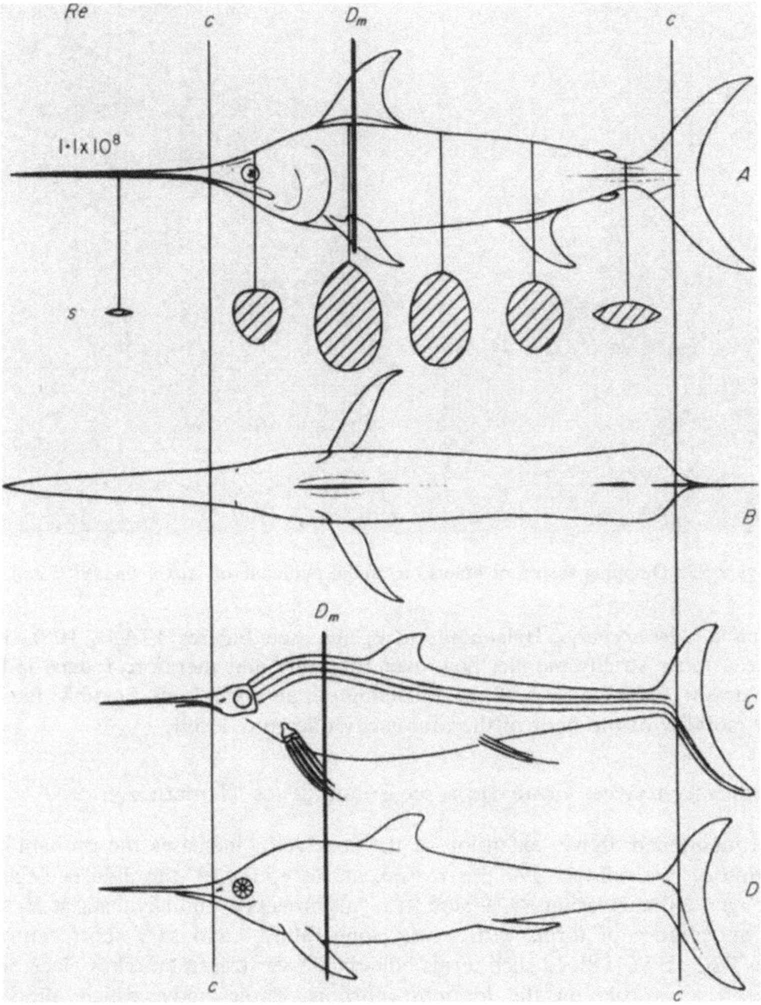

Figure 98. A, B, Xiphias gladius L., (*A*) side and (*B*) top view, $L_a = 356.0$ cm; *C, D, Eurhinosaurus longirostris* Jaeger, (*A*) after Huene (1956, with modifications), and (*D*) author's reconstruction, $L_a = 500$ cm.

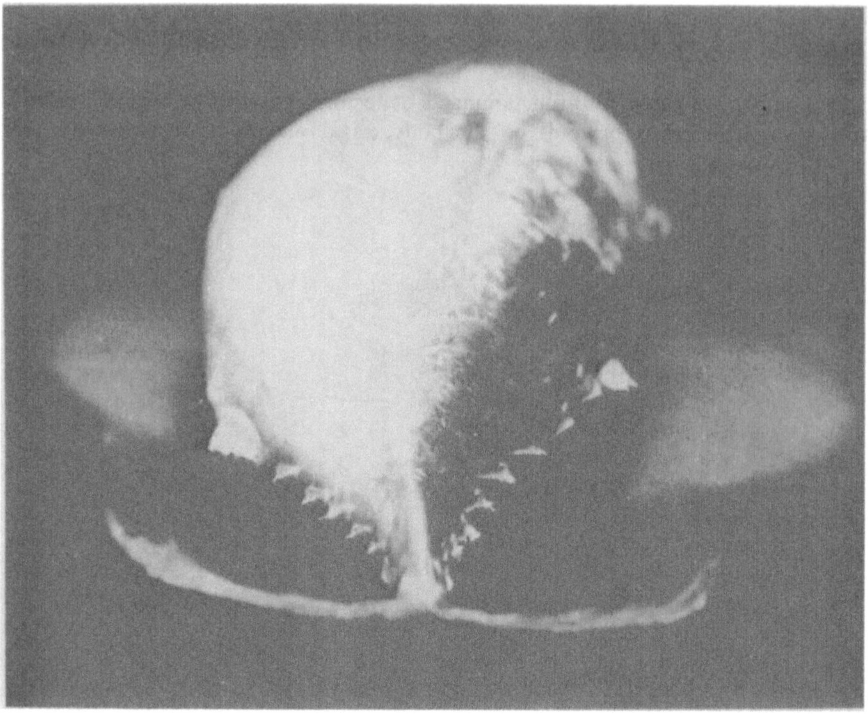

Figure 99. Damping screen of finlets on caudal peduncle of *Auxis thazard* (Lac.).

(*Rhincodon, Dermochelys,* Balaenopteridae, etc.; see Figures 17A,B, 101). These produce a more strictly parallel flow over the body and therefore reduce induced drag (Aleyev 1963a, 1966a,b). In Balaenopteridae such folds possibly facilitate greater mobility of the floor of the oral cavity (Nemoto 1959).

2. Boundary-layer Transition to Turbulence (Turbulization)

As was mentioned above, extension of the contractor increases the probability of separation on the diffuser. For this reason, at $Re > 5.0 \times 10^6$ the diffuser begins to grow longer with increasing *Re* (Figure 91). This process eventually leads at $Re > 10^7$ to the appearance of forms with a very long diffuser and very short contractor (Figures 92G,H, 96–98). In such forms boundary-layer transition takes place on the contractor, as a rule on the foremost portion of the body, which offers two advantages: first, it excludes absolutely the possibility of laminar boundary-layer separation; second, and this is most important, according to the laws of hydromechanics it allows thickening of the turbulent boundary layer over a greater part of the body surface, thus reducing tangential stress of friction τ, as a result of which the total friction drag is reduced (Aleyev 1970b). Thus there appears an absolutely distinct complex of adaptations aimed at reducing drag based on turbulent flow over the entire body, which leads to the appearance of a number of characteristic features in the morphology of nektonic animals.

The initial stages of this process are manifested by the occurrence at $Re > 10^7$ of a relative shortening of the contractor, which is characteristic of many Scombridae (*Thunnus albacora, Th. thynnus, Acanthocybium solandri*, etc.), Coryphaenidae and Cetacea (*Globicephalus, Tursiops, Physeter, Orcinus, Balaenoptera*, etc.; see Table 14). There are no grounds to suppose that in Scombridae the 'corset' of large scales situated behind the gill slits induces boundary-layer transition (Walters 1962) since we find this corset also in animals with typical laminarized body shapes (*Auxis, Sarda*, etc.). Rather, in Scombridae, just as in cetaceans, transition of the boundary layer in the downstream part of the contractor at $Re > 10^7$ occurs as a result of its general instability at these *Re*. According to calculations (Kermack 1948; Bainbridge 1961), the state of the boundary layer on whales is on the whole turbulent.

In its most developed form we see reduction of hydrodynamic drag though great extension of the diffuser and complete boundary-layer transition in comparatively

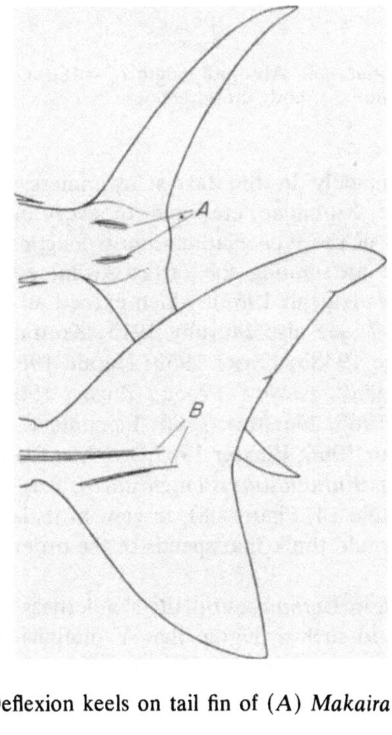

Figure 100. Deflexion keels on tail fin of (*A*) *Makaira* and (*B*) *Lamna*.

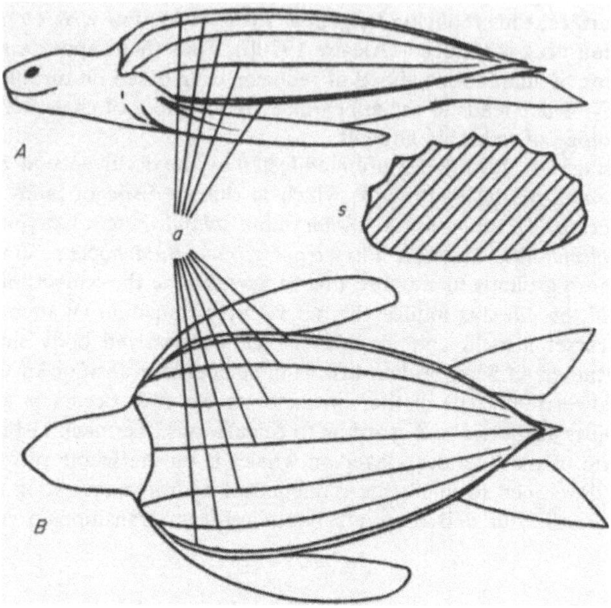

Figure 101. Dermochelys coriacea L. Absolute length $L_a = 160$ cm. *A*, side view; *B*, top view, *a*, longitudinal keels on armour; *s*, body cross-section.

few nektonic animals, namely in the fastest swimmers: in fishes of the group Xiphioidae (Istiophoridae, Xiphiidae, etc.) and, by every indication, in some rather specialized ichthyosauras of the type *Eurhinosaurus longirostris* (Aleyev 1970b).

The group Xiphioidae are among the fastest swimmers, with speeds reaching 30–36 m/s (Walford 1937; Nursall 1962), which exceed all known figures for other nektonic animals (Table 7; see also Murphy 1915; Krettmann 1932; Gawn 1948; Scheffer 1950; Bainbridge 1958b; Fraser 1959; Herald 1962; Slijper 1962; Walters 1962, 1966; Weaver 1962; Aleyev 1963a; Zuyev 1964a, 1966; King 1964; Sukhanov 1964; Focke 1965; Fierstine 1966; Fierstine & Walters 1968; Lang & Norris 1966; Lang & Pryor 1966; Blaxter 1967, 1969, and others). The general body shape of the ichthyosaurus *Eurhinosaurus longirostris* is very close to that of fishes of the Xiphioidae group (Table 14, Figure 98); in view of its large size, up to 6 m long (Tatarinov 1964b), one would think that speeds of the order of 15–20 m/s were well within its reach.

Both in Xiphioidae and in *Eurhinosaurus* the slenderness ratio markedly increases and the diffuser extends to such a degree that *Y* diminishes to 0.20–0.31. At the same time special adaptations emerge for boundary-layer control on the contractor, facilitating a boundary-layer already turbulent on the foremost part of the body:

from convex, the upper and lower surfaces of the head become concave, and there appears a **pointed rostrum**. This is formed by the upper jaw or, more seldom, by both jaws, its length measured from its apex to the frontal point of the effective body length L_c is in various cases from 14 to 45% of L_c (Table 14, Figures 92G,H, 96–98).

In a number of cases, which pursuing large prey, the rostrum, or, as it is frequently called, the sword, is used for piercing the victim, evidence of which are numerous occasions when fishes pierced by the sword have been found in the stomachs of various Xiphioidae species, including such fast swimmers as large tuna and other Scombroidei, as well as *Coryphaena*, *Lutianus*, etc. (Grey 1926, 1928; Conrad & La-Monte 1937; Farrington 1937, 1942; Gregory & La-Monte 1947; June 1951; Morrow 1951, 1952; Hubbs & Wisner 1953; Anonymous 1955; Royce 1957; Fitch 1958; Wisner 1958; Talbot & Penrith, 1962, and others). It has also been pointed out (Günther 1880; Voss 1956; Sheer 1963: Budylenko 1973) that *Xiphias* and *Istiophorus* are capable of aiming lateral blows at their victims with the sword. At the same time many authors (Günther 1880, Grey 1926; Gudger 1940; Farrington 1942; Moore 1950; Morrow 1951; Smith 1956; Wisner 1958, and others) have noted the normal fastness of fishes with a broken or bent rostrum, as well as cases when large fishes without any trace of having been pierced were found in the stomachs of Xiphioidae, i.e. they had simply been seized with the jaws (Bandini 1933; Heilner 1953; Ovchinnikov 1966b). This indicates that the use of the rostrum for catching prey is not a necessity. Numerous reports of 'attacks' by representatives of Xiphioidae on various objects moving in the water, ships, pontoons and boats (Smith 1956, and others) as well as sharks (Gordon 1935) and *Balaenoptera* whales (Jonsgard 1959, 1962; Brown 1960, and others) are explained by accidental collisions of the predator with these objects while pursuing pilot fishes (Barsukov 1960a; Aleyev 1963a).

The basic function of a rostrum is hydrodynamic (Smith, 1956; Wisner 1958; Barsukov 1960a; Walters 1962; Aleyev 1963a, 1965b, 1970b; Sheer 1963; Ovchinnikov 1966a, 1967a, 1968; 1970, and others), and consists of controlling the boundary layer, with the primary aim of reducing friction drag (Aleyev 1970b).

The head of Xiphioidae represents on the whole a generator of turbulence, which, morphologically, is due to 1. the presence of concave regions on its surface and 2. the presence of supercritical roughness on the rostrum. These peculiarities in the body structure of Xiphioidae make the boundary layer already completely turbulent at the beginning of the contractor, which decisively affects the entire flow pattern.

The concave regions of the frontal part of the head of Xiphioidae facilitate boundary-layer transition on account of more intensive mixing of the layer's outer and inner (near-wall) parts: the fluid particles moving rapidly over the concave regions of the head are capable of penetrating to the surface of the body under the effect of centripetal forces, whereas the slowly moving fluid particles, on the contrary, are thrown away from this surface (Walters 1962).

The rostrum of Xiphioidae itself also induces turbulence (Ovchinnikov 1966a,b), which is facilitated by supercritical roughness on it in the form of very small tubercles, folds and spinelets (Aleyev 1970b).

The rostrum of *Xiphias* is a horizontal blade of elliptical cross-section sharpened on both edges and pointed at the end: its general shape really does resemble a two-edge sword (Figures 98A,B). The length of the sword in adult individuals comprises about 40–45% of L_c. The microrelief on the upper and lower sides of the *Xiphias* rostrum is practically the same and consists of a complicated system of very tortuous and branched longitudinal tuberous crests and crescents (Figures 102A,B). This relief is ideally adapted for boundary-layer transition. According to our examinations (Aleyev 1970b) the height of the elements making up the roughness on the *Xiphias* rostrum for a fish's length L of about 200 cm comes to about 0.10–0.50 mm, reaching a maximum at about one third of its length, measuring from the apex (Table 16, Figure 103A). It is apparently for this reason that it is impossible when assessing the turbulizing action of the *Xiphias* rostrum, to draw an analogy with a smooth plate (Ovchinnikov 1966a).

The rostrum of members of the Istiophoridae, of which we examined *Istiophorus, Tetrapturus* and *Makaira,* has a round or nearly round cross-section, forming a pointed spike, its length in different species being from 14 to 30% of L_c (Figures 96D,E, 97). The roughnesses of the upper and lower surfaces of the Istiophoridae rostrum differ fundamentally. The upper surface has only a very low roughness, which is more-or-less the same in different species and formed by the minutest granular protuberances, which in adult fishes are usually not higher than 0.20 mm and only in *Makaira* reach 0.35 mm on the very tip of the rostrum (Table 16, Figures 102D,E, 103B,C, 104). The ventral and partially the lateral surfaces of the Istiophoridae rostrum have considerably large, sharp, conical spinelets, i.e. skin teeth, whose height on various parts of the rostrum varies in the different species from 0.35 to 0.75 mm, achieving a clearly defined maximum at the end of the foremost third of the rostrums' length (Table 16, Figures 102D–F, 103B,C, 104).

A comparison of the cross-sectional size of the protuberances forming the roughness on the rostrum with the height k_{cr} of the critical roughness for each point of the rostrum surface at swimming velocities of 1, 3 and 10 m/s (Aleyev 1970b) shows that the rostrum roughness in *Xiphias* becomes supercritical at speeds faster than 1 m/s and in the Istiophoridae at speeds of 3–10 m/s on the dorsal surface and at speeds of 1–3 m/s on the ventral side (Table 16, Figures 103, 104). These speeds are certainly below the swimming speeds of Xiphioidae fishes, from which it follows that the rostrum roughness in all of them is practically always supercritical and causes intensive transition of the boundary layer. The Xiphioidae rostrum should be regarded primarily as a carrier of supercritical roughness whereby this roughness is brought out far ahead, so that the boundary layer is fully turbulent even before it begins to flow over the compact part of the body, being already turbulent on the rostrum, whose surface area is relatively small. This causes, in accordance with the laws of hydromechanics already outlined, a thickening of the turbulent boundary layer and a reduction of the tangential stress of friction τ over the greater part of the

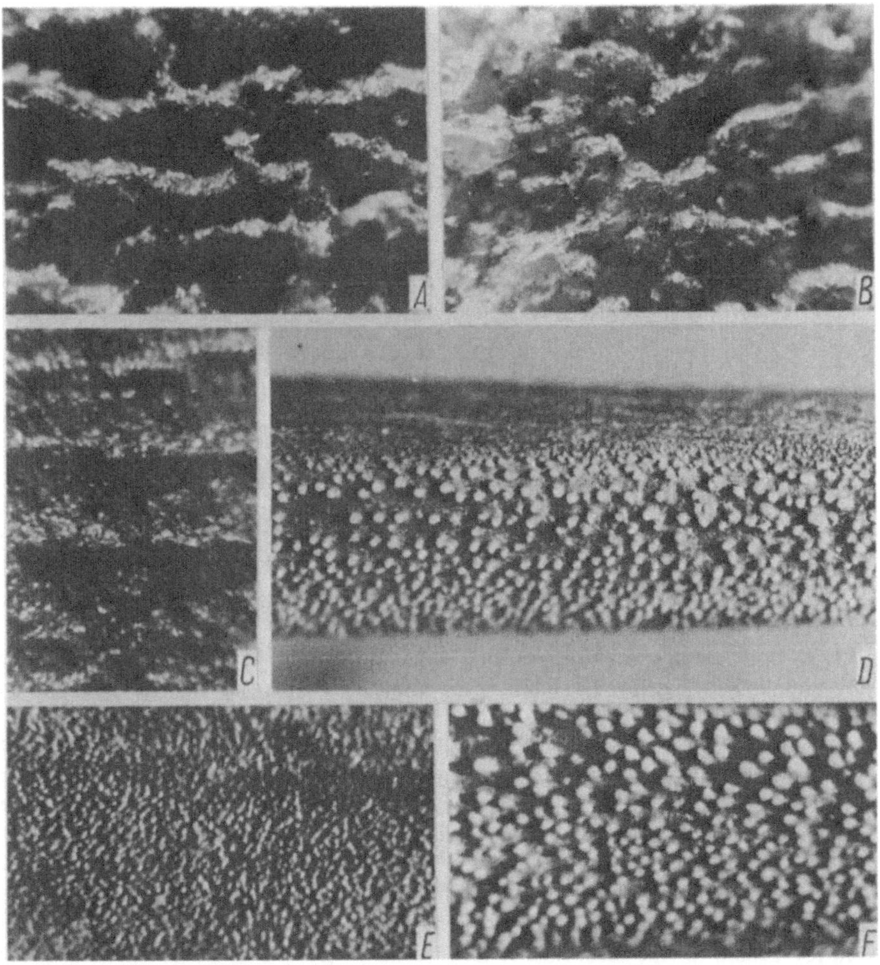

Figure 102. Microrelief on surfaces of rostrum and body of Xiphioidae. *A-C*, *Xiphias gladius* L., $L_a = 200.0$ cm. *A*, dorsal side of rostrum, magnification ×56; *B*, ventral side of rostrum ×56, *C*, lateral surface of body, ×24, longitudinal orientation of microrelief (left to right in photo) clearly visible. *D-F*, *Makaira ampla* (Poey), $L_a = 200.0$ cm. *D*, general side view of rostrum individual cuticular teeth visible, differences in degree of roughness on dorsal and ventral rostrum surfaces clearly distinguishable; *E*, dorsal side of rostrum, ×8; *F*, ventral side of rostrum, ×8.

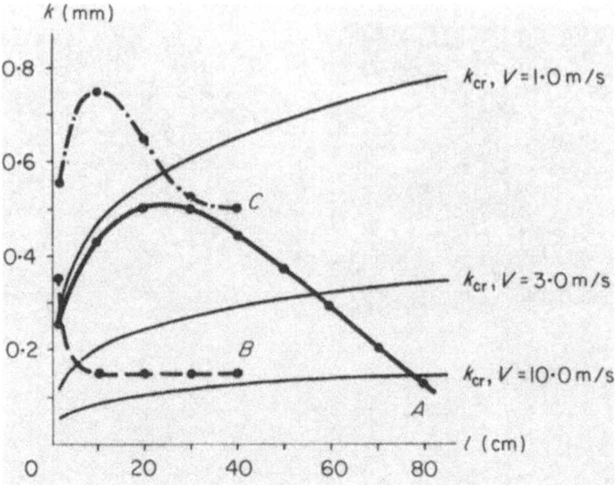

Figure 103. Height k of roughness elements on rostrum in *Xiphias gladius* L., $L_a = 200.0$ cm (*A*, upper and lower rostrum surfaces), and *Makaira ampla* (Poey), $L_a = 200.0$ cm (*B*, upper; *C*, lower rostrum surface, compared with height k_{cr} of critical roughness for swimming speeds of 1, 3 and 10 m/s, l, length of rostrum, counting from tip (cm).

body surface, which leads to a general reduction of the total friction drag. The view (Ovchinnikov 1966a) that boundary-layer transition on the *Xiphias* rostrum leads to a general increase in friction drag and is useful only in so far as it reduces form drag is groundless. Nor are there any grounds for asserting (Ovchinnikov 1966a) that *Xiphias* and *Istiophorus* have poorly streamlined bodies: experimental research we conducted showed that, according to the value of C_{Dp}, *Xiphias* and *Istiophorus* are analogous to the majority of fast-swimming fishes (Table 17).

There is a small longitudinal fold in the body surface of adult Xiphioidae, its height and width in different cases being from 0.1 to 1.0 mm. In Istiophoridae these folds are formed by longitudinally stretched acicular scales, similar to those described (Aleyev 1963a) for the most mobile forms of Carangidae (Figure 142); in Xiphiidae they appear as minute coriaceous crests (Figure 102C). The function of all this longitudinal folding in Xiphioidae, just as in other fishes, is to adjust the flow in the boundary layer, i.e. to prevent the development of excessive turbulent pulsations in it (Burdak 1968, 1969a, and others), and is examined in greater detail below. In young Xiphioidae the surface of the body is covered with scales carrying all kinds of spinelets and turbercles, acting most probably as laminarizing elements. With age all these structures regress, which corresponds to transition to completely turbulent flow over the body. In large *Xiphias* individuals the scales are functionally absent (Aleyev 1963a), though small scales concealed in the skin have been reported (Koval 1972) in fishes more than 2.5 m long. In ontogeny, however, roughness on the rostrum of

Table 16. Critical roughness height k_{cr} and maximum height k of roughness elements on rostrum of Xiphioidae.

Species, nektonic type and absolute length L_a (cm)	Distance from tip of rostrum (cm)	k (mm) dorsal surface	k (mm) ventral surface	k_{cr} (mm) at various swimming speeds V: $V=1$ m/s	$V=3$ m/s	$V=10$ m/s
Xiphias gladius L. (EN), $L_a=200.0$	1	0.25	0.25	0.26	0.11	0.05
	10	0.42	0.43	0.47	0.20	0.08
	20	0.50	0.50	0.55	0.24	0.10
	30	0.49	0.50	0.61	0.27	0.11
	40	0.44	0.43	0.66	0.29	0.12
	50	0.39	0.36	0.69	0.31	0.13
	60	0.29	0.30	0.72	0.32	0.13
	70	0.20	0.20	0.75	0.33	0.13
	80	0.10	0.15	0.78	0.34	0.14
Istiophorus platypterus (Show & Nodder) (EN) $L_a=44.5$	1	0.05	0.30	0.26	0.11	0.05
	5	0.05	0.05	0.39	0.17	0.07
	10	0.05	0.10	0.47	0.20	0.08
Istiophorus platypterus (Show & Nodder) (EN) $L_a=180.0$	1	0.18	0.35	0.26	0.11	0.05
	10	0.18	0.60	0.47	0.20	0.08
	20	0.18	0.52	0.55	0.24	0.10
	30	0.16	0.48	0.61	0.27	0.11
Tetrapturus belone Raf. (EN) $L_a=175.0$	1	0.16	0.45	0.26	0.11	0.05
	10	0.15	0.58	0.47	0.20	0.08
	20	0.15	0.50	0.55	0.24	0.10
Makaira ampla (Poey) (EN) $L_a=200.0$	1	0.35	0.65	0.26	0.11	0.05
	10	0.15	0.75	0.47	0.20	0.08
	20	0.15	0.65	0.55	0.24	0.10
	30	0.15	0.52	0.61	0.27	0.11
	40	0.15	0.50	0.66	0.29	0.12

Xiphioidae develops progressively (Table 16), which is evidence of the progressive development of the rostrum's turbulizing function.

Along with its turbulizing function, being a macrostructure, the Xiphioidae rostrum redistributes the dynamic pressure over the body surface. A comparison of the dynamic pressure distributions we obtained (Table 13) for an aerodynamically smooth model of *Xiphias gladius* (Figure 86, continuous curve) and for the same model with the rostrum replaced by a conventional short conical fairing (Figure 86, dashed curve) shows that on the model without a rostrum pressure changes along the longitudinal axis of the body are sharper, while the minimum pressure is lower (Aleyev 1970b), which is an indirect indication of higher form drag. This we established also experimentally: for the model with a rostrum $C_{Dp}=0.004$, for the

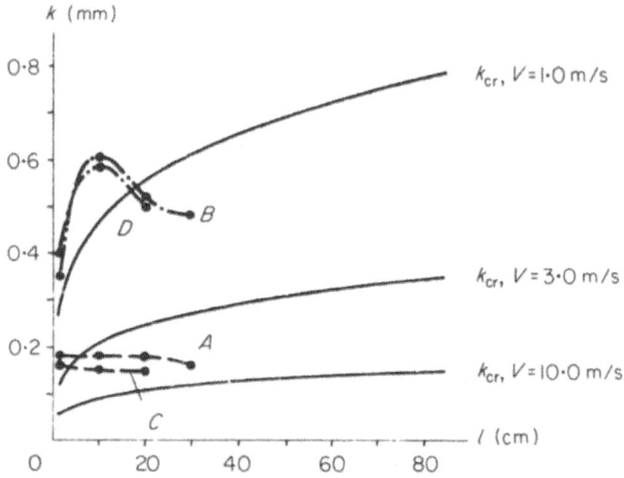

Figure 104. Height k of roughness elements on rostrum in *Istiophorus platypterus* (Show and Nodder), $L_a = 180.0$ cm (*A*, upper; *B*, lower rostrum surface), and *Tetrapturus belone* Raf., $L_a = 175.0$ cm (*C*, upper; *D*, − lower rostrum surface), compared with height k_{cr} of critical roughness for swimming speeds of 1, 3 and 10 m/s. *l*, length of rostrum, counting from tip (cm).

model without a rostrum $C_{Dp} = 0.005$ (Table 17). Positive pressure gradients on the surface of the frontal part of the Xiphioidae head are directly connected with the presence of concave regions on it: in the absence of a rostrum, when the head has no concave surfaces, no positive pressure gradients occur (Table 13, Figure 86).

It has been suggested (Aleyev 1963a, 1965b) that, when Xiphioidae move near the surface of the water at high speeds of the order of 20–30 m/s, hydrodynamic drag is probably reduced on account of cavitation and of some part of the body finding itself in the air. This was not confirmed experimentally when completely immersed models of *Xiphias gladius* and *Istiophorus platypterus* were tested in a cavitation tunnel, nor when a *Xiphias* model was tested in a free-surface flow in a trough with the dorsal fin sticking out of the water (Aleyev 1970b).

Apart from Xiphioidae and the specialized ichthyosaurs mentioned earlier (*Eurhinosaurus*, etc.), as well as the sharks and Acipenseridae examined above, individual representatives of a number of other nektonic groups (*Pristiophorus*, *Pristis*, Hemirhamphidae, Polyodontidae, *Monodon*, etc.) also have as part of the head a rostrum extending forward from the mouth slit, but there is no indication that it fulfils any special hydrodynamic function. In distinction from Xiphioidae and *Eurhinosaurus*, none of these animals are fast swimmers. In such cases the appearance of a rostrum is connected with various adaptations involving feeding and defence.

Thus the saw-like rostrum of *Pristiophorus* and *Pristis* is used for tearing the silt

Table 17. Nektonic animals' C_{Dp} values according to model trials

Species and nektonic type	Absolute length L_a of animal (cm)	C_{Dp}	Reynolds number: Live specimen Re	Reynolds number: Model Re_m
Cryptocleidus oxoniensis Phill. (XN)	335.0	0.009	1.5×10^7	10^7
Acipenser güldenstädti colchicus V. Marti (BN)	150.0	0.007	4.5×10^6	4.5×10^6
Chelonia mydas (L.) (XN)	154.0	0.007	3.0×10^6	3.0×10^6
Eudyptes chrysolophus Brandt (XN)	69.6	0.006	5.0×10^6	5.0×10^6
Pagophoca groenlandica (Erxl.) (XN)	197.0	0.006	10^7	10^7
Prionace glauca (L.) (EN)	403.0	0.005	4.0×10^7	4.0×10^7
Abramis brama (L.) (BN)	75.0	0.005	8.0×10^5	8.0×10^5
Mola mola (L.) (EN)	200.0	0.005	2.0×10^6	2.0×10^6
Stenopterygius quadriscissus Quenst. (EN)	210.0	0.005	2.1×10^7	1.2×10^7
Arctocephalus pusillus (Schreb.) (XN)	200.0	0.005	1.6×10^7	1.2×10^7
Eubalaena glacialis (Bonnatt.) (EN)	1400.0	0.005	6.0×10^7	9.5×10^6
Physeter catodon L. (EN)	1800.0	0.005	1.1×10^8	1.2×10^7
Squalus acanthias L. (EN)	120.0	0.004	4.2×10^6	4.2×10^6
Salmo trutta labrax Pall. (BN)	98.8	0.004	1.2×10^7	1.1×10^7
Esox lucius L. (BN)	103.4	0.004	10^7	10^7
Odontogadus merlangus merlangus (L.) (BN)	49.7	0.004	10^6	10^6
Mugil auratus Risso (BN)	41.4	0.004	1.5×10^6	1.5×10^6
Mugil cephalus L. (BN)	70.0	0.004	4.2×10^6	4.2×10^6
Sphyraena barracuda (Walb.) (EN)	130.0	0.004	2.0×10^7	1.4×10^7
Serranus scriba (L.) (BN)	30.0	0.004	8.0×10^5	8.0×10^5
Thunnus alalunga (Bonnat.) (EN)	89.0	0.004	1.6×10^7	1.2×10^7
Thunnus thynnus (L.) (EN)	196.0	0.004	4.0×10^7	1.2×10^7
Xiphias gladius L. (EN)	356.0	0.004	1.1×10^8	1.5×10^7
Makaira indica (Cuv. et Val.) (EN)	450.0	0.004	1.6×10^8	1.5×10^7
Istiophorus platypterus (Show & Nodder) (EN)	183.0	0.004	6.1×10^7	10^7
Eurhinosaurus longirostris Jaeger (EN)	500.0	0.004	7.5×10^7	1.1×10^7
Balaenoptera physalus (L.) (EN)	2500.0	0.004	2.1×10^8	1.5×10^7
Phocoena phocoena (L.) (EN)	128.0	0.004	1.4×10^7	1.1×10^7
Delphinus delphis ponticus Barab. (EN)	178.5	0.004	2.7×10^7	1.2×10^7
Loligo vulgaris Lam. (EN)	36.0	0.003	1.5×10^6	1.5×10^6
Symplectoteuthis oualaniensis (Less.) (EN)	45.2	0.003	2.0×10^6	2.0×10^6
Clupea harengus pallasi Val. (EN)	36.0	0.003	8.0×10^5	8.0×10^5

Table 17. (contd.)

Species and nektonic type	Absolute length L_a of animal (cm)	C_{D_p}	Reynolds number: Live specimen Re	Reynolds number: Model Re_m
Alosa kessleri kessleri (Grimm) (EN)	50.0	0.003	10^6	10^6
Hirundichthys rondeletii (Cuv. et Val.) (EN)	24.0	0.003	2.5×10^6	2.5×10^6
Pomatomus saltatrix (L.) (EN)	99.0	0.003	6.0×10^6	6.0×10^6
Trachurus mediterraneus ponticus Aleev (EN)	56.0	0.003	3.5×10^6	3.5×10^6
Coryphaena hippurus L. (EN)	181.0	0.003	3.4×10^7	1.2×10^7
Scomber scombrus L. (EN)	50.0	0.003	5.0×10^6	5.0×10^6
Sarda sarda (Bl.) (EN)	73.0	0.003	10^7	10^7
Auxis thazard (Lac.) (EN)	52.1	0.003	5.0×10^6	5.0×10^6
Scomberomorus commersoni (Lac.) (EN)	120.0	0.003	3.0×10^7	1.2×10^7
Acanthocybium solandri (Cuv. et Val.) (EN)	150.0	0.003	3.2×10^7	1.2×10^7

and striking prey through lateral movements of the head (Suvorov 1948; Arambourg & Bertin 1958; Lindberg & Legeza 1959).

The Hemirhamphidae are fishes inhabiting the surface layers of the water, as is indicated, in particular, by their very low median line (Aleyev 1960a); their enlarged and wide lower jaw has an aerodynamically smooth surface and fulfils the function of the lower square of a pelagic trawl, preventing the catch (of planktonic animals) from escaping downwards, which is most probable in the surface layers of the water (Figures 105A,B). Given the laminarized body shape of Hemirhamphidae and the fact that $Re < 5.0 \times 10^6$ the elongation of their lower jaw cannot be aimed at fulfilling the function of the rostrum of Xiphioidae.

The spatula-like horizontal rostrum of *Polyodon* fulfils the function of the upper square of a benthic trawl (Nikolsky 1954), preventing the catch from escaping upwards, which in benthic conditions, where *Polyodon* catches its prey (planktonic animals; Bertin 1958e), is most probable (Figures 105C,D).

The rostrum-like tusk of the male *Monodon monoceros*, the left tooth of the upper jaw extended forwards horizontally, is the main weapon for inflicting frontal blows when breaking through ice covering the water, and also probably for repelling predators (Tomilin 1957).

3. Sucking Away the Boundary Layer

Sucking away the boundary layer as a means of preventing its separation is encountered among nektonic animals in cephalopods only. Here sucking away the

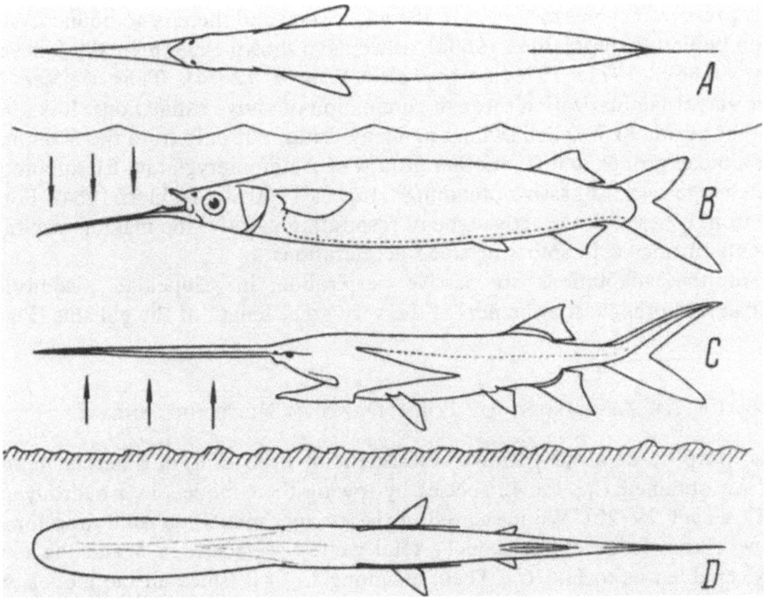

Figure 105. Hemirhamphus commersonii Cuv., $L_a = 38.0$ cm (*A*, top view; *B*, side view), and *Polyodon folium* Lac., $L_a = 56.3$ cm (*C*, side view; *D*, top view). Arrows indicate direction of food objects' escape at moment of capture.

boundary layer is connected with the intake of water into the mantle cavity through the mantle slit, necessary for respiration and the operation of the hydrojet propulsor; such suction has the further purpose of preventing boundary-layer separation.

4. Other Ways of Preventing Boundary-layer Separation

The adaptations of this type which have not already been examined are of a more particular character.

It is assumed (Walters 1962) that in fishes with the passive type of respiration (Scombridae, etc.) the head represents an analogue of a leading-edge slat, since the mouth and gill slits are constantly open and accelerated streams of water from the gill cavities are constantly flowing into the boundary layer. This leads to continuous displacement of the old, low-energy boundary layer by a new, high-energy layer created by the water flowing out of the gill slits, thus helping to prevent boundary-layer separation.

Moreover, during the passive type of respiration in such fishes as the Scombridae, the water constantly flowing through the gill slits prevents the appearance of areas of very low pressure beyond the edges of the gill covers, and there is no boundary-layer transition behind them (Burdak 1969a), as we have shown experimentally (Aleyev & Ovcharov 1969, 1971, 1973a,b; see also Figures 72–74). Thus we see here boundary-layer laminarization through elimination of active exhalations. It is for this reason that during high-speed swimming many fishes, not only from the Scombridae and Xiphioidae groups, but from other groups of Actinopterygii and Elasmobranchii as well, change over to passive breathing (Hughes 1960a,b; Wahlert 1964). For the same reason, fishes with the active type of respiration usually stop making respiratory movements during rapid spurts or steep accelerations.

One of the adaptations to passive respiration in Clupeidae, Scombridae, Xiphioidae and other fast swimmers is the very great length of the gill slits (Figures 96–98).

5. Comparative Assessment of Form Drag in Nektonic Animals

With the purpose of a comparative assessment of form drag in different nektonic animals we obtained C_{Dp} for 42 species by towing their models in a hydrodynamic channel (Aleyev 1972b). We measured in these experiments the total drag force F, which, with knowledge of the model's total wetted surface area S and the towing speed V, enabled us to find C_D. Then, assuming $C_{Di} = 0$ (since all the models were made for the case of movement by inertia) and knowing C_{Df}, which was found by the method generally accepted in hydromechanics, i.e. by measuring the total wetted surface area S of the models (Schlichting 1956), we found C_{Dp} from formula (2).

To carry out these experiments we used the same series of models as when finding $\bar{p}$ (Figure 106, etc.), with the addition of a number of new models made by the methods described earlier and based on the author's direct measurements of the animals; the models of *Eurhinosaurus* and *Cryptocleidus* were reconstructed by the author on the basis of data in the literature (Abel 1912; Huene 1956; Tatarinov 1964k).

The models were towed fully immersed on a transverse support at a depth of about 0.6 m in a hydrodynamic channel 40 m long and 3 m wide with a water depth of about 2.5 m (Figure 106). The towing speed in different cases varied from 0.6 to 12.5 m/s, Re_m having a value between 8.0×10^5 and 1.5×10^7 which in most cases corresponded to the maximum Re for the live animal. When experimental conditions precluded the attainment of the maximum Re characteristic of the given species (whales, *Xiphias*, *Istiophorus*, etc.), towing was carried out at $Re_m \geqslant 9.0 \times 10^6$, which corresponded in this case to the similarity region, where the influence of Re on C_D is completely or practically absent. Measurement of the force F under the conditions $Re_m = Re_{max}$ and $Re_m \geqslant 9.0 \times 10^6$ gives, naturally, identical results, owing to which all the C_{Dp} values we obtained are comparable. To measure the force F we used tensometric sensors and recorded their readings on an H-700 oscillographer. From twelve to fifteen towings were carried out for each model.

It should be stressed that the C_{Dp} values for the models are identical to those for

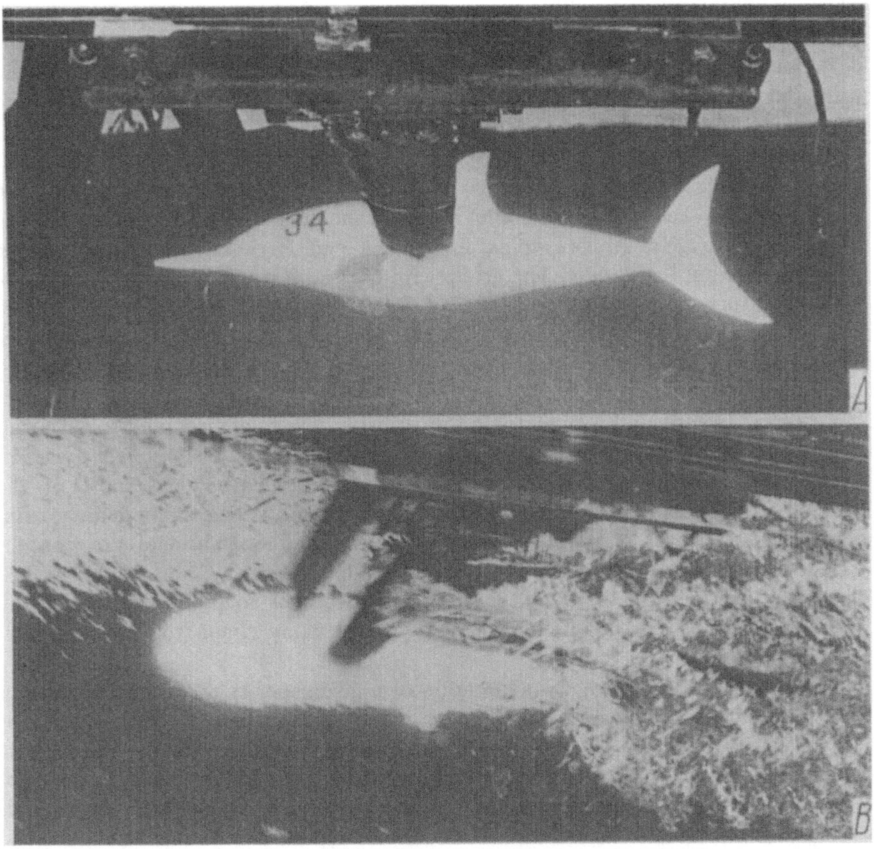

Figure 106. Models of (*A*) *Stenopterygius quadriscissus* Quenst. (starting position) and (*B*) *Coryphaena hippurus* L. (towing).

their live prototypes, since C_{Dp} depends only on the shape of the body being investigated, which in the case of each model is an exact replica of its living prototype.

Table 17, where the individual nektonic species are arranged in order of diminishing C_{Dp}, shows that in the series obtained in this manner the animals' ecology undergoes regular variation. At the beginning of the series we have the least mobile forms and at the end the most active swimmers, both those that are on the move most of the time and those which perform separate sharp spurts. Characteristic of the majority of the nektonic animals examined, 37 species out of 42, are low C_{Dp} values, from 0.003 to 0.005, which points to well-streamlined bodies and low form drag, which is advantageous to all nektonic animals. We find higher C_{Dp} values (0.006-0.009) only in species whose body form is hydrodynamically less perfect owing to the development of adaptations not connected with locomotion and conducive to contact with the bottom (*Acipenser*) or land (*Cryptocleidus, Chelonia, Eudyptes* and *Pagophoca*); eunektonic forms are absent in this range.

Within the C_{Dp} range 0.003–0.005 one sees noticeable ecological differences among nektonic animals according to the value of C_{Dp}. Thus, in the group with $C_{Dp} = 0.005$ we see animals swimming at low or medium speeds (*Prionace, Abramis, Mola, Stenopterygius, Arctocephalus, Eubalaena* and *Physeter*); the fastest pelagic swimmers are not found here. The group with $C_{Dp} = 0.004$ has none of the very slow swimmers, such as *Mola, Abramis* or *Eubalaena*, though we still see comparatively fast benthonektonic forms (*Salmo, Esox, Odontogadus, Mugil* and *Serranus*); predominating in this group are fast eunektonic species (*Sphyraena, Thunnus*, Xiphioidae, *Eurhinosaurus* and Delphinidae). Finally the group with $C_{Dp} = 0.003$ is represented exclusively by eunektonic forms: fishes and squids adapted to sustained movement, both comparatively slow (*Clupea* and *Alosa*) and fast (*Loligo, Symplectoteuthis, Hirundichthys, Pomatomus, Trachurus, Coryphaena* and most of the Scombridae). Thus all the best pelagic swimmers (Teuthoidea, Sphyraenidae, Exocoetidae, Trachurus, Coryphaenidae, Xiphioidae and Scombridae) are concentrated, as should be expected, in the two last groups, in the C_{Dp} range 0.003–0.004. The value of C_{Dp} can, therefore, serve as a measure of adaptation to movement, directly reflecting the adaptation of the general body shape to reducing hydrodynamic resistance.

One can assess the degree of adaptation of a given species to various regimes of locomotion from the character of the variation in the force F of resistance to movement in the speed-up period. The added mass and, accordingly, the starting effort are least in species adapted to the performance of sudden spurts: in this case the starting effort F_2 is not more than 3–5 times the drag force F_1 which corresponds to uniform movement (Figure 107E). In forms adapted to sustained movement at comparatively low accelerations, F_2 may be 9–10 times F_1 (Figures 107A,B). Nektonic animals adapted to more varied regimes of locomotion occupy positions between these two extremes (Figures 107C,D).

Few attempts have been made so far to determine the value of C_D in living nektonic animals directly. The most objective method of finding C_D is by filming the animals when they are moving by inertia (Lang & Pryor 1966; Semenov 1969; Aleyev & Kurbatov 1973; Kurbatov 1973; Mordvinov 1973). Determining C_D by calculation (Pyatetsky 1970; Pyatetsky & Kayan 1972a,b, and others), by measuring the drag force F on animals tethered by a thread in an oncoming stream (Matyukhin 1973; Matyukhin & Turetsky 1972, and others) or by the method of additional drag (Grushanskaya & Korotkin 1973) provides results that are not accurate enough.

We determined (Aleyev & Kurbatov 1973; Kurbatov 1973) C_D for several nektonic animals, various fishes, the dolphin *Phocoena phocoena* and the seal *Pagophoca groenlandica*, by means of an original 'Scopa' photoelectric automatic

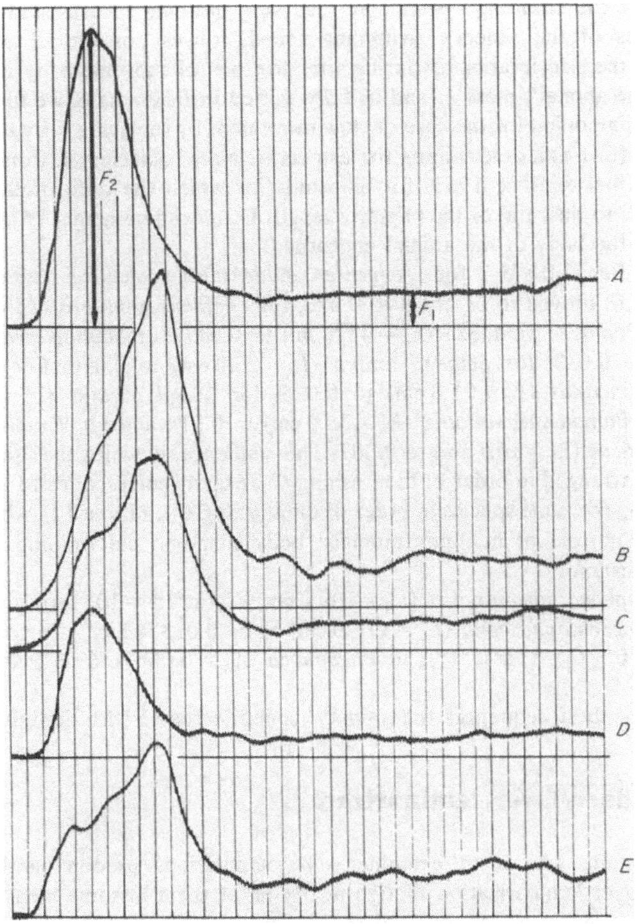

Figure 107. Oscillograms of force **F** resisting movement of different animal models during initial stretch of towing. Towimg speed 2.5–3.0 m/s throughout. *A*, *Phocoena phocoena* (L.); *B*, *Auxis thazard* (Lac.); *C*, *Mugil auratus* Risso; *D*, *Salmo trutta labrax* Pall.; *E*, *Sphyraena barracuda* (Walb.). See text.

scanning system (Aleyev & Kurbatov 1972, 1974), capable of lasting and stable scanning of the animals while they are moving about freely in a hydraulic channel (Figure 31). The system's movable platform is constantly over the object being watched, travelling at the same speed and acceleration, the magnitudes of which are measured and recorded by an oscillographer; plan filming of the object with a 'Konvass' camera is carried out simultaneously, the camera being mounted on the platform of the scanning system. By analysing the ciné records and synchronous oscillograms of the object's swimming speed, it was possible in every case to determine the deceleration a during the stretches of movement by inertia. Then, knowing the animal's mass m and its fully wetted surface area S, we found C_D from formula (1) modified for the case of slow movement by replacing F by the equivalent expression $(m+k)a$, determining the associated mass coefficient k from nomograms (Kochin, Kibel & Rose 1963) for ellipsoids for which the ratio of the three axes corresponds to the ratio of the effective length L_c, maximum height H and maximum width I of the body of the animal concerned.

C_D at $Re = 2.0 \times 10^5$ for *Acipenser güldenstädti colchicus* (absolute length $L_a = 68.0$ cm) proved to be equal to 0.076, for *Diplodus annularis* ($L_a = 20.0$ cm) to 0.55, for *Puntazzo puntazzo* ($L_a = 43.5$ cm) to 0.047, for *Squalus acanthias* ($L_a =$ 61.5 cm) to 0.033, for *Sciaena umbra* ($L_a = 25.0$ cm) to 0.030, for *Odontogadus merlangus euxinus* ($L_a = 21.8$ cm) to 0.025, for *Mugil auratus* ($L_a = 27.5$ cm) to 0.017, for *Pomatomus saltatrix* ($L_a = 38.0$ cm) to 0.014 and for *Trachurus mediterraneus ponticus* ($L_a = 16.7$ cm) to 0.007. This sequence, in which the fishes examined have been arranged in order of diminishing C_D, exactly corresponds to that obtained by arranging the same species in order of diminishing C_{Dp} (Table 17), which points to the important role of nektonic animals' body shape in determining their hydrodynamic resistance.

For *Pagophoca groenlandica* ($L_a = 108.0$ cm) C_D at $Re = 10^6$ proved to be 0.032 and for *Phocoena phocoena* ($L_a = 116.0$ cm) to be 0.013, which also corresponds to the results for C_{Dp} (Table 17): in *Pagophoca* $C_{Dp} = 0.006$ and in *Phocoena* $C_{Dp} =$ 0.004.

In each of these sequences the animals' speed increases with diminishing C_D.

4. Boundary-layer laminarization

It now remains for us to examine some adaptations predominantly aimed at boundary-layer laminarization though nearly all of them in some measure or other help also to prevent boundary-layer separation.

1. Controlling Boundary-layer Flow by Means of Rigid Microrelief on the Body Surface

We have already examined a number of adaptations aimed at controlling flow in the boundary layer and connected mostly with the macrostructure of nektonic animals. Now, however, we examine adaptations based on a complex of rigid microstructures on the body surface fulfilling a special hydrodynamic function.

The body surface of most nektonic animals is aerodynamically smooth, which is conducive to the preservation of boundary-layer laminarity and the reduction of friction drag (Aleyev 1974). This is so in fishes (Aleyev 1963a), Teuthoidea (Zuyev

1964b) and dolphins (Babenko & Surkina 1969; Khadzhinski 1972); an aerodynamically smooth surface is formed also by the armour plates of Chelonioidea, the disks and scales of Hydrophidae, the feather coat of Sphenisciformes and the hair coat of most of the pinnipeds (with the exception of Odobenidae). Aerodynamically smooth too, probably, was the surface of the skin of Ichthyosauria without armour.

The microrelief on the body surface of nektonic animals is usually made up of well-streamlined structures, examples of which are the ctenoids and some other formations on the scales of fishes and also the feathers of Sphenisciformes and the hairs of the pinnipeds. All these and similar structures on the body surface have a decisive effect on the character of the boundary-layer flow, most often fulfilling the function of laminarization.

A. Hydrodynamic functions of fish scales

One of the varied hydrodynamic functions of fish scales consists of ensuring a certain smoothness of the body surface and preventing the appearance of transverse folds on the sides of the fish (Walters 1962, 1963; Aleyev 1963a; Wahlert & Wahlert 1964; Kudryashov & Barsukov 1967a,b; Burdak 1968, 1969a–c, 1970, 1972, 1973a–c, Kudryashov 1969a,b; Kobets & Komarova 1971; Zayets 1972, and others). This is confirmed by the fact that in ontogeny scales initially appear on the sides of the body (Aleyev 1957b, 1963a; Burdak 1957, 1968, 1969a and others), where the emergence of folds is most probable, and also by the fact that in fishes swimming with insignificant body undulations scales usually disappear (Kudryashov & Barsukov 1967b). However, the relief on the scaly surface fulfills more complicated hydrodynamic functions. In this connection we note first of all the imbricate character of the scales in the overwhelming majority of extant fishes, which has great hydrodynamic significance (Aleyev 1963a) as a means of ensuring a certain unsteadiness of the boundary layer. It is not accidental that scales sometimes appear even in aquatic mammals (*Castor*) and even in Teuthoidea (*Lepidoteuthis grimaldi*). The elements of body-surface microrelief in various fishes have a rather diverse structure (Hertwig 1874, 1876, 1879, 1882; Klaatsch 1890; Hofer 1890; Ryder 1892; Goodrich 1907; N. Rosen 1913, 1914; Carter 1919; Roule 1924; Gnadeberg 1926; Sewertzov 1932; Budker 1938; Daget 1952; Kerr 1952; Bertin 1958c,d; Walters 1962; Aleyev 1963a; Wahlert & Wahlert 1964; Burdak 1968, 1969a–c, 1970, 1972, 1973a–c, 1974, and others). However they are fundamentally quite similar, since they always consist of some prominent processes, spines, crests, etc., i.e. structures directly controlling the boundary layer (Aleyev 1963a; Kudryashov & Barsukov 1967a,b; Burdak 1968, 1969a, and others).

The hydrodynamic functions of the skin microrelief of fishes and ichthyoids have been thoroughly studied by Burdak (1968, 1969a–c, 1970, 1972, 1973a–c, 1974, and others), who showed that the scaly covering of ichthyoids and fishes constitutes a complicated morphological structure aimed at controlling the flow in the boundary layer, and that the most characteristic feature of this structure is **the presence on the**

body surface of longitudinally oriented relief. This general concept of the hydrodynamic function of the skin of ichthyoids and fishes has been confirmed experimentally (Aleyev & Ovcharov 1969, 1971, 1973a,b, and others).

Morphologically the surface relief on the scales and armour of ichthyoids and fishes may be built up of various kinds of ctenoid structures, e.g. spines, dents and tubercles, or of longitudinally oriented crescents and crests with a smooth or angular cross-section, which may be collectively described by the term 'run-off grooves' (Figures 108–110). The run-off grooves on the scales become clearly visible if one draws on photographs of fish scales lines similar to isohypses, i.e. connecting points situated at an equal distance from the lower surface of the scales (Burdak 1972). These isohypses may be easily traced as the contours of individual plates embedded in the scaly covering (Figure 109). In some cases the run-off grooves on the scales occur in combination with longitudinal rows of more isolated often conical elevations which supplement the relief formed by the grooves (Figure 110). Functionally, all these details of the relief are very similar: they all represent elements of a laminarizor and are aimed at reducing or eliminating turbulent pulsations in the boundary layer (Burdak 1968, 1969a–c, 1970, 1972, 1973a–c, 1974, and others).

Longitudinally oriented relief on the scales and armour is very common in the most diverse, including the most ancient, groups of ichthyoids and fishes, both nektobenthic and nektonic. It is found on the placoid scales of Thelodonti (*Thelodus, Lanarkia* and *Phlebolepis*; Traquair 1899a,b; Gross 1947; Bystrov 1949; Obruchev 1964a), on the armour and individual tesserae of Heterostraci (*Tesseraspis, Aphataspis, Traquairaspis, Archegonaspis, Tolypelepis* and others; Kiaer 1932; Tarlo 1962; Obruchev 1964a), on the armour and scales of Osteostraci (*Timanaspis, Tannuaspis* and others; Obruchev 1956; 1964a; Kossovoi & Obruchev 1962), on the placoid scales of Acanthodei (*Nostolepis, Archaeacanthus, Gomphodus, Haplacanthus* and others; Gross 1947; Novitskaya & Obruchev 1964) and Elasmobranchii (on the crowns of skin dents in most sharks; Bigelow & Schroeder 1948; Glikman 1964), on the cosmoid scales of Sarcopterygii (Holoptychioidei, Eusthenopteridae, Rhizodontidae, Coelacanthida and others; Obruchev 1947; Jarvik 1950; Schaeffer 1952; Kondratyeva & Obruchev 1955; Obrucheva 1955; Orvig 1957; Millot & Anthony 1958; Vorobyeva 1963; Vorobyeva & Obruchev 1964), on the ganoid scales of Palaeonisci (Matveyeva 1958; Berg, Kazantseva & Obruchev 1964) and Lepisosteiformes (Goodrich 1907; Bertin 1958c; Danilchenko 1964a) and on the elasmoid (ctenoid and cycloid) scales of Actinopterygii (many Cyprinidae, Mugilidae, Perciformes and others; Goodrich 1907; Bertin 1958c; Burdak 1968, 1969a–c, 1970, 1972, 1973a,b, and others). The longitudinal relief on scales and armour consists in every case of a complex of well-streamlined structures (Figures 108–110) functionally similar to the diffuser grid or turbulence screen, which are successfully used in aero- and hydromechanics for reducing turbulence by breaking up the vortices (Burdak 1968).

It has been shown (Burdak 1968, and others) that in various fishes and Agnatha, including the most archaic fossil forms (Thelodonti, etc.), the development of

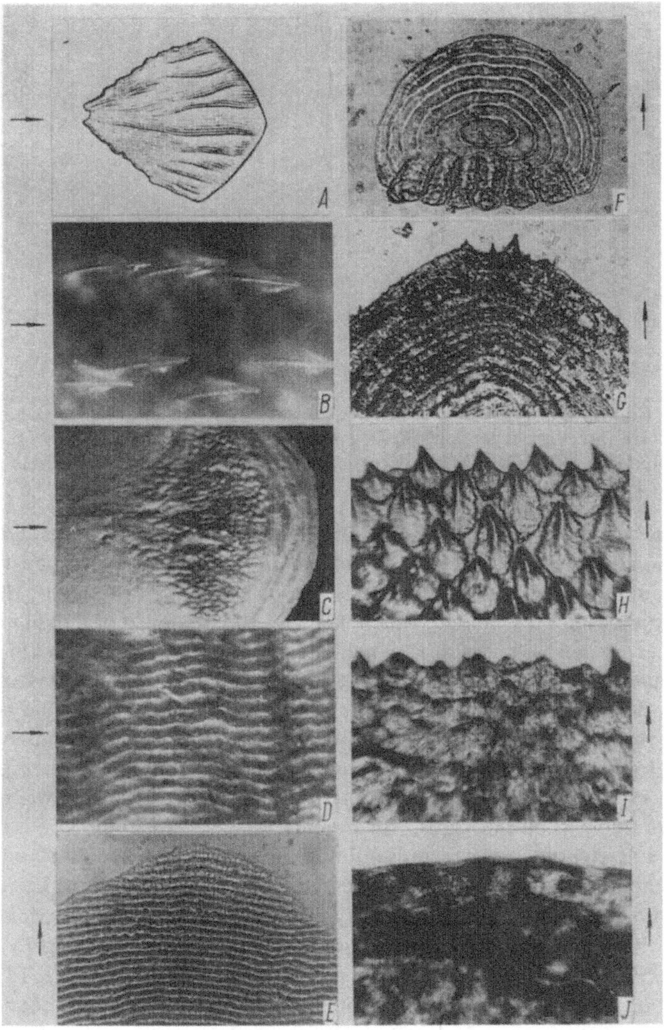

Figure 108. Microrelief on surface of scales of (*A*) *Haplacanthus persensis* Gross, (*B*) *Squalus acanthias* L., (*C,D*) *Neoceratodus forsteri* Krefft (*C*, general view of exposed part of scales showing large longitudinal relief in central area; *D*, small longitudinal relief in peripheral area of exposed part of scales), (*E*) *Pomatomus saltatrix* (L.), (*F, G*) *Mugil saliens* Risso and (*H-J*) *Mugil cephalus* L. Absolute length of fishes (cm): *B*, 61.0; *C,D*, 75.0; *E*, 14.6; *F,G*, 1.6; *H*, 57.0; *I*, 58.5; *J*, 70.0. Magnification: *A*, ×56, *B*, ×8; *C*, ×2.3; *D*, ×8; *E*, ×56; *F-J*, ×32. Arrows indicate direction of stream. *A*, after Gross (1947); *B-J*, after Burdak (1969a, 1970, 1972, 1973b).

Figure 109. Run-off grooves (dashed lines) on fish scales. Isohypses drawn as continuous lines. Arrows indicate direction of stream. *A*, — *Pomatomus saltatrix* (L.), length of fish to end of vertebral column $L = 70.0$ cm; *B*, *Leuciscus idus* (L.), $L = 36.5$ cm. After Burdak (1972).

longitudinal relief is suppressed on certain areas of the body in one and the same way. No longitudinal relief at all is found over the forebody; it appears at some distance from the anterior end of the body and its height, determined by the length k of the ctenoids, increases caudad, which indicates the existence of a definite relationship between k and k_{add}, since k_{add} increases in the same direction (Figures 111, 112). In the fishes we investigated $k > k_{add}$ (Figure 112); with an increase in Re both k_{add} and k decrease. The line dividing the areas of cycloid and ctenoid scales on a fish's body is always situated somewhat forward of the border between the contractor and diffuser (Figure 111), i.e. ctenoid scales appear on the downstream part of the contractor, where accelerations in the outer flow are already low and where, because of this, favourable conditions arise for boundary-layer transition.

Cycloid and ctenoid scales represent two interconnected stages of one and the same adaptation, leading into one another, owing to which they must be regarded as subtypes of one and the same elasmoid type (Burdak 1970). The functional connection between cycloid and ctenoid scales is particularly clear in ontogeny, as shown (Burdak 1957, 1969a, 1970) in the example of *Mugil* (Figures 108F–J, 113). During swimming in obviously subcritical regimes, when the threat of boundary-layer transition does not yet arise, i.e. at $Re < 10^4$, the scales remain cycloid (Figures 108F, 113A–C). In the range $10^4 \leqslant Re \leqslant 10^5$ areas of turbulent flow appear on the fish's body (Aleyev & Ovcharov, 1969, 1971, and others; see also figures 72–75) and, accordingly, at $Re = 8.0 \times 10^3$ the formation of the ctenoid apparatus begins (Figures 108G, 113D–F), concluding at $Re \approx 2.0 \times 10^5$ (Figures 108H, 113G–M). It should be noted that ctenoids form first of all on those areas of the body where the boundary layer is most strongly influenced by turbulizing factors: behind the fins and the gill slits (Figures 113, 114). This confirms the assessment of the role of ctenoids as

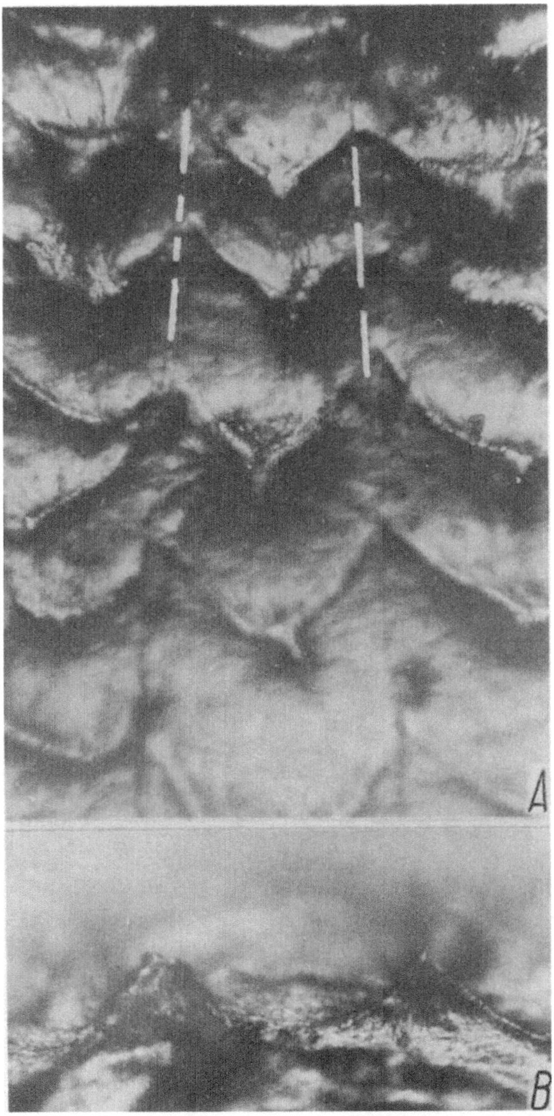

Figure 110. Run-off grooves on scales of *Carassius auratus gibelio* Bloch. Fish length to end of vertebral column $L = 10.7$ cm. *A*, microrelief of scale surface in plan, run-off grooves marked with dashed lines; *B*, same microrelief photographed laterally along the longitudinal axis of the fish, i.e. streamwise, showing clearly the semicircular profile of the run-off groove formed by the two rows of conical protuberances shown in plan in frame *A*. After Burdak (1972).

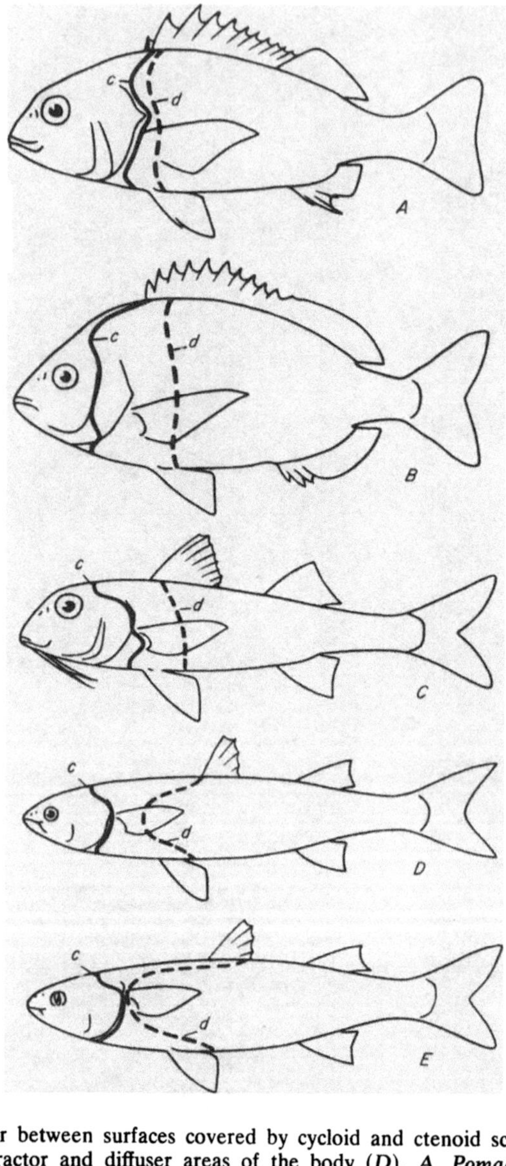

Figure 111. Border between surfaces covered by cycloid and ctenoid scales (*C*) and border between the contractor and diffuser areas of the body (*D*). *A*, *Pomadasis guoraca* Russ., average length (2 specimens) to end of vertebral column $L = 38.5$ cm; *B*, *Diplodus annularis* (L.), 10 specimens, $L = 15.2$ cm; *C*, *Mullus barbatus ponticus* Essip., 10 specimens, $L = 17.6$ cm; *D*, *Mugil auratus* Risso, 10 specimens, $L = 28.5$ cm; *E*, *Mugil cephalus* L., 10 specimens, $L = 36.0$ cm. After Burdak (1968).

laminarizing elements, 'combing' the boundary layer (Burdak 1969a). At $Re \approx 3.5 \times 10^6$ the ctenoid apparatus of grey mullets begins to regress (Figure 108I), which at $Re \approx 5.0 \times 10^6$ ends in its complete destruction, i.e. the scales again become cycloid (Figure 108J); this corresponds to a situation where the effective reduction of turbulence by means of ctenoids becomes no longer possible. Thus we observe in ontogeny changes in scale structure of a cyclic character, the ctenoid stage lasting from $Re \approx 10^5$ to $Re = 5.0 \times 10^6$ (Burdak 1970).

A similar picture of cyclic changes in the structure of the scaly covering has been established for phylogeny (Burdak 1968, 1969c): most often, longitudinal relief on the scales occurs at *Re* from 10^4 to 10^7. The occurrence of longitudinal relief on the scales in ontogeny and phylogeny in one and the same range of Reynolds numbers undoubtedly points to the development of such relief being wholly determined by its

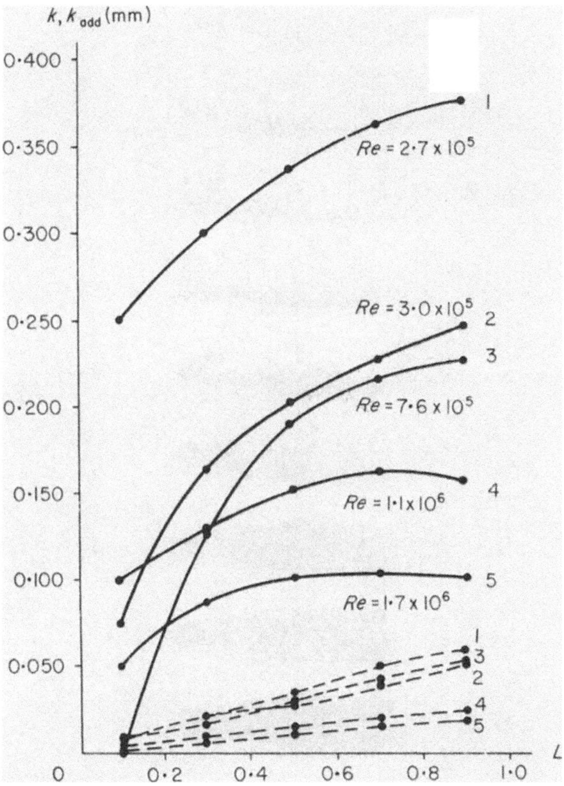

Figure 112. k (continuous lines) and k_{add} (dashed lines) versus *L*. 1, *Mullus barbatus ponticus* Essipov; 2, *Diplodus annularis* (L.); 3, *Pomadasis guoraca* Russ.; 4, *Mugil auratus* Risso; 5, *M. cephalus* L. After Burdak (1968).

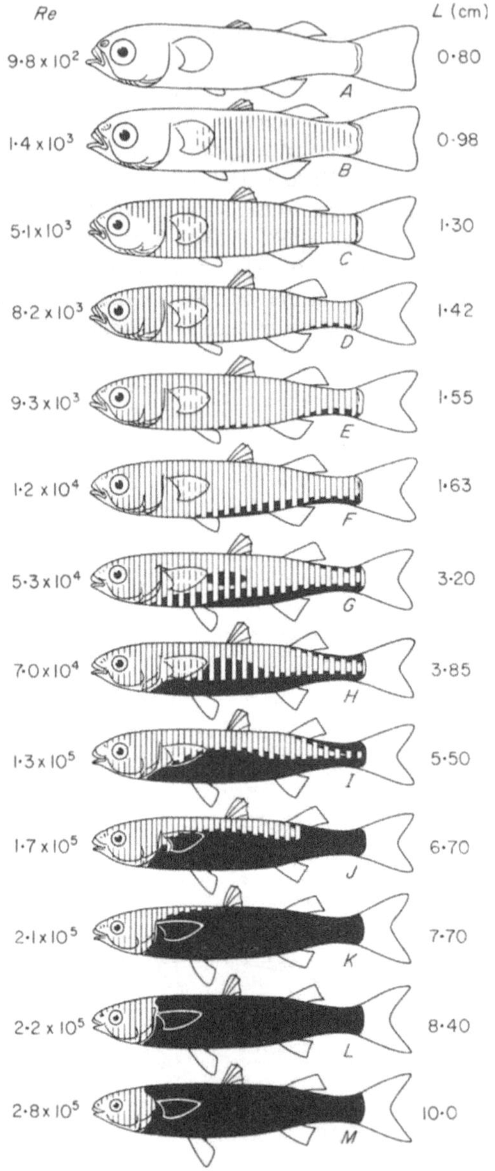

Figure 113. Development of scales in *Mugil saliens* Risso. ▭, no scales; ||||||, cycloid scales, **I I I I**, cycloid and ctenoid scales; ■, ctenoid scales. *Re* = Reynolds number, *L* = length of fish to end of vertebral column. See text. After Burdak (1969a).

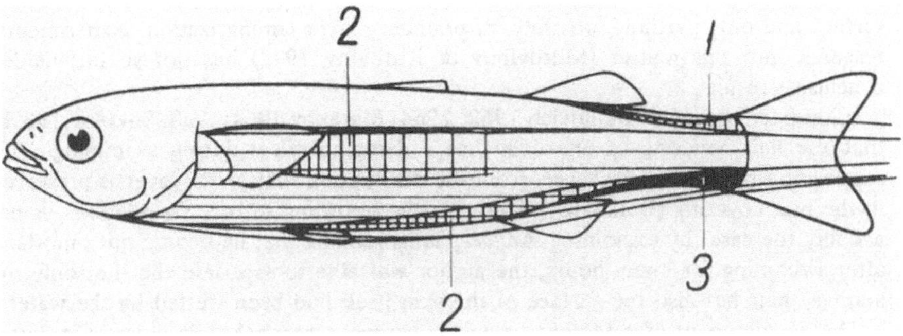

Figure 114. Position of potential turbulization zones. Intensity of hatching corresponds to intensity of the effect of turbulizing factors; figures indicate number of superimposed zones of turbulization. After Burdak (1969a).

hydrodynamic function. It is obvious that at $Re < 10^4$ no laminarizor is needed on the scales as in this case no threat of boundary-layer transition arises on such well-streamlined bodies as fishes. At higher *Re* the reduction of turbulent pulsations in the boundary layer is ensured by special adaptations, one of which is longitudinal relief on the scales. However, at $Re > 10^7$ the preservation of laminarity of the boundary layer or an appreciable reduction of its turbulence proves impossible even by means of special adaptations, and in these circumstances fishes adapt to swimming in a turbulent regime. At the same time, the presence of longitudinal microfolds on the body of *Xiphias* (Koval 1972; see also Figure 102C) and large sharks (Bigelow & Schroeder 1948; Zayets 1972, 1973a) is evidence that in a number of cases fishes preserve the hydrodynamic function of longitudinal relief even at $Re > 10^7$, which fully conforms to the conclusion (Burdak 1970, 1973a,b) on the universality of the hydrodynamic function of longitudinal relief on the body surface in various groups of ichthyoids and fishes.

The various types of scale in ichthyoids and fishes, placoid, cosmoid, ganoid and elasmoid, correspond to the various stages of these animals' adaptation to the nektonic mode of life, in the course of which the scales gradually lose their initial defensive function and acquire a new, hydrodynamic one. The concluding step in the development of the hydrodynamic function of the scales in the group Chondrichthyes was the appearance of placoid scales with a well-defined crown, bearing clear-cut longitudinal crests (as in *Sphyrna, Carcharhinus,* etc.), and in the Osteichthyes group the appearance of fine elasmoid (i.e. cycloid and ctenoid) scales, fulfilling diverse hydrodynamic functions; in both these cases the scales were already completely devoid of a protective function (Burdak 1973b).

B. Hydrodynamic functions of feather and hairy coverings

Apparently, the hydrodynamic function of feather and hairy coverings is comparatively insignificant, consisting mainly of ensuring a certain smoothness of the body surface and only partially, possibly, of boundary-layer laminarization. Experimental research into this matter (Mordvinov & Kurbatov 1972) has not so far yielded conclusive results.

It has been assumed (Belkovich 1962, 1964; Marakov 1964; A. S. Sokolov 1969b) that the hair covering of Otariidae has a damping effect during swimming. This supposition proceeded, however, from the assumption that an air layer is preserved in the hair covering (Belkovich 1962), which, according to our observations, is not actually the case: by examining *Arctocephalus pusillus* that had come out onto land after swimming for many hours, the author was able to ascertain that not only the animals' hair but also the surface of the skin itself had been wetted by the water.

The development of a feather or hair covering in xeronektonic animals is determined not only by their degree of adaptation to the nektonic mode of life, but also by the conditions under which these animals stay on land: the hydrometeorological conditions in their breeding grounds, the character of the substrata, etc. Because of this, it is impossible to do away with feather and hair coverings in the xeronektonic stage, particularly in higher latitudes.

The degradation of the hair covering in mammals when they go over to the nektonic form of life is due on the one hand to the impossibility of preserving the air layer in the hair covering when the animal stays for long in the water, which destroys the heat insulation it provides (Sokolov 1965), and on the other to the need to ensure certain hydrodynamic functions that can be performed by naked skin only.

2. Boundary-layer Control by Elastic Deformation of the Body Surface

A. Skin of nektonic animals as a damping complex

Everything indicates that one of the means of controlling the boundary-layer flow over nektonic animals is elastic deformation of the body surface during swimming, especially directed at damping turbulent pulsations. So far the existence of such a mechanism has not been proved experimentally for any nektonic animal owing to the great difficulty of carrying out the appropriate experiments. However, on the basis of morphological examination of the skin of various nektonic animals, the existence of such a mechanism has been demonstrated for cetaceans (M. O. Kramer 1957, 1960a–c, 1962, 1964, 1965; V. E. Sokolov 1960, 1965a,b; Tomilin 1962b,c, 1965, and others), and assumed for fishes with a reduced scaly covering, such as Scombridae (Walters 1962; Aleyev 1963a), for Cephalopoda (Zuyev 1966, 1969) and for some others. It has also been suggested (Burdak 1972) that there might be a similar damping mechanism based on the epithelium of the elasmoid scales of fishes.

Of the greatest interest in this regard, among nektonic animals whose skin is not covered by scales, feathers or hair, are the cetaceans, whose skin structure has **specific features obviously aimed at boundary-layer control**. Without dwelling on the generally known details of the histoanatomy of the skin of cetaceans, which have recently been thoroughly investigated precisely because of their possible hydrodynamic importance (V. E. Sokolov 1953, 1955, 1960a,b, 1962a,b, 1965a,b, 1971; Sokolov & Kuznetsov 1966; Sokolov, Kuznetsov & Rodionov 1968; Sokolov, Kalashnikova & Rodionov 1971; Naaktgeboren 1960; Purves 1963; Palmer & Weddel 1964; Surkina 1968, 1970, 1971a,b, and others), we shall note that the anti-turbulence properties of the skin of cetaceans are due to all its morphological features acting together. The complete absence of hairs permits direct contact between the aerodynamically smooth epidermal surface (Figure 115A) and the flowing stream, which is necessary for the skin to perform its anti-turbulence function. Firm adhesion of the dermis and epidermis, ensured by the penetration of dermal papillae into the cells of the lower layer of the epidermis (Figures 115A,B), prevents shifts in the epidermis due to friction against the flowing stream in contact with it (V. E. Sokolov 1955). The upper epidermal layer, which is not perforated by the cells of the dermal papillae and in dolphins is about 0.5 mm thick, forms a comparatively dense elastic membrane and transmits without distortion all boundary-layer pressure pulsations to the underlying layer of the epidermis, which is capable of elastic deformation since the dermal papillae which permeate it are made of looser and more hydrated tissue, including fat cells (Figures 115A–D). The papillary layer in dolphins is about 1.0 mm thick and is capable under excessive pressure of being compressed perpendicular to the skin's surface, whereas at points of reduced pressure it expands in the same direction. Such compression or expansion is accompanied by deformation of the elastic structures in the papillary and subpapillary layers of the dermis, primarily its fat cells (M. O. Kramer 1957, and others; Anonymous 1966; Tomilin 1965, and others), during which energy is exchanged between the external medium and the skin, resulting in damping of the incipient turbulent pulsations. The overall elasticity of the dermis is increased on account of bundles of collagen and elastin fibres woven into it in every direction (Figures 115A,C). Individual bundles of these fibres extend upwards forming dermal crests entering the epidermis (Figure 115C) and oriented, on the whole, longitudinally (Figure 116A); papillae rise from these crests (Figures 115A,B). The blood filling the capillaries passing through the centre of a papilla can be regulated, thereby altering the viscoelastic properties of the papillary layer of the dermis and the skin as a whole (Babenko, Kozlov & Pershin 1972 and others). The longitudinal dermal crests with rows of papillae on them sometimes form visible longitudinal relief on the body surface (Figure 116B), which is thought (Purves 1963) also to facilitate boundary-layer laminarization. The space between the collagen and elastic fibres is filled with elastic fatty tissue, the amount of which increases from the subpapillary layer to the dermis (Figure 115A); in some cases fat is contained in the papillae. Both the epidermis and dermis become thicker and stronger on the frontal surfaces of the body, where the pressure of the water during movement is higher; the network of collagen fibres in such areas is thicker and denser and the dermal papillar are more strongly developed. A skin muscle (*m. panniculus carnosus*) stretches along the lateral and ventral surfaces of the cetacean body from the lower jaw and eye to the

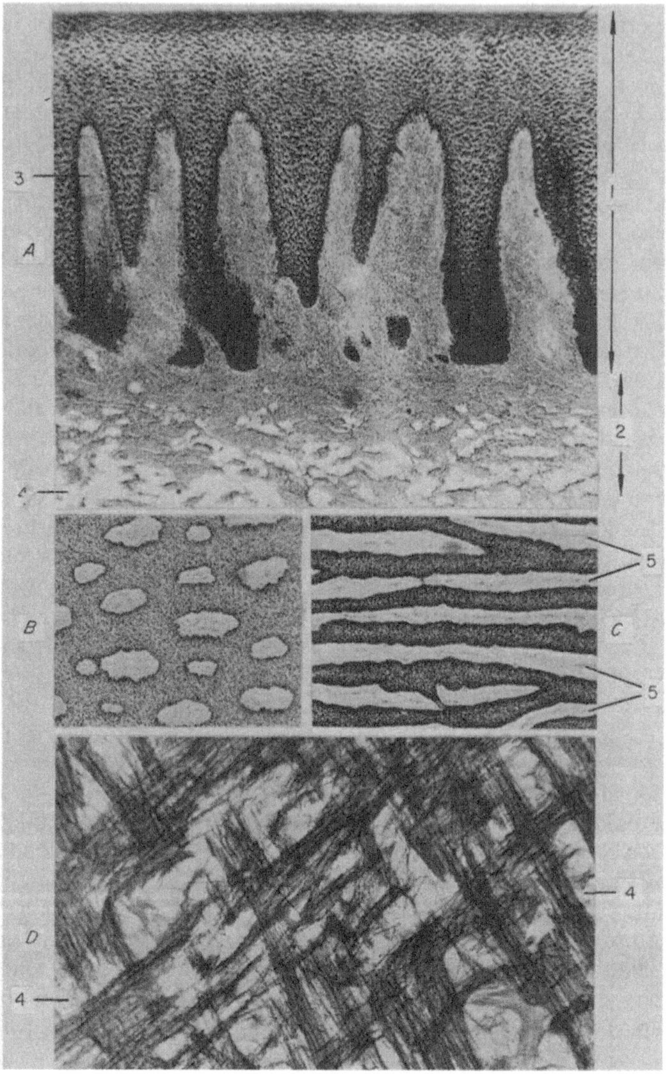

Figure 115. *A-C*, *Tursiops truncatus* (Montagu) (absolute length $L_a = 205.0$ cm); *D*, *Phocoena phocoena* (L.) ($L_a = 110.0$ cm). Skin on lateral part of body level with dorsal fin. *A*, section perpendicular to body surface; *B-D*, tangential sections. *B*, at level of middle dermal papillae; *C*, at level of base of dermal papillae (dermal tori visible running along body); *D*, at level of subpapillary dermal layer (square cells formed by bundles collagen fibre visible). 1, epidermis; 2, reticular dermal layer; 3, dermal papillae; 4, fatty tissue; 5, dermal crests.

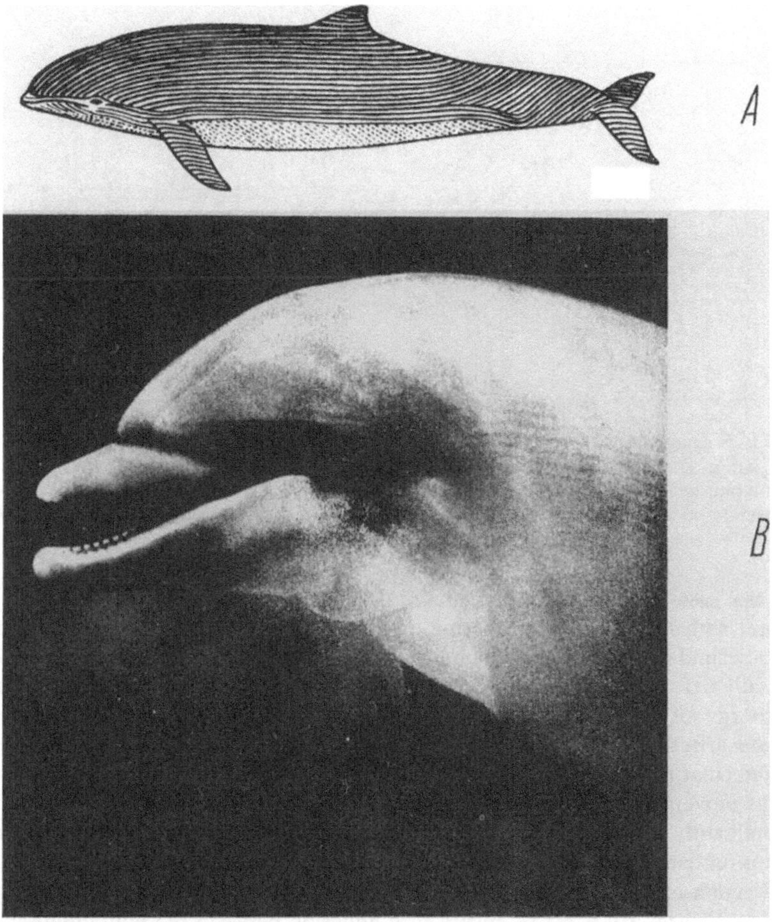

Figure 116. *A*, direction of dermal crests in skin of dolphin (*Phocoena phocoena* (L.); after Purves 1963, modified); *B*, longitudinal relief formed by these crests on skin surface (*Tursiops truncatus* (Montagu); after Essapian 1955).

anal region at the border between the dermis and fatty tissue, and the overall contractions of this are capable of changing the general elasticity of the skin.

To judge from everything, all the anti-turbulence properties of the skin of cetaceans play an important role in the general system of adaptations aimed at reducing hydrodynamic resistance. On the basis of analysis of the skin structure of dolphins, or possibly (Petrova 1970) independently of it, a number of artificial damping coverings have been created by M. O. Kramer (1964, and others).

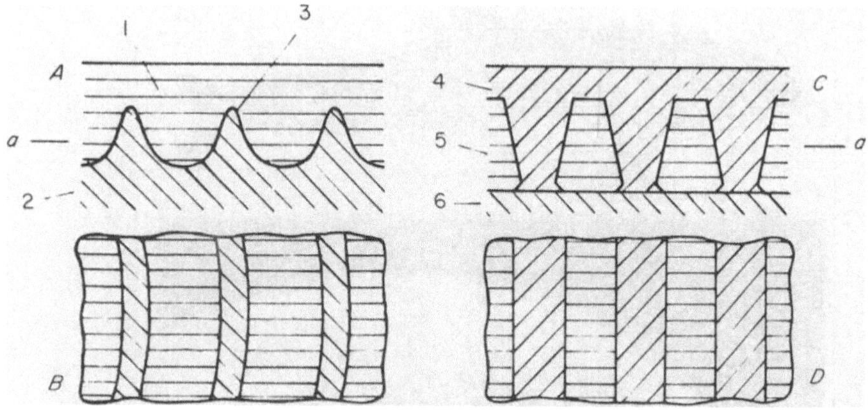

Figure 117. Diagrams of (*A*,*B*) structure of sclerite apparatus on elasmoid scales of *Rutilus rutilus rutilus* (L.) and (*C*,*D*) Kramer's ribbed cover. *A*,*C*, vertical section; *B*,*D*, horizontal section along *aa*. 1, epithelium; 2, bone tissue of scale; 3, sclerite ribs; 4, ribbed diaphragm; 5, intercostal space filled with viscous fluid; 6, rigid sheath of model or ship. After Burdak (1972).

In the case of elasmoid fish scales it is assumed (Burdak 1972) that sclerites, together with the epithelium covering them, function as a damping complex, fundamentally similar to the rib-like damping covering of Kramer (Figure 117), where an elastically deformable epithelium covers the scales and serves as the agent absorbing the energy of turbulent pulsations in the boundary layer. Whenever excessive pressure arises at some point in the boundary layer, the epithelium is compressed perpendicular to its surface; this is accompanied by the deformation of epithelial cells and the movement of viscous cellular juices parallel to the surface of the epithelium, predominantly along the sclerite lunules. The mechanism of this energy exchange remains fundamentally the same whatever the arrangement of the sclerites relative to the direction of the flowing stream. The ribs of the sclerites, being set at an angle to the flow (Figures 108E, 117), ensure firm adhesion between the epithelium and the scales and act at the same time as stiffeners, increasing the structural rigidity of the whole scale (Burdak 1972).

B. Hydrodynamic significance of mobile skin folds

A number of hypotheses have posed the existence of a mechanism reducing the hydrodynamic resistance of nektonic animals by damping turbulent flow and by shifting the incipient vortices back to the rear end of the body by means of mobile skin folds on the body surface (Essapian 1955; M. O. Kramer 1957, 1960a,b, 1962, 1965; Kudryashov & Barsukov 1967a,b; Barsukov 1969; Kudryashov 1969a; Zayets

1973b, and others). However, in no case has this supposed useful effect been proved experimentally owing to the high complexity of staging relevant experiments (Aleyev 1970a), while the very existence of mobile skin folds has so far been demonstrated instrumentally (Essapian 1955) only for Delphinidae.

Experimental studies carried out by this author (Aleyev 1970a, 1973c,d 1974), far from confirming the existence of the supposed useful hydrodynamic effect, have led to the reverse conclusion, showing that in Delphinidae **the mobile skin folds on the body surface** is not an adaptation for reducing hydrodynamic resistance, but merely a parasitic feature on account of which this resistance increases. It is precisely for this reason that, despite the abundance of ciné records on dolphins, a significant ciné record of travelling waves on the surface of a dolphin's body can be seen in one study only (Essapian 1955). Naturally, the structure of the skin of dolphins and other cetaceans is directed at preventing this parasitic effect, and if it does occur during swimming at maximum speeds (Essapian 1955), this is merely evidence that the skin of the dolphin is not so perfectly adapted to movement at such extreme speeds after all.

Indeed, it has been established by means of underwater filming of dolphins in an oceanarium (Essapian 1955) that wavelike folds in the skin's surface perpendicular to the direction of the animal's movement appear on the ventral and lateral surfaces of the body at moments of steep acceleration and during swimming at maximum speed (Figures 118–120). These folds are as a rule immobile, and only in adult animals and at the highest swimming speeds do they tend to move caudad. Such travelling folds appear in two cases: 1. during short sharp spurts lasting not more than 1 s and 2. during sustained swimming at high speeds. In the latter case one observes a series of cycles in the formation of the folds, individual cycles lasting only about 2 s. After local disturbances in the flow have vanished, i.e. when the dolphin stops or slows down, the folds on the surface of the skin disappear owing to the elastic elements it contains.

In the opinion of the majority of researchers (V. E. Sokolov 1962a; Sokolov, Bulina & Rodionov 1969; Babenko & Surkina 1969; Babenko, Gnitetsky & Kozlov 1969; Surkina 1971 and others), travelling deformation waves appear in the skin of the dolphin as a result of the contraction of certain trunk and skin muscles. So far, however, this has not been confirmed experimentally.

Meanwhile, we have demonstrated (Aleyev 1970a, 1973c,d, 1974) that in humans, to be more precise, in women aged between 17 and 30, an absolutely analogous

---

Figure 118. Wavelike folds in skin appearing during rapid swimming. *A*, dolphin *Tursiops truncatus* (Montagu) during spurt when seizing prey (after Essapian 1955); *B*, swimmer (subject 16) when towed at speed of 3.0 m/s, consecutive cine-record frames showing moving folds on ventral side of trunk, filmed by author.

Figure 119. Wavelike folds in skin appearing on entry into water. *A*, dolphin *Tursiops truncatus* (Montagu) (after Essapian 1955); *B,C,* swimmers entering water with arms forward. *B*, subject 18; *C*, subject 24, consecutive frames of cine-record showing moving folds on ventral side of trunk. Filmed by author.

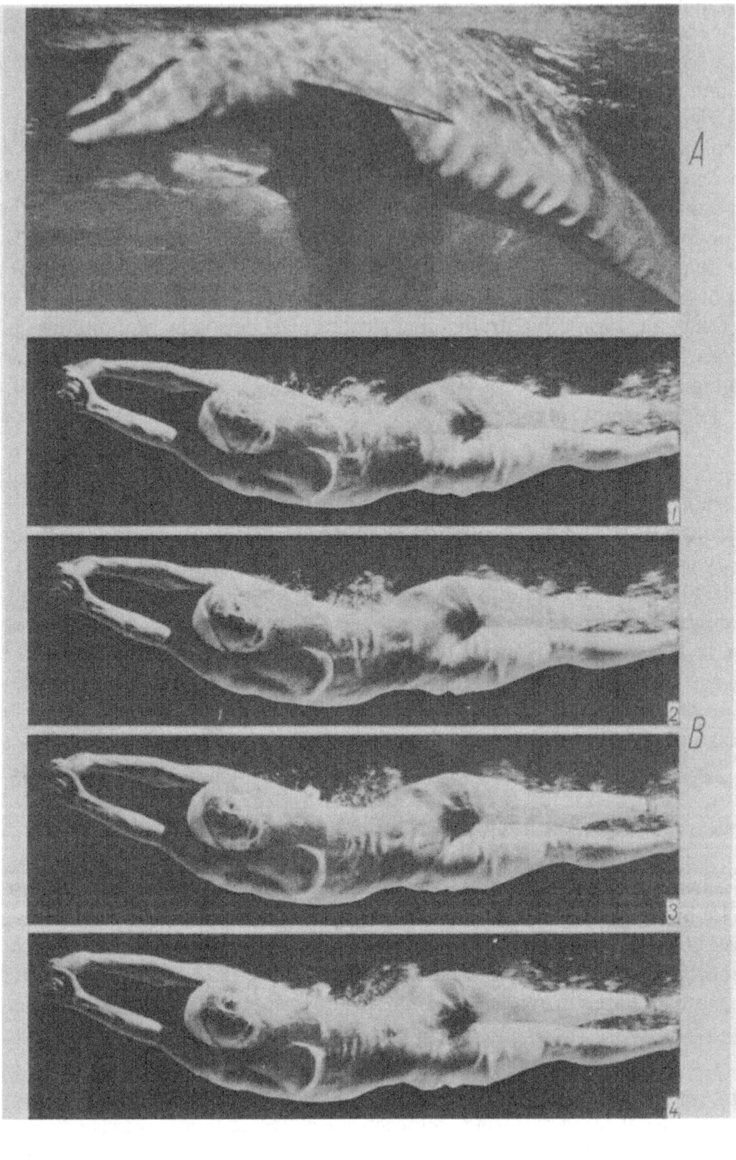
A
1
2
B
3
4

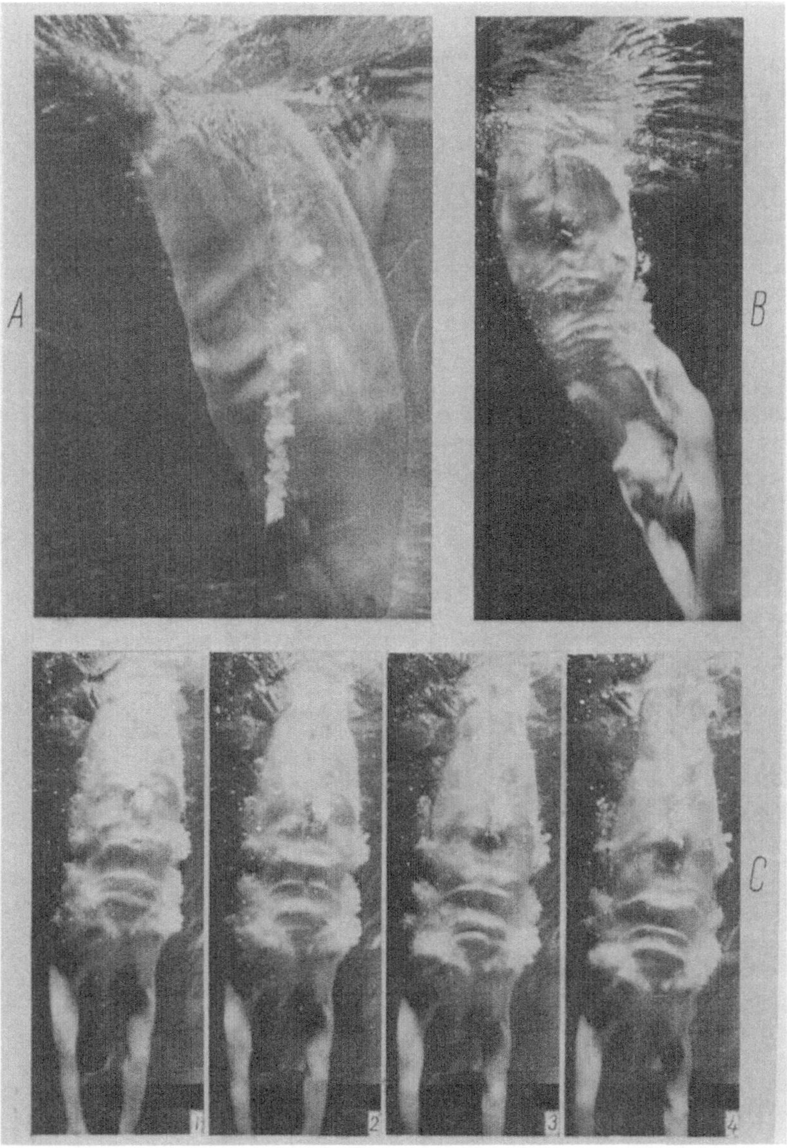
A
B
C
1
2
3
4

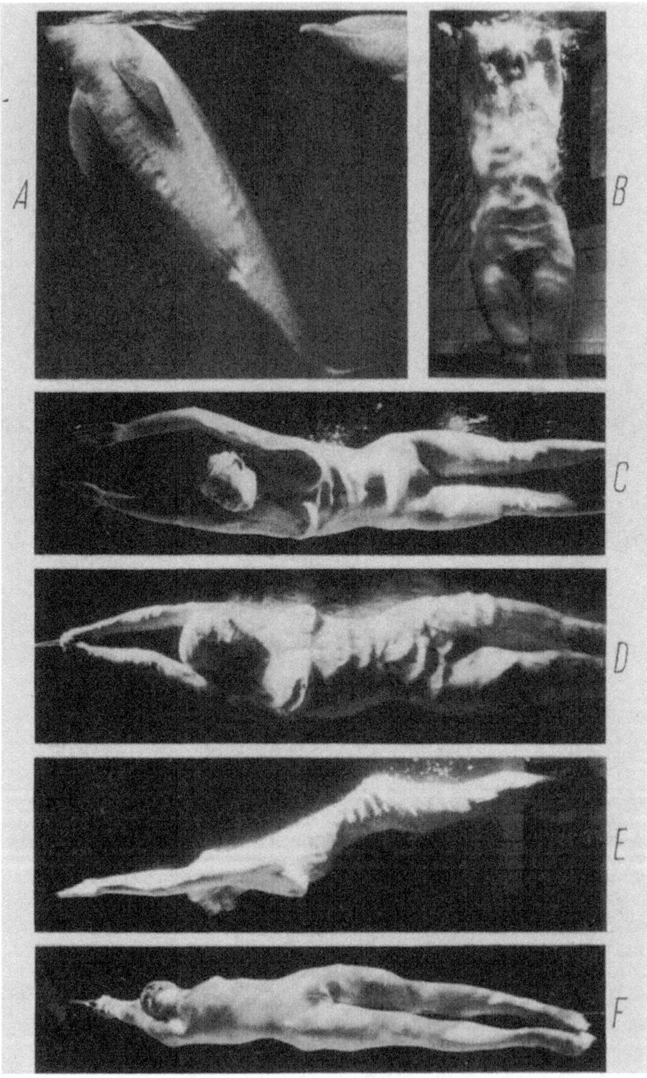
A
B
C
D
E
F

mobile deformation of the skin occurs passively, under the effect of only external hydrodynamic forces. This helps one to understand the mechanism of the appearance of mobile folds on the skin of dolphins and other nektonic mammals.

Women are similar in body size to average-sized dolphins of the *Delphinus* type. For women 160–170 cm tall swimming with arms stretched forward at a speed of 2.0–4.0 m/s, the range of Reynolds numbers is about 3.0×10^6–9.0×10^6, which is entirely inside that most usual for dolphins of the *Delphinus* type (3.0×10^6–1.8×10^7). Just as in dolphins, the general body contours of women are typically smooth, which is due to the specific development of the bones and muscles and the presence of a comparatively thick subcutaneous fatty layer (Bammes 1964, and others), the thickness of which for the greater part of the body is usually from 1 to 4 cm, which is close to the range for dolphins (usually 3–6 cm). Both in dolphins and in women, locomotor muscles lie under this layer of fatty tissue. As far as this particular problem is concerned the body surface of the typical woman may be considered to a sufficient approximation hairless, which is characteristic also of dolphins. All this is to say that woman, as a physical analogue, satisfies in this case all the basic conditions of the biohydrodynamic experiment. To this it should be added that within the trunk and limbs humans have no muscles capable of moving the skin; the subcutaneous *m. platysma* situated on the neck ends at the level of the second rib. The human body is especially valuable in this respect, since no wavy skin folds can appear on the trunk and limbs on account of the contraction of muscles.

With the purpose of clarifying whether mobile wavy skin folds could appear under the action of external hydrodynamic forces alone, we carried out experiments with professional women swimmers (Aleyev 1970a, 1973c,d, 1974). The test subjects, professional sportswomen and masters of sport, numbering 40 and aged between 17 and 28, were from 154 to 168 cm tall. The hairs on the trunk and thighs in all the subjects were few and their pattern of fat deposition and general build close to normal. During the experiments the completely nude subjects swam or were towed (by means of an electric winch) completely immersed at speeds of 2.0–4.0 m/s at a depth of 0.5–1.5 m below the surface of the water, the depth of the pool being about 2.0–2.5 m, and also jumped or dived into the water and rapidly pushed themselves away from the wall of the pool with their feet. Continuous underwater filming of the subjects with a 'Konvas' camera went on throughout the experiment. Along with this a number of **functional-morphological investigations** were carried out. The thickness of the fatty layer, the length of the transverse skin perimeters and the skin elasticity on various parts of the body were measured for all the subjects. Skin elasticity was

---

Figure 120. A-E, different cases of skin deformation under action of hydrodynamic forces; *E,* measurement of resistance forces by means of towed dynamometer. *A,* dolphin *Tursiops truncatus* (Montagu) during spurt (after Essapian 1955); *B,* swimmer (subject 20) on entry into water feet first; *C,* swimmer (subject 24) during spurt while swimming dolphin stroke; *D,* swimmer (subject 11) towed at speed of 3.0 m/s, flow visualized by silk streamers; *E,* swimmer (subject 30) pushing away from wall; *F,* swimmer (subject 38) in hydrodynamically neutral suit during measurement of resistance forces by towed dynamometer. Filmed by author.

measured by a special tensometric elastometer of original design (Figure 121), whose readings were recorded by an H-102 oscillographer. The skin elasticity J_n in section n was expressed as the magnitude j_n of the sagging caused by the application of a standards normal force to a standard area of the skin's surface divided by the maximum sag $j_{\max}$ found for the given subject:

$$J_n = j_n j_{\max}^{-1} \tag{31}$$

To investigate the relation between the nature of the developing mobile skin folds and the characteristics of the flow, the latter was visualized in a number of experiments by gluing silk streamers onto the subjects' bodies (Figure 125).

It was established that, during swimming, during sharp pushes away from the walls of the swimming pool, during jumps and dives into the water, during movement fully immersed at speeds greater than 1.5 m/s, as well as during towing of the subjects (fully immersed) at speeds of 2.0 and 4.0 m/s, clearly visible, comparatively large wavy skin folds appeared on the trunks and thighs of all of them. These folds were on the whole perpendicular to the direction of the subject's movement, and on each section of the body, perpendicular to the direction of the flow, which could be assessed from the position of the silk streamers (Figures 118–120). These folds had nothing to do with the relief formed by the muscles, since at equal speeds of movement they took shape with equal intensity during active swimming and during passive gliding, when the locomotor musculature is completely at rest (during jumps into the water and towing). Besides, rapid filming of the same subjects in air while performing the crawl and dolphin strokes on a special trainer imitating the loads arising during swimming, as well as when performing various gymnastics, proved quite objectively that no skin deformation waves appear in these cases (Figure 122). That the nervous system has nothing whatever to do with the formation of folds in the skin is proved, particularly, by the fact that the direction of travel of the skin deformation waves was wholly determined by the direction of the flow, the waves moving with equal intensity caudad, during swimming, towing or dives into the water head first, and craniad, during jumps into the water feet first (Figures 118–120).

The principal areas where skin deformation waves appeared on the subjects were 1. the ventral and lateral surfaces of the trunk below the mammary glands, 2. the dorsal surface of the trunk below the waist and 3. the entire surface of the thighs (Figures 118–120, 123). A study of the distribution of dynamic pressure on the surface of a rigid model of subject no. 2 (Aleyev 1970a) was carried out in a biohydrodynamic channel. This showed that all the principal sites where skin deformation waves appeared are behind the line M of minimum dynamic pressure both in women and in dolphins (Figure 123), i.e. in this respect too the conditions for the formation of deformation waves on the body surfaces of dolphins and women turn out to be the same.

The appearance of wavy skin folds and the time for which they existed were the same in dolphins and women. Both in swimmers and in dolphins folds appeared in two cases: 1. during sharp spurts and 2. during sustained swimming at speeds above

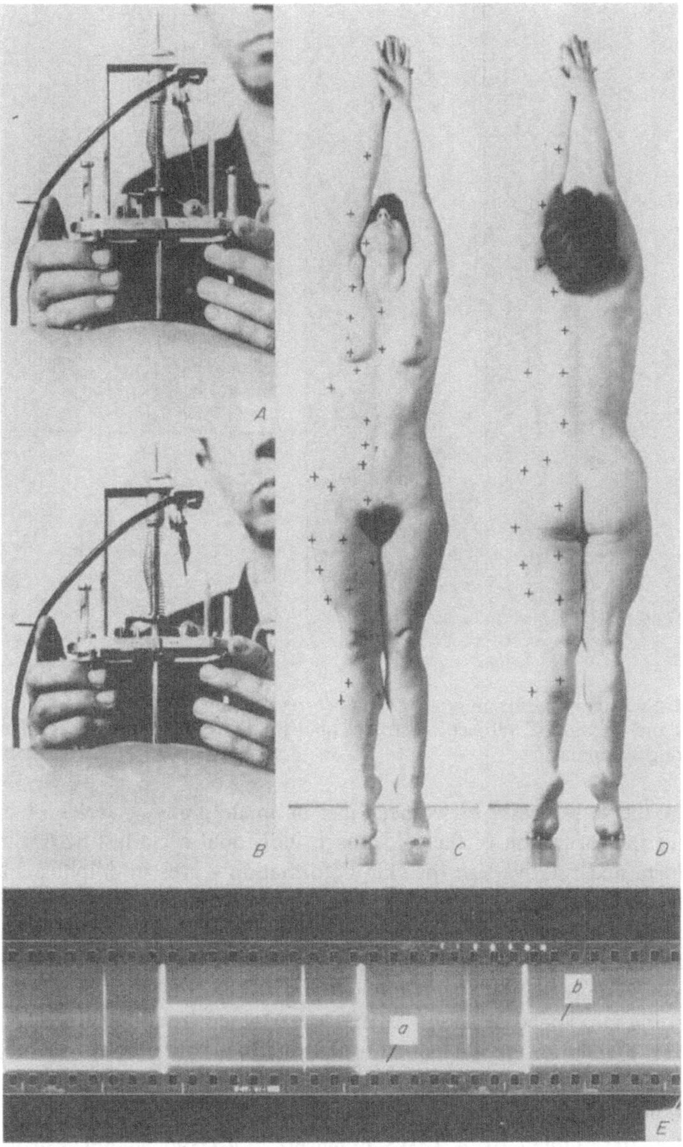

Figure 121. Measuring body elasticity by means of elastometer. *A,B*, elastometer in use (*A*, rod in zero position; *B*, rod in working position); *C,D*, location of points examined (subject 16); *E*, oscillogram of body elasticity (*a*, zero point; *b*, working point).

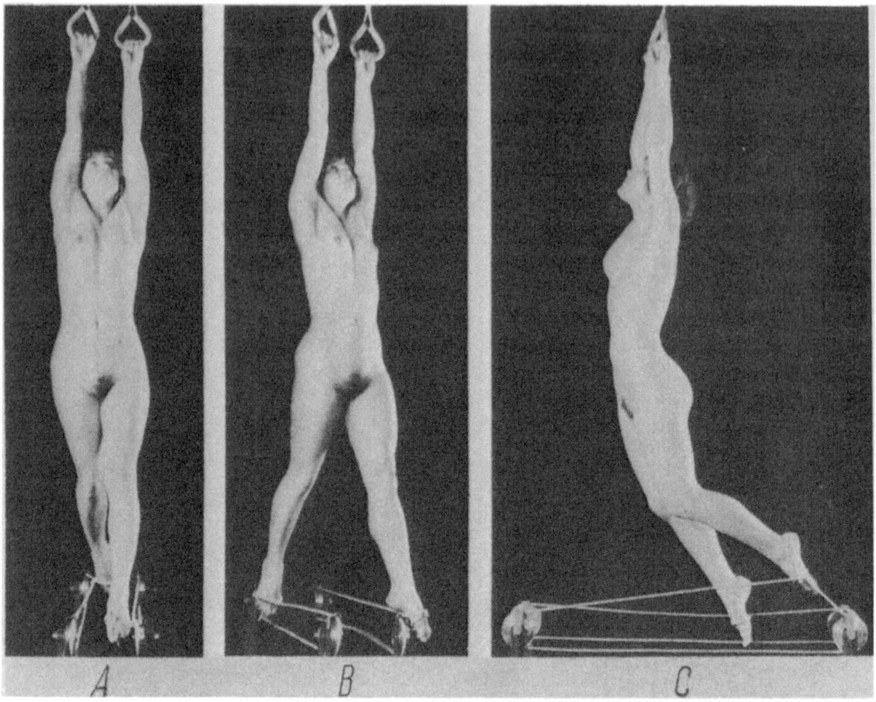

Figure 122. Swimmers performing movements corresponding to the crawl (*A*, subject 27; *B*, subject 19) and dolphin (*C*, subject 27) strokes on a special trainer imitating the dynamic loads arising during swimming.

1.5 m/s. In the latter case, in women, just as in dolphins, a series of cycles was observed in the formation of the folds, each individual cycle lasting not more than 1.5 s. In form and relative size the skin deformation waves in dolphins and women were fairly similar (Figures 118–120). The speed of movement of the deformation waves over the subjects' skin was close to their speed of swimming, which fits in with the appearance of these waves being entirely due to the flowing stream; thus the speed at which deformation waves moved over the skin of subject no. 2 when she was swimming at a speed of about 2.0 m/s, determined from film frames, was about 1.8 m/s. The maximum mobility of skin deformation waves, both in women and in dolphins, was observed at maximum speeds.

The frequency of appearance of skin deformation waves in all the subjects, expressed as the number n_d of waves appearing in a given section n of the body in 1 s, like their amplitude a_d, is directly dependent on the swimming speed, the thickness of the subcutaneous fatty layer and the length of the transverse perimeter and the elasticity of the skin. The thickness of the fatty layer and the length of the transverse perimeter of the skin determine the degree of skin elasticity, directly dependent on which is the frequency at which skin deformation waves are formed, as can be seen from the nature of the curves $n_d = f(J_n)$ and $a_d = f(J_n)$ (Figure 124). The noticeable streamwise reduction in the thickness of the fatty layer, the transverse perimeter of the skin and skin elasticity on the distal areas of the thighs in women and at the base of the caudal peduncle in dolphins leads to severe damping and destruction of the deformation waves in the skin there (Figure 123).

By the same methods as for women, we obtained data on the thickness of the subcutaneous fatty layer, the transverse perimeters of the body and skin elasticity for the dolphins *Tursiops truncatus* and *Phocoena phocoena.* Comparison of these data with the results (Essapian 1955) on the development of mobile folds on the skin of dolphins indicates that the dependence of the development of mobile folds on the functional-morphological properties of the skin is on the whole identical in dolphins

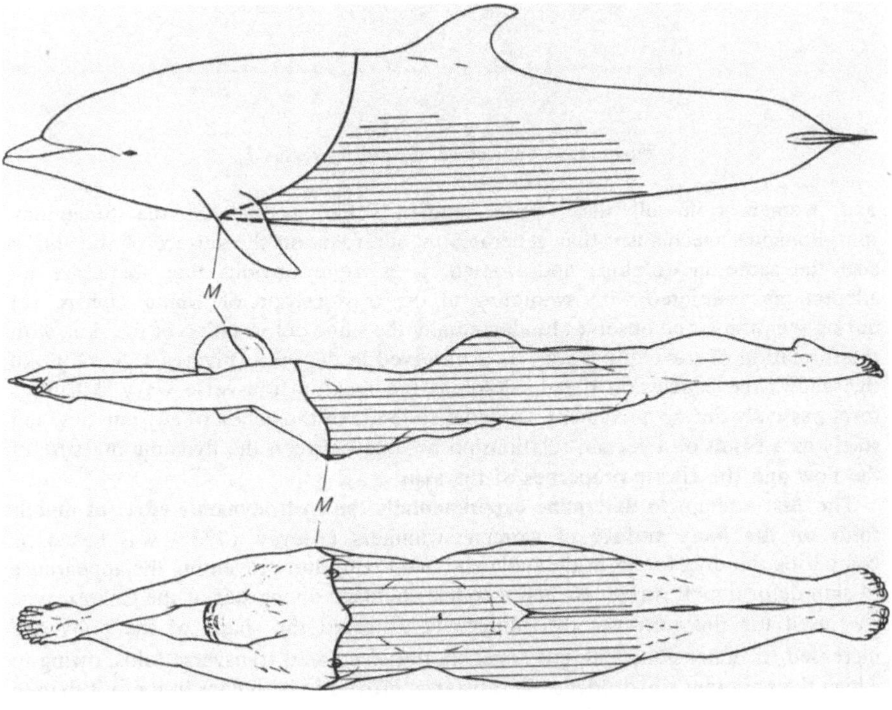

Figure 123. Lines *M* of minimum dynamic pressure on the body of the dolphin *Tursiops truncatus* (Montagu) and of woman (subject 2) according to results of testing models. Hatching shows basic areas where sking deformation waves appear on body surface under the action of hydrodynamic forces.

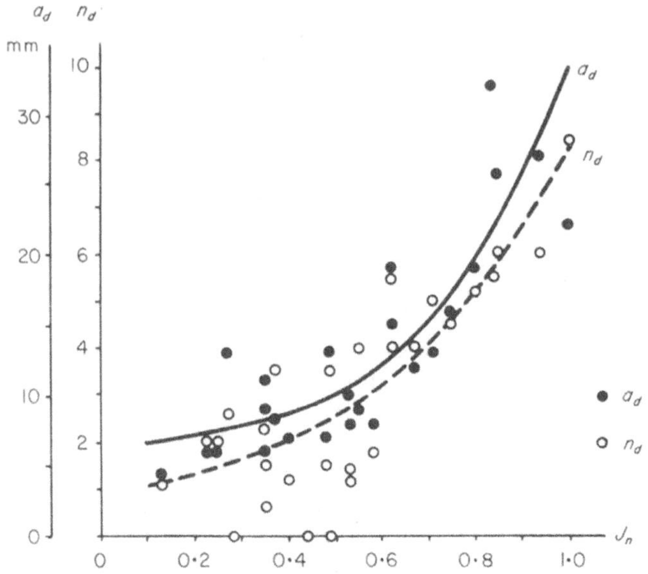

Figure 124. Subject 16. a_d and n_d versus J_n.

and women from all the aspects studied. This shows that the functional-morphological mechanism that generates mobile folds on the surface of the skin is also the same in dolphins and women. It is quite obvious that there are no adaptations associated with swimming in the organization of *Homo sapiens*, yet during swimming one observes fundamentally the same deformation of the skin, with the formation of travelling waves, as is observed in dolphins (Figures 118–120). All this allows the conclusion that in dolphins too moving transverse wavy skin folds form passively during movement, i.e. independently of the action of any muscles, and solely as a result of a certain relationship holding between the dynamic pressure of the flow and the elastic properties of the skin.

The first attempt **to determine experimentally the hydrodynamic effect of mobile folds on the body surface** of women swimmers (Aleyev 1973c) was based on comparing the drag forces in the nude and in a diving suit preventing the appearance of skin deformation waves. At first a coarse standard diving suit of the Calipso type was used for this purpose, this noticeably distorted the shape of the body and increased its dimensions and, moreover, in places created transverse folds, owing to which the swimmer's hydrodynamic resistance proved to be higher in the suit than in the nude (Aleyev 1973c). These tests made obvious the need to make a special, hydrodynamically reliably neutral suit that would not distort the body shape of the subjects, and also to devise a more accurate method of measuring the resistance forces.

Later, with the purpose of assessing the hydrodynamic effect of mobile folds by means of a special tensometric towed dynamometer of special design, we made (Aleyev 1974) measurements and produced oscillographs of the resistance forces when the subjects were towed nude and in special hydrodynamically neutral suits (Figure 120; see also frontispiece). Individually fitted suits consisting of thin, elastic, tightly clinging, synthetic fabric were made for 10 swimmers. Keeping the body shape absolutely intact, these swim-suits had no transverse seams and covered practically the whole body (Figure 125), and during towing prevented deformation waves from forming on the surface of the body; by stretching uniformly over all parts of the body they stabilized their shape (Figure 120). Examination of the elasticity of the body in such a suit by means of an elastometer showed that the surface elasticity on all areas of the body (Figure 121) was proportionally reduced. This could be assessed from the changes in the j_n: when measured in a suit these came to 51–69%, on average 62.1%, of similar measurements made in the nude. Accordingly, the intensity of body surface deformations under the effects of hydrodynamic forces, expressed by the values of n_d and a_d, during towing in a swim-suit was reduced nearly to zero, so that the values of n_d and a_d, found from ciné records, were not more than 10%, and more often under 5%, of the corresponding magnitudes obtained when the same subjects were towed at the same speed in the nude. On this basis, and for the purposes of the present investigation, it could be accepted with adequate accuracy that the suit completely eliminated deformation of the skin.

The hydrodynamic neutrality of the fabric of which the suits were made was preliminarily established experimentally by measuring and making oscillographs of the resistance force on a 100 cm long ellipsoid with a polished surface with and without the fabric covering it. The tests showed that at towing speeds $V \leqslant 3.5$ m/s and $Re \leqslant 3.5 \times 10^6$ the fabric did not increase the hydrodynamic resistance of the ellipsoid. Therefore the swimmers were towed in swim-suits at $V = 2.0$ m/s with $3.2 \times 10^6 \leqslant Re \leqslant 3.4 \times 10^6$.

These experiments (Aleyev 1974) showed that deformation of the body surface under the action of hydrodynamic forces increased hydrodynamic resistance, i.e. constitutes a parasitic effect: in the absence of body-surface deformation, i.e. when a swimmer was towed in a neutral suit, the resistance F_1 always proved to be noticeably lower than that when she was towed in the nude (F). F_1 for the ten subjects at V 2 m/s was from 91.2 to 96.2% of F, so that on average F_1 was 93.9% of F.

As for the cutaneous muscle in cetaceans, its function is not to generate deformation waves in the skin, as is usually assumed (Surkina 1971a, and others), but precisely the reverse: to prevent, through its general contraction, parasitic deformation of the skin at moments of steep acceleration and during uniform swimming at maximum speeds, which is never for very long. When our subjects were towed at a speed of about 4 m/s, they experienced in a number of cases painful sensations on body areas corresponding to those of maximum development of mobile skin folds in dolphins, in which such painful sensations probably appear also and which they eliminate by general contraction of the cutaneous muscle.

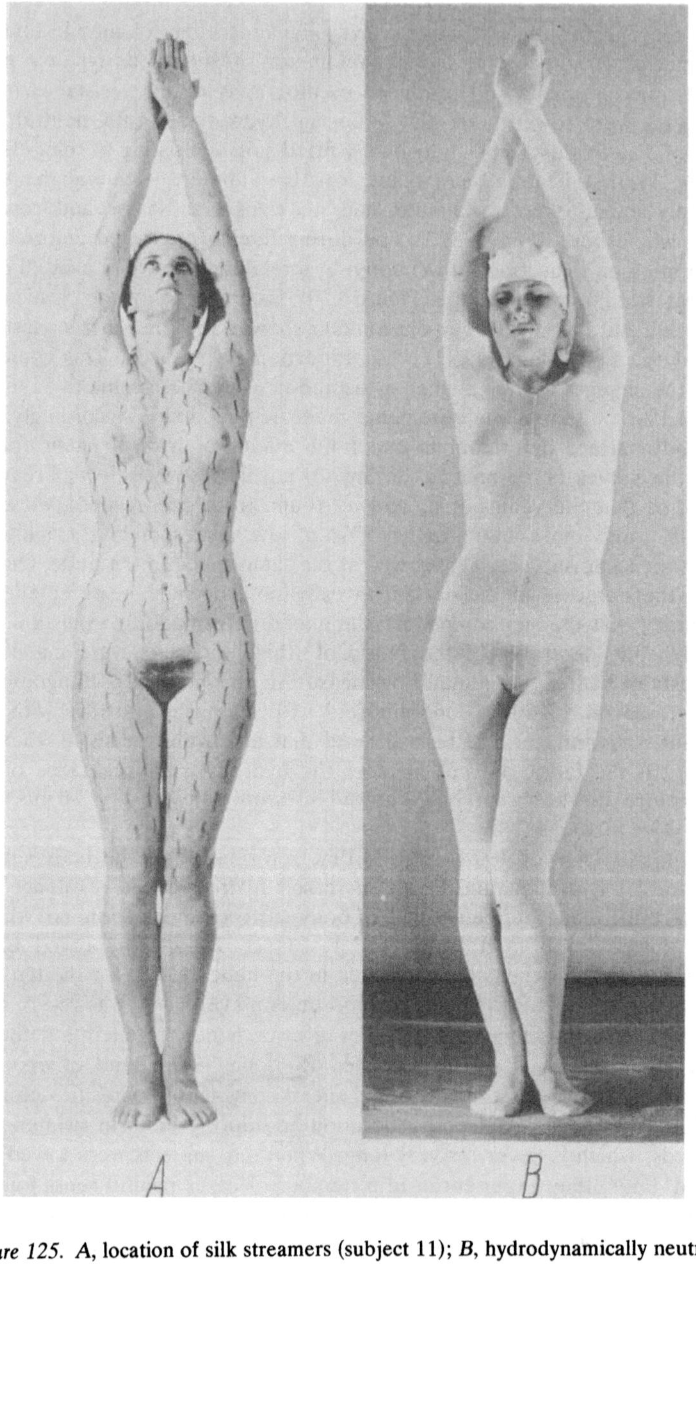

Figure 125. *A*, location of silk streamers (subject 11); *B*, hydrodynamically neutral suit (subject 31).

The skin of cetaceans is richly innervated (Palmer & Weddel 1964; Agarkov & Ferents 1967; Agarkov & Khadzhinski 1970; Khomenko 1970a,b, and others) and readily responds to various types of tactile stimulation. This is connected, however, not so much with its hydrodynamic function (V. E. Sokolov 1962), as with its function as a sense organ (Palmer & Weddel 1964; Khomenko 1970a,b, and others), which is of enormous importance from various aspects of interspecific relations. The clear-cut connection between the respiratory rhythm and the emergence of definite areas of the animal's body into the air characteristic of cetaceans (Tomilin 1957) undoubtedly requires such very complicated skin innervation. By the way, for man, just as for cetaceans, abundant skin innervation as compared with all other primates has been specifically reported (Nestrukh 1970), as well as a large number of blood vessels in the skin, which is associated with the degradation of the hair covering.

All the above leads to the conclusion that, not only in Delphinidae, but in other nektonic animals as well, mobile skin folds on the surface of the body is by no means a useful adaptation and does not help to reduce hydrodynamic resistance.

3. Boundary-layer Control Methods not Connected with Body Surface Relief and its Variations

Among the methods of boundary-layer control characteristic of nektonic animals and not immediately connected with body surface relief and its variation, the three that are of the most general importance are 1. excretion of mucilage by the skin glands, 2. warming the boundary layer and 3. hydrophobic skin coverings.

The role of mucous excretions by the skin glands in reducing hydrodynamic resistance is essential in primary nektonic animals, fishes and cephalopods, and possibly in Sagittoidea as well. The significance of mucilage in ensuring smoothness of the body surface in fishes, rejected by some investigators (Richardson 1936; Gero 1952), is undoubtedly great (Zenkevich 1944; Kudryashov & Barsukov 1967a,b; Burdak 1968; Kalugin & Merkulov 1968; Kozlov & Pyatetsky 1968; Pyatetsky & Savchenko 1969; Whitear 1970; Rosen & Cornford 1971; Belyayev & Koval 1972; Chaikovskaya & Sedykh 1972; Koval 1973, and others). It has been established that fish mucilage reduces frictional drag through boundary-layer laminarization (Kobets 1969; Kobets, Zavyalova & Komarova 1969) by inducing anomalously low viscosity in the boundary layer because its viscosity is lower than that of water (Merkulov & Khotinskaya 1969, and others).

Action of mucilage as a factor facilitating hydrodynamic smoothness of the body surface has likewise been reported for Teuthoidea (Zuyev 1964b). It is assumed (Surkina, Uskova & Momot 1972) that the skin excretions of dolphins also have some influence on the boundary-layer flow; this, however, is rather doubtful.

In some cases **warming of the boundary layer** facilitates its laminarization, for this reduces the viscosity of the water layer immediately adjacent to the nektonic animal's body, on account of which the retardation of the boundary layer is reduced and its laminarity thereby maintained.

We can speak with greatest confidence of this effect with regard to warm-blooded cetaceans (Parry 1949c; Belkovich 1965; Babenko & Surkina 1969; Babenko, Gnetetsky & Kozlov 1970, and others), Measuring the skin temperature of *Phocoena phocoena* during swimming showed (Parry 1949c) that the warmest areas of the body surface, where the temperature is 2–3 °C higher than on neighbouring areas, correspond to the most probable sites of boundary-layer transition.

More problematic is such a thermal effect in the boundary layer of fishes. It is assumed (Walters 1962) that it occurs in Scombridae.

The body surface of a number of nektonic animals (cetaceans, etc.) is **hydrophobic** It has been supposed (Tomilin 1962a, 1965) that in cetaceans some importance with respect to reducing hydrodynamic resistance is attached also to the hydrophobic properties of the skin. However, calculations (Kobets 1968) show that for an animal of length 3 m swimming at a speed of 20 m/s the relative gain due to hydrophobicity with a turbulent boundary layer is a mere 0.04% or so. It follows, then, that hydrophobicity of the body surface does not appreciably influence the hydrodynamic resistance of nektonic animals.

5. Hydrodynamic resistance and swimming speeds

An examination of the flow around nektonic animals, the structure and performance of nektonic propulsors and the development of nektonic body shapes in phylogeny and ontogeny allows certain conclusions to be reached regarding the form of the dependence of V_m on L_a in swimming animals (Figure 126).

Shuleikin (Shuleikin, Lukjanova & Stas 1937, 1939; Shuleikin 1953, 1965, 1966) noted that, with increasing length, the maximum speeds aquatic animals can achieve increase too, since resistance to movement grows in proportion to the square of the animal's linear dimensions, while body volume, and consequently the mass of muscular tissue, grows in proportion to their cube. Having throughout the large length range $L_a > 1.5$ m only one experimental point (for the coalfish whale), Shuleikin concluded on the basis of his theoretical calculations that this dependence is valid for aquatic animals of any size, '. . . from the huge whales and down to unicellular organisms' (Shuleikin 1953, p. 898). However our investigations, based on a great deal of varied experimental material concerning different groups of nektonic animals (Tables 7, 8, 14 and material from Parin 1971c), showed that this dependence is valid only for $L_a \leqslant 450$ cm, whereas in the case $L_a > 450$ cm it is reversed, i.e. with increasing length in this range one observes no longer an increase, but a decrease in the maximum speed V_m which can be attained (Figure 126).

It follows from the above that no single, universal relation $V_m = f(L)$ valid for all nektonic animals exists. The formula suggested by Shuleikin (1968), according to which $V_m = 125\ L^{1/3}$ cm/s for all fishes and cetaceans longer than 150 cm, cannot actually be used for determining the maximum speed of these animals even in the range $L_a \leqslant 450$ cm, since it completely fails to take into account their specific

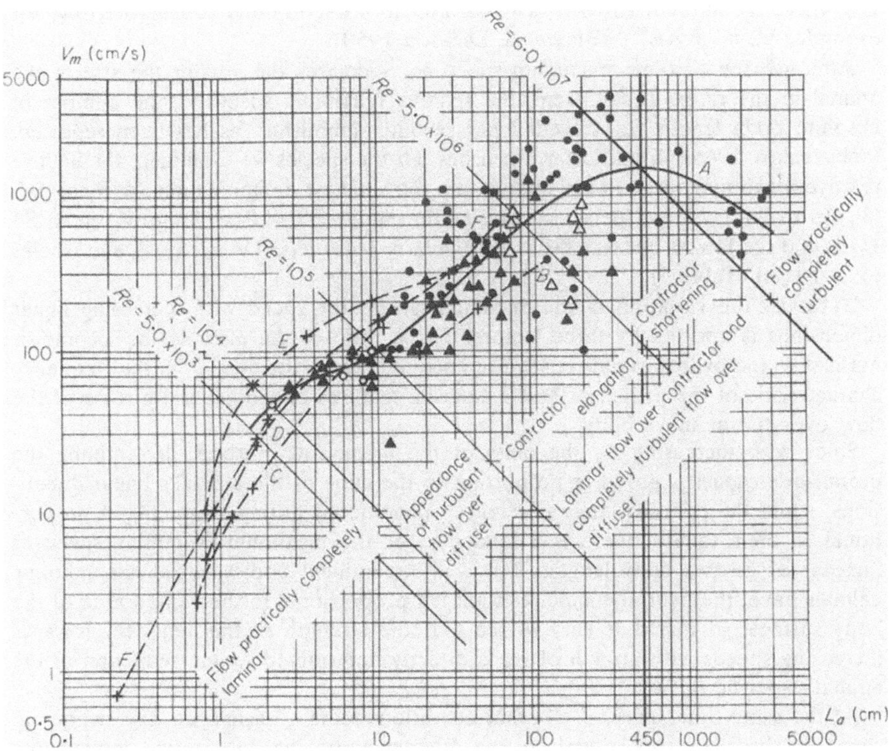

Figure 126. V_m versus L_a for nektonic animals. *A*, all nekton, phylogeny; *B-F* ontogeny. *B*, *Squalus acanthias* L.; *C*, *Carassius auratus* (L); *D*, *Lebistes reticulatus* Peters; *E*, *Mugil saliens* Risso; *F*, *Trachurus mediterraneus ponticus* Aleev. ●, ○, ▲, △, phylogeny (●, eunekton; ○, planktonekton; ▲, benthonekton; △, xeronekton); +, ontogeny, eunekton and benthonekton.

features. Thus, according to Shuleikin's formula, the maximum speed of animals 150 cm long is 668 cm/s, whereas in fact *Thunnus albacora* of this length reaches a speed of 2070 cm/s (Walters & Fierstine 1964), *Acanthocybium solandri* reaches 2140 cm/s (Walters & Fierstine 1964), while *Acipenser güldenstädti colchicus* reaches 300 cm/s (Table 7). According to the same formula the maximum swimming speed at a length of 300 cm is 836 cm/s, whereas *Thunnus thynnus* of this length attains a speed of 2500 cm/s (Nursall 1962), *Xiphias gladius* attains 3610 cm/s (Nursall 1962), while the swimming speed of *Physeter catodon* 18 m long does not exceed 600 cm/s (Tomilin 1962a). There is no point in quoting more examples of this kind; the impossibility of determining the maximum swimming speed of an animal on the basis of its length alone is clear enough to every biologist.

With regard to individual fish species, Bainbridge (1960, 1961) found that, within the range $L_a \leqslant 450$ cm, the maximum speed V_m a fish can achieve is proportional to

L^{α}, where in different cases α is from 0.65 to 1.09. For the oceanic herring, for example, $V_m = 7.6 \times L^{0.94}$ (Blaxter & Dickson 1959).

Although the absolute maximum speed V_m increases the greater the size of the animal in the range $L_a \leqslant 450$ cm, the relative maximum speed V_r, the number of absolute body lengths L_a travelled per second, diminishes, as has been reported, (Saburenkov 1966), in particular, for fishes. Of the species we examined, the highest relative maximum speeds were attained by the smallest nektonic animals, flying fish (V_r up to 24 L_a/s) and pelagic young of the mullet 3.0–4.0 cm long (V_r up to 34 L_a/s), and the lowest by large fishes, particularly *Huso huso* ($V_r = 2$ L_a/s), and whales ($V_r < 1$ L_a/s) (Table 7).

Naturally the variation in the maximum swimming speed with increasing linear dimensions is affected by three factors: 1. the ratio of the mass of the locomotor muscles to the hydrodynamic resistance encountered by the animal, 2. the frequency characteristic of the propulsor and 3. specific features of the general picture of the flow over the animal's body.

Since with increasing L_a the mass of the locomotor muscles, determining the propulsor's capacity, grows in proportion to the cube of the animal's linear dimensions, while the hydrodynamic resistance encountered during swimming is proportional to their square, there is a tendency for the maximum swimming speed to increase at greater body lengths. For well-streamlined bodies, which all nektonic animals have, the hydrodynamic resistance is proportional to the wetted area of the body surface, so that it is easy to see that the strength of this tendency towards increasing speeds with growth of L_a is directly determined by the reduction in the animal's specific surface.

At the same time, since, if all other conditions remain unchanged, the swimming speed is directly proportional to the frequency of the locomotive movements performed by the propulsor, and this frequency on the whole diminishes with an increase in the animal's linear dimensions both in phylogeny and in ontogeny, there is a tendency for the maximum swimming speed to decrease with increasing body length. The decrease in the working frequency of the axial undulatory propulsive mechanism the greater the size of the nektonic animal can be clearly seen both when comparing different species (Pershin 1970) and in ontogeny (Bainbridge 1958a, 1960), while in the nektonic range of *Re* the propulsor's operating frequency diminishes by a factor of more than ten with increasing L_a. Thus, for example, though in flying fishes (Exocoetidae), most of which are 20–30 cm long, the propulsor can operate at frequencies up to 50–70 c/s (Edgerton & Breder 1941; Bertin 1958a, and others), in the 3–10 cm long fry of the grey mullet (*Mugil*) it can perform, according to our ciné records, at frequencies up to only 30 c/s, and in 110–250 cm long dolphins (*Phocoena, Delphinus* and *Tursiops*), according to the same sources and our ciné recordings, at not more than 10 c/s, while in the nektonic giants, the whales (Pershin 1970), this frequency apparently does not even exceed 2 c/s.

The actual swimming speed in every specific case is determined by the interaction of these two opposite tendencies, the result of which is affected in considerable measure by a third factor: the general characteristics of the flow over the animal's body, which in turn change essentially with increasing L_a and Re.

An analysis of data on the swimming speeds of nektonic animals (Figure 126) leads to the conclusion that the highest absolute speeds can be achieved with some mean absolute linear dimensions and some mean frequency of the propulsor. Conversely, combinations of extreme values of these quantities, minimal linear dimensions combined with a maximal operating frequency of the propulsor and an Re at the very bottom of the nektonic range, or maximal linear dimensions combined with a minimal frequency of the propulsor and an Re extremely high for nekton, do not create optimal conditions for a rise in swimming speeds. It is clear that at low Re swimming speeds are limited by high specific surface area and at high Re by the low operational frequency of the propulsor.

The specific features of the flow, and particularly the state of the boundary layer, essentially influence the speed of nektonic animals whatever the value of Re, as becomes clear from a comparison of the experimental curve $V_m = f(L_a)$ (Figure 126) with our experimental material on the flow over the animals' bodies (Aleyev & Ovcharov 1969, 1971, 1973a,b, and others).

In the planktonic range and in the lower part of the nektonic range, i.e. for $L_a \lesssim 2$ cm and $Re < 10^4$, there takes place a rapid increase in V_m with growth of L_a, which corresponds to the preservation of practically completely laminar flow over the animals; this may be clearly traced in the ontogeny of nektonic fishes (Figure 126, curves C–F). For 2 cm $\lesssim L_a \lesssim 10$ cm and $10^4 \leqslant Re \leqslant 10^5$, there occurs a steep drop in the rate of increase of V_m, which corresponds to the appearance and gradual extension of an area of turbulent flow on the diffuser (Figure 126, curves A,C–F); at the upper end of this range the boundary-layer flow is turbulent over practically the entire diffuser (Aleyev & Ovcharov 1969, 1971). Then, for L_a from about 10 to 80–100 cm and $10^5 \leqslant Re \leqslant 5.0 \times 10^6$, the distribution of the laminar and turbulent flow scarcely changes, i.e. laminar flow is preserved on the contractor and turbulent flow on the diffuser (Figure 126), although the contractor extends with increasing Re (Figures 92A–D). Accordingly, because of the comparatively high operational frequencies of the propulsor, a high, more-or-less steady growth rate of V_m is preserved with growth of L_a: the slope of the curve $V_m = f(L_a)$ for nekton hardly changes in this range (Figure 126, curve A). For $80 \lesssim La \leqslant 450$ cm and $5.0 \times 10^6 \leqslant Re \leqslant 6.0 \times 10^7$ there takes place a progressive shortening of the contractor (Figure 92E–H) and, consequently, a shortening of the laminar section and extension of the turbulent section of the flow, owing to which the growth rate of V_m, against the background of continuing decrease of the propulsor's operating frequency, falls progressively and at $L_a = 450$ cm becomes on average (for nekton) equal to zero (Figure 126, curve A). For $L_a > 450$ cm and $Re > 6.0 \times 10^7$ the further turbulization of the boundary layer, on the contractor too, results in almost completely turbulent flow over the body and, along with the diminishing frequency of propulsive movements, leads to lower values of V_m: situated in this interval is the descending stretch of the curve $V_m = f(La)$ (Figure 126, curve A).

The presence of an ascending and a descending portion of the curve $V_m = f(L_a)$ is responsible for the different directions in the ranges $L_a \lesseqgtr 450$ cm of certain morphological changes in nektonic animals with increasing linear dimensions in phylogeny and ontogeny: whereas in the range $L_a < 450$ cm an increase in L_a corresponds to adaptation to movement at ever increasing speeds, as can be seen in the examples of fishes (Aleyev 1963a) and Teuthoidea (Zuyev 1966), this is not so in the range $L_a > 450$ cm, an example of which is the ontogenetic development of whales (Chepurnov 1968).

Comparing the curves A in Figures 91 and 126, it is interesting to note 1. that at $Re = 10^5$ there begins both a region of predominantly laminarized body shapes (Figure 91) and the region of considerable and more-or-less steady growth of V_m (Figure 126, curve A), 2. that $Re = 5.0 \times 10^6$ corresponds to both a maximum in the curve $Y = f(Re)$ (Figure 91) and the upper border of the region of intensive growth of V_m (Figure 126, curve A), and 3. that at $Re = 6.0 \times 10^7$ the predominantly laminarized body shapes are replaced by predominantly non-laminarized shapes (Figure 91) and simultaneously the descending portion of the curve $V_m = f(L_a)$ begins (Figure 126, curve A). All these coincidences undoubtedly indicate deep connections between definite functional-morphological features and both elements of the overall picture of the flow over nektonic animals and the swimming speeds they can attain. They point particularly to very great significance in a nektonic animal's hydrodynamics of the features of body structure that are reflected by the index Y, as has been noted earlier (Aleyev 1962).

All this may provide the answer to the question of why nektonic cephalopods, fishes, ichthyosaurs, dolphins, sirenians and pinnipeds are on the whole characterized by $L_a < 450$ cm: at these sizes, within the L_a range of about 4 to 5 m they are capable of achieving the fastest speeds attainable by swimming animals, which without doubt ensures for them ecological advantages in the pelagic zone when escaping from enemies and catching prey. Only the whales, by adapting to feeding on relatively small objects, trapped by means of straining, have considerably overstepped the threshold $L_a = 450$ cm, which both allowed them to increase the absolute dimensions of the straining apparatus and also diminished considerably the number of their potential enemies, of which they are in effect free. Predatory forms with $L_a > 450$ cm, such as *Eurhinosaurus* or *Physeter*, are very few and on the whole are not characteristic of eunekton.

In the range $L_a > 450$ cm, with the growth of L_a one observes in nektonic animals a tendency towards maximum retardation of the inevitable decrease in V_m, which is well illustrated by the dependence of C_5 on Re for eunekton (Figure 48): the value of C_5, showing the degree of concentration of the propulsive action in the rearmost (fifth) elementary section of the body (Aleyev 1969a, 1973a,b), remains extremely high even in whales (Figure 48, area *W*). It is for this reason that we find in all the large late (Cretaceous) ichthyosaurs (Ichthyosauridae and Stenopterygiidae) and in all the whales scombroid-type propulsors, which are known (Aleyev 1963a) to facilitate the maximum preservation, all other conditions being unchanged, of high speeds of locomotion.

6. The phylogeny of adaptations towards reducing resistance to movement

The tendency to reduce hydrodynamic resistance in phylogeny during transition to the nektonic mode of life is among the very first nektonic features to appear. Already in nektobenthic and nektoxeric animals the body, as a rule, is well streamlined. The development of these adaptations reaches, as we have seen in all nektonic forms, a rather advanced stage, being central among the adaptations associated with movement.

On the whole one sees a later appearance in phylogeny of the most specifically nektonic features, special adaptations for controlling the boundary layer being the most important among them.

The basic characteristic of the progressive development of this set of adaptations common to all nektonic groups is the constantly increasing complexity of the individual adaptations making up the set. It is the general rule that with higher relative and absolute speeds of locomotion the complex of adaptations reducing hydrodynamic resistance occupies a more and more dominant position in the overall system of the organism's adaptations and has an ever more determining influence on the general course of the phylogenetic development of the species, which is explained by a progressively increasing share of the total energy of the organism being spent on locomotion.

It should also be noted that most of the adaptations in the group under examination have multiple functions, which corresponds to the close relationship between all the elements of the flow over a nektonic animal's body and the multiple process of controlling it.

VI. CONTROLLING MOVEMENT

1. General

Movement control includes stabilization and changing the direction of movement, maintenance of equilibrium and furnishing the necessary positive and negative accelerations, i.e. speeding up and braking. The performance of all these functions is connected with the creation of moments M_k rotating the body of the nektonic animal about axes passing through its centre of gravity. For a moment M_k to arise it is necessary to apply a single force or a pair of forces away from the centre of gravity. The greater the applied force (or the greater the lesser of a pair of forces) and the greater its distance from the centre of gravity (or the greater the distance between the vectors corresponding to a pair), the greater the **rotating moment** will be

$$M_k = \pm F l_0, \tag{32}$$

where F is the magnitude of the force applied (or the lesser of a pair) and l_0 is the distance from the vector representing the force to the centre of gravity (or the distance between the vectors representing a pair). The signs $\pm$ indicate that the direction of rotation may be different. Let us call a moment M_k negative when it resists a change in the body's position, i.e. restores it to its former position (a stabilizing moment for rotation about one of the transverse axes and a righting moment for rotation around the longitudinal axis). Accordingly, we shall call positive a moment facilitating a change in body position (a turning moment for rotation about any of the transverse axes and a rolling moment for rotation around the longitudinal axis). Thus, in the case of an animal's forward movement, we shall regard a negative moment M_k as indicating dynamic stability and a positive moment as indicating dynamic instability. Slowing down does not, as a rule, require the creation of rotating moments, however, when additional resistance forces are created by deflecting the fins there appear reciprocally damping rotating moments. In some cases braking consists precisely of a sharp change in the direction of movement.

When the force F is created by setting some surface at an angle to the stream, its magnitude is directly proportional to the area of that surface (be it a fin or the trunk of the animal itself) and the speed at which the animal is moving (or, when this is zero, the speed at which the fin is moving). When the force F is created by the action of a hydrojet propulsor its magnitude is determined by the mass of water ejected per unit time.

Movement control is in all cases connected with the creation of temporary rotating moments and is provided 1. by bending the body and 2. by changing the position of the fins or the funnel of a hydrojet propulsor.

A capacity to bend the body is possessed in some degree or another by all nektonic animals. In Testudinata the capacity to bend the longitudinal axis of the body is preserved in the cervical and caudal parts, in Sphenisciformes in the cervical portion only and in Ostraciidae in the caudal portion only. In cephalopods the mantle is comparatively less flexible, while they have very flexible arms. The capacity to bend the body is considerable not only in anguilliform, ribbon-like and snake-like nektonic animals (Anguilliformes, Trichiuridae, Hydrophidae, etc.; Figure 127D), but also in forms with a more compact and relatively short body, such as, in particular, the great majority of fishes and all Sagittoidea, Cetacea, Sirenia and Pinnipedia (Figures 127C,E, 128); the body of the majority of nektonic animals is capable of forming at the instant of turning an arc of angle 90–180° (Figures 127, 128).

Bending the body is in most cases the best possible way to create the above-mentioned moments of rotation essential for stabilization and changing the direction of movement. Fins and flippers in these cases are more often merely additional means of creating the necessary moments. Of great importance in this regard for cephalopods is changing the position of the funnel.

Conversely, maintaining equilibrium and slowing down is connected in the majority of cases mainly with action of the fins, and bending of the body here either does not take place at all or is of secondary importance only. However, in some cases, when movement of the fins is restricted, for example, in Elasmobranchii and partially in cetaceans, and also during swimming at very high speeds, slowing down is achieved mainly by a very sharp turn, owing to which the nektonic animal's entire body finds itself situated perpendicular to the direction of the inertia force, which facilitates a steep rise in the total drag and damping of the inertia force. In cephalopods braking in every case is very much facilitated by changing the position of the funnel.

In all cases the creation of positive accelerations is associated with only one particular mode of performance of the propulsor, i.e. concerns the complex of adaptations already examined in chapter IV.

An essential role in the general system of morphological adaptations directed at movement control belongs in the majority of nektonic animals to the fins and flippers, which perform the functions of rudders, stabilizers, keels, braking panels, etc. In different nektonic animals all these fin functions are rather varied both in the degree of their development and in the details of their morphological structure (Duges 1905; Houssay 1909a,b, 1912; Schlesinger 1911a,b; Schmalhausen 1916; Gray 1933f, 1937; Harris 1936, 1937a,b, 1938, 1947, 1953; Oehmichen 1938; Vasnetsov 1941, 1948; Chabanaud 1943; Baron 1950; Svetovidov 1952; Sleptsov 1952; Aleyev 1955, 1957a–c, 1958a–c, 1959a–d, 1963a, 1969c; Aleyev & Vodyanitsky 1966; Bourdelle & Grasse 1955; Petit 1955; Huene 1956; Rose 1957; Tomilin 1957; Bertin 1958a; Monod 1959; Wickler 1960; Nachtigall 1961; Klausewitz 1962; Akimushkin 1963; Tatarinov 1964e; Khozatsky & Yuryev 1964;

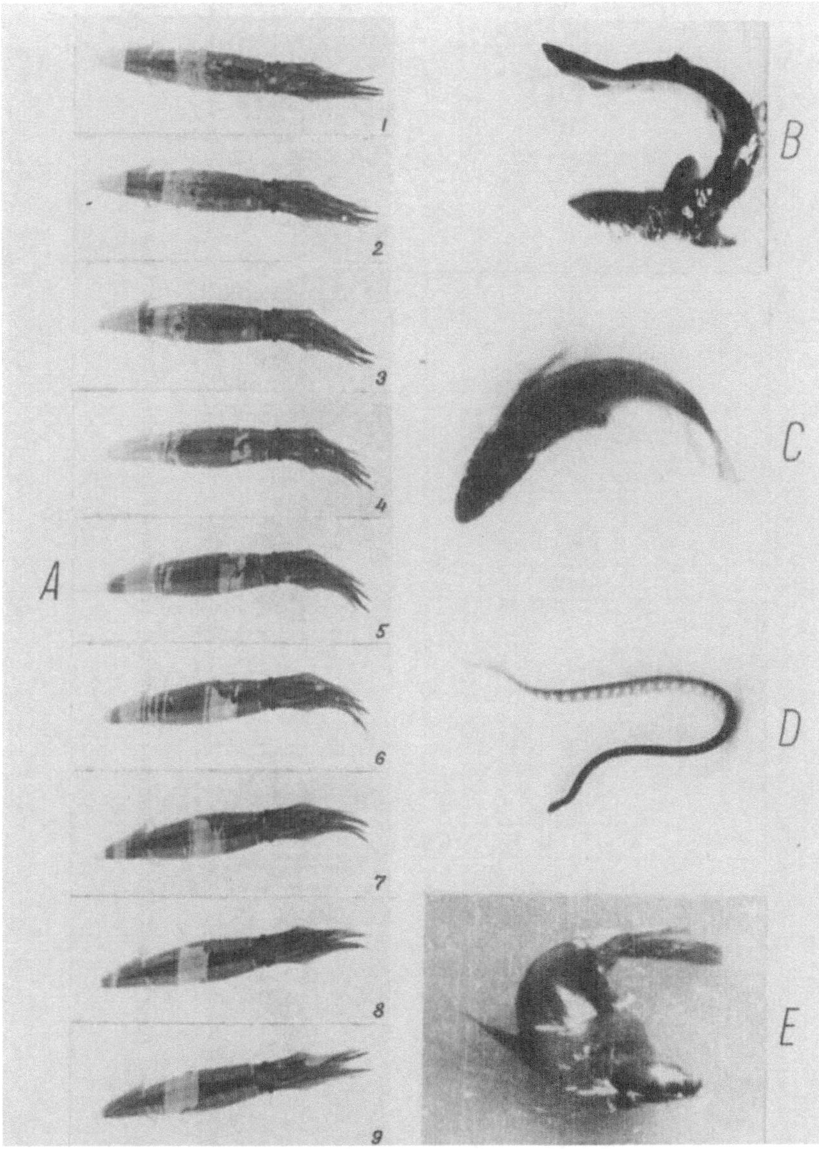

Figure 127. Body bending in nektonic animals when turning. *A*, *Todarodes pacificus* (Steenst.) (cine-record of left turn through 20°, filmed by B. V. Kurbatov); *B*, *Squalus acanthias* L.; *C*, *Diplodus annularis* (L.); *D*, *Enhydrina schistosa* (Daud.); *E*, *Arctocephalus pusillus* (Schreb.) (filmed by Yu. E. Mordvinov).

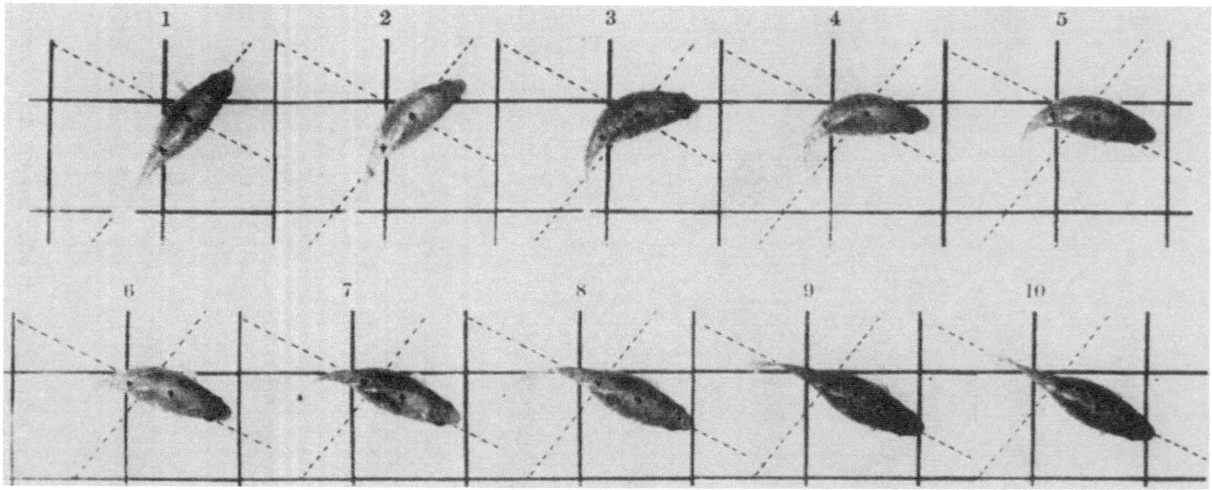

Figure 128. Goldfish (*Carassius auratus* (L.) during an 80° right turn. Film (after Gray 1933f).

Zuyev 1966; Felts 1966; Alexander 1967; Ovchinnikov 1966c, 1967b; Mordvinov 1968; Muller 1968; Niiler & Wite 1969; Ovcharov 1974, and others). However, despite all this functional variety, **the development of fins and flippers in phylogeny and ontogeny** has a number of common features, explainable by the fact that the functioning of all of them is based on the common task of creating additional resistance forces (Aleyev 1957a, 1963a, and others). Of the most general interest among these special features is the regular diminution, in phylogeny and ontogeny, of the fins' relative dimensions as swimming speeds increase.

Since the hydrodynamic resistance of any fin performing the function of a rudder, stabilizer, balancer, braking element or locomotive organ is directly proportional to the square of the speed V of the stream flowing over it, it is obvious that with an increase in this speed the fin will be able to perform its former function at a smaller relative size, owing to which its relative size will decrease. This trend, though initially reported for fishes (Aleyev 1958b), is valid for other nektonic animals as well (Aleyev 1963a). According to this dependence, for nektonic animals, with increasing size of individuals, which, as was indicated in chapter IV, is accompanied in the range $L_a < 450$ cm by increasing swimming speeds, there is both in ontogeny and in phylogeny a decrease in relative fin size. In phylogeny, as a rule, this is sharper the greater the difference in size between individuals of the related forms concerned, and the greater their ecological and morphological similarity, as, in particular, has been reported for fishes (Aleyev 1958a, 1963a). Not conforming to this dependence are apparently only the mantle fins of Teuthoidea (Zuyev 1966), probably because these fins are meant to act during slow swimming only, whereas during rapid movement they are invariably folded around the mantle, forming a fairing for the mantle end of the body.

In the ontogeny of fishes, fin development in the typical case can be divided into two periods (Aleyev 1958b, 1963a): a period of relative growth and a period of relative diminution. This may easily be seen either in the changes in the relative sizes of the longest rays of a given fin (Aleyev 1958b, 1963a), or – with greater accuracy – in the changes in the fin's relative area S_p, which can be found from the formula

$$S_p = s_p^{1/2} L_c^{-1}, \tag{33}$$

where s_p is the fin's absolute area when maximally extended. By designating S_p values for the respective fins by symbols generally accepted for them, we shall obtain for the pectoral fins S_{pp}, for the pelvic fins S_{pv}, for the dorsal fins S_{pd1}, S_{pd2}, etc., for the anal fins S_{pa1}, S_{pa2}, etc., and for the caudal fin S_{pc} (Table 18).

The relative fin growth in the first period serves to improve its performance. After the appearance of a fin its relative size increases until it reaches the size necessary for its optimal functioning at the particular swimming speeds of the fish in question. The moments when such a point (the point of equilibrium) is reached for the different fins of one and the same fish occur at different body lengths, since the different fins do not start to grow simultaneously and develop at different rates. This can be seen from Table 18 for the examples of *Mugil* and *Trachurus*. Further growth of the fish within

Table 18. Changes in relative fin area S_p with age in nektonic fishes

Species and nektonic type	Length L of fish to end of vertebral column (cm)	S_{pp}	S_{pv}	S_{pd1}	S_{pd2}	S_{pa}	S_{pc}
Mugil saliens	0.60	0.11	0.02	0.04	0.07	0.07	0.21
Risso (BN)	0.73	**0.12**	0.04	0.04	0.08	0.08	**0.23**
	1.4	0.11	0.06	0.06	**0.08**	**0.09**	0.21
	4.8	0.10	**0.08**	**0.07**	0.08	0.08	0.19
	36.0	0.09	0.07	0.06	0.07	0.07	0.17
Trachurus medi-	0.69	0.10	0.02	0.06	0.13	0.12	0.18
terraneus	0.78	**0.14**	0.03	0.06	0.15	0.14	0.22
ponticus Aleev	1.0	0.13	0.07	0.10	**0.16**	**0.14**	**0.22**
(EN)	1.5	0.12	0.08	**0.10**	0.15	0.13	0.21
	2.8	0.11	**0.09**	0.10	0.14	0.13	0.20
	6.0	0.11	0.09	0.09	0.13	0.12	0.16
	16.1	0.10	0.08	0.09	0.13	0.12	0.15
	43.3	0.10	0.07	0.08	0.11	0.10	0.14

the range $L_a < 450$ cm and the related growth in swimming speeds produce a continuous decrease in the optimal relative size of all the fins.

Exceptions to this rule, which is common to almost all fishes, are explained in every individual case by the specific features of ecological changes according to age. Thus, for example, in *Spicara* the dorsal and pelvic fins increase in size with age because of the increase in their keel function to enhance manoeuvrability of older fishes. This phenomenon is explained by the sexual inversion characteristic of *Spicara*: when young the fish is female but later on turns into a male (Lozano Cabo 1953, and others), which in the process of spawning is more mobile and needs greater manoeuvrability than the female (Aleyev 1963a).

In other nektonic animals, ontogenetic changes in absolute body size are not so great as in fishes and the changes in the relative size of the fins and flippers with age are, therefore, not so marked.

A functional-morphological analysis of the complex of adaptations connected with movement control entails many difficulties. From the functional aspect the overall system of fins associated with movement control is in itself in most cases very complicated. No less complicated is the constant interaction of this system with the nektonic animal's trunk. Each fin and each group of fins develops and functions not as a self-sufficient part of some aggregate, but primarily as an element of a functionally integral complex. In this connection the fins of any nektonic animal should be regarded as a whole first of all as a single and complicated system which, though its morphological structure may vary, is characterized by a definite topography of fin functions common to the great majority of nektonic animals (Aleyev 1957a). This is why amputated fins are so easily compensated for, a fact which accounts for the main shortcoming of most of the former methods of investigating fin functions, which consisted mainly of the amputation of individual fins or preventing them from functioning in some other way and then observing the defects appearing in the animal's movements (Duges 1905; Vasnetsov 1941, 1948; Mordvinov, 1968, and others). These methods, though allowing assessment in some measure of the functions of the fins, fail to explain most of the specific features of their locations on the animal's body. These characteristics can only be understood when the nektonic animal is regarded as a single, indivisible system and observed during the simultaneous action of all its numerous elements (Aleyev 1957a, 1958c, 1959a, 1963a, 1969c).

When studying adaptations associated with movement control, two methods assume particular significance: 1. observing the swimming and fin performance of animals in natural (field and aquarium) conditions and 2. modelling certain functional aspects of the general system of adaptations under review.

We carried out direct observations of the swimming and the functioning of the fins of more than a hundred species of nektonic animals belonging to the following genera: *Sagitta, Squalus, Huso, Acipenser, Sprattus, Clupea, Sardina, Clupeonella, Alosa, Engraulis, Salmo, Esox, Umbra, Nannostomus, Hemigrammus, Gymnocorymbus, Rutilus, Leuciscus, Leucaspius, Alburnus, Alburnoides, Chalcalburnus, Blicca, Abramis, Pelecus, Tinca, Brachydanio, Barbus, Carassius, Cyprinus, Hipophthalmichthys, Silurus, Amiurus, Anguilla, Belone, Odontogadus, Gadus, Gasterosteus, Gambusia, Xiphophorus, Mollienisia, Lebistes, Zeus, Mugil, Atherina, Sphyraena, Serranus, Stizostedion, Perca, Pomatomus, Trachurus, Sciaena, Corvina, Diplodus, Puntazzo, Spicara, Pterophyllum, Chromis, Crenilabrus, Symphodus, Labrus, Ophidion, Gymnammodytes, Scomber, Sarda, Trichogaster, Macropodus, Betta, Echeneis, Gobius, Aphya, Enhydrina, Chelonia, Eretmochelys, Caretta, Pygoscelis, Eudyptes, Phoca, Arctocephalus, Odobenus, Delphinus, Phocoena, Tursiops* and some others. The observations were carried out predominantly in the aquaria and pools of the Institute of the Biology of the Southern Seas of the Soviet Union, Ukrainian Academy of Science, as well as in the field: in the Black and Azov Seas and various inland waters of the USSR. Numerous films were also used, which allowed the number of species investigated to be considerably increased, to include forms which the author was unable to observe directly (*Rhincodon, Carcharodon, Manta, Chaetodontidae, Ostracion, Balistes, Physeter, Balaenoptera,* etc.).

The methods used for modelling individual aspects of the adaptations associated with movement control are described later, when the individual adaptations concerned are examined.

2. Stabilization and changing direction of movement

1. General principles

The capacity of any streamlined body to maintain and change its direction of movement is basically determined by the position of two points on that body's longitudinal axis: **the centre of inertia**, determined taking account of the added mass, and **the centre of dynamic pressure**, i.e. the point of application of the total hydrodynamic force. It is known (Aleyev 1963a) that when the centre of inertia is situated craniad relative to the centre of dynamic pressure the state of a moving animal is dynamically stable, but when the centre of inertia is situated caudad relative to the centre of dynamic pressure its state of motion becomes dynamically unstable. According to this, all other conditions being unchanged, the closer the centre of inertia is to the anterior (in the direction of movement) end of the animal's body and the further caudad is the centre of dynamic pressure, the higher is its dynamic stability and vice versa. Stabilization of the direction of movement calls for a certain dynamic stability of a streamlined body, whereas an imperative condition for changing the direction of movement is, on the contrary, its dynamic instability. In other words, manoeuvrability of a streamlined body constitutes a property incompatible with its dynamic stability, as has already been noted with regard to swimming animals (Harris 1936; Aleyev 1963a). Therefore adaptations associated with stabilization of the direction of movement and adaptations associated with the manoeuvrability of nektonic animals act in opposition.

When, owing to some chance circumstance, a nektonic animal finds itself situated at an angle to its direction of movement, a righting (stabilizing) moment returns it to the usual position in which its longitudinal axis is parallel to the direction of movement. The functions of stabilizers are performed in nektonic animals by surfaces situated behind the centre of inertia: both the trunk itself, and the fins and other appendages (Houssay 1909a,b, 1912; Schlesinger 1911b; Abel 1912; Schmalhausen 1916; Harris 1936, 1947, 1953; Vasnetsov 1941; Sleptsov 1952; Aleyev 1957a, 1959b, 1963a, 1969c; Burdak 1957; Bertin 1958a; Tomilin 1962a; Zuyev 1966; Mordvinov 1968, and others). The most stable is the locomotion of nektonic animals whose body is built on the principle of a 'feathered arrow', i.e. whose stabilizing fins are shifted towards the posterior end of the body (Schlesinger 1911b; Harris 1936; Aleyev 1963a).

During a turn, the body of a nektonic animal, through the action of a turning moment, is deliberately placed at an angle to the direction of movement, after which forward movement continues in the new direction. Changing the direction of movement in the great majority of cases is associated with bending of the animal's longitudinal axis so that a transverse hydrodynamic force acts on the anterior end of the body, its moment effecting the body's rotation about one of its transverse axes (Gray 1933f; Harris 1936; Aleyev 1963a).

A general rule for nektonic animals is unequal adaptation to effecting manoeuvres in the dorso-ventral and lateral planes: in each of the systematic groups of nektonic animals there is a certain tendency to replace all maneouvres by some single, **preferred manoeuvre** which is better ensured morphologically and is performed with the greatest ease, as has already been noted for fishes (Aleyev 1958a, 1963a). Thus, according to this writer's observations, animals with an axial undulatory propulsor always prefer to manoeuvre in the plane of the body's propulsive undulations, which is made easier by the locomotor fins movements in this plane and the (usually) higher body flexibility; accordingly, fishes, amphibians and reptiles prefer the lateral manoeuvre, whereas Sagittoidea, cetaceans and sirenians prefer the dorso-ventral manoeuvre. Animals with a paddling propulsor, Testudinata, Sphenisciformes and pinnipeds (and also, judging by everything, Sauropterygia), prefer the dorso-ventral manoeuvre since it is better ensured morphologically on account of the lateral arrangement of the limbs, which function as paddles and rudders simultaneously. Thanks to the great mobility of the funnel and extremely high flexibility of the bundle of arms in Teuthoidea, they apparently have no strongly preferred manoeuvre; however, it may be assumed that the dorso-ventral manoeuvre is performed with greater ease than the lateral one owing to the presence of the laterally situated mantle and arm fins and the capacity to bend the mantle end of the body predominantly in the dorso-ventral plane. It has been reported (Zuyev 1966) that in the squids *Symplectoteuthis, Ancistroteuthis* and *Illex* the angle at which the mantle end of the body is bent in the dorso-ventral plane comes to 20–30°, being 15–20° in the horizontal plane; in *Loligo* these figures are 10–15° and 10° respectively.

As one can see from the above, the presence of a preferred manoeuvre in nektonic animals is due in every case to specific features of the type of locomotion and the structure of the propulsive mechanism.

Morphologically, the capacity to stabilize and change the direction of movement is determined mainly by four factors: first, the position of the animal's centre of gravity relative to the ends of the body: second, the shape of the longitudinal projection of the body onto a plane perpendicular to the plane of the preferred manoeuvre, and the position of the centre of this projection relative to the ends of the body (below we shall call this longitudinal projection for brevity the effective longitudinal projection of the body); third, the degree of body flexibility and the presence of passive stabilizers and rudders; and fourth, the presence and strength of active stabilizers and rudders.

Shifts of the centre of gravity along the body's longitudinal axis make the centre of inertia shift in the same direction. Therefore a craniad shift of the centre of inertia enhances stabilization of the direction of movement, whereas a caudad shift increases the animal's manoeuvrability.

The shape of the effective longitudinal body projection is one of the crucial factors determining the location of the centre of dynamic pressure. Shifting the centre of this projection along the longitudinal axis makes the centre of dynamic pressure shift in the same direction. Therefore a backwards shift of the centre of the effective longitudinal projection, all other conditions being unchanged, enhances stabilization of the direction of movement, whereas shifting the centre of this projection forwards, conversely, diminishes the animal's dynamic stability and increases its manoeuvrability.

The animal's body flexibility and the presence of passive stabilizers and rudders determine whether or not the transverse hydrodynamic forces necessary for stabilizing and changing the direction of movement can be created passively. i.e. merely as a result of the nektonic animal's forward movement on account of certain parts of the body surface or certain fins being at an angle to the flow. Such a method of creating stabilizing and rotary cross-forces is characteristic mostly of animals adapted to sustained or nearly continuous, predominantly fast movement, such, for instance, as most of the eunektonic sharks and all Scombridae, Xiphiidae, Istiophoridae, cetaceans, etc.

The action of active stabilizers and active rudders is based on the active creation of hydrodynamic cross-forces, which presupposes either special fin movements or the functioning of a hydrojet propulsor with its funnel at an angle to the animal's longitudinal axis. Such methods of creating hydrodynamic cross-forces are characteristic mostly of slower swimmers as well as of nektonic animals which swim in separate sharp bursts or with frequent stops when there is no forward movement, such, for instance, as nearly all benthonektonic fishes, xeronektonic Chelonioidea and *Trichechus*, the eunektonic *Mola*, etc.

2. Modes of stabilization and changing direction of movement

Turning in **Teuthoidea** is effected predominantly by changing the direction of the water jet ejected from the funnel by turning its free end, and also by bending the longitudinal axis of the body: its mantle end and, particularly, the bundle of arms, whose action as rudders and stabilizers is enhanced by the presence on them of lateral fins (Figure 127A). An analysis of cine records of swimming *Todarodes pacificus* showed that the manoeuvre begins with deflexion of the mantle end of the body and the bundle of arms in the direction of the turn, which creates a tendency for the animal to move along an arc (Figure 127A, frames 2–7); once the manoeuvre is over, the body's longitudinal axis straightens out again (Figure 127A, frames 8, 9). When the mollusc is swimming slowly, of considerable importance in performing lateral manoeuvres may be the intensity and direction of undulations of the mantle fins, as has been reported for Sepioidea (Jaeckel 1958). The mantle fins and the arms are at opposite ends of the body, so when set at an angle to the stream and by virtue of their own movements, they create a large torque (Akimushkin 1963; see also Figures 129B,C). Thus the rudder system of Teuthoidea consists of frontal and rear rudders.

In **Sagittoidea** changes in the direction of movement are achieved by bending the body, although in planktonektonic forms the fins have been morphologically divided to perform the functions of frontal rudders, stabilizers and propulsive organs (Figures 131C, 137A).

Fishes turn in the lateral plane by means of an asymmetric propulsive wave with an unusually large amplitude and a small speed of propagation: the muscular contractions pass along the side towards which the turn is being made (Gray 1933f, 1937). The angular speed of the turn can be regulated by the degree of asymmetry of the propulsive wave and its speed, as well as by active movements of the pectoral fins. When turning in the horizontal plane a fish first bends the anterior half of the body in the direction of the turn. At this moment the fins situated on the anterior half of the

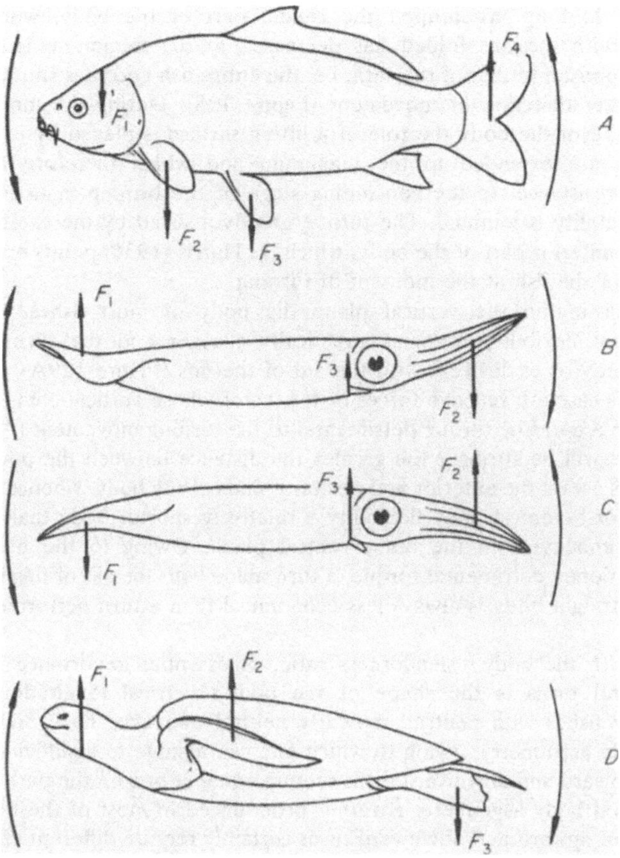

Figure 129. Dorsoventral turning mechanism in nektonic animals with low dorsoventral body flexibility. *A*, *Cyprinus carpio carpio* L.; *B,C*, *Loligo pealei* Lesueur; *D*, *Chelonia mydas* L., F_1, F_2, F_3, F_4, vertical components of forces acting in vertical plane. Curved arrows indicate direction of turn.

body are folded while the caudal fin and other fins on the posterior half of the body are fully extended. Thus a lifting surface is formed by the posterior part of the body, fixing the position of the body's caudal part, whereas its anterior half, conversely, may be moved with maximal ease in the direction of the turn, right or left, since its resistance to transverse movements at this moment is minimal. After the anterior part of the body has been deflected through the necessary angle to the right or left, the fins on it, which are capable of increasing the area of its vertical longitudinal projection, extend, while the fins on the posterior part of the body, including the caudal fin, fold up. Thereupon the caudal part of the body, whose transverse resistance with the fins folded has decreased to the minimum, is moved in the direction opposite to that of the turn, i.e. the entire fish becomes situated at an angle to the former direction of movement (Figure 128). During the movement of the posterior half of the body the role of a lifting surface is played by its anterior half, whose fins are extended to the maximum and which therefore has maximum cross-flow resistance. In the concluding stage of the turning manoeuvre the fish's dynamic stability is minimal. The turn is greatly assisted by the extended unpaired fins on the anterior part of the body, which, as Harris (1936) points out, increase the instability of the fish at the moment of turning.

When turning in the vertical plane the body of most fishes, owing to low dorso-ventral flexibility, remains practically straight and the turn is performed predominantly or exclusively with the aid of the fins (Figure 129A). Acting on the body in this case are reactive forces of the water whose vertical components form a couple with a parasitic torque detrimental to the turning movement (Aleyev 1963a). This torque will be stronger the greater the distance between the points where the cross-forces act on the anterior and posterior ends of the body, whence it follows that with low dorso-ventral body flexibility a relatively shorter body makes it easier to perform manoeuvres in the dorso-ventral plane. Owing to the presence of the above-mentioned detrimental torque, a turn made with the aid of fins with a straight or nearly straight body is always less economical than a turn performed by bending the body.

Along with the body's slenderness ratio, of essential importance in performing dorso-ventral turns is the shape of the body's vertical longitudinal projection. Eunektonic fishes with neutral or nearly neutral buoyance have only mild dorso-ventral body asymmetry, owing to which one can assume to a sufficient approximation that upward and downward turns require equal efforts on the part of the fish. As dorso-ventral body asymmetry is rather pronounced in most of the benthonektonic species, here upward and downward turns certainly require different efforts. According to Vasnetsov (1948), fishes which particularly frequently make turns in the vertical plane with no forward movement or very slow forward movement usually have a deep, disk-like body (Chaetodontidae, *Abramis*, *Blicca* and others; Figure 51D) or a shape close to a semicircle with its convex edge upwards (*Cyprinus*; Figure 39D); a strongly convex dorsal contour of the body with a straight or nearly straight ventral contour makes it very easy for the fish to turn downwards along an arc. Moreover lateral compression of the body facilitates passage through narrow gaps in the submarine landscape.

In some eunektonic fishes ecologically closely connected with the surface of the water, the upper profile of the body becomes straighter while the lower becomes more convex, which facilitates upward turns along an arc; in this case sharp dorso-ventral body asymmetry is usually accompanied by an upper mouth, essential for grasping food from below, and relatively enlarged pectoral fins, which are also very important for making turns in the vertical plane (*Pelecus*).

In fishes the main horizontal rudders assisting upward and downward turns are the pectoral fins (Figures 129A). The pelvic fins are used less as rudders, however this function of these fins increases when they are situated closer to the anterior end of the body, i.e. with their removal from the centre of gravity (Schmalhausen 1916; Harris 1934, 1936, 1937a,b, 1938, 1947, 1953; Vasnetsov 1941; Aleyev 1957a, 1963a, and others). During turns in the dorso-ventral plane, the pectoral fins and the posterior parts of the dorsal and anal fins, as well as the caudal fin, act as active rudders creating their own vertical hydrodynamic forces facilitating turning movements in many Actinopterygii, particularly in slow swimmers. The pectoral fins during this manoeuvre make downward or upward strokes. The active rudder action of the dorsal and anal fins behind the centre of gravity consists, as Vasnetsov (1941, 1948) demonstrated, of deflecting the fin rays simultaneously to the right or left, as a result of which the dorsal fin creates a downward force and the anal fin an upward force; this is characteristic, in particular, of many benthonektonic Cyprinidae (*Abramis, Blicca, Carassius, Cyprinus, etc.*). The rudder action of the caudal fin during turns in the dorso-ventral plane consists of creating upward or downward forces, whose mechanism has already been examined in chapter III.

In nektonic **reptiles** turns are made both by bending the body and also by the action of the fins.

Observations by this author of the swimming of the sea snake *Enhydrina schistosa* indicate that Hydrophidae turn in any direction wholly on account of bending of the body, through the passage along the body of an asymmetric propulsive wave of unusually large amplitude. By changing the degree of bending the snake can change its direction of movement immediately by 180° (Figure 127D). Despite the very high body flexibility of *Enhydrina* one still observes a clear tendency to replace dorso-ventral turns by lateral ones: the lateral compression of the body makes for much greater lateral flexibility.

The high slenderness ratio of the body in Mosasauridae doubtless ensured for them very high flexibility, owing to which members of this group turned mainly on account of bending of the body, though the fins also took part. Body flexibility was apparently nearly as high in Metriorhychidae, Mesosauria and early Ichthyosauria of the *Cymbospondylus* type (Figure 10B). In the more specialized later ichthyosaurs, similar to *Stenopterygius, Macropterygius* and *Eurhinosaurus* (Figures 10D,E, 98C), the body grew denser and less flexible, owing to which the fins began playing a greater role in the process of manoeuvring.

Of great importance for Sauropterygia when turning was their long flexible neck, which acted as a frontal rudder. In addition, turns were very much facilitated by flipper-like limbs and, to a lesser degree, by the tail. A similar stabilizing and rudder system was characteristic of Placodontia, in which the role of the neck and tail in stabilization and changing the direction of movement was directly proportional to their relative sizes in the various forms.

The armour-clad body of Testudinata is absolutely inflexible. Observations of swimming *Chelonia mydas, Eretmochelys imbricata, Caretta caretta* and *Emys orbicularis* show that in all turtles, both nektonic and nektoxeric, the transverse hydrodynamic forces necessary for stabilization and changing direction of movement are almost entirely created by the flippers and neck only. During turns the neck in all cases bends in the direction of the turn, while the flippers create the necessary transverse forces both passively (during faster movement) and actively (during slow movement), with the foreflippers performing at a much higher pace than the hind ones (Figure 129D). Turtles, being slow swimmers, create the necessary stabilizing and turning forces in the main actively. During rectilinear swimming the hind flippers of nektonic turtles function as stabilizers.

In **Sphenisciformes** the transverse hydrodynamic forces necessary for stabilization and changing the direction of movement are created by both pairs of limbs and the flexible neck. Observations by this author of a swimming *Eudyptes chrysolophus* show that in penguins the stabilizer-rudder complex is based on the same pattern as in turtles: the neck and legs in every case are deflected in the direction of the turn.

The rudder system of **Hesperornithes** was, by all indications, close to that of the penguins, with the only difference that, in the absence of forelimbs (Figure 60), the rudder action of the head and flexible neck was of still greater importance.

In **cetaceans** all turns are carried out both by bending the body and on account of passive functioning of the pectoral and caudal flippers. Our observations of swimming dolphins (*Delphinus, Tursiops* and *Phocoena*) show that turns in the vertical plane are produced by dorso-ventral bending of the body and by setting the pectoral fins at a positive or negative angle of attack, while turns to the right and left are produced by lateral bending of the body. The same methods of performing manoeuvres have been described also for whales (Sleptsov 1952; Tomilin 1957, 1962a, and others). The functions of stabilizers in cetaceans are performed mainly by the laterally compressed caudal stem and the tail fin as well as the dorsal fin (when present).

The stabilizer-rudder system of **sirenians** is on the whole morphologically and functionally analogous to that of cetaceans. However, the paired fins of sirenians are capable of more varied movements than those of cetaceans (Petit 1955), thanks to which they act not only as passive rudders but as active rudders as well.

In **pinnipeds**, just as in cetaceans, all turns are made both by bending the body and neck and on account of passive and active functioning of the flippers (Ray 1963; Mordvinov 1968, 1972; see also Figure 127E). However, as the author's observations of swimming *Pagophoca groenlandica, Phoca caspica* and *Arctocephalus pusillus* kept in the marine aquaria of the Institute of the Biology of the Seas of the Southern Soviet Union show, the role of bending of the body in Otariidae in this regard is much greater than in Phocidae and still greater than in Odobenidae. Functioning as stabilizers in pinnipeds are the hind flippers.

3. Passive dynamic stability

Using methods developed earlier (Aleyev 1957a, 1958c, 1959a, 1963a, and others), we have found the first numerical indices characterizing the degree of development of some of the morphological adaptations functionally associated with stabilization and changing the direction of movement. Morphologically, the development of adaptations associated with the passive creation of stabilizing and turning forces may be described, according to the above, by the relative position on the longitudinal axis of the body of the animal's centre of gravity (c) and the centre of the effective longitudinal projection of the body (o). This, in turn, can be expressed as the ratio K of the distance l between c and o to the body's effective length L_c (Aleyev 1957a):

$$K = \pm l L_c^{-1} \tag{34}$$

The point o, the centre of the effective longitudinal projection, may easily be found as the centre of a piece of cardboard the same shape. The signs ± indicate that K, which we shall call the index of passive dynamic stability, changes sign depending on the relative position of c and o (Figure 130): positive values of K correspond to the centre of the effective longitudinal projection being in front relative to the centre of gravity and vice versa. Accordingly, on algebraic increase in the index K, all other conditions being unchanged, corresponds to a reduction of dynamic stability, i.e. a greater capacity of the animal for manoeuvring through the action of passive rudders, and vice versa.

The size and shape of the effective longitudinal body projection are variable in most nektonic animals, mainly because of their capacity to unfold and fold up various

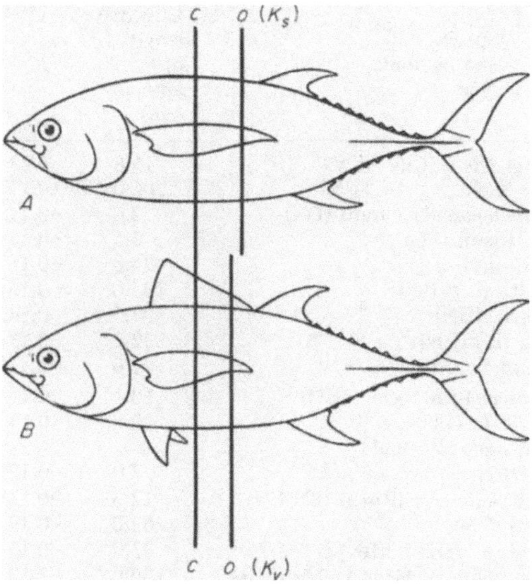

Figure 130. Effective longitudinal projections of *Thunnus alagunga* (Bonnat.). *A*, during rectilinear movement; *B*, during turn in horizontal (lateral) plane. *C*, centre of gravity; *o*, centre of projection. See text.

fins or flippers, thus changing also the position o. The rearmost position of o corresponds to rectilinear movement by inertia, when the fins on the forebody are folded up to the utmost while the stabilizing fins on the posterior part of the body, if present, are extended to the maximum; let us designate the value of K for this case by the subscript s (K_s). The foremost position of o is observed during the concluding phase of a turning movement, when the fins on the anterior half of the body are extended to the maximum and those on the posterior part are folded; let us designate the value of K corresponding to this case by the subscript v (K_v). Thus we obtain two indices, K_s and K_v, found from one and the same formula (34), with the magnitude of l in each case corresponding to the distance from the centre of gravity to the centre of the effective longitudinal projection. When drawing the effective longitudinal projection all the fins, limbs, tentacles (arms) and other appendages were depicted in their positions during rapid rectilinear movement by inertia (for a projection corresponding to K_s) or in the concluding phase of a sharp turn during rapid movement (for a projection corresponding to K_v).

The values of K_s and K_v found by this author for 63 nektonic species are given in Table 19 with the individual species arranged in order of increasing K_v, i.e. in order of diminishing passive dynamic stability.

Table 19. K_s and K_v values in nektonic animals.

Species and nektonic type	Effective length L_c of animal (cm)	K_s	K_v
Hirundichthys rondeletii (Cuv. et Val.) (EN)	15.8	−0.23	−0.23
Mola mola (L.) (EN)	200.0	−0.21	−0.21
Gambusia affinis holbrooki Girard (BN)	2.0	−0.21	−0.20
Aphya minuta (Risso) (PN)	3.5	−0.18	−0.17
Labrus viridis L. (BN)	24.8	−0.16	−0.15
Mustelus mustelus (L.) (BN)	103.0	−0.15	−0.15
Acipenser sturio L. (BN)	97.5	−0.14	−0.14
Prionace glauca (L.) (EN)	132.0	−0.13	−0.13
Squalus acanthias L. (EN)	101.0	−0.13	−0.13
Atherina mochon pontica Eichw. (EN)	10.1	−0.13	−0.13
Stromateus fiatola L. (EN)	19.6	−0.14	−0.13
Clupeonella delicatula delicatula (Nordm.) (PN/EN)	7.0	−0.12	−0.12
Sprattus sprattus phalericus (Risso) (PN/EN)	12.0	−0.12	−0.12
Lichia amia (L.) (EN)	62.0	−0.13	−0.12
Samo trutta labrax m. fario L. (BN)	23.0	−0.13	−0.12
Acipenser güldenstädti colchicus V. Marti (BN)	130.0	−0.12	−0.12
Acipenser stellatus Pall. (BN)	111.3	−0.11	−0.11
Esox lucius L. (BN)	121.7	−0.11	−0.11
Chaetodon striatus L. (BN)	11.0	−0.11	−0.11

Table 19. K_s and K_v values in nektonic animals (*contd.*)

Species and nektonic type	Effective length L_c of animal (cm)	K_s	K_v
Sciaena umbra L. (BN)	39.1	−0.12	−0.11
Mugil auratus Risso (BN)	41.2	−0.12	−0.11
Seriola dumerilli (Risso) (EN)	20.8	−0.12	−0.11
Belone belone euxini Günth. (EN)	68.0	−0.11	−0.11
Oncorhynchus nerka (Walb.) (BN)	47.0	−0.11	−0.11
Onchorhynchus kisutch (Walb.) (BN)	55.7	−0.11	−0.11
Puntazzo puntazzo (Cetti) (BN)	44.0	−0.10	−0.10
Engraulis encrasicholus ponticus Alex. (EN)	14.0	−0.11	−0.10
Alosa kessleri pontica (Eichw.) (EN)	25.0	−0.10	−0.10
Oncorhynchus gorbuscha (Walb.) (BN)	47.0	−0.10	−0.10
Gymnammodytes cicerellus (Raf.) (BN)	9.9	−0.09	−0.09
Odontogadus merlangus euxinus (Nordm.) (BN)	41.6	−0.11	−0.09
Spicara smaris (L.) (EN)	15.0	−0.11	−0.09
Sphyraena sphyraena (L.) (EN)	56.0	−0.09	−0.09
Sphyraena barracuda (Walb.) (EN)	130.0	−0.09	−0.09
Clupea harengus harengus L. (EN)	28.0	−0.08	−0.08
Zeus faber pungio (Val.) (BN)	25.3	−0.12	−0.08
Trachurus mediterraneus ponticus Aleev (EN)	44.9	−0.09	−0.08
Megalaspis cordyla (L.) (EN)	41.5	−0.10	−0.08
Thunnus alalunga (Bonnat.) (EN)	68.0	−0.10	−0.08
Anguilla anguilla (L.) (EN)	90.0	−0.07	−0.07
Sardina pilchardus (Walb.) (EN)	15.6	−0.08	−0.07
Pomatomus saltatrix (L.) (EN)	94.1	−0.07	−0.07
Auxis thazard (Lac.) (EN)	40.3	−0.08	−0.07
Scomberomorus commersoni (Lac.) (EN)	121.0	−0.08	−0.07
Istiophorus platypterus (Show & Nodder) (EN)	122.0	−0.09	−0.07
Xiphias gladius L. (EN)	243.0	−0.07	−0.07
Coryphaena hyppurus L. (EN)	71.9	−0.10	−0.06
Scomber scombrus L. (EN)	34.1	−0.06	−0.05
Sarda sarda (Bl.) (EN)	76.9	−0.06	−0.05
Tetrapturus belone Raf. (EN)	120.0	−0.06	−0.05
Pagophoca groenlandica (Erxl.) (XN)	160.0	−0.05	−0.05
Loligo vulgaris Lam. (EN)	26.5	−0.05	−0.05
Symplectoteuthis oualaniensis (Less.) (EN)	24.1	−0.05	−0.05
Enhydrina schistosa (Daud.) (EN)	95.0	−0.04	−0.04
Chelonia mydas (L.) (XN)	144.0	−0.05	−0.03
Phocoena phocoena (L.) (EN)	128.0	−0.04	−0.03
Delphinus delphis ponticus Barab. (EN)	178.5	−0.04	−0.03
Caretta caretta (L.) (XN)	126.0	−0.05	−0.02
Tursiops truncatus (Montagu) (EN)	325.0	−0.03	−0.02
Eudyptes chrysolophus Brandt (XN)	69.6	−0.02	−0.01
Arctocephalus pusillus (Schreb.) (XN)	200.0	−0.02	−0.01
Eretmochelys imbricata (L.) (XN)	71.0	−0.03	0.00
Trichiurus lepturus L. (EN)	149.0	0.00	0.00

The 34 species in the first part of the series, the range $-0.23 \leq K_v \leq -0.09$, consist of all the benthonektonic species examined except one, i.e. 17, or 50% of the 34, all the planktonektonic species, of which there are 2, i.e. 6%, and 15 eunektonic species, i.e. 44%. Here are found all the species examined of the 'sagittal' type, which are adapted to rapid rectilinear acceleration, i.e. to sharp bursts (*Hirundichthys, Salmo, Esox, Mugil, Belone, Oncorhyncus* and *Sphyraena*), and the majority of the slow-swimming forms, in which the necessary manoeuvrability is achieved mainly through the operation of active rudders (whose action the indices K_s and K_v do not take into account). This first half of the series includes most of the small species with a maximum known length L_c less than 50 cm; these constitute 19 out of the 34, i.e. 56%.

In the second half of the series, $-0.08 \leq K_v \leq 0.00$, containing 29 species, we find all the xeronektonic species examined, 6 out of the 29, i.e. 21%, most of the eunektonic species, 22 out of the 29, i.e. 76%, and one benthonektonic species, constituting 3%. Clearly evident in this range is a marked preponderance of fast swimmers adapted to active pursuit of prey, in whom the necessary manoeuvrability is mainly or exclusively achieved through the action of a passive rudder system. This half of the series includes all the investigated members of the Teuthoidea, Scombridae s.l., Xiphiidae, Istiophoridae, Coryphaenidae, Pomatomidae, Sphenisciformes, Otariidae, Delphinidae, and of the Carangidae *Megalaspis* and *Trachurus*, etc., i.e. in effect all the fastest swimmers of the pelagic zone. Here we also see the elongated eel-like (*Anguilla*), snake-like (*Enhydrina*) and ribbon-like (*Trichiurus*) forms, whose high manoeuvrability is well known, as well as forms with a long flexible neck having the role of a frontal rudder, enhancing the animal's high manoeuvrability (*Pagophoca, Arctocephalus*, Chelonioidea and *Eudyptes*). This second half of the series is obviously dominated by large forms: only nine of the 29 species here, constituting only 31%, include individuals whose maximum known length does not exceed 50 cm.

Apart from the natural changes in the animals' ecology in the series under consideration, we see on the whole that with increasing K_v their dimensions also increase, which is accompanied, according to the dependence described earlier (see chapter V), by increasing mean swimming speeds.

Table 19 shows that the inequality $K_s < K_v$ holds for the majority of nektonic animals. This corresponds to a decrease in dynamic stability at the moment of turning on account of extension of the fins situated in front relative to the centre of gravity (Figure 130), the point of which is clear from the above. More seldom we encounter equality: $K_s = K_v$.

Thus, in the course of adaptation to sustained rapid movement the animals' passive dynamic stability, on the whole, diminishes, which affords a greater possibility of manoeuvring through the action of passive rudders, thereby rationalizing the process of movement control, which may be quantitatively characterized by the increase in the indices K_s and K_v.

It goes without saying that increased manoeuvrability of nektonic animals does not imply that their rectilinear movement becomes insufficiently stabilized. It should be borne in mind that, apart from the location of an animal's centre of gravity and the centre of its effective longitudinal projection, the degree of its dynamic stability is influenced also by a whole series of other factors, principally various active movements on the part of the animal itself.

Morphological adaptations aimed at ensuring a definite level of dynamic stability consist primarily of the development of special fins or other elements situated craniad and caudad relative to the centre of gravity and as close as possible to the anterior and posterior ends of the body respectively. These fins or other elements expand at the right moment to create the essential additional forces of resistance on the anterior or posterior end of the body facilitating a change in dynamic stability as required. Important in this regard too is the shape of the trunk, whose effective longitudinal projection may be broader mostly in front of or behind the centre of gravity, which will affect the position of its centre. Figure 131 shows typical examples of the body shape and the location of fins in nektonic animals adapted to different modes of locomotion.

Critical remarks have been made (Zuyev & Kudryashov 1968) with regard to passive dynamic stability indices suggested by this writer (Aleyev 1957a, 1958c, 1959a, 1963a). These remarks, however, cannot be described as objective. The assertion that Aleyev '... identifies the centre of pressure with the geometric centre of the fish's vertical profile' (Zuyev & Kudryashov 1968, p. 1062) is utterly groundless. Aleyev does not identify these two concepts, but merely points out '... that the location of the centre of pressure (*o*) depends in some measure on the location of the centre of the fish's longitudinal projection, and, in particular, that a forward shift of the centre of projection facilitates a forward shift of the centre of dynamic pressure (*o*) as well' (Aleyev 1963a, p. 177). Equally groundless is these critics' assertion that when developing the methods under review '... Yu. G. Aleyev proceeds from the wrong premise of the constancy of the location of the fish's centre of pressure during a turn' (Zuyev & Kudryashov 1968, p. 1062). In the work quoted by his critics Aleyev, on the contrary, stresses that 'the area of longitudinal projection and its shape, and, consequently, also its centre, change depending on the degree of flexion of the fish's body and the degree of expansion of its fins' (Aleyev 1963a, p. 177). On the basis of the above-mentioned unfounded assertions, these critics come to the conclusion "... that the methodology suggested by Yu. G. Aleyev for calculating 'indices of dynamic stability' of fishes fails to reflect the essence of the phenomenon it claims to investigate" (Zuyev & Kudryashov 1968, p. 1062). However, Aleyev (1963a) specially stresses that the indices obtained '... cannot characterise the true magnitude of a fish's dynamic instability, but only indicate the peculiarities of its structure, important for ensuring a certain dynamic stability' (Aleyev 1963a, p. 180). Well, this seems to clarify the matter.

Lacking experimental material of their own, and merely proceeding from their theoretical models and the ciné records found in Gray's work (1933f) showing the turning movements of a goldfish, which, as Ovcharov noted (1974), they analysed

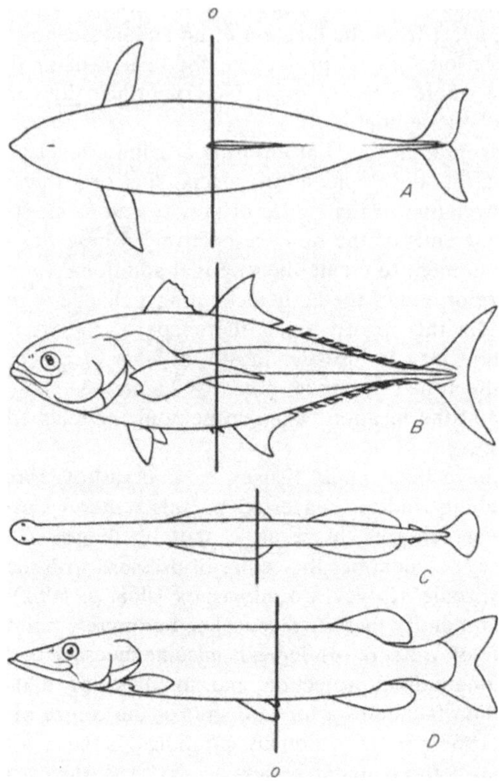

Figure 131. Effective longitudinal projections of nektonic animals adapted to different modes of locomotion. *o*, centre of projection. *A*, *Phocoena phocoena* (L.), effective body length $L_c = 128.8$ cm; *B*, *Megalaspis cordyla* (L.), $L_c = 41.5$ cm; *C*, *Sagitta setosa* Müll., $L_c = 2.2$ cm; *D*, *Esox lucius* L., $L_c = 121.7$ cm. *A*,*B*, sustained rapid swimming with constant manoeuvering; *C*,*D*, movement predominantly in separate spurts.

carelessly and interpreted wrongly, Zuyev & Kudryashov (1968) arrived at the erroneous conclusion that the body of the fish at the moment of turning is in a dynamically unstable state. However, it has been shown experimentally (Ovcharov 1974), by observing the pattern of the flow and analysing film records, that when the fish turns its body remains in a dynamically stable state that fully corresponds to the magnitudes of the passive dynamic stability indices quoted above (Tables 19, 20).

To assess the manoeuvrability of dolphins Maslov (1970) proposed a dimensioness index that takes into account the animal's mass, length and strength, as well as the speed and linear degree of manoeuvring and the time in which the manoeuvre is performed. However, this index completely fails to take into account any morphological features of the animal associated with ensuring this or any other level of manoeuvrability.

As the sizes of nektonic animals increase **in ontogeny**, the swimming speeds they can develop increase in the range $L_s < 450$ cm, owing to which the same changes take place as in the series examined earlier with regard to phylogeny: passive dynamic stability diminishes, while the animal's capacity to manoeuvre through the action of passive rudders increases, which creates the same energetic advantages as in phylogeny (Table 20, Figure 132). This process is all the more pronounced the greater the differences between the minimal and maximal dimensions and swimming speeds of the individuals concerned in ontogeny, the causes of which are clear. It is most pronounced in fishes and least pronounced in cetaceans, pinnipeds and Sphenisciformes (Table 20).

Morphologically, the increasing manoeuvrability of nektonic animals with age is ensured by the progressive development of the same adaptations as in phylogeny, which facilitate a shift of the centre of the effective longitudinal projection forwards and a shift of the centre of gravity backwards. This is achieved mainly by the progressive development and a forward shift of the fins situated in front of the centre of gravity and, simultaneously, by a decrease in the relative sizes of fins situated behind the centre of gravity, as well as a backward shift of the maximum transverse cross-section of the body (Figure 132). A forward shift of most of the fins in ontogeny has been reported for a great many fishes (Fage 1920; Lebour 1921; Ford 1930; Francois 1955; Aleyev 1957a, and others) and is undoubtedly the general rule for fishes (Aleyev 1963a); these shifts are due to the unequal relative growth of individual parts of the body (Ford 1930; Desbrosses 1945; Aleyev 1957a, 1963a) and the decrease in the relative size of the cranium (Aleyev 1957a, 1963a). In many cases the change in the location of a fin takes place in opposite directions relative to the ends of the body and relative to the spinal column, as is observed, for example, in *Atherina, Odontogadus, Corvina* and *Squalus* (Figure 133). This shows that the changing location of the fins is not merely due to the unequal growth of the various sections of the spinal column, but is directly dependent on the specific functions of the fins and the development of these functions with age. The phenomenon of unequal longitudinal growth of the individual sections of the spinal column as such should be regarded rather as a means of effecting the functionally necessary shift of the fins.

In fishes, in the majority of cases there takes place in ontogeny more-or-less complete division of the dorsal fin into two or three autonomous parts, performing the functions of rudder, stabilizer and propulsor respectively. Likewise the anal fin divides into two portions, fulfilling, respectively, the functions of stabilizer and propulsor. This differentiation takes place on account of slowing down or cessation of the relative growth of the fin rays situated on the borders between the fin's segregrating portions, and on account of the functional shift of its various parts in different directions along the body's longitudinal axis (Aleyev 1957a, 1963a). In the

Table 20. Variation of K_s and K_v in nektonic animals with age.

Species and nektonic type	Effective length L_c of animal (cm)	K_s	K_v
Mustelus mustelus (L.) (BN)	41.0	−0.17	−0.17
	103.0	−0.15	−0.15
Squalus acanthias L. (EN)	21.8	−0.16	−0.16
	40.6	−0.14	−0.14
	101.0	−0.13	−0.13
Atherina mochon pontica Eichw. (EN)	11.6	−0.19	−0.19
	8.1	−0.13	−0.13
Clupeonella delicatula	4.3	−0.12	−0.12
delicatula (Nordm.) (PN/EN)	7.0	−0.12	−0.12
Salmo trutta labrax m. fario L. (BN)	7.0	−0.18	−0.17
	10.5	−0.17	−0.16
	13.5	−0.15	−0.14
	23.0	−0.13	−0.12
Acipenser stellatus Pall. (BN)	24.5	−0.17	−0.17
	36.5	−0.14	−0.14
	111.3	−0.11	−0.11
Mugil auratus Risso (BN)	1.8	−0.19	−0.18
	20.1	−0.15	−0.14
	41.2	−0.12	−0.11
Oncorhynchus nerka (Walb.) (BN)	2.34	−0.20	−0.20
	47.0	−0.11	−0.11
Oncorhynchus kisutch (Walb.) (BN)	2.4	−0.20	−0.20
	55.7	−0.11	−0.11
Alosa kessleri pontica (Eichw.) (EN)	8.0	−0.13	−0.13
	25.0	−0.10	−0.10
Oncorhynchus gorbuscha (Walb.) (BN)	2.57	−0.21	−0.21
	47.0	−0.10	−0.10
Odontogadus merlangus euxinus	5.2	−0.15	−0.15
(Nordm.) (BN)	41.6	−0.11	−0.09
Spicara smaris (L.) (EN)	5.1	−0.12	−0.11
	15.0	−0.11	−0.09
Trachurus mediterraneus	1.0	−0.15	−0.15
ponticus Aleev (EN)	1.5	−0.14	−0.14
	2.8	−0.13	−0.12
	6.0	−0.12	−0.11
	17.2	−0.11	−0.10
	44.9	−0.09	−0.08
Pomatomus saltatrix (L.) (EN)	4.8	−0.16	−0.13
	10.5	−0.15	−0.12
	15.1	−0.13	−0.11
	20.2	−0.11	−0.10
	20.8	−0.11	−0.10
	50.8	−0.07	−0.07
	94.1	−0.07	−0.07

Table 20 (*contd.*)

Species and nektonic type	Effective length L_c of animal (cm)	K_s	K_v
Xiphias gladius L. (EN)	37.8	−0.13	−0.11
	343.0	−0.07	−0.07
Scomber scombrus L. (EN)	12.1	−0.11	−0.10
	16.0	−0.08	−0.07
	23.7	−0.06	−0.05
Phocoena phocoena (L.) (EN)	89.2	−0.04	−0.04
	128.0	−0.04	−0.03

great majority of cases the differentiation of unpaired fins in fishes is completed at a length of less than 2–3 cm, i.e. before the nektonic stage of ontogeny begins (Figure 95), and much more seldom during later stages of development (Figure 132C).

In a number of cases, the different mobility of the males and females of some nektonic animals determines the appearance of **sexual dimorphism** in adaptations associated with stabilization and changing the direction of movement; this is characteristic, in particular, of many fishes. Thus, for example, it is known (Nikolsky 1940; Suvorov 1948, and others) that in many cases the males (particularly in the carp family) have relatively longer paired fins, which can be attributed to their greater mobility during spawning.

The spawning dress of the males of *Oncorhyncus* – the formation of a hump on the anterior part of the back and elongation of the jaws – facilitates a shift of the centre of the effective longitudinal projection forwards, resulting in greater manoeuvrability of the fish, which is of great importance for the very mobile behaviour of the male in the spawning grounds (Figure 134). In the females of *Oncorhynchus*, whose behaviour during spawning is much more passive, the changes during breeding are much less pronounced (Aleyev 1963a, 1964b, 1969b).

Similar features increasing the manoeuvrability of the male are observed in *Coryphaena hippurus*: in the male the upper contour of the head's profile rises more steeply and the dorsal fin begins somewhat closer to the anterior end of the body, thanks to which the position of the centre of the effective longitudinal projection is further forward than in the female (Figure 134). In distinction from *Oncorhynchus*, these features in *Coryphaena* are permanent.

It should be noted that sezual dimorphism associated with increased male manoeuvrability is observed in fishes that spawn in small groups, when the male swims around the female repulsing other males, which requires much higher manoeuvrability than that of the female (*Salmo*, most Cyprinidae, etc.). In fishes that spawn in shoals (Clupeidae, Carangidae, Scombridae, etc.) these features of sexual dimorphism are usually not pronounced. Characteristic of some and perhaps many such fishes

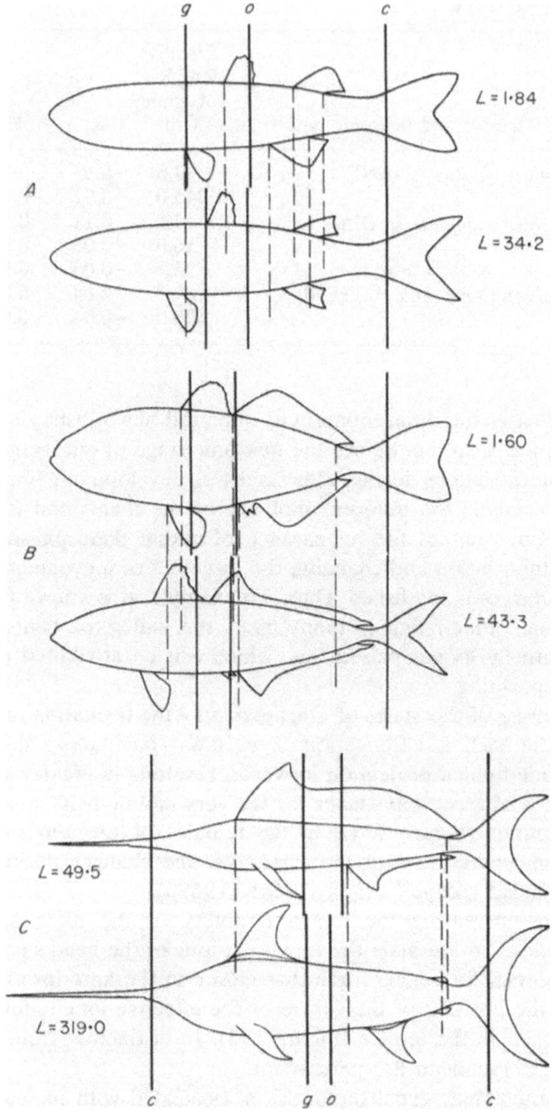

Figure 132. Variation in location of centre of gravity and centre of effective longitudinal projection in the ontogeny of (*A*) *Mugil auratus* Risso, (*B*) *Trachurus mediterraneus ponticus* Aleev and (*C*) *Xiphias gladius* L. Vertical lines pass through (g) centre of gravity (*o*) centre of projection and (*c*) ends of effective body length L_c. L = length of fish to end of vertebral column (cm).

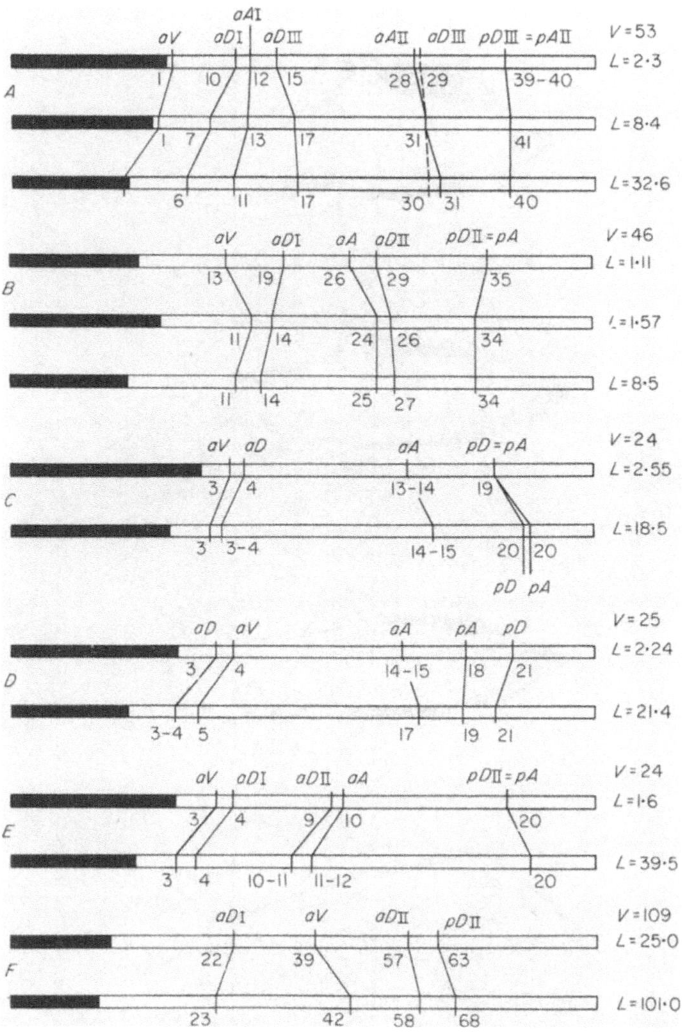

Figure 133. Diagram of fin shifts relative to ends of body and individual vertebrae in the ontogeny of (*A*) *Odontogadus merlangus euxinus* (Nordm.), (*B*) *Atherina mochon pontica* Eichw., (*C*) *Diplodus annularis* (L.), (*D*) *Corvina umbra* (L.), (*E*) *Trachurus mediterraneus ponticus* Aleev and (*F*) *Squalus acanthias* L. Shaded horizontal strips designate the cranium, unshaded horizontal strips the vertebral column. *aD*, beginning of base of dorsal fin; *aA*, beginning of base of anal fin; *aV*, beginning of base of pelvic fins; *pD*, end of base of dorsal fin; *pA*, end of base of anal fin; figures by lines through vertebral column at the beginning or end of a fin base designate the number of the vertebra through which the line passes; *V* = number of vertebrae; *L* = fish's length to end of vertebral column (cm).

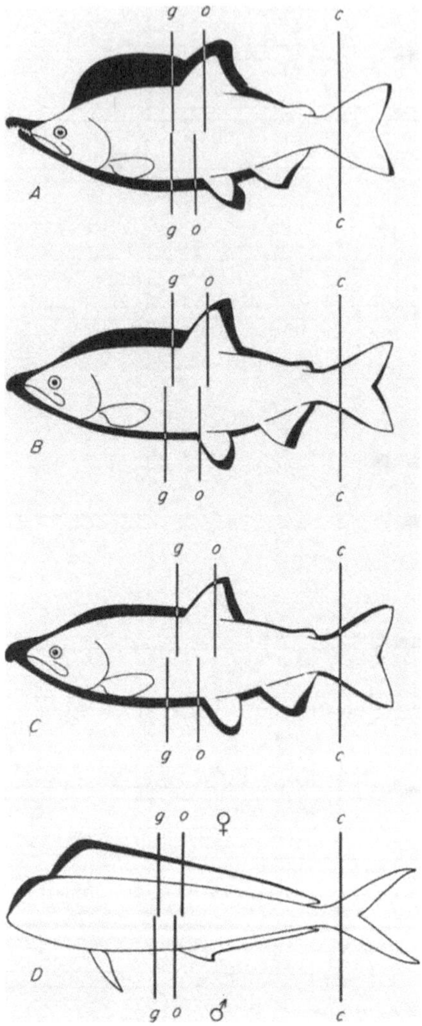

Figure 134. A, Oncorhynchus gorbuscha (Walb.); *B, Oncorhynchus nerka* (Walb.); *C, Oncorhynchus kisutch* (Walb.); *D, Coryphaena hyppurus* L. *A-C*, males without (unshaded profile) and with (shaded profile) breeding dress. Fish length to end of vertebral column without and with breeding dress respectively: *A*, 41.1 and 41.1 cm; *B*, 30.8 and 50.2 cm; *C*, 42.5 and 54.9 cm. *g*, centre of gravity; *o*, centre of effective longitudinal projection; *c*, end of vertebral column; above, for male without breeding dress; below, for make with breeding dress (projection centre shifts forwards). *D*, female (unshaded profile) and male (shaded profile). *g, o, c*, as for *A-C*; above, for female; below, for male (centre of projection closer to anterior end of body).

(Tokarev 1949; Aleyev 1952) is a 'stratified' structure of the spawning shoals; the males keeping to a 'layer' above (as in *Trachurus*) or below (as in *Clupeonella*) the females. After spawning the roe rises or sinks respectively, passing through the layer of milt so that fertilization takes place. In this case the individuality of the males during spawning is to a considerable degree lost, and therefore they hardly strive to approach a particular female, owing to which the mobility of the males is reduced and the related morphological peculiarities aimed at increasing manoruvrability disappear.

3. Maintaining equilibrium and braking

Maintaining equilibrium, which consists of eliminating or preventing the appearance of undesirable roll and yaw differences, and braking, which consists of creating negative acceleration, are closely related in nektonic animals since deflexion and active movement of fins and flippers in order to create moments necessary for preserving equilibrium inevitably increase the animal's hydrodynamic resistance.

During rapid forward movement, the fins (flippers) acting as rudders and stabilizers create transverse forces which suffice to maintain the animal in the desired position even at insignificant deflexions from the body, provided that they are set at least at a small angle to the direction of movement. With a drop in the speed of the flow over the animal it is compelled to extend the stabilizing fins more and more. When, finally, the speed is reduced so much that even with the paired fins extended to the maximum the transverse forces they create prove inadequate for maintaining equilibrium, these fins have to move autonomously in order to create the necessary transverse forces. During intentional retardation of forward movement (braking) which is effected not by a sharp turn but while continuing in the initial direction of movement, all fins are extended to the maximum, as a result of which the total drag steeply increases, primarily because vortex formation begins over the pectoral fins (deflected from the body), as has been shown (Aleyev & Ovcharov 1973b) experimentally (Figure 74). At the same time the fully extended fins undoubtedly prevent rotation of the animal's body about its longitudinal axis, thereby facilitating the preservation of equilibrium. It is this that provides the close relationship between preserving equilibrium and braking in nektonic animals.

Ensuring equilibrium by setting certain passive rudders at an angle to the direction of forward movement is the simplest way of doing so. The simplest way of braking that requires no special braking system is to change sharply the direction of movement, thereby steeply increasing the total drag. This mode of braking is the main or only possible one for nektonic animals whose fins have limited mobility (Sagittoidea, Elasmobranchii, Acipenseridae, Cetacea, etc.) or very small relative dimensions (Trichiuridae, most of the Anguilliformes, etc.), or when there are no paired fins at all (Hydrophidae, many Anguilliformes); this has already been reported for the Elasmobranchii and Acipenseriformes (Schmalhausen 1916; Harris 1953, and others).

Ensuring equilibrium and braking by special fin movements or operating a hydrojet propulsor is more complex.

In Teuthoidea the transverse forces necessary for maintaining equilibrium are created by undulation of the mantle fins (Akimushkin 1963; Zuyev 1966) and by the hydrojet propulsor, which, when performing in unison, can create practically any rotational moment. Braking is effected predominantly by turning the outer end of the funnel through 180° so that the hydrojet propulsor creates a braking force; at this moment the arms are extended like a fan and the mantle fins are also extended, thereby increasing the resistance still more (Zuyev 1966).

In fishes, nektonic reptiles, Sphenisciformes and pinnipeds with sufficient mobility of the paired fins and flippers (the majority of the Actinopterygii, Ichthyosauria, Sauropterygia, Placodontia, Testudinata, Sphenisciformes and Pinnipedia), preservation of equilibrium and braking, according to the author's observations of the movement of diverse fishes (Elasmobranchii and Osteichthyes), sea turtles (*Chelonia*, *Eretmochelys* and *Caretta*), penguins (*Eudyptes*) and pinnipeds (*Phoca*, *Pagophoca* and *Arctocephalus*), are achieved during slow swimming predominantly through active movements of all the above fins or flippers, and during rapid swimming predominantly or exclusively through simple deflexion of the paired fins (flippers) away from the body, i.e. by placing them at an angle to the stream. This is also the way in which these functions are performed in sirenians, in whom the pectoral flippers possess considerable mobility (Petit 1955).

The stronger balancing and braking functions of the pectoral fins in Actynopterygii are facilitated both by the vertical orientation of their bases (Schmalhausen 1916; Gregory 1928; Harris 1937a, 1953; Vasnetsov 1941; Chabanaud 1943; Aleyev 1963a, and others) and also by the shift of their bases upwards to the level of the centre of gravity, which prevents the emergence of a parasitic pitching moment during deflexion of these fins (Schmalhausen 1916; Harris 1937a; Aleyev 1963a, and others). An intensification of these functions of the pelvic fins in Actinopterygii takes place during the movement of these fins towards the anterior end of the body (Harris 1937a,b; Aleyev 1963a), which helps to reduce undesirable pitching moments arising during their deflexion. The pectoral and pelvic fins of Actinopterygii represent a single balancing-braking system which as a whole creates no parasitic torque, since the moments created by these fins cancel each other (Harris 1937a; Aleyev 1963a; see also Figure 135).

Equilibrium during the movement of most fast eunektonic fishes and cetaceans is maintained mostly or exclusively by passive stabilizing action of the pectoral fins (flippers), while braking is effected, as a rule, by a sharp turn, which the author was able to see for himself when observing the swimming of different Scombridae and dolphins (*Tursiops*, *Delphinus* and *Phocoena*). These and similar animals encounter scarcely any obstacles and are therefore hardly capable of braking sharply while maintaining the initial direction of movement.

A peculiar mode of braking is observed in *Istiophorus*: extension of the huge dorsal fin of these fishes steeply increases frictional drag. There are no grounds to

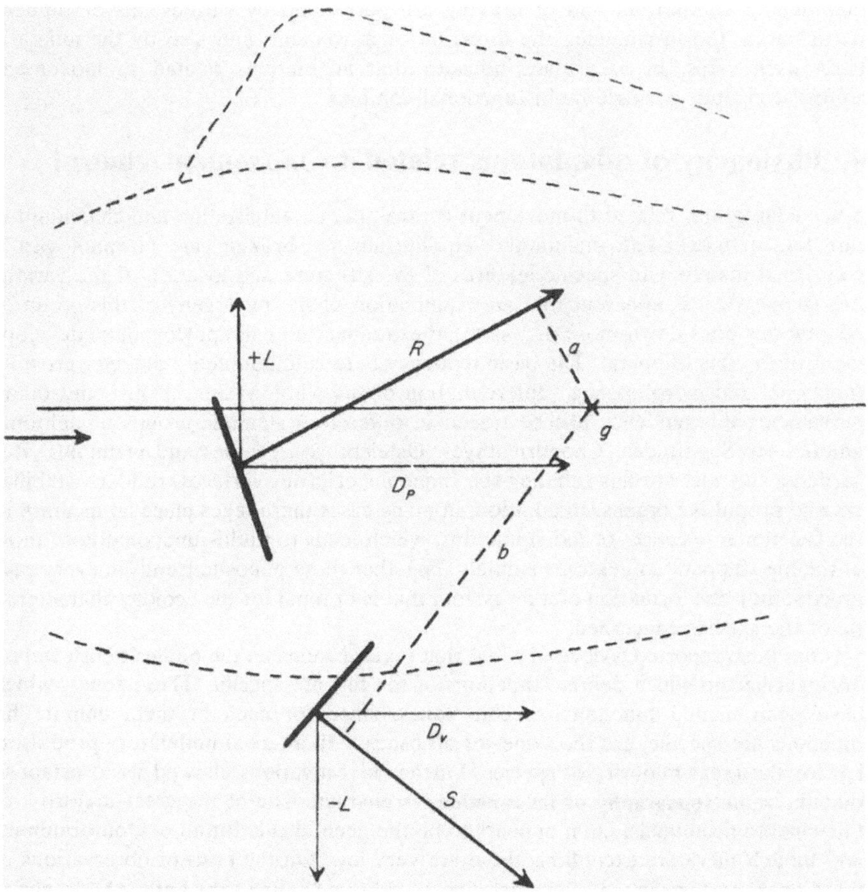

Figure 135. Diagram of operation of braking system formed by paired fins of fishes of the perch family. *R*, *S*, drag forces acting on pectoral and pelvic fins; $+L$, D_p, $-L$, D_v their components; *g*, centre of gravity; *a*, distance from vector *R* to *g*; *b*, distance from vector *S* to *g*. If $Ra = Sb$, the system of four paired fins produces braking only (Harris 1937).

believe that the dorsal fin of *Istiophorus* serves to ensure partially high manoeuvrability (Ovchinnikov 1967b) since *Istiophorus* needs such manoeuvrability no more than other fast-swimming predators and, most important, adequate manoeuvrability can be ensured at the high swimming speeds characteristic of this genus by relatively very small rudder fins.

As the above shows, nektonic animals have no special morphological adaptations aimed exclusively at the preservation of equilibrium and braking. The functions of maintaining equilibrium and of braking are performed by various fins or flippers participating too in changing the direction of movement, and also by the animal's body itself. This, in particular, indicates that adaptations related to movement control constitute a single multi-functional complex.

4. Phylogeny of adaptations related to movement control

Since adaptations related to movement control, i.e. to stabilization and changing the direction of movement, maintaining equilibrium and braking, are morphologically concerned mainly with specific features of the structure and location of the various fins (flippers), it is apparent that an examination of the phylogeny of this group of adaptations boils down more or less to an examination of the phylogenetic development of the fins (flippers). The basic tendency here is undoubtedly the ever growing functional and morphological differentiation of the whole system of fins controlling movement; this tendency can be traced in different systematic groups of nektonic animals – in Sagittoidea, Chondrichthyes, Osteichthyes, reptiles and mammals. Regarded as distinct are fins fulfilling the functions of lifting surfaces, rudders, stabilizers and propulsive organs. In addition, in many cases there takes place an increase in the functional 'valence' of individual fins, which leads to multi-functionality of most of the fins (flippers) in nektonic animals. Together these opposite trends in every case precondition the formation of a fin system that is optimal for the ecology characteristic of the species concerned.

It has been reported (Aleyev 1957a) that several zones on the bodies of fish can be distinguished to which definite functions of the fins are specific. These zones, which have been named functional-specific zones, since for each of them definite fin functions are specific, are the same for all fishes with an axial undulatory propulsor, i.e. for the great majority of species. Further investigations allowed us to establish that the same **topography of fin functions** is characteristic of the great majority of other nektonic animals, i.e. it appears to be the general rule for all nektonic animals, and though there are exceptions these are very few. On the basis of observations of fin (flipper) performance in the most diverse nektonic animals the author has reached the conclusion that inherent in nektonic animals is a common topography of fin functions, determined by the hydrodynamic features of the individual functions. That is, on the body of a nektonic animal there are three functional-specific zones, each of which is characterized by definite fin functions. In the caudal direction these zones are as follows:

(I) Zone of frontal rudders, balancers and lifting surfaces.
(II) Zone of stabilizers.
(III) Zone of rear rudders, balancers, lifting surfaces and propulsive organs.

The functions of zone I, i.e. the functions of the frontal rudders, balancers and lifting surfaces supporting the anterior end of the body, are associated with the creation of moments changing the position of the body's longitudinal axis. These functions are carried out by fins placed as close as possible to the anterior tip of the body and, accordingly, removed to the maximum from the centre of gravity, which maximizes the rotational moments resulting from the transverse forces created by these fins. During uniform or accelerating rectilinear movement all these fins (flippers), when they do not function in addition as propulsive organs, are as a rule folded up, i.e. pressed to the body, and remain extended only in animals in which their mobility is notably restricted. This mostly happens either as a result of adaptation of these fins to act as lifting surfaces (pectoral fins of sharks and sturgeons) or as a result of adaptation to very fast swimming, when the stability of the anterior fins must be greater, which is achieved at the cost of restricting their mobility (the first dorsal and pectoral fins in *Xiphias* and *Makaira indica*).

In Teuthoidea the zone I functions are performed by the mantle fins, which, when extended, can act as active rudders and balancers, and when the mantle end of the body is bent, can act as passive rudders. In ichthyoids and fishes, the functions of zone I are ensured by the pectoral fins, which act both as frontal rudders and balancers and as lifting surfaces, and in many cases also by the anterior (more or less separate) portion of the dorsal fin when it is close to the anterior tip of the body (*Coryphaena*, *Auxis*, *Sarda*, *Megalaspis*, etc.). In this case, when the frontal dorsal fin is extended during turning it has to withstand the action of transverse forces, owing to which its rays are usually thickened, frequently (Perciformes, Mugilidae, etc.) turning into firm spines; this gives the blade of the fin the necessary transverse stability. In fishes whose pelvic fins are situated on the thorax, these fins also act as frontal rudders, as a result of which their first rays frequently become less flexible. In nektonic reptiles and in pinnipeds, cetaceans and sirenians the functions of zone I are performed by flippers on the forelimbs.

In Sagittoidea, Hydrophidae and Sphenisciformes the functions of zone I are not performed by fins since the anterior part of the body (or the entire body) has greater flexibility, thanks to which the necessary transverse forces on the anterior end of the body can be created by bending it. In Sagittoidea the function of frontal rudders may however be partiallv performed by the frontal pair of lateral fins.

The function of zone II, i.e. the function of stabilizers, is the creation of transverse forces behind the centre of gravity preventing deflexion of the body's longitudinal axis from the direction of movement. This function is performed, accordingly, by the fins situated behind the centre of gravity. The location of zone II on the animal's body is determined by two contradictory factors. On the one hand, it is advantageous to shift the stabilizing fins backwards to the utmost, thereby removing them from the centre of gravity. On the other hand, these fins must be located on that part of the body where the animal's median plane does not undergo any noticeable periodic deflexions during performance of the propulsor. Accordingly, the position of the stabilizing fins in nektonic animals partially depends on the type of propulsive mechanism. In animals with a paddling or hydrojet propulsor, as well as in some species with a peripheral undulatory propulsor, these fins are situated close to the posterior end of the body (the arm fins of Teuthoidea, the posterior flippers of

Otariidae, the tail fin of *Zeus*, etc.). In animals with an axial undulatory propulsor, whose body axis constantly bends during active movement, the stabilizing fins are usually located immediately behind the centre of gravity, where the amplitude of the propulsive wave is still small, which guarantees comparatively small deflexions of the fin blades from the animal's median plane during swimming movements. Such a location of the stabilizing fins is characteristic of Sagittoidea, fishes swimming by means of an axial undulatory propulsor, Ichthyosauria and cetaceans; however, when these animals move by inertia, the locomotor (tail) fin, situated at the rearmost point of the body, also has a stabilizing function. In a number of fishes this contradictory location of the stabilizing fins leads to a considerable elongation of their rays, which permits a considerably shift of the centre of the fin blades backwards, facilitating a higher stabilizing effect. A classical example of this arrangement can be seen in the tuna fishes (*Thunnus alalunga, Thunnus albacore*, etc.; Figure 137). Stabilizing fins act both during rectilinear movement and during turning, when they create additional forces preventing transverse drift of the animal under inertia forces; this is why these fins are always extended to some degree.

Stabilization in Sagittoidea is performed mainly by the posterior pair of lateral fins situated behind the centre of gravity. In Teuthoidea this function is performed by the arms and arm fins. In fishes stabilization is carried out by the posterior portions of the dorsal fin, situated immediately behind the centre of gravity, as well as by the anal fin or its morphologically separated anterior portion, also situated immediately behind the centre of gravity. The front rays of stabilizing fins are frequently thickened to ensure greater transverse stability of these fins. When a fish has two dorsal fins (as in most of the Perciformes), in the great majority of cases the posterior one acts as a stabilizer and the anterior one as a frontal rudder. When there are three dorsal fins (as in Gadinae), the first acts as a frontal rudder, the second as a stabilizer and the third as a propulsive organ. In fishes with a long anal fin, its frontal part, consisting of elongated rays (as in *Trachurus*), acts as a stabilizer and its rear part as a propulsor. When there are two separate anal fins (as in Gadinae) the front one is a stabilizer and the rear one a propulsor. In most cases the dorsal and anal stabilizing fins of fishes are situated one above the other and are more-or-less equal in shape and size, which ensures the creation of stabilizing forces symmetric about the body's longitudinal axis, i.e. the absence of any parasitic torque.

The function of a stabilizer in Ichthyosauria was performed by the dorsal fin and the fins formed by the posterior limbs, while in cetaceans it is performed by the dorsal fin only. In Sphenisciformes, Chelonioidea, Dermochelyoidea, Otariidae and partially in Phocidae and Odobenidae, the role of a stabilizer is played by the posterior limbs, however in members of the last two groups the basic function of these limbs is propulsive. In some Chelonioidea and a number of other nektonic reptiles stabilization is partially performed by the tail.

In Hydrophidae stability is ensured not by the fins but by the very high slenderness ratio of the body. Neither are there any stabilizing fins in sirenians nor in some, mostly slow swimming, cetaceans (*Balaena, Monodon, Delphinapterus* and *Lissodelphis*).

The functions of zone III, i.e. the functions of rear rudders, balancers, lifting surfaces and propulsive organs, are the creation of transverse forces on the posterior end of the body (rudders balancers and lifting surfaces) and the creation of a propulsive force (locomotor fin). The rudders and balancers function best when removed to the utmost from the centre of gravity, the reasons for which have been examined above. The locomotor fin operates as a propulsor with the highest efficiency, as we have seen in chapter IV, when it is situated at the rearmost point of the body. In fishes, not only the locomotor fin itself, which directly creates the propulsive force, must be close to the rearmost point of the body, but also the functionally related specialized finlets, i.e. derivatives of the posterior portions of the dorsal and anal fins which prepare the stream flowing into the propulsor area, forming a damping screen. Posterior rudders and balancers function during turns (rudders and balancers) and in the absence of forward movement (balancers), while lifting surfaces and propulsive organs function during active forward movement; when the animal is moving by inertia all the fins in zone III act as stabilizers. Accordingly, the fins in zone III are usually extended both during forward movement and in its absence.

The functions of zone III in Sagittoidea are performed by the tail fin, which is the main propulsor. In Teuthoidea the functions of zone III are carried out by the arm fins, which, by the way, function also as rear rudders. In fishes the functions of zone III are performed by the posterior portions of the dorsal and anal fins and by the tail fin. A balancing function of all these fins is especially characteristic of slow-swimming, predominantly benthonektonic Actinopterygii; in fast-swimming fishes it is reduced to the minimum or is absent. Rudder and propulsive functions of these fins are characteristic of the great majority of fishes. However we find a variant of this system of providing zone III functions in the eunektonic *Mola*, in which the tail fin is a stabilizer, while the propulsive action is performed by dorsal and anal fins close to the rearmost point of the body. In nektonic reptiles zone III functions are performed by the tail fin when present, or by a tail of the usual reptile type, even if it is devoid of a special fin, as well as by the flippers formed by the hind limbs. The tail fin carries out the functions of zone III in all cetaceans and sirenians. In Phocidae and Odobenidae the zone III functions are performed by the hind flippers. In Sphenisciformes and Otariidae the propulsive function of zone III is practically absent; the functions of balancers and rear rudders are carried out by the hind limbs. A peculiar way of ensuring zone III functions occurred in the xeronektonic Hesperornithes: the functions of propulsive organs, rear rudders and balancers were performed by the hind limbs, whose hydrodynamically active elements were set close to the rearmost point of the body.

It follows from all the above that the functional-specific zones are the same in nektonic animals with fins and flippers with the most diverse structure and locations: the arrangement of these body zones of a nektonic animal is always the same (Figures 136, 137). This constancy of the functional topography of the fins is an expression of the fact that the external morphological organization of nektonic animals is subject

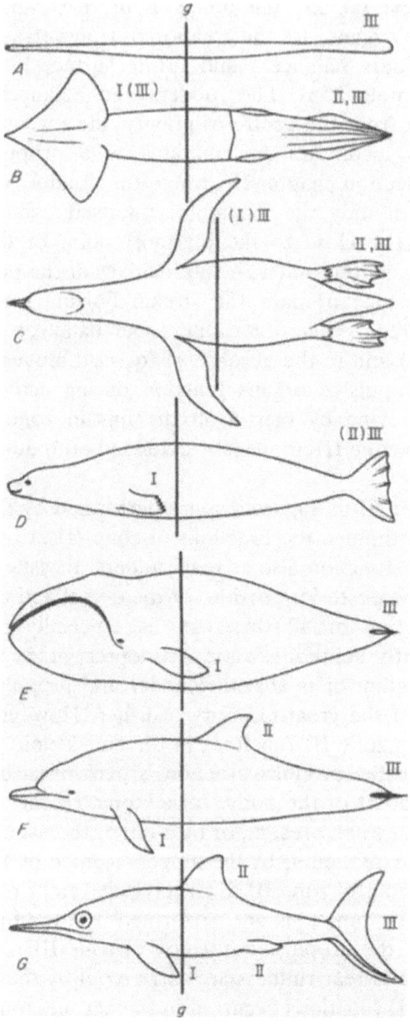

Figure 136. Fins labelled according to the various functional-specific zones (Roman numerals) in nektonic animals. Bracketed numbers mean that the function of that zone is not the chief one of the fin concerned. g, centre of gravity. With the purpose of displaying the fins more clearly, some animals are depicted laterally and some in plan. *A*, *Enhydrina schistosa* (Daud.) (side view), $L_a = 90.0$ cm; *B*, *Symplectoteuthis oualaniensis* (Less.) (plan view), $L_a = 24.1$ cm; *C*, *Eudyptes chrysolophus* Brandt (plan view), $L_a = 69.7$ cm; *D*, *Pagophoca groenlandica* (Erxl.) (side view), $L_a = 197.0$ cm; *E*, *Balaena mysticetus* L. (side view), $L_a = 1500$ cm; *F*, *Delphinus delphis ponticus* Barab. (side view), $L_a = 178.5$ cm; *G*, *Stenopterygius quadriscissus* Quenst. (side view), $L_a = 210.0$ cm.

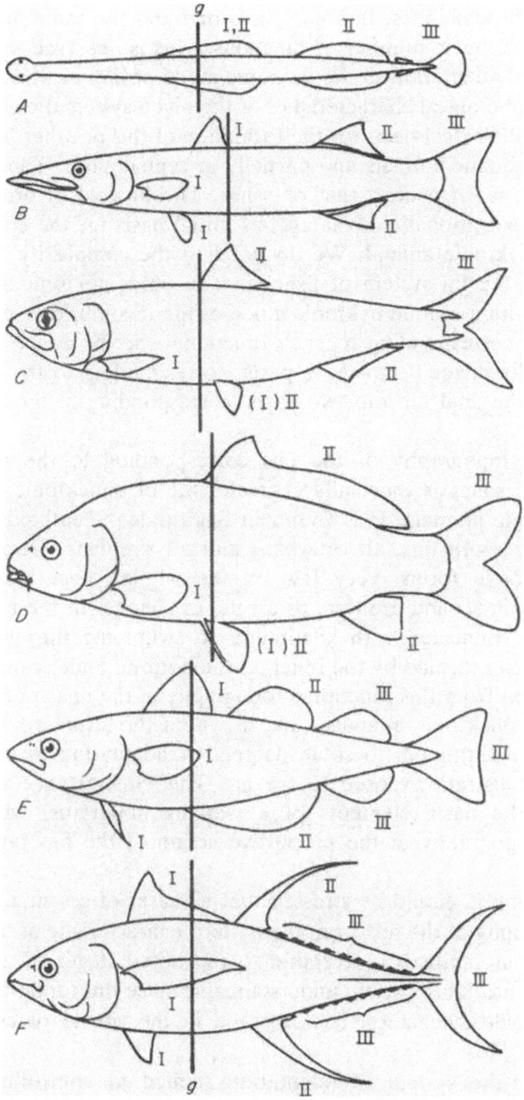

Figure 137. Fins labelled according to different functional-specific zones in nektonic animals. *A, Sagitta setosa* Müll. (plan view), $L_a = 2.0$ cm; *B, Sphyraena barracuda* (Walb.) (side view), $L_a = 150.0$ cm; *C, Alosa kessleri pontica* (Eichw.) L. (side view), $L_a = 27.9$ cm; *D, Cyprinus carpio carpio* L. (side view), $L_a = 50.0$ cm; *E, Odontogadus merlangus euxinus* (Nordm.) (side view) $L_a = 47.1$ cm; *F, Thunnus albacora* (Lowe) (side view), $L_a = 135.0$ cm. Symbols as in figure 136.

to the laws of hydrodynamics. In some cases one and the same fin or part of a fin carries out not one, but a number of functions. This is observed most frequently in the initial stages of adaptation to the nektonic mode of life, as well as in cases when, owing to the morphological characteristics of the given systematic group, there is not an adequate morphological basis for the formation of this or other fins, which is so in many secondary aquatic animals and partially in cephalopods. The most plastic and diverse fin system is, of course, that of fishes. The fin fold of primary vertebrates proved to be an exceptionally advantageous initial basis for the construction of the fin system of a nektonic animal. We do not find the complexity and evolutionary responsiveness in the fin system of fishes in any other nektonic animal. In highly organized fishes with a mobile nektonic mode of life (Scombridae, etc.), there is as a rule a special fin corresponding to each functional-specific zone. Accordingly, the dorsal fin is usually divided into three parts, corresponding to the first, second and third zones, and the anal fin into two parts, corresponding to the second and third zones.

The functional topography of the fins corresponding to the above-mentioned functional-specific zones is especially characteristic of eunektonic, planktonektonic and benthonektonic animals. It is found in Sagittoidea, Teuthoidea, all fishes, all eunektonic reptiles with fins, all cetaceans and all sirenians. Deviations from this pattern in the above forms, very few on the whole, most often occur among benthonektonic animals and are due, as a rule, to changes in the topography of the propulsive action connected with adaptation to swimming through a complicated labyrinth of obstacles formed by the relief on the bottom, underwater plants or coral. Essential deviations from this functional topography of the fins are found, judging by everything, in xeronektonic animals only, in whom the structure of the propulsive apparatus is still determined, to some degree, by adaptation to movement on the surface of solid substrata exposed to the air. This circumstance usually results in preservation of the basic elements of a walking apparatus, which hinders the formation of a topography of the propulsive action of the fins typical of nektonic animals.

In this connection it should be stressed that all cases of essential deviation of the functional topography of the fins from the pattern characteristic of the overwhelming majority of nektonic animals correspond to the initial stages of adaptation to the nektonic mode of life. This is quite understandable since this topography, determined by the laws of hydromechanics, has emerged in the course of adaptation to the nektonic mode of life.

On the whole, the system of adaptations aimed at controlling movement is provided more and more with specialized fins during the changeover to the eunektonic mode of life. Among the individual fin functions, the phylogenetically most ancient is undoubtedly the propulsive function, which is ensured by special fins in one way or another in all nektonic animals. Following it, according to everything, there appeared the whole complex of zone I functions, i.e. the functions of frontal rudders, balancers and lifting surfaces; all these functions are most important elements of the functional complex associated with movement control. Zone I functions are performed by special fins in the great majority of nektonic animals. The next in importance and, one would think, in phylogenetic age is the function of stabilization; this too is ensured by fins at comparatively early stages of adaptation to the nektonic mode of life, but, at the same time, much later than the functional complex of zone I. This order of appearance of individual functional-morphological fin complexes in the phylogeny of nektonic animals corresponds to the role of these complexes in the general system of morphological adaptations aimed at creating and controlling forward movement.

VII. CAMOUFLAGE AND DEFENCE

1. General

Whereas the development of adaptations associated with movement (chapters III–VI) has the strongest influence on the morphological organization of nektonic animals, second place in this respect goes to the complex of adaptations associated with camouflage and defence against enemies.

Camouflage may have either of two purposes. First, it may be used by the victim, and second, by the predator lying in wait for or stealing up on its prey. The most varied adaptations associated with camouflage are found in nektonic animals dwelling in the upper, well-illuminated layers of the water, down to depths of about 500 m, which corresponds to the greatest diversity of ecological conditions being at these depths. Cryptic elements of body shape are practically absent in nektonic animals which always or for most of their life cycle dwell at depths greater than 500 m, i.e. in the weakly illuminated and aphotic zones, while in colouring, along with trends of simplification, there appear very specialized features aimed at preserving concealment; along with this luminescent organs appear, which vary both morphologically and functionally.

Adaptations associated with defence may be interpreted very widely indeed. A capacity for fast movement, camouflage, high resolving power of sensory organ receptors, the presence of solid armour, all kinds of spines, acanthas, etc., the presence of repelling structures and elements of colouring, poisonousness in combination with a memorable colouring, and many other things are all adaptations associated with defence against enemies. However the means of defence enumerated are not all of equal value. For example, it is obvious that a capacity for fast movement, camouflage and a high resolving power of sensory organ receptors may not be due only to the development of adaptations associated with defence. At the same time there do exist structural features associated with defence alone: all kinds of ossification of the skin, spines, acanthas, etc. These are the adaptations we shall examine in this chapter. Other features which in some cases may also be regarded as partial adaptations associated with defence against enemies are examined in part in chapter VIII.

Below we examine principally morphological aspects of adaptations associated with camouflage and defence, since these facilitate in greatest measure the convergence of nektonic animals belonging to different systematic groups.

2. Camouflage

1. Methods of Camouflage in Nektonic Animals

Three basic groups of adaptations associated with camouflage can be established for nektonic animals: 1. high body transparency, 2. cryptic features of body shape and 3. cryptic colouring.

The general development of adaptations associated with camouflage changes regularly in nektonic animals with changing depth of habitat and changing Reynolds number. Among the extant deep-water nektonic fauna, consisting of Sagittoidea, Teuthoidea and fishes, the morphological features associated with camouflage are on the whole more uniform than in nektonic animals dwelling in the upper, illuminated layers of the water. In the latter, with increasing size of individuals and, therefore, increasing swimming speeds, we observe a regular variation in the basic principles of camouflage which, judging by everything, is absent or nearly absent in deep-water species.

The smallest representatives of nekton populating the upper, well-illuminated water layers, in particular, nearly all the planktonektonic forms, are to a high degree transparent. Body transparency significant for concealment is maintained in Sagittoidea and Teuthoidea up to $Re \approx 5.0 \times 10^5$ and in Osteichthyes usually up to $Re \approx 5.0 \times 10^4$ but in exceptional cases (for example in planktonektonic Comephoridae) up to $Re \approx 2.0 \times 10^5$, i.e. on average transparency is maintained in nektonic animals up to about $Re \approx 10^5$, which corresponds to the border between planktonekton and eunekton. With an increase in the size of the animals, securing high optical transparency of the body grows more and more complicated. For this reason there begin to develop camouflage adaptations based on the effect of cryptic colouring and on the effect of full or selective reflection of light by the body surface, i.e. on the mirror effect, which is ensured by peculiarities of body shape, and a high reflectivity of its surface. Cryptic features of body shape in combination with cryptic colouring remain the basic form of camouflage in nektonic animals belonging to different groups on average till *Re* reaches values of the order to 10^6–10^7. However, with a further increase in the animals' dimensions and the related growth of average swimming speeds (see chapter IV), the process of adaptation to movement sets ever more rigid demands on body shape, so that an essential alteration of the shape of the body's cross-section aimed at camouflage despite the necessity of optimizing hydromechanical parameters becomes less and less possible. At $Re > 10^7$ cryptic features of body shape disappear completely and the function of furnishing concealment is taken over entirely by colouring.

With increasing size and swimming speeds of nektonic animals the number of their potential enemies diminishes; thus escaping from predators moves increasingly into first place among means of protection, whereas the importance of camouflage become more and more secondary.

2. Body Transparency as a Means of Camouflage

High optical transparency of the body is especially characteristic of planktonic and planktonektonic forms and occurs among only the smallest eunektonic species and only in exceptional cases. Camouflage on account of high optical transparency of the body is characteristic in some measure of representatives of three nektonic groups: Sagittoidea, Teuthoidea and Actinopterygii.

In Sagittoidea high body transparency is a very characteristic feature. However in nektonic Teuthoidea and Actinopterygii concealment is secured wholly or partially on account of optical transparency only in the planktonic young, and, in the adult state, in planktonektonic species (Enoploteuthidae, Salangidae, Gonostomidae, Steronoptychidae, Sudidae, Myctophidae, etc.). Both in Teuthoidea and in fishes concealment is usually achieved through a certain combination of body transparency and cryptic colouring, and more seldom on account of body transparency alone (Salangidae).

3. Cryptic Features of Body Shape

Features of body shape associated with camouflage are found among members of two nektonic groups: Actinopterygii and Reptilia. In the most general case these features produce a shape of the body's cross-section which eliminates the betraying shadow on the animal's ventral side. This type of camouflage occurs in comparatively small nektonic animals dwelling as a rule in the uppermost, well-illuminated water layers. Some benthonektonic Actinopterygii have other cryptic features of body shape, not associated with the body's cross-section.

A. Camouflage and the cross-sectional shape of the body

Many non-deep-dwelling Actinopterygii, and also Hydrophidae from among the reptiles, have cryptic features in the shape of the body's cross-section. The shape of the body's cross-section is determined in these animals, as we have already noted, primarily by the development of adaptations associated with movement. Along with this, however, the cross-sectional shape of the body in Actinopterygii, and also in many Hydrophidae, is always partially determined too by the development of adaptations associated with camouflage: as a rule, the shape of the body's cross-section helps to eliminate the betraying shadow on the animal's belly (Aleyev 1960b).

As is known from hydro-optics (Shuleikin 1933; Timofeyeva 1951), light brightness in water, which diminishes with depth, does not approach complete uniformity in different directions: it tends merely to some steady distribution in the different directions. The polar diagram of this steady brightness distribution in the different directions is close to a circle with the pole shifted towards the surface of the water, i.e. upwards, relative to the centre of the circle (Figure 138I). Accordingly, in the most general case a nektonic animal is illuminated mainly by a converging cone-like

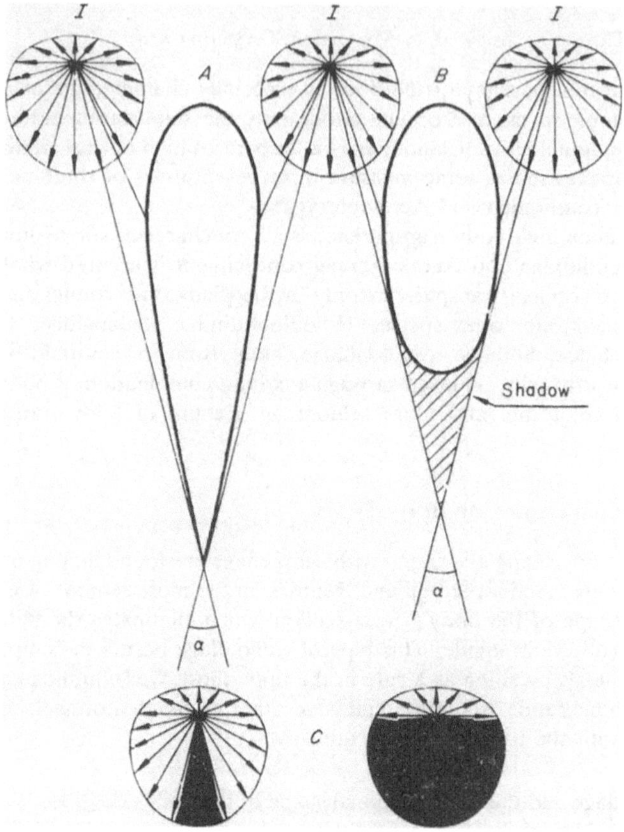

Figure 138. Diagram of (*A*) elimination of betraying shadow by ventral keel and (*B*) appearance of betraying shadow on ventral surface of body in the absence of a ventral keel. *I*, polar diagram of stationary distribution of brightness for media with a low absorption factor, *C*, rays illuminating all body surfaces facing downwards; shaded areas correspond to light rays in directions incapable of illuminating downward-facing surfaces. See text.

bundle of rays incident from above and composed of a mixture of direct and diffuse light (Figure 138). The weaker diffuse light incident on the animal from the surrounding space outside this cone does not illuminate it from beneath with the same intensity and therefore a betraying shadow appears on the ventral side of the body, emphasizing the its depth, i.e. volume.

Most expedient for nektonic animals which may be observed from above, from the side and from beneath is the complete elimination of betraying shadows. Therefore many nektonic animals have a body cross-section that maximizes elimination of the betraying shadow on the ventral part of the body.

In many eunektonic, planktonektonic and, more seldom, benthonektonic fishes, both marine and freshwater, the abdomen is compressed into a sharp-edged longitudinal ribor **ventral keel** (in the majority of the Clupeidae; among the Cyprinidae, in *Scardinius, Aspius, Alburnus, Alburnoides, Erythroculter, Culter, Pelecus,* etc.; among the Stromateidae, in *Stromateus,* etc.; see Figure 66D). It has been shown (Aleyev 1960b) that, when combined with a light, silvery or golden colouring and high reflectivity (i.e. a glittering, sometimes mirror-like surface) of the lower half of the body, a ventral keel constitutes an adaptation associated with camouflage. An absolutely analogous keel performing the same cryptic function is found on the abdomen of eunektonic Hydrophidae (Figure 12C), in which its presence just as in fishes, is combined with a light yellow or greenish colouring of the sides and ventral part of the body. The ventral keel of fishes may be formed by special keel scales (Clupeidae) or may appear as a skin growth (*Pelecus*). However in fishes the cryptic function of a ventral keel is performed not only by a special structure known by this term, but also by less severe pointedness of the lower edge of the body, which is characteristic to some degree of nearly all nektonic fishes (Carangidae, Scombridae, etc.). Both in fishes and in Hydrophidae, the ventral keel serves to eliminate the betraying abdominal shadow appearing when the animal is illuminated from above. As Figure 138B shows, a rounded abdomen makes this shadow inevitable, since in the absence of a ventral keel, all surfaces facing downwards can be illuminated only by rays that are directed upwards (Figure 138C), which according to the polar diagram of the diffusion (Figure 138I) provide the least luminosity. In the presence of a ventral keel, this shadow is absent or at least strongly reduced (Figure 138A) since the keel fills up, as it were, the region of the shadow; in this case all the surfaces facing downwards are no longer illuminated only by upward rays, but also by the majority of downward rays (Figure 138C), which according to the polar diagram of the diffusion (Figure 138I) provide the highest luminosity.

The cryptic effect created by a ventral keel has been modelled (Aleyev 1963a). Two wooden slabs were shaped for this purpose such that the cross-section of one was identical to that shown in Figure 138A, while the cross-section of the other was identical to that shown in Figure 138B. These models were illuminated from the 'dorsal' side through a mat screen with uniformly distributed light sources behind it. To the sides of and below the models there were screens reflecting light on to them from beneath. The models were photographed from beneath against the background of the bright upper screen, which radiated light (Figure 139C). By changing the strength of the light sources and the arrangement of the lower screens, we could easily achieve a situation in which the slab shaped to the cross-section of a fish with a ventral keel (Figure 138A) became practically invisible, disappearing against the background of the bright upper screen (Figure 139A). However the slab shaped to the cross-section of a fish without a ventral keel (Figure 138B) remained clearly visible whatever the strength of the light sources or the position of the screens, owing to the presence of a sharp betraying shadow on the 'belly' (Figure 139B).

Thayer (1909) reported the 'counter-shadow' phenomenon in the colouring of

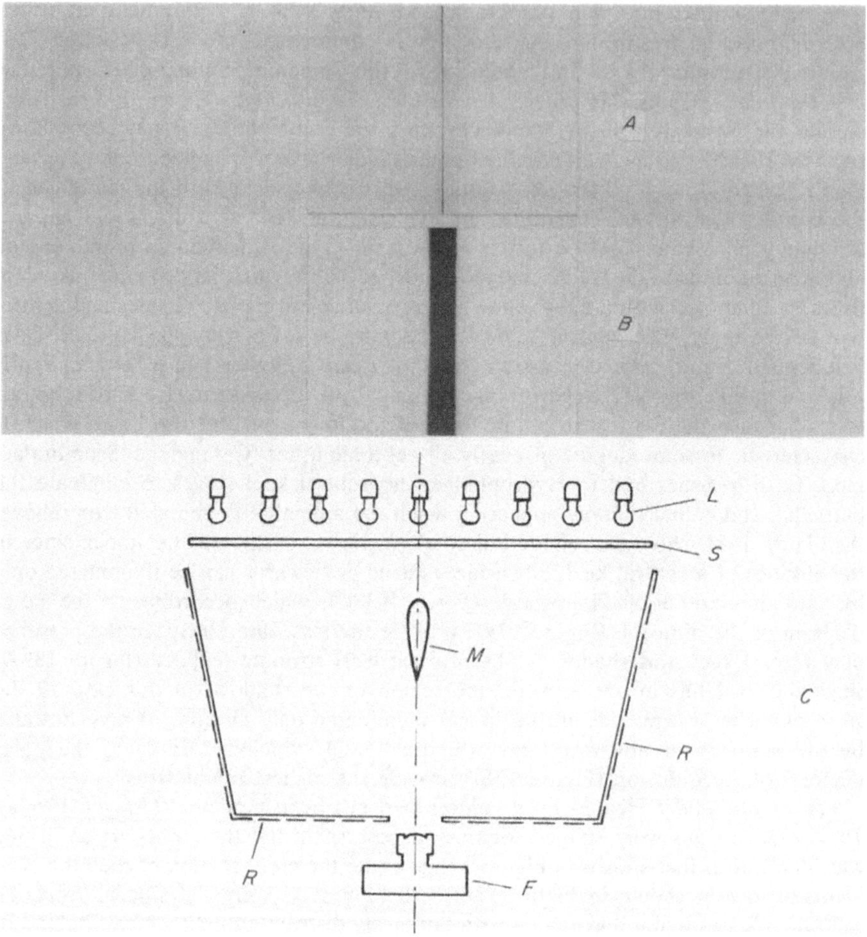

Figure 139. Bottom views illuminated from above of models of fish bodies (*A*) with a ventral keel (*Pelecus cultratus* (L.) and (*B*) without a ventral keel (*Salmo trutta labrax m. fario* L.). In the presence of a ventral keel (*A*) the model practically merges with the light upper background. *C*, diagram of installation used to model cryptic function of ventral keel. *M*, model; *L*, light source; *S*, mat screen; *R*, reflecting screens; *F*, camera.

fishes and noted that it does conceal the shadow. The ventral keel represents an adaptation with fundamentally the same purpose, since it gives an observer the illusion that the body of the nektonic animal lacks volume by eliminating the shadow emphasizing depth. In distinction, however, from Thayer's counter-shadow, the cryptic function of a ventral keel consists not of concealing the betraying shadow but of its elimination, i.e. a ventral keel is a more perfect adaptation effective during illumination of various intensities.

With brighter illumination under which the background (above or to the side) becomes lighter than the shadow on the white belly of the nektonic animal, which may take place in the uppermost, well-illuminated layers of the water, Thayer's counter-shadow proves unable to eliminate the betraying effect of volume introduced by the shadow. In this case, to camouflage the shadow on the belly by means of colouring would require such a strong darkening of the sides that when viewed sideways the animal would, though 'flat', be clearly-visible as a dark silhouette against a light background; the animal would also be clearly visible from beneath, despite the whiteness of the belly. Therefore, when the shadow on the ventral side of the animal becomes darker than the background, a new, more effective adaptation is added to Thayer's counter-shadow camouflage system which is better capable of creating a cryptic effect whatever the illumination, i.e. a ventral keel. It is for this reason that a ventral keel is most characteristic of fishes dwelling in the uppermost, surface layer of the water, among which we find all the cases of the strongest keel development (*Gasteropelecus, Pelecus, Culter, Hemiculter,* most of the *Clupeonella,* etc.); dwelling in the surface layers also are Hydrophidae. The degree of ventral-keel development in nektonic animals is, of course, determined also by the degree of water transparency, which determines the light diffusion and absorption factor and the shape of the polar diagram of the diffusion, which directly affects the degree of sharpness of the ventral keel required to produce the cryptic effect necessary in the given conditions.

A well-developed ventral keel is, as a rule, an attribute of small nektonic animals, which need camouflage for protection against predators most of all: in fishes such a keel occurs predominantly in the *Re* range 5.0×10^3 to 10^6, in Hydrophidae mostly in the *Re* range 10^5 to 5.0×10^6.

It is apparent that with an increase in the angle γ at the apex of a ventral keel its cryptic action will inevitably diminish, disappearing altogether at $\gamma = 180°$. Accordingly, in fishes whose adult forms have a well-developed ventral keel, the body's height (H) to width (I) ratio increases in ontogeny as this ensures reduction of the angle γ. Thus, in *Sprattus sprattus phalericus* with body lengths L to the end of the vertebral column of 2.6, 6.0 and 10.0 cm, the ratio H/I comes to 1.80, 1.96 and 2.02 respectively, and in *Clupeonella delicatula delicatula* with $L=2.7$, 4.3 and 6.6 cm, $H/I=2.22$, 2.43 and 2.50 respectively. The changes in the ventral keel in the ontogeny of fishes correspond to the ecological changes at various stages of development. A keel is frequently present in young pelagic fish but disappears in adult stages which are to some extent ecologically associated with the bottom (*Latridopsis,* etc.).

B. Other features of body shape associated with camouflage.

The other features of body shape in nektonic animals associated with camouflage are of a more specialized character. Such features are characteristic principally of benthonektonic animals, predominantly fishes. Many benthonektonic fishes have elongated fin rays and all kinds of processes on the skin resembling elements of underwater vegetation (*Zeus, Alectis, Anthias, Hipposcarus, Callyodon*, etc.; see Figure 140A). Also, the general shape of the body sometimes changes (to varying extents) for the sake of concealment in small slow-swimming benthonektonic forms; a case in point is the freshwater fish *Monocirrhus polyacanthus*, which is very similar to a floating leaf of a tree (Figure 140B).

4. Cryptic Colouring

Cryptic colouring is characteristic of the great majority of nektonic animals, the only exceptions being a few absolutely transparent planktonektonic forms.

Without repeating here the general theses concerning the cryptic variation in colouring of animals (Cott 1950), let us note that the variation in nektonic animals' cryptic colouring should be considered mainly from two aspects: first, in connection with changes in depth of habitat; second, in connection with stronger ecological contact with various kinds of solid substrata. In eunektonic and planktonektonic animals dwelling in the body of the water, changes in colouring are due almost

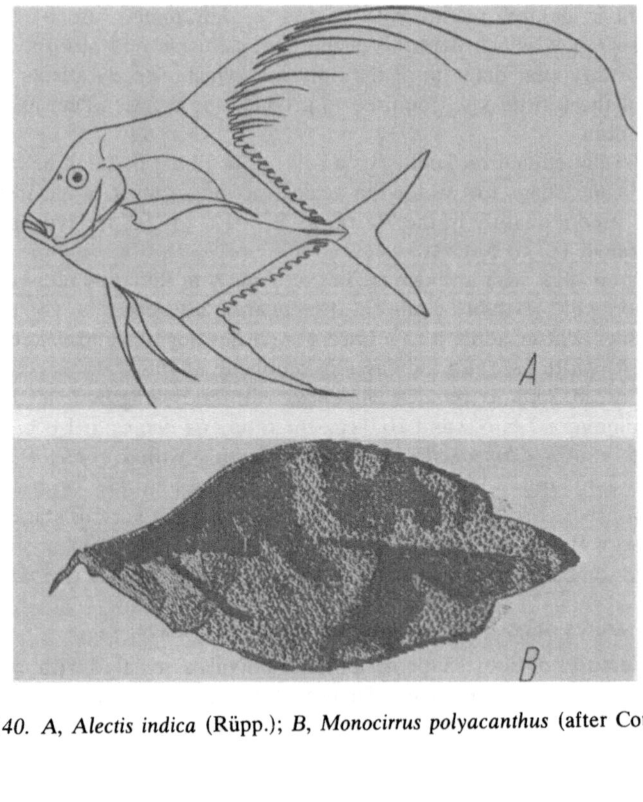

Figure 140. A, Alectis indica (Rüpp.); *B, Monocirrus polyacanthus* (after Cott 1950).

entirely to the first factor only, whereas in benthonektonic and xeronektonic species changes may be due to either.

A. Changes in cryptic colouring with depth of habitat

The changes in the colouring of aquatic animals with changing depth of habitat are common knowledge (Günther 1877; Joubin 1893; Brauer 1906, 1908; Chun 1910; Murray & Hjort 1912; Parker 1930; Suvorov 1948; Zernov 1949; Cott 1950; Bertin 1958f,g; Marshall 1958; Akimushkin 1963; Nikolsky 1963, and others). One of the special cases of such changes consists of those in nektonic animals. The reason underlying the changing colouring of animals with depth is the tendency for preservation of the cryptic function of colour with the changing illumination.

In well-illuminated, near-surface water layers, blue and green hues predominate in the spectral composition of the light as well as in the colouring of eunektonic and planktonektonic animals, which produces sufficiently intensive light reflection, creates the impression of light colouring and, in these conditions of good illumination, enhances the necessary cryptic effect. The phenomenon of the counter-shadow is clearly evident in the colouring of all eunektonic and planktonektonic animals inhabiting the upper water layers, and sudden transition from one colour to another is not typical in this case. This, however, does not rule out the presence of not very contrasting patterns in the colouring of the upper side of the body, imitating the play of light on the surface of the water. Such, for example, are the patterns made up of blurred and more clear-cut dark stripes on the dorsal part of the body in *Scomber* and *Sarda*, the light and dark spots on the body in *Rhincodon* and *Carcharhinus*, etc. In addition, the even colouring of eunektonic animals is in a number of cases broken up by individual contrasting spots and stripes; these are elements of so-called shoal colouring, which helps the orientation of the animals in a shoal. On the whole, the colouring of eunektonic and planktonektonic species gives the impression of being quiet, soft and by no means mottled.

In the colouring of eunektonic animals dwelling in the surface zone itself, there appear elements that imitate certain features of the water surface. A classical example of this phenomenon is found in the dolphins *Delphinus delphis*, on the sides of whose bodies sea waves are 'drawn' (Figure 141); when these animals protrude from the water an impression of 'white horses' is created. A similar cryptic significance may perhaps be attached to the white tips of the dorsal and tail fins of some sharks, particularly *Carcharhinus albimarginatus*.

The preponderance of blue hues in the colouring of nektonic animals is maintained till depths of the order of 150 m (Murray & Hjort 1912; Marshall 1958, and others), subsequently giving way to a predominance of gray and red tones. In this twilight zone concealment is achieved through comparatively dark colouring, which is why red hues appear: since red rays are absent at these depths, red pigment seems black. Also, counter-shadows become less pronounced because of the diminishing overall illumination.

At depths greater than 500 m only negligible traces of sunlight corresponding to

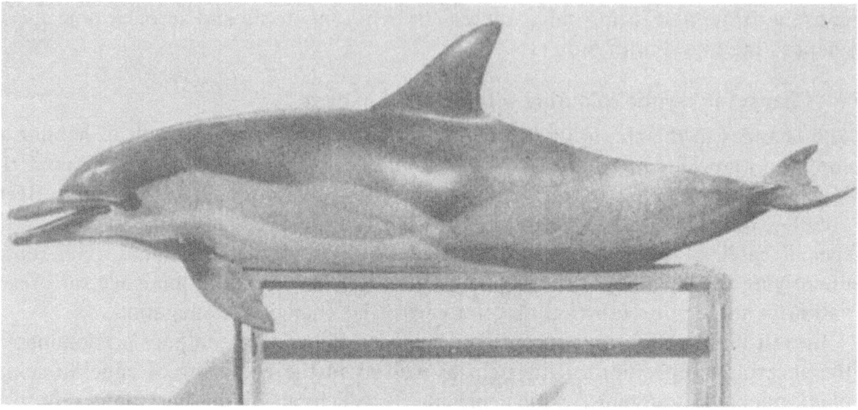

Figure 141. Cryptic colouring of *Delphinus delphis ponticus* Barab.

the blue portion of the spectrum, penetrate, so that dominating the colouring of the animals are red and black tones (Joubin 1893; Brauer 1906, 1908; Murray & Hjort 1912; Parker 1930; Marshall 1958; Akimushkin 1963, and others), which in these conditions create the maximum cryptic effect, precluding any reflection of light from the body surface. Counter-shadows in the colouring are now absent.

Along with changes in the colouring of deep-water nektonic animals, various luminescent organs appear in many deep-water Teuthoidea and fishes (Lendenfeld 1887, 1905; Joubin 1893; Brauer 1906, 1908; Chun 1903, 1910; Murray & Hjort 1912; Ege 1918, 1931; Schmidt 1918; Tåning 1918; Berry 1920; Suvorov 1948; Bertin 1958h; Marshall 1958; Akimushkin 1963, and others). The function of bioluminescence in deep-water nektonic animals is far from always clear; apparently, in various cases it may differ vastly. The possibility that under certain conditions bioluminescence may have a certain cryptic significance has not been ruled out.

On the whole, the variation in the colouring of nektonic animals with depth is a strictly regular phenomenon, based on the equally regular variation in the spectral composition of the light and the overall illumination with depth.

B. Changes in cryptic nektonic colouring with increasing ecological contact with solid substrata

With stronger ecological contact with a solid substratum, i.e. the bottom or benthic plants, floating ice, floating plant material or the land, elements of the animals' colouring creating a cryptic effect in the proximity of the substratum are enhanced.

The colouring of forms ecologically associated with floating ice or dwelling near ice floes in some cases has no marked features associated with the proximity of the ice (*Boreogadus saida*), whereas in other cases such features are strongly pronounced

(*Delphinapterus* and *Monodon*). Undoubtedly, white or nearly white animals swimming among ice floes at the surface of the water are imitating the ice.

In benthonektonic species, the general tone, character and pattern of the colouring approximate on the whole those of the bottom or the underwater vegetation or coral growths among which the given species lives. Quieter and rather uniform in the higher latitudes, brighter and more mottled in warm and tropical seas, the colouring of benthonektonic animals is distinguished overall by considerable variety, which corresponds to that of the appearance of solid substrata and underwater vegetation on the bottom; i.e. the habitats of these nektonic animals. It is quite impossible to describe the entire range of colouring in benthonektonic animals, particularly benthonektonic fishes (Bertin 1958g; Herald 1962; Smith & Smith 1963, and others), nor is it necessary, since this would take too much time while hardly being of any use for revealing the general characteristics of benthonektonic animals, for what is general here is merely a tendency towards maximum similarity between colouring and the appearance of the surrounding relief. Moreover, all these specific features of the colouring of benthonektonic fishes are well known and adequately described in various regional ichthyological and other reference books.

As for xeronektonic animals, their colouring is better adapted to producing camouflage on a solid substratum in the air the longer the periods the animals remain there. Thus, in Chelonioidea, which spend only a few days a year on the shore, the colouring is not adapted at all to producing camouflage on land. In the pinnipeds, however, we frequently come across colouring that is equally well adapted to producing a cryptic effect in the water and on land, an example of which is the final phase in the colouring of *Pagophoca*: when in the water this seal resembles pieces of broken ice, the dark elements of its colouring in this case looking like gaps between small ice floes; when on the ice, the light areas merge with the background of the substratum and the dark elements look like darker patches on the uneven ice relief (Figure 67).

The colouring of Sphenisciformes, despite their lasting stays on solid substrata, is adapted mainly to ensuring concealment when the animals are in the water, which is probably due in considerable measure to the practically complete absence of enemies on land.

3. Defence against enemies

Adaptations associated with defence against enemies, interpreted in the above-mentioned narrow sense, are on the whole uncharacteristic of nektonic animals. In eunektonic and planktonektonic species they are exceptionally rare, occurring somewhat more frequently in benthonektonic and xeronektonic forms.

We do not find adaptations of this type among Sagittoidea, while in Teuthoidea they are limited, apparently, to the ejection of ink-like fluid which disorients the predator.

In fishes adaptations associated with defence are limited mainly to the appearance of various kinds of rigid covering, i.e. armour, plaques, scutes, etc., as well as all kinds of prickly processes, i.e. spines, sharp-pointed rays, etc., all of which are characteristic mainly of benthonektonic species (Aleyev 1963a).

Among extant nektonic fishes, **protective armour** formed by rigid skin elements is best developed in Lepisosteidae, Gasterosteiformes, many Tetrodontiformes and some others. This armour is comprised of a number of non-adhering bony plaques or plates densely pressed to each other (*Gasterosteus* and *Balistes*), or of large rhombic ganoid scales (*Lepisosteus*), or of a continuous consolidated rigid covering (Ostraciidae; Figure 51B).

A **scaly covering** consisting of fine elasmoid (cycloid or ctenoid) scales does not actually create any protection against the most common enemies, i.e. large predators which swallow their prey whole. An elasmoid scaly covering may give some protection against small predators and parasites, but, as Barsukov (1960b) noted, there is too little concrete data to confirm this role of the scales.

In the great majority of cases the rigid skin elements lose their protective function altogether and acquire a new function, associated with propulsion. Instead of on heavy armour, protection now depends on the animal's high speed.

All kinds of **prickly appendages** occur as means of defence in benthonektonic fishes, and much more seldom, in eunektonic fishes. Most often they appear as thickened and pointed rays of the first dorsal, anal or pelvic fins, either connected by a membrane (*Perca, Stizostedion, Hapalogenis,* etc) or set separately, i.e. already stripped of the function of fin support alone (Gasterosteidae, *Lichia, Chorinemus santipetri,* etc.; Figure 142A), and sometimes combined spines on the operculum (*Epinephelus, Sebastes, Sebastodes,* etc.). In Balistidae the pelvic fins too have turned into spines (Monod 1959).

In rare cases the protective spines may be formed neither by fin rays nor by spines on the operculum. Cases in point are the spinose bony plaques in *Zeus* (Figure 51D), the spines on the armour of *Acanthostracion* (Figure 51B) and the spines scattered all over the body in Diodontidae. In Acanthuridae there is a strong folding spine on each side of the caudal peduncle (Figures 142E–G); in many species situated at the centre of a bright spot, which as Cott (1950) notes, helps to attract the predator to this, most dangerous part of this prey's body.

Among the adaptations associated with defence in some nektonic reptiles are rigid armour plates, comprised either of individual scales and plates (Hydrophidae, Teleosauridae, Mosasauridae, etc.) or monolithic formations (Testudinata, Cyamodontoidei, etc.). However, in nektonic reptiles as a whole, one sees a regular reduction of such protective armour in all groups for the same reasons as in nektonic fishes. An important protective adaptation in Hydrophidae is poisonousness.

Sphenisciformes, pinnipeds, cetaceans and sirenians have no special adaptations for protection against enemies in the above-mentioned sense. *Monodon* probably uses its tusk for defence against large predators (Tomilin 1957), but its main function is different (see chapter VI).

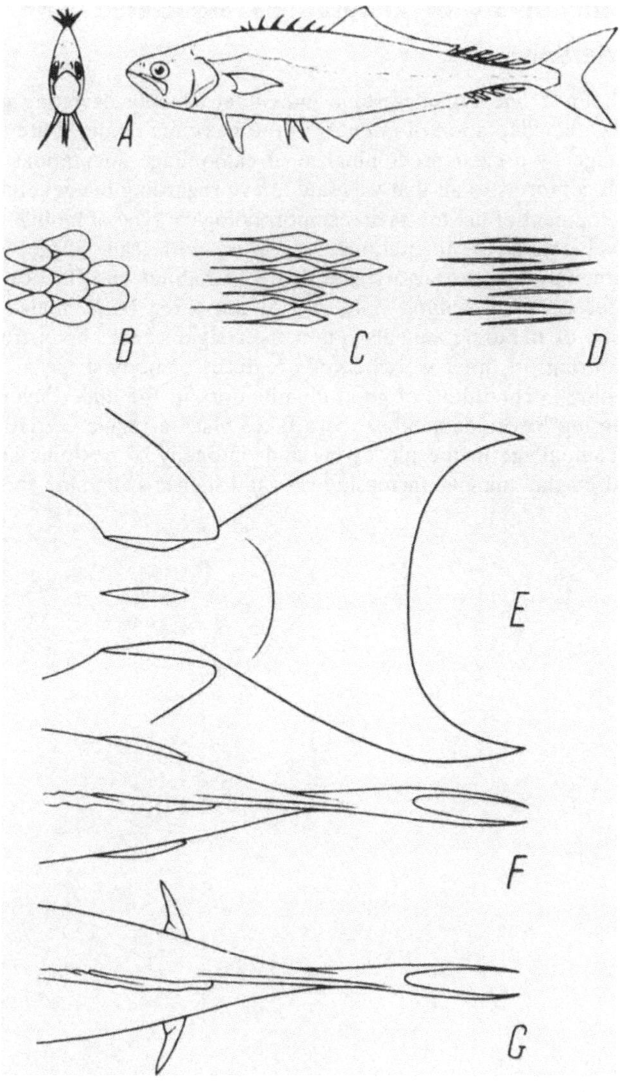

Figure 142. A-D, Chorinemus (*A, Ch. santipetri* Cuv. et Val.; *B*, scales of *Ch. tala* (Cuv. et Val.); *C*, scales of *Ch. tolooparch* (Rüpp.); *D*, scales of *Ch. lysan* (Forsk.); *E-G, Paracanthurus hepatus* (L.) (*E, F*, spines folded; *G*, unfolded).

4. The phylogeny of adaptations associated with camouflage and defence

On the whole, if one excludes some individual, steeply deviating groups, such as Testudinata, the adaptation of animals to the nektonic mode of life is characterized at every stage by a clear predominance of camouflage adaptations over defensive ones, which conforms to all that was said above regarding the general characteristics of the development of nekton as an ecomorphological type of mobile aquatic animal.

The development of adaptations associated with camouflage is regulated in phylogeny mainly by two factors: the depth of habitat and the Reynolds number. With greater depth of habitat only two of the three basic modes of camouflage characteristic of nektonic animals retain their significance: body transparency and cryptic colouring. In other words cryptic features of body shape are found only in forms dwelling in conditions of good illumination, in the upper layers of the water. With increasing Reynolds number there takes place a regular variation in the basic means of camouflage in the phylogeny and ontogeny of nektonic animals, directly determined by the animals' increasing size and higher swimming speeds.

VIII. OTHER ADAPTATIONS

1. General

All the other adaptations in nektonic animals are of less importance in the formation of the morphological features common to all these animals than the complexes of adaptations associated with propulsion and camouflage, examined in Chapters III–VII. All those adaptations not associated with propulsion or camouflage are very different in animals belonging to different systematic groups, which is due both to differences in the initial morphological organization which served as a basis for their development and also to differences in ecology already present in the nektonic stage of phylogeny. Accordingly, a detailed examination of these other adaptations would inevitably result in an outline of generally known zoological material pertaining to individual systematic groups of animals, which is not our aim and which we rejected right from the beginning. We shall therefore merely touch briefly on those basic adaptations that are important in the formation of morphological features common to all nektonic animals. Of the greatest interest in this regard are adaptations associated with the reception and transmission of information and the capture of food, which we shall examine first; along with this, we shall note also some specific adaptations not associated with the above-mentioned functions.

2. Reception and transmission of information

Reception of information from outside and transmission of information to other individuals of one's own species are in all animals among the most important functions. Most general and ecologically most significant for the majority of nektonic animals are reception and transmission of information along two channels, the optical and acoustical.

1. Optical Channel

Vision is one of the most important senses of nektonic animals. It is absent only in a very few deep-water fishes (Brauer 1906, 1908, and others), and among cephalopods, in the deep-water *Cirrothauma* (Akimushkin 1963). The overwhelming majority of nektonic animals receive information along the optical channel which is vital both in the well-illuminated enviroment of the surface layers and in the conditions of nearly absolute darkness of the abyssal depths, broken by bioluminescence only.

The importance of optical information differs in the different systematic groups of nekton. Thus, for example, for Teuthoidea the optical receptor is basic and irreplaceable: when the optical organs are out of action the animals die of hunger, as has been shown experimentally (Akimushkin 1963). Contrary to this, some nektonic fishes and cetaceans are capable of finding food even when the optical receptor is useless, which is proved by instances of these animals feeding at night, by cases when blind but well fattened animals have been caught, and by direct experimental data (Andriashev 1955; Tomilin 1965, and others). In spite of these differences, the optical receptor is undoubtedly one of the most important in the general system of receptors in nektonic animals as a whole. The two aspects most common in the morphology of the optical receptors of animals belonging to different groups are evident: 1. the degree of structural complexity and relative size of the eyes and 2. their location on the body.

Eye structure in all nektonic animals is fairly complex. Despite the comparatively low general morphological organization of Sagittoidea, their eyes have a comparatively complex structure (Burfield 1927). The complex structure of the eyes of cephalopods, particularly Teuthoidea, is well known (Chun 1903, 1910; Abel 1916; Kondakov 1940; Krymgolts 1958; Akimushkin 1963, and others); for complexity of eye structure they are superior to all other invertebrates, which is doubtless associated with adaptation to fast swimming. We find well-developed eye structure also in fishes and other nektonic vertebrates: reptiles, birds and mammals (Matthiessen 1886; Putter 1903; Brauer 1906, 1908; Rochon-Duvigneaud 1940, 1943, 1950, 1954, 1958; Mann 1946; Suvorov 1948; Bourdelle & Grassé 1955; Frechkop 1955; Petit 1955; Terentyev 1961; Pilleri 1964; Rozhdestvensky 1964a; Tatarinov 1964 e.g. and others).

Eye size in nektonic animals varies greatly (Table 21). Under good illumination the eyes, as a rule, develop normally, and in most nektonic animals the horizontal diameter of the eyes is 2–10% of the effective body length (Table 21). Eye atrophy associated with diminution of illumination is characteristic, in particular, of deep-water fishes. It is known (Joubin 1893; Brauer 1906, 1908; Chun 1910; Murray & Hjort 1912; Zernov 1949; Marshall 1958; Rochon-Duvigneaud 1958; Bertin 1958f, and others) that with greater depth of habitat and lower illumination the eyes at first relatively increase, which is particularly characteristic of semi-deep-water species, living in waters still reached by faint sunlight (the majority of Teuthoidea, and among the fishes, *Sebastes, Myctophum, Argyropelecus, Sternoptyx*, etc.; Table 21, Figure 143A). The further reduction of light usually results in a considerable functional and morphological reduction of the eyes and a decrease in eye size in aphotic-zone dwellers (*Cyclothone*, etc.; Table 21, Figure 143B). In other cases the eyes remain huge and telescopic, adapting to catch the very faint bioluminescent light. Of interest in this connection is the great enlargement of the eyes in *Anguilla* when it transfers to great depths during the spawning period (Schmidt 1912, 1925). In benthonektonic fishes dwelling in turbid continental waters of low transparency, the role of vision in the general system of sense organs is frequently reduced, resulting in a decrease in

Table 21. Relative horizontal eye diameter *o* in nektonic animals

Species and nektonic type	Effective body length, L_c (cm)	*o* (%) L_c
Enhydrina schistosa (Daud.) (EN)	90.0	0.5
Lapemis curtus (Shaw) (EN)	99.0	0.5
Balaenoptera musculus (L.) (EN)	2500	0.6
Physeter catodon L. (EN)	1600	0.6
Huso huso (L.) (BN)	180.0	1.1
Tursiops truncatus (Montagu) (EN)	350.0	1.1
Sagitta setosa Müll. (PN)	2.2	1.3
Anguilla anguilla (L.) (EN)	106.0	1.3
Acipenser stellatus Pall. (BN)	108.5	1.4
Phocoena phocoena (L.) (EN)	128.8	1.5
Cyclothone microdon (Günth.) (PN)	4.9	1.7
Arctocephalus pusillus (Schreb.) (XN)	200.0	1.8
Eudyptes chrysolophus Brandt (XN)	69.6	2.0
Prionace glauca (L.) (EN)	132.5	2.0
Trichiurus lepturus L. (EN)	150.0	2.0
Pagophoca groenlandica (Erxl.) (XN)	197.0	2.2
Chelonia mydas (L.) (XN)	130.5	2.3
Eretmochelys imbricata (L.) (XN)	119.0	2.5
Scomberesox saurus (Walb.) (EN)	24.6	2.8
Sphyrna zygaena (L.) (EN)	195.0	3.0
Remora remora L. (EN)	33.3	3.0
Mustelus mustelus L. (BN)	109.0	3.1
Esox lucius L. (BN)	117.7	3.2
Makaira indica (Cuv. et Val.) (EN)	180.0	3.3
Istiophorus platypterus (Show & Nodder) (EN)	120.0	3.3
Mugil auratus Risso (BN)	34.8	3.5
Coryphaena hippurus L. (EN)	105.8	3.6
Xiphias gladius L. (EN)	220.0	4.0
Aphya minuta Risso (PN)	3.46	4.0
Clupea harengus harengus L. (EN)	34.0	4.0
Thunnus alalunga (Bonnatt.) (EN)	106.0	4.1
Eurhinosaurus longirostris Jaeger (EN)	360.0	4.1
Auxis thazard (Lac.) (EN)	40.3	4.3
Abramis brama (L.) (BN)	36.9	4.5
Salmo trutta labrax m. fario L. (BN)	24.1	4.8
Mola mola (L.) (EN)	110.0	5.0
Trachurus mediterraneus ponticus Aleev (EN)	35.0	5.2
Stenopterygius quadriscissus (Quenst.) (EN)	177.0	5.3
Pelecus cultratus (L.) (EN)	25.8	5.4
Boreogadus saida (Lepechin) (BN)	21.0	5.9
Atherina mochon pontica Eichw. (EN)	10.7	5.9
Melanogrammus aeglefinus (L.) (BN)	27.2	7.3
Gasterosteus aculeatus L. (BN)	6.9	7.5
Hirundichthys rondeletii (Cuv. et Val.) (EN)	15.7	7.6
Alectis indica (Rüpp.) (BN)	20.4	7.9
Lebistes reticulatus Peters (BN)	1.8	7.9
Chaetodon striatus L. (BN)	13.8	8.0

Table 21 (contd.)

Species and nektonic type	Effective body length, L_c (cm)	o (%) L_c
Loligo vulgaris Lam. (EN)	26.5	8.0
Leucaspius delineatus (Heck.) (PN)	5.2	8.1
Zeus faber L. (BN)	25.4	8.3
Symplectoteuthis oualaniensis (Less.) (EN)	24.1	9.0
Caranx equilla Temm. et Schl. (EN)	20.3	9.8
Sebastes marinus mentella Travin (BN)	22.7	11.5
Myctophum punctatum Raf. (PN)	3.9	12.2
Pterophyllum scalare Cuv. et Val. (BN)	6.0	12.3
Myctophum rissoi (Cocco) (PN)	5.0	19.0

relative eye size (Table 21; *Huso, Acipenser*). Eye atrophy due to low illumination is always compensated for by the other sense organs being stronger, particularly the tactile, taste and lateral line organs (Rochon-Duvigneaud 1958; Marshall 1958; Bertin 1958f, and others).

In small nektonic forms the eyes are comparatively larger than in large forms. This is due to the fact that for normal functioning of the eye as an optical instrument it is unnecessary for its absolute size to exceed a certain limit in the ecology characteristic of the given species (Aleyev 1963a). This trend may easily be traced both for nekton as a whole and within individual systematic groups, Teuthoidea, Osteichthyes, Cetacea, etc., as can be seen from Table 21, where the individual species are arranged in order of increasing relative eye size.

In small nektonic animals whose eyes occupy a considerable part of the head, as in most Teuthoidea and many small fishes, a direct relation is observed between eye

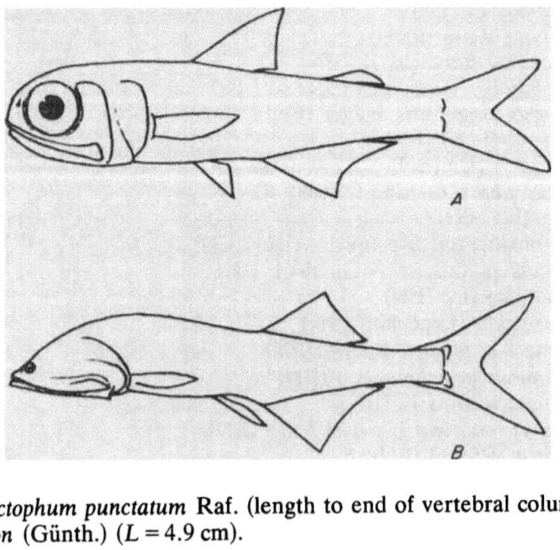

Figure 143. A, Myctophum punctatum Raf. (length to end of vertebral column $L = 3.9$ cm); *B, Cyclothone microdon* (Günth.) ($L = 4.9$ cm).

diameter and body height (Murray & Hjort 1912; Aleyev 1963a). With smaller body height the vertical eye diameter is more and more severely limited, owing to the necessity of preserving overall streamlining of the head. It is for this reason that in nektonic animals with elongated anguilliform, ribbon-like and similar body shapes the relative eye size is usually small; in some cases it comes to only about 0.5% of the body length (Table 21; *Enhydrina, Lapemis*). For the same reason, some fishes and Teuthoidea with an elongated body and relatively large eyes have a horizontally extended eyeball (Aleyev 1963a; Zuyev 1966). In the ontogeny of fishes relative eye size at first increases and then diminishes again; this is the general rule for egg-laying species and takes place because of changes in the absolute dimensions of the fish and its relative body height with age (Aleyev 1963a). In the early stages of development, when the larval body is still very slender, eye size is limited by the height of the body and increases together with body height. Upon reaching some absolute magnitude optimal for the species' ecology at the given stage of development, the eye's absolute growth is retarded and its relative diminution begins, owing to the continuing growth of the fish (Figure 144). This is observed both in bottom-dwelling fishes, in whom the importance of the visual receptor relatively diminishes in the adult state, for example in some carp (Kryzhanovsky 1956), and also in pelagic fishes, in whom the visual receptor is the chief one at any age (Aleyev 1963a).

In viviparous fishes, particularly sharks (egg-laying species included) and Cyprinodontiformes, as well as in fishes whose larvae hatch from the egg with all the fins already formed (Gobiidae), the increase in relative eye size usually ends in the

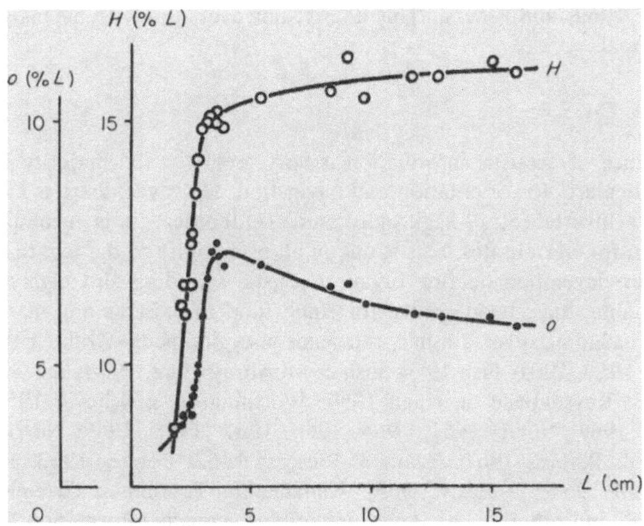

Figure 144. Dependence of horizontal diameter *o* of eye and maximum body height *H* on fish length *L* to end of vertebral column in *Engraulis encrasicholus ponticus* Alex.

embryonic period, and in post-embryogenesis there occurs only a decrease in relative eye size. Thus our studies show that the horizontal eye diameter in newborn *Squalus acanthias* having an effective body length $L_c = 20$ cm is about 4.5% of L_c, while at $L_c = 49.0$ cm it is about 3.8% and at $L_c = 108.8$ cm only 2.9% of L_c. A similar picture is seen also in Teuthoidea (Zuyev 1966).

The changes in relative eye size with age in nektonic reptiles, birds and mammals are insignificant, owing to the small changes in body size with age.

Thus, when the relative body height is not so small as to considerably limit their size, the eyes, both in ontogeny and phylogeny, become relatively smaller as the size of the animal increases.

The eyes of nektonic animals in the typical case are situated laterally, on the sides of the head, at a level near to the longitudinal axis of the body. In this case their fields of vision above and below the body are symmetric, which is important for receiving optical information from all the surrounding space. In addition, this position of the eyes reduces to the minimum the possibility of overlapping of the fields of vision of the right and left eyes, owing to which the field of **binocular vision** in nektonic animals is usually small or absent altogether. In fishes, for example, it comes to only 28–30° (Nikolsky 1954). Nor is it large, judging by everything, in Teuthoidea (Akimushkin 1963). Cetaceans have monocular vision only (Tomilin 1965). A wider field of binocular vision is characteristic of seals (Walls 1942).

Apart from through the eyes, some nektonic animals may receive limited optical information by means of a peculiar **pineal organ** (*gl. pinealis*), characteristic, for example, of some tuna fishes (Rivas 1953; Holmgren 1958; Magnusson 1963; Ovchinnikov 1969, and others). This ability, however, is by no means common in nektonic animals.

2. Acoustic Channel

The importance of acoustic information is very great for the majority of nektonic animals, particularly for orientation and navigation. However, there is every indication that for invertebrates – Sagittoidea and Teuthoidea – it is incomparably less essential than for vertebrates. This is due in all probability to the fact that the latter have a better developed hearing organ. Acoustic sounding and highly developed acoustic signaling have been known for some time to exist among many nektonic vertebrates, including diverse fishes, cetaceans and pinnipeds (Griffin 1950; Kellogg 1953, 1958, 1959, 1961; Fish 1954; Fish & Mowbray 1962, 1970; Schevill, Watkins & Ray 1963; Reysenbach de Haan 1957; Worthington & Schevill 1957; Tavolga 1960, 1964, 1967; Norris 1961, 1964, 1966, 1967, 1968, 1969; Norris, Prescott, Asa-Dorian & Perkins 1961; Evans & Prescott 1962; Poulter 1963, 1965, 1966, 1967; Tomilin 1964; Protasov 1965; Romanenko, Tomilin & Artemenko 1965; Lockley 1968, and others). The important role of acoustic information for nektonic animals is due (Tomilin 1965, and others) to the sound-conducting properties of water: the high speed and considerable range of sound propagation in water, which make the acoustic channel particularly suitable for remote underwater communication.

Despite the very great ecological importance for nektonic animals of acoustic information and acoustic sounding, the development of morphological adaptations associated with receiving such information and generating sounds contains comparatively little that is common to all nektonic animals. Taking nekton as a whole, we shall find no common morphological characteristics associated with the functions under examination. Taking nektonic vertebrates only, such common features are limited morphologically to a comparatively few generally known facts, of which we ought to mention at least two: 1. the absence of auricles (or their more or less strong reduction in secondary aquatic mammals) and 2. a tendency to enhance the sound-receiving function of the cranial bones.

In fishes, nektonic reptiles and Sphenisciformes the absence of an external ear is primary; in nektonic mammals we observe different stages of external-ear reduction from a comparatively early stage in Otariidae to the final phase in Odobenidae, Phocidae, Sirenia and Cetacea. The reduction of the external ear in nektonic animals is due to the need to maximize the smoothness of the surface of the anterior part of the body, as well as to the high sound conductivity of water, which means that special 'sound locators' can be dispensed with.

Perception of sound through the cranial bones is also primary in fishes and secondary in all other nektonic vertebrates. Yet the latter perceive sound mainly not via the bones of the cranium but through the auditory meatus; this is presumed to be so even in cetaceans (Tomilin 1957), whose auditory meatus is very narrow and in most cases tightly closed by an auditory plug, capable, however, of transmitting sound.

Other morphological adaptations in nektonic animals associated with the reception and generation of sound are rather varied even among the vertebrates, and actually suggest no ideas as to the paths of the formation of a convergent similarity among nektonic animals as a specific ecomorphological type.

Of considerable interest is **the adaptation for directional emission of a sound signal** in cetaceans, which is of great importance during acoustic steering.

In Odontoceti sounds are generated when air is exhaled through the nasal duct, and also during the circulation of air in a complicated system of air sacs in the nasal passage, with the air not necessarily expelled into the environment (Lawrence & Schevill 1956; Romanenko, Tomilin & Artemenko 1965; Tomilin 1965, and others). The contraction of special muscles makes the air sacs in the nasal passage change their shape and volume, on account of which the sound's characteristic frequency changes (Romanenko, Tomilin & Artemenko 1965; Tomilin 1965). Dolphins (*Tursiops* and *Delphinus*) have three pairs of air sacs: vestibular, tubular and premaxillar. These are all connected to the nasal passage (Figure 145), which allows the animal to issue different types of sound simultaneously.

Wood (1954) expressed the opinion that the fatty pad on the dolphin's head plays the role of an acoustic lens facilitating directional propagation of the sound issued by

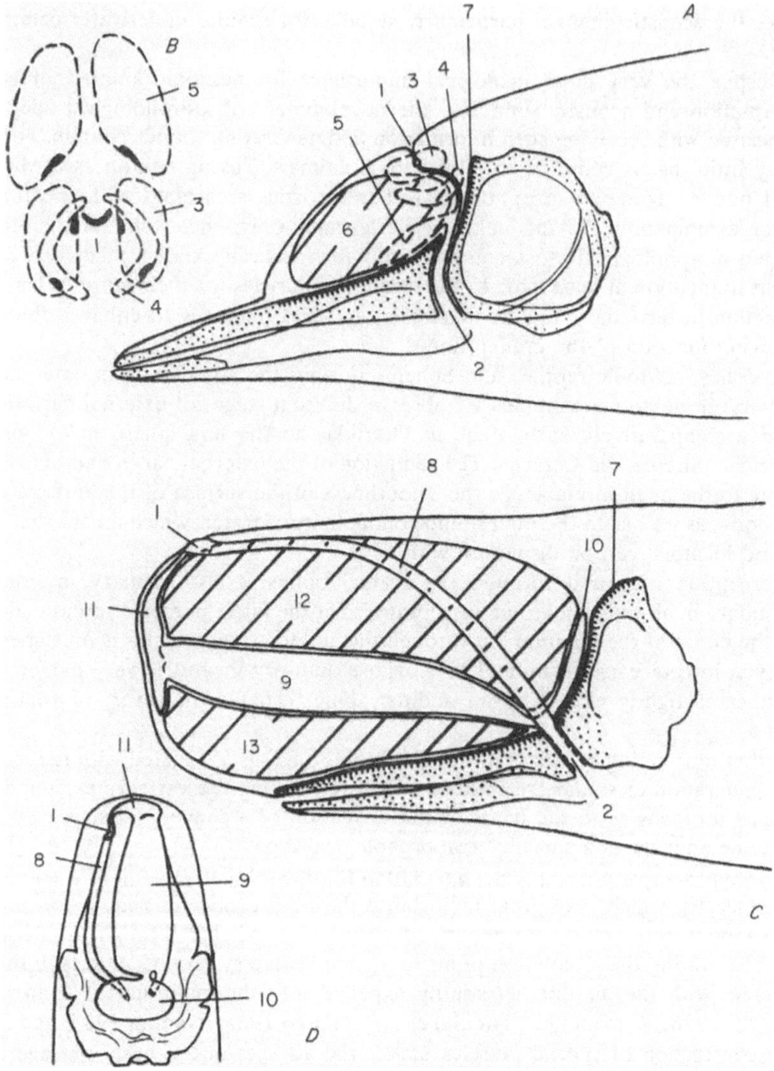

Figure 145. Diagram of sound-emitting apparatus in (*A,B*) *Delphinus delphis* L. and (*C,D*) *Physeter catodon* L. *A*, *C*, sagittal section; *B*, *D*, plan view of air sacs. 1, nostril; 2, bony nasal duct; 3, vestibular air sac; 4, tubular air sac; 5, premaxillary air sac (shaded parts in (*B*) are where air sacs join nasal duct); 6, fatty pad (acoustic lens); 7, sound focusing surface of cranium; 8, left nasal passage; 9, right nasal passage; 10, frontal air sac; 11, distal air cavity; 12, 13, upper and lower spermaceti sac (acoustic lens). After Sleptsov (1952) and Romanenko, Tomilin & Artemenko (1965), modified.

the animal. Later, Evans & Prescott (1962), by passing air under pressure through the larynx and nasal passage of severed dolphin heads (*Stenella* and *Tursiops*), established certain directionality in the propagation of the sound, and tentatively attributed it to the functioning of the dolphins' fatty pad as an acoustic lens together with the focusing effect of the frontal cranial surfaces. By using still subtler methods, Romanenko, Tomilin & Artemenko (1965) obtained directional sound diagrams for a severed head of *Delphinus delphis ponticus* (Figure 146), which confirmed the capacity of dolphins for directional sound propagation; it turned out that the directionality of the sound generated by the dolphin's head was greater the higher its characteristic frequency (Figure 146). An investigation by the same authors of the directional diagrams obtained for an intact dolphin head and for the cranium without soft tissues made it possible to establish that the directionality of sound propagation is affected both by the cranial bones and by the soft tissues (fatty pad), yet the main role in this respect is played by the cranium. Thus the functional significance of the fatty pad of Odontoceti and of the convex shape of the frontal surface of their cranium was established: combined, these structures form a sound-emitting reflector which is undoubtedly of great importance for acoustic steering.

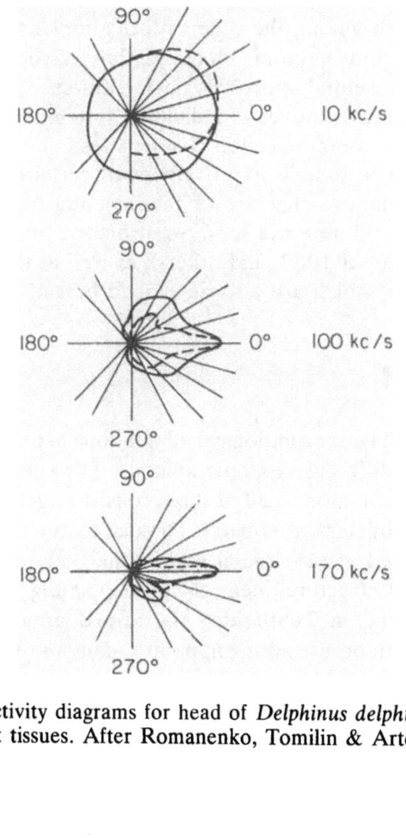

Figure 146. Sound directivity diagrams for head of *Delphinus delphis ponticus* Barab. cranium alone; cranium with soft tissues. After Romanenko, Tomilin & Artemenko (1965, modified).

In this light, evident also is the purpose of the huge spermaceti pad of *Physeter*: its main function is undoubtedly that of an acoustic lens which amplifies and enhances the directionality of the sound emitted by the animal (Figures 145C,D) which is particularly important at greater depths, where the action of the visual receptor is completely excluded.

The fatty pad of the Odontoceti doubtless fulfils in some measure the role of a hydrostatic organ too, which is generally characteristic of all fatty deposits, including those of cetaceans (see chapter III).

In Mystacoceti, instead of air sacs connected with the nasal passage, there is a large throat sac opening into the upper part of the larynx. It is supposed (Tomilin 1965) that forcing air out of this sac through the larynegeal slit helps to produce the low-frequency sounds characteristic of these cetaceans, which they use for echo-piloting.

The great relative size of the brain in cetaceans is probably explainable (Tomilin 1968) by the need to process quickly a large amount of acoustic information.

3. Other Information-receiving and Communication Channels

It is difficult to distinguish among the other information-receiving and communication channels used by various nektonic creatures those associated with the development of more or less essential morphological features common to all nektonic animals. In occasional representatives of individual systematic groups, these or other information-receiving and communication channels assume rather great importance, creating prerequisites for a notable improvement of certain morphological adaptations. Among such adaptations that are of fundamental importance for individual nektonic groups, one should note the seismo-sensory system, characteristic of fishes alone (Disler 1960; Dijkgraaf 1962, and others), as well as the chemo-receptive and chemo-signalling systems, which are also developed best of all in fishes.

3. Capture of food

Like the methods of feeding, morphological adaptations associated with food capture are extremely varied in different nektonic animals. The specific features associated with the capture of food common to all of them consist of generally known structural features of animals with bilateral symmetry. Of such features we point out only two. First, the mouth is located at the cranial end of the body, where the eyes too are situated. Second, special structures near the oral opening help to retain prey; in Sagittoidea these are ceta, in Teuthoidea the armed arms and tentacles, and in vertebrates jaws with teeth or a straining apparatus composed of gill rakers (in fishes) or of special horny plates (in Mystacoceti).

Most characteristic of eunektonic and planktonektonic animals, irrespective of peculiarities in feeding, is a terminal situation of the mouth, which is found equally in

Sagittoidea, Teuthoidea and all vertebrates, and is due to the fact that they live in a three-dimensional domain where food objects occur with equal probability above and below them.

Apart from these characteristics, whose functional foundations are obvious, we could note with regard to the morphology of the group of adaptations under examination only less widespread features, common to the nektonic representatives of individual animal types, i.e. nektonic Mollusca, Chaetognatha and Chordata.

Nor do we observe any significant features in nektonic representatives of Mollusca, i.e. in nektonic Teuthoidea, that distinguish them fundamentally from their non-nektonic relatives. However, in nektonic Teuthoidea and the faster nektobenthic Sepioidea we observe some relative shortening of the tentacles as compared with the slower nektobenthic Octopoda (Zuyev 1966).

In planktonic and nektonic Chaetognatha, the adaptations associated with grasping food are on the whole identical, which is basically due to the very insignificant difference in the sizes of individuals of the former and the latter.

Among the Chordata, nektonic forms are present only in the vertebrate group. In Vertebrata adaptation to the nektonic mode of life creates some common structural traits in the dental system. Teeth, whenever they are present, are most often of a single type (fishes, Ichthyosauria, Odontoceti, etc.) and serve as a rule only for securing prey. Biting off pieces from a food object in water is complicated by its changing position and therefore occurs in eunektonic forms only when this object is rather big and cannot be swallowed whole; such, for example, are cases when sharks and killer whales (*Orcinus*) bite pieces out of large animals (whales, dolphins, seals, etc.). Essential modifications of the dental system of nektonic vertebrates occur almost exclusively in cases when the food objects are situated on the bottom and are incapable of fast movement or nearly immobile.

The numerous variations in the relative size and structure of the jaw apparatus and dental system in different nektonic animals are all entirely determined by features of their ecology. What all these variations represent is not a basis for the convergence of nektonic forms, but mainly a basis for their divergence, and we shall therefore not dwell on them, all the more so because they are described exhaustively in zoological literature.

4. Other adaptations

Among all the other adaptations it is difficult to set apart anything that contributes in some essential way to the convergence of nektonic animals belonging to different systematic groups. Adaptations associated with respiration, digestion, excretion, reproduction and a number of other vital functions, though frequently changing in the course of adaptation to the nektonic mode of life, hardly ever create a basis for the convergence of nektonic animals belonging to different systematic groups. A description of all these adaptations can be found in the relevant sections of zoology for Teuthoidea, Sagittoidea and nektonic vertebrates.

Among the reproductive adaptations common to representatives of individual systematic groups of nektonic animals, one may point to the emergence of planktonic eggs in nektonic Teuthoidea and Osteichthyes (though such eggs are not peculiar to the nektonic members of these groups alone), and to viviparism in sharks and Ichthyosauria (Branca 1908a,b). Both are directed at freeing the process of reproduction from any connection with some solid substratum, which should be regarded as a fundamental stage of adaptation to living permanently in a mass of water. Of equal importance is the capacity to give birth to young in the water and to perform coitus while swimming, which developed in Ichthyosauria and Cetacea.

Of great importance for all secondary aquatic animals, as regards adaptation to the nektonic mode of life, is the capacity to rest and sleep in the water. From the morphological point of view this capacity is mostly due to the development of different kinds of hydrostatic adaptation.

Another adaptation to the nektonic mode of life is aggregative behaviour, which to one degree or another is characteristic of the majority of nektonic animals. There is no doubt that in at least some cases (perhaps in most cases) movement in a dense shoal is more advantageous energetically than swimming alone. This is explained primarily by the emergence of a common added mass, which allows more intensive secondary utilization of the kinetic energy of the boundary layer than is possible in the case of solitary swimming. This is why speeds that are unattainable by a solitary nektonic animal may be attainable in a shoal, as has been repeatedly noted earlier (Parin 1968, and others). One can see from this why aggregative behaviour is particularly characteristic of smaller nektonic animals. At the same time it is obvious that the ecological significance of aggregative behaviour is not restricted to energetic advantages alone.

PART 3. ORIGINS AND ECOLOGICAL DIVERGENCE OF NEKTON

Having examined the most important adaptations in nekton, it seems to be possible not only to give a detailed description of individual modifications and ecomorphological classes of nekton, but also to examine the questions of nekton's origins and the common features of the formation of nekton in different types of bodies of water. This pair of questions, whose discussion makes up the third part of this book, are of interest not only as general problems in zoology and nektonology, but also on a wider, evolutionary plane.

IX. ECOMORPHOLOGICAL CLASSES OF NEKTON AND THEIR ORIGINS

1. General

The ecomorphological classification of nekton was discussed in chapter I, where definitions and brief characteristics of the four ecomorphological classes of nekton were given as well as the class definitions of other ecomorphological types of bionts related to nekton. In the present chapter, on the basis of the above analysis (chapters III–VIII) of the basic nektonic adaptations, we characterize the nektonic classes in greater detail and examine their origins, without which it is impossible to go into the characteristics of the emergence and development of nekton as a whole for particular bodies of water, which are the subject of chapter X. Below, when describing individual classes of nekton, we shall avoid repeating the overall characteristics of nekton already described in preceding chapters, and shall endeavour to stress only those which are specific to and characteristic of the class in question, and indeed make up its distinctive features.

Naturally, owing to the incompleteness of available paleontological material and the lack of ecological data pertaining to the majority of fossil forms, in many cases a group or species of extinct animals cannot be placed in a particular nektonic class with such certainty as can extant animal forms. Therefore, when dealing with fossil forms, we shall often have to be content with mere assumptions. When writing the following sections of this chapter the author used a number of cumulative paleontological files and studies of a more particular character, which to avoid unduly burdening the text will not be cited in every case (these were Fraas 1891, 1902, 1910, 1911; Traquair 1899a,b; Woodward 1901; Abel 1907, 1912, 1916, 1922; Huene 1916, 1923, 1956; Wiman 1922; Stensio 1932; White 1935; Kuhn 1937; Watson 1937; Romer 1939, 1956; Berg 1940, 1955; Berg, Kazantseva & Obruchev 1964; Berg & Obruchev 1964; Westoll 1944; Gross 1947; Davitashvili 1949; Matveyeva 1958; Terentyev 1961; Vorobyova & Obruchev 1964; Vyushkov 1964; Glikman 1964; Danilchenko 1964a,b; Danilchenko & Yakovlev 1964; Dementyev 1964; Kazantseva 1964; Konzhukova 1964a–i; Maleyev 1964a–e; Novitskaya & Obruchev 1964; Novozhilov 1964a–d; Obruchev 1964a–c; Obruchev & Kazantseva 1964a,b; Obrucheva 1959; Rozhdestvensky 1964a,b; Sukhanov 1964; Tatarinov 1964a–t; Khozatsky & Yuryev 1964; Chudinov 1964a–d; Shishkin 1964).

2. Ecomorphological classes of nekton

Benthonekton are nektonic animals swimming at $Re > 5 \times 10^3$ which are ecologically connected with the bottom or with the immersed surface of some floating solid substratum (floating ice, vegetable material, etc.) but not necessarily with land or some floating solid substratum jutting out into the air. The ecological bonds between benthonektonic animals and a solid substratum develop mainly 1. because food is to be found on the surface of the substratum, within the substratum (e.g. in the ground) or in the labyrinth formed by the substratum (e.g. within thickets of subaquatic plants), and 2. because of use of the shelter provided by the substratum (growths of underwater vegetation, coral, rock crevices, etc.). Ecological connections with the substratum are essentially the same whether it is located on the bottom or floating (immersed surfaces of drifting ice, vegetable material, etc.). Thus, for example, the connections between nektonic animals and immersed surfaces of drifting ice, coastal rocks, various floating vessels or piles driven into the ground, etc., are absolutely identical. There is no reason, therefore, to draw distinctions between nektonic animals ecologically connected with solid substrata on the bottom or freely drifting; the former and the latter are both benthonekton and are characterized by the development of the same adaptations.

Represented among benthonekton are nine classes of animals: Cephalopoda, Diplorhina, Monorhina, Placodermi, Acanthodei, Chondrichthyes, Osteichthyes, Reptilia and Mammalia. The great majority of benthonektonic species are primary aquatic animals (representatives of the first seven of the above-listed classes) and only benthonektonic reptiles and mammals, comparatively few overall, are secondary aquatic animals.

Among the cephalopods some Belemnoidea are by every indication benthonekton. There probably exist benthonektonic forms also among extant Teuthoidea, particularly among Loliginidae, however the ecology of the majority of Teuthoidea has not been sufficiently well studied to permit certainty in this regard. In addition, some representatives of Sepioidea ecologically connected with the bottom, but spending most of the time suspended in the interior of the water, may also be placed among benthonekton.

Of Diplorhina, probably benthonektonic were individual representatives of Pteraspidida (*Pteraspis, Podolaspis, Doryaspis*, etc.); it has not been ruled out, however, that they were nektobenthic (see chapter II).

Of Monorhina, belonging among benthonekton were probably many representatives of the subclass Anaspida (Birkeniae).

Placodermi were on the whole a benthic group; belonging among benthonekton were probably only very few members of this class, namely some representatives of Pachyosteida, particularly those belonging to the families Synaucheniidae, Oxyosteidae and Leptosteidae.

Among Acanthodei, if not predominant, benthonektonic forms were undoubtedly very common. Judging by the morphology of these fishes (thick body, broad cranium, frequently flattened ventral surface of body), many of them were still nektobenthic forms (Diplacanthida and Gyracanthida). However, forms more laterally compressed and with a more graceful body that had lost (or nearly lost) the spines between the pectoral and pelvic fins were in most cases no longer benthonektonic (Ischnacanthida and Acanthodida).

Of the Chondrichthyes, the benthopelagic sharks, which are ecologically connected with the bottom, should be placed among benthonekton, particularly Triakidae (*Mustelus*, *Triakis*, etc.).

Among Osteichthyes benthonektonic forms occur in the majority of orders. For example, among Sarcoperygii one may point to such groups as Eusthenopteridae and Latimeriidae, and among the Actinopterygii to *Acipenser*, *Huso*, *Blicca*, *Barbus*, *Cyprinus*, *Carassius*, *Gadus*, *Boreogadus*, Mugilidae, Polynemidae, Pomadasyidae, Chaetodontidae, Balistidae, Monacanthidae, etc.

The benthonektonic reptiles include some forms of Ichthyosauria ecologically connected with the bottom, in particular, mollusc-eating Omphalosauridae with crushing teeth. All Acrochordinae, which have completely lost all ecological ties with land, are also benthonektonic. The mollusc-eating Mosasauridae (*Globidens*, etc.) may have been benthonektonic too, however it has not been ruled out that they, like the extant Laticaudinae, preserved an ecological connection with land, i.e. were xeronektonic.

The mammals which are benthonektonic consist of some sirenians that have completely lost ecological ties with land (*Dugong*) and bottom-feeding cetaceans (*Eschrichtius*, *Platanista*, etc.).

Morphologically, benthonektonic animals are on the whole characterized by neutral buoyancy. Neutral or close-to-neutral buoyancy is one of the most essential features distinguishing benthonektonic animals from the close nektobenthic forms, whose buoyancy, as a rule, is markedly negative. We find among benthonektonic animals all three types of nektonic propulsor: undulatory (axial and peripheral), paddling and hydrojet. High-speed swimmers moving at speeds in excess of 10 m/s are very few among them, and accordingly well-streamlined body shapes are rare. All the comparatively few cases of fast-swimming benthonektonic animals are either predators of the ambush type, such as *Esox*, or rheophilic forms, such as some *Salvelinus* or *Salmo trutta m. fario;* all these animals are adapted predominantly for performing separate rapid spurts. The pursuing-type of predator, capable of swimming at speeds above 10 m/s, does not occur among benthonektonic forms. The body shape of many benthonektonic animals has specific features functionally associated with the maintenance of contact with a solid substratum. Frequent among such features is less convexity of the ventral as compared with the dorsal contour of the body, which, combined with a low location of the mouth, is particularly characteristic of bottom-feeding forms and is directly aimed at facilitating grasping food from the bottom and turning downwards, i.e. towards the bottom. Another frequent feature of body shape in benthonektonic animals is strong lateral compression of the body, which turns it into a vertically oriented disk, making it easier for the animal to pass through narrow vertical gaps in benthic relief formed by aquatic vegetation, coral, etc., as well as making it easier to turn in the vertical plane, frequently performed when approaching or leaving the bottom. A disklike body shape is characteristic of many benthonektonic fishes dwelling among coral or underwater plants. Living among the complicated labyrinths of obstacles formed by the bottom relief induces improvement of the manoeuvrability of benthonektonic animals. Along with features already mentioned, characteristic of benthonektonic animals in this respect are highly active rudder and braking systems morphologically supported by various fins or a hydrojet apparatus; special attention should be drawn in this connection to the extremely high mobility of the pectoral fins in most benthonektonic Osteichthyes in comparison with typical eunektonic forms in this class, such as Clupeidae, Scombridae or Xiphioidae. On the whole, benthonektonic animals use the nektonic type of camouflage, manifested primarily in the general cross-sectional shape of the body. However the colouring of benthonektonic animals clearly has specific features associated with dwelling amidst bottom relief or among elements of floating solid substrata (e.g. floating vegetable material), which is manifested in the development of diverse colouring adapted to providing concealment in a particular environment, i.e. near a particular substratum (Aleyev 1973a).

Benthonekton are one of the primary, most ancient forms of nekton. Benthonektonic Monorhina (Anaspida) and Acanthodei (Ischnacanthida) existed already in the Silurian, from which it follows that benthonekton as a whole are certainly not younger than the Silurian.

Originally, benthonekton usually evolved from nektobenthic forms. The great majority of the most ancient Paleozoic benthonektonic forms originated directly from nektobenthic ancestors. Exactly such, one must presume, is the origin of benthonektonic Belemnoidea (from the Carboniferous period); also undoubtedly descended from nektobenthic forms are the benthonektonic representatives of Diplorhina (Devonian), Monorhina (Silurian-Devonian), Placodermi (Devonian) and Acanthodei (Silurian-Permian), and early benthonektonic Chondrichthyes (Devonian). The only benthonektonic Osteichthyes which may be assumed probably to be descended from nektobenthic forms are some primitive and in the main also Paleozoic groups; such, in particular, from among Sarcopterygii are benthonektonic Rhipidistia (Devonian-Permian) of the type of Devonian *Eusthenopteron*, and from among Actinopterygii, benthonektonic representatives of Palaeonisci (Devonian-Cretaceous) and Acipenseriformes (from the lower Lias).

In some cases benthonektonic forms probably evolved from nektoplanktonic ones, but the available material is too scanty for a more conclusive opinion in this respect.

A number of extant benthonektonic Chondrichthyes and Osteichthyes originate from eunektonic ancestors. Such, for example, are the benthonektonic sharks of the type Triakidae, the benthonektonic representatives of Carangidae, etc. Also descended from eunektonic forms are probably benthonektonic Teuthoidea (if any exist).

Some benthonektonic forms, but apparently comparatively few, originate from xeronektonic ones. A case in point are the Triassic mollusc-eating Ichthyosauria from the Omphalosauridae group, as well as some extant sirenians that have lost all ecological ties with land (*Dugong*), and bottom-feeding cetaceans (*Eschrichtius* and *Platanista*).

Both benthic and planktonic stages may occur in the ontogeny of benthonektonic animals; here development most often follows either the pattern: benthos – plankton – benthonekton or the pattern: plankton – benthonekton. An example of the former can be seen in the demersal roe, planktonic larvae and early young fry of benthonektonic Cyprinidae. In this case ecological bonds with the bottom appear at $Re < 5.0 \times 10^3$ (*Carassius, Cyprinus*, etc.). An example of the latter pattern is presented by benthonektonic fishes with planktonic roe and planktonic larvae, in which ecological bonds with the bottom also appear at $Re < 5.0 \times 10^3$ (*Diplodus annularis, Sciaena umbra*, etc.). Ontogeny in addition often follows the pattern: euplankton – planktonekton – benthonekton. Examples are provided by such benthonektonic animals as *Odontogadus* or *Mugil*, for which it has been reported (Burdak 1957, 1959, 1964, 1968, 1969b) that the young fry remains palagic not only at $Re \leqslant 5.0 \times 10^3$, but also at $Re > 5.0 \times 10^3$, i.e. have a planktonektonic stage. During a prolonged pelagic stage when ecological bonds with the bottom appear at $Re > 10^5$ in fishes ecologically similar to the above-mentioned *Odontogadus* and *Mugil*, development may follow the pattern plankton (or benthic forms if the roe is demersal) – planktonekton – eunekton – benthonekton. In some cases we observe in benthonektonic animals direct development via the pattern: benthonekton – benthonekton, for instance in some sirenians (*Dugong*).

Planktonekton unites nektonic animals that have no obligatory ecological connection with a solid substratum and swim in the range $5.0 \times 10^3 < Re \leqslant 10^5$. Planktonektonic animals are, therefore, eupelagic animals.

Three classes of animals are represented in planktonekton: Cephalopoda, Sagittoidea and Osteichthyes. Thus all planktonektonic animals are primary aquatic animals.

The cephalopod species belonging among planktonekton are relatively small. These consist principally of some extant Teuthoidea with a nektonic body shape and swimming in the range $5.0 \times 10^3 < Re \leqslant 10^5$. Such, for example, are some Neoteuthidae and Brachioteuthidae, most Enoploteuthidae, some Octopoteuthidae, Lycoteuthidae, occasional Histioteuthidae and Chiroteuthidae. By every indication, planktonektonic also are some small eupelagic Sepioidea in which $U \leqslant 0.40$ and $Re \leqslant 10^5$, such as Sepiolidae (e.g. *Rossia pacifica*, etc.). It is probable that in addition planktonektonic animals occurred among extinct Belemnoidea. Some eupelagic Octopoda from the Cirroteuthoidea group with $U \leqslant 0.40$ and $S_0 \leqslant 4.50$ for which the condition $Re \leqslant 10^5$ holds may also be planktonektonic.

Planktonektonic Sagittoidea comprise not only large species, for example the genus *Flaccisagitta*, but also smaller ones in which length of individuals exceeds 2.0 cm, in particular *Sagitta setosa*. It has not been ruled out that for some particularly large Sagittoidea, such as *Flaccisagitta hexaptera* or *Flaccisagitta gazellae*, $Re > 10^5$; i.e. they could, according to this parameter, be eunekton. However, since all Sagittoidea swim in separate rapid bursts, speeds satisfying the condition $Re > 10^5$ are, in fact, only instantaneous, i.e. they are maintained for a very short time: not more than 1 s in *Sagitta setosa*, judging by the author's observations. It should be assumed on this basis that even for the largest Sagittoidea, reaching 10 cm in length, the average maximum swimming speeds maintained for longer than 1 s hardly satisfy the condition $Re > 10^5$. With this in mind, as well as the presence of planktonic features in the general morphological organization of Sagittoidea, which are very clearly pronounced even in the largest, it should be concluded that even these biggest representatives of the group cannot be placed among eunekton, but are merely planktonektonic.

The planktonektonic Osteichthyes consist of small pelagic forms on average less than 10 cm long, such as Myctophidae, Salangidae, some pelagic Cyprinidae (*Leucaspius*), Gobiidae (*Aphya*), etc.

The most essential morphological characteristic of planktonekton is the presence of both nektonic and planktonic features. Although nektonic, forms descended from plankton (such as Sagittoidea) preserve and forms originating from eunekton or benthic animals (such as Myctophidae or *Aphya* respectively) acquire some clearly planktonic features, primarily a planktonic type of camouflage, based on high optical transparency of the body. The presence of a number of planktonic features in planktonektonic animals, particularly planktonic camouflage, is explainable by their small linear dimensions: at *Re* from 5.0×10^3 to 10^5 and with a nektonic body shape, their length is usually less than 10 cm. The buoyancy of planktonektonic animals is as a rule close to neutral, a characteristic, as we have seen, of both nektonic and most planktonic forms. We find among these animals all three types of nektonic propulsor, undulatory (axial and peripheral), paddling and hydrojet, through the paddling type occurs comparatively seldom (i.e. in some small Osteichthyes) and is never the main nor, moreover, the only one. The main propulsor in all planktonektonic animals is of either the undulatory type (Sagittoidea, Osteichthyes and some Cephalopoda) or the hydrojet type (most Cephalopoda). The small length of planktonektonic animals (on average less than 10 cm) and their relatively low swimming speeds (2 m/s maximum) lead to the development of adaptations associated with reducing hydrodynamic drag relevant to swimming in the laminar regime. A manifestation of this is the absence of high-speed body shapes in these animals, and in planktonektonic Osteichthyes, the absence of ctenoid scales, which are intended, as we have seen, to suppress vortex formation. It should be noted in this connection that the cycloid stage in the development of ctenoid scales, characteristic of the ontogeny of all fishes with ctenoid scales (Burdak 1968, 1969a,b; 1970, and others), corresponds precisely to the planktonektonic stage of development, whose upper limit is defined by the condition $Re = 10^5$, which coincides (Burdak 1968, 1969a, and others) with the moment when the ctenoid apparatus starts functioning. Life in open water predetermines the absence of special adaptations in planktonektonic animals facilitating a manoeuvre in some definite direction; their manoeuvrability is on the whole good (Aleyev 1973a).

Like benthonekton, planktonekton represent one of the primary, most ancient nektonic forms. Among the earliest planktonektonic forms are possibly representatives of Chaetognatha, known in the fossil state (*Amiskwia* from among Archisagittoidea) from the Cambrian (Tokioka 1965a). Planktonektonic Osteichthyes and Cephalopoda are younger: the former have most probably existed since the Devonian (Palaeonisci) and the latter since the Carboniferous or Permian period (Belemnoidea).

Planktonektonic Sagittoidea originate from euplanktonic forms. Planktonektonic Cephalopoda are descended from benthonekton and eunekton. The majority of extant planktonektonic Osteichthyes, in particular such groups as Myctophidae and Salangidae, are descended from eunekton; more seldom, planktonektonic Osteichthyes are of benthonektonic origin, for example some Gobiidae (*Aphya*).

A euplanktonic stage, so far as can be ascertained from the ecology of extant forms, is inevitable in the ontogeny of planktonektonic animals. In planktonektonic animals with pelagic eggs and young fry, such as all Sagittoidea, most Cephalopoda and Osteichthyes, development follows the pattern euplankton-planktonekton. When there is a benthic stage (demersal eggs) in the ontogeny, development follows the pattern benthos-euplankton-planktonekton; such a sequence can be seen, for example, in some Cyprinidae (*Leucaspius*).

Eunekton consist of nektonic animals that have no obligatory ecological connections with any solid substratum and swim in the regime $Re > 10^5$. Therefore eunektonic animals, like planktonektonic ones, are eupelagic animals.

Seven classes of animals are represented among eunekton: Cephalopoda, Placodermi, Acanthodei, Chondrichthyes, Osteichthyes, Reptilia and Mammalia. Thus we see in eunekton both primary aquatic animals (Cephalopoda, Placodermi, Acanthodei, Chondrichthyes and Osteichthyes) and secondary ones (reptiles and mammals).

Of the cephalopods, some Belemnoidea were undoubtedly eunektonic; most Teuthoidea too are eunektonic. Everything indicates that some eupelagic Octopoda from the Cirroteuthoidea group for which the inequalities $U \leqslant 0.40$, $S_0 \leqslant 4.50$ and $Re > 10^5$ hold are also eunekton; however, as the ecology of these forms has not yet been adequately studied, it is impossible to make more definite conclusions in this regard. It may well be that the Cirroteuthoidea group contains planktonektonic forms only.

Among Placodermi, forms with a wedge-like body cross-section tapering downwards, such as some Synaucheniidae (*Synauchenia*) and Oxyosteidae (*Oxyosteus*), may have been eunektonic.

Among Acanthodei, probably only few forms were eunektonic. According to the morphology of these fishes (Novitskaya & Obruchev 1964, and others), it is highly probable that eunektonic species were present in the Acanthodida group. In particular, some representatives of Acanthodidae (*Acanthodes*) which were characterized by a deeply emarginate tail fin, additional branches of the lateral line on the ventral surface of the body and the absence of intermediate spines on the belly were according to every indication eunektonic.

Among Chondrichthyes, all the eupelagic Elasmobranchii, mostly sharks and only a few rays (Mobulidae), which underwent a secondary transition to pelagic life, are eunektonic. There are no eunektonic forms among extant Holocephali; available paleontological material indicates that they were absent also among extinct representatives of this group.

Among Osteichthyes, there are, according to every indication, no eunektonic Sarcopterygii. Conversely, however, eunektonic forms are very numerous among Actinopterygii, where they are found in most orders. To list all the eunektonic Actinopterygii would be futile, so great is their number. By way of example, one may name such very characteristic and predominantly or exclusively eunektonic groups as Albulidae, Elopidae, Megalopidae, Clupeidae, Engraulidae, Pomatomidae, Caranginae, Coryphaenidae, Alepisauridae, Trichiuridae, Gempylidae, Scombridae s.l., Xiphiidae, Istiophoridae, Molidae, etc. Molidae, which are sometimes placed among plankton (Davis 1955; Parin 1968) are objectively classed as eunekton here from the results of an ecomorphological analysis of members of this group. At $U<0.40$, $S_0<4.50$ and $Re>10^5$ eupelagic Molidae have a comparatively well-streamlined body shape characterized by $C_D=0.005$, which coincides the values of this index for such eunektonic animals as *Prionace*, *Phocoena* and *Delphinus*. The peripheral undulatory propulsor of Molidae is not specific per se to any nektonic class thus its presence does not place an animal in a specific nektonic class; this type of propulsor is common, for instance, among eunektonic cephalopods.

Eunektonic reptiles include most of the Ichthyosauria (with the exception of benthonektonic Omphalosauridae), the eupelagic Palaeophidae and some Hydrophidae (Hydrophinae, giving birth to their young in the water, for example *Microcephalopsis gracilis*). There is every indication that Metriorhynchidae, which had probably completely lost (Terentyev 1961) the capacity to climb out onto the shore, and also most of the predatory Thalattosauria and Mosasauridae were eunektonic, though it has not been ruled out that some of them still preserved ecological connections with land (as regards reproduction), i.e. were merely xeronektonic. Nor do we exclude the presence of eunektonic forms among Sauropterygia, since it has been suggested (Tatarinov 1964f) that they were viviparous.

Among the mammals only cetaceans (with the exception of benthonektonic forms) are eunektonic. As for sirenians, they all feed on the off-shore aquatic vegetation, both completely immersed and semi-immersed, i.e. they are ecologically connected with the bottom or with land and are, accordingly benthonektonic or xeronektonic.

Most vividly manifested in the morphology of eunekton are all the specific features of nekton as a particular ecomorphological type. The buoyancy of eunektonic animals is in the great majority of cases neutral or close to neutral. Of the three types of nektonic propulsor, most typical of eunektonic animals is the undulatory propulsor, either axial (for all groups except Cephalopoda) or peripheral (Cephalopoda and some fishes, particularly Molidae, Mobulidae, etc.). The hydrojet propulsor is characteristic of cephalopods alone. The paddling propulsor, though it does occur in eunektonic forms, for instance in many fishes, is on the whole uncharacteristic of eunektonic animals and in any case is merely accessory. The main propulsor in eunektonic animals always is either an axial undulatory or hydrojet propulsor. The high incidence of the axial undulatory propulsor among eunektonic animals is due to its special advantages for building up the relative and absolute power of the locomotor apparatus, which for eunektonic forms is of decisive importance in view of their highly mobile mode of life in the majority of cases. It is among eunektonic animals that we find the fastest and most sturdy swimmers in the pelagic zone, many having swimming speeds in excess of 10 m/s (the majority of eunektonic species over 1 m long), quite a number having speeds in excess of 20 m/s (e.g. many large tunas, the Xiphioidae, etc.), and a very few even having speeds reaching 30–33 m/s (Xiphioidae). In most cases these speeds correspond to $Re > 10^6$ (the great majority of cephalopods, fishes, reptiles and cetaceans), in many cases to $Re > 10^7$ (large Teuthoidea, large tuna fishes, Coryphaenidae, large sharks, most Ichthyosauria and Mosasauridae, all cetaceans, etc.), occasionally even to $Re > 10^8$ (*Xiphias;* probably some particularly large predatory Ichthyosauria, i.e. *Macropterygius, Cymbospondylus, Eurhinosaurus,* etc.; Mosasauridae, and whales). Accordingly, it is in eunektonic animals that we find the most perfect body shape as regards hydrodynamic properties, characterized by $C_{Dp} \leqslant 0.005$ and adapted, according to the ecology of the species concerned, to movement at the above values of Re. All other adaptations directed at boundary-layer control and the reduction of hydrodynamic drag also reach the highest perfection in eunektonic animals. The manoeuvrability of eunektonic animals on the whole is high, which corresponds to their generally high level of mobility. The nektonic type of camouflage is characteristic of all eunektonic animals, being provided, as a rule, in comparatively small species by both a cryptic body shape and cryptic colouring, but in larger forms (sharks, large tuna fishes, Xiphioidae, Ichthyosauria, cetaceans, etc.), predominantly or exclusively by cryptic colouring (Aleyev 1973a).

In comparison with benthonekton and planktonekton, eunekton are a younger form of nekton. This is proved by the fact that eunekton are not directly descended either from benthic forms or from plankton, but evolved from some other nektonic form: benthonekton, planktonekton or xeronekton. This, in turn, is due to the fact that, in the transition of planktonic, benthic or terrestrial fauna to the eunektonic mode of life, an intermediate benthonektonic, planktonektonic or xeronektonic phase is unavoidable. Of these intermediate forms, xeronekton are the youngest (see below). Apparently, eunekton on the whole emerged in the Devonian period, according to the time of appearance of the first eunektonic fishes in the groups Placodermi, Acanthodei, Chondrichthyes and Osteichthyes.

Eunektonic cephalopods originate from benthonektonic or planktonektonic ancestors. Eunektonic Placodermi, Acanthodei and Chondrichthyes evolved from benthonektonic forms. The majority of eunektonic Osteichthyes are descended from benthonekton, though it is possible that the immediate ancestors of some eunektonic species were planktonektonic. All eunektonic reptiles and cetaceans originate from xeronekton.

In the ontogeny of eunektonic animals we see in most cases benthic and planktonektonic stages, which are common, in particular, in Cephalopoda and Osteichthyes. In such cases development usually follows either the pattern benthos-euplankton-planktonekton-eunekton, as, for example, in fishes with demersal roe and planktonic larvae and young fry (*Clupea*, etc.), or the pattern euplankton-planktonekton-eunekton, as in fishes with planktonic roe and larvae (Scombridae, etc.). In some cases, when the young fry upon hatching swim in the regime $Re > 10^5$, eunektonic animals develop directly according to the pattern eunekton-eunekton, as is observed in viviparous eunektonic sharks, Ichthyosauria and cetaceans. In small viviparous eunektonic forms whose young on emergence swim in the regime $5.0 \times 10^3 < Re \leqslant 10^5$, development follows the pattern planktonekton-eunekton; examples of this are found among small eunektonic sharks (*Euprotomicrus bispinatus*), among small viviparous Hydrophidae and among small Ichthyosauria.

Xeronekton are nektonic animals ecologically connected with land and swimming in the regime $Re > 5.0 \times 10^3$.

Four classes of animals are represented in xeronekton: one primary aquatic, amphibians, and three secondary aquatic, reptiles, birds and mammals. Thus xeronekton consist of vertebrates only.

The presence of xeronektonic forms among amphibians can not as yet be considered as proved. At the same time there are no grounds to deny the existence of xeronektonic forms among some extinct groups of amphibians, primarily Ichthyostegalia. At the present stage of knowledge about extinct amphibians the question remains open, i.e. the existence of xeronektonic primary aquatic animals is problematic.

Xeronektonic reptiles include some Pleurosauria (*Pleurosaurus*, etc.), all or nearly all the Sauropterygia (with the exception of eunektonic forms, if any exist), Placodontia, Mesosauria, some Pleurosternidae (Desmemydinae), all Thalassemydidae, Apertotemporalidae, Chelonioidea, Dermochelyoidea, most Hydrophidae (except eunektonic Hydrophinae) and all Teleosauridae. Some Thalattosauria, Mosasauridae and Palaeophidae were probably xeronektonic too, as their process of reproduction was connected with land to some degree. In addition, the presence of primitive xeronektonic forms in some other reptile groups has not been ruled out, particularly in such groups as Rhynchocephalia (Claraziidae), Choristodera, Aigialosauridae, Dolichosauridae, Cholophidia and Phytosauria.

Only comparatively few birds are xeronektonic. These are the Cretaceous Hesperornithes and Sphenisciformes (Impennes), which appeared in the Oligocene.

Of the mammals, all pinnipeds and some sirenians are xeronektonic. Xeronektonic forms occurring among sirenians include *Trichechus*, which (Petit 1955) are capable of climbing onto the beach and into shallow littoral waters, where they lie and move about in a semi-immersed state; everything indicates that *Hydrodamalis gigas* were xeronektonic too, since they were closely linked with shallow littoral waters. Also, xeronektonic forms must have existed among the ancestors of cetaceans.

The morphological characteristics of xeronekton are determined above all by the development of adaptations directed at ensuring ecological contact with land or some substitute for it. The buoyancy of xeronektonic animals is either close to neutral or markedly positive; the latter is observed, for instance, among pinnipeds during periods of increased fatness, which usually coincide with the beginning of a terrestrial period in their life cycle and are directly due to the storage of energy in the form of subcutaneous blubber deposits (when on land pinnipeds do not feed). Body shapes are rather varied among xeronektonic animals: we observe here snake-like, turtle-like, crocodile-like, seal-like and other forms, which on the whole corresponds to the comparatively early stage of adaptation of all these animals to the nektonic mode of life, and reflects the phylogenetically different quality of the animals that make up xeronekton. In very many xeronektonic forms (Sauropterygia, Placodontia, Testudinata, Pinnipedia, etc.) the body is more convex above than beneath, and has a more or less flattened ventral surface, which is directed at easing terrestrial locomotion, which is rather difficult for most xeronektonic animals. At the same time, such a body shape facilitates its conversion into a lifting surface, creating additional lift during swimming, which, as we observed in chapter III, is of especial value when the animal descends to great depths. One of the most common propulsive mechanisms in xeronektonic animals is the paddling type of propulsor (Sauropterygia, Placodontia, Testudinata, Pinnipedia, etc.). This is undoubtedly explained by its capacity to ensure movement both in the water and on land, where the 'leg-paddles' function to some extent as elements of the walking apparatus. Along with this, frequently occurring in xeronektonic animals is the undulatory type of propulsor, – axial (Pleurosauria, Mesosauria, Palaeophidae, Hydrophidae, Teleosauridae, Sirenia, etc.) or pseudo-axial (Phocidae and Odobenidae). No very fast swimmers are found among xeronektonic forms: swimming speeds never exceed 10 m/s and are usually much lower. Nor do we find, accordingly, any body shapes particularly perfect in hydrodynamic terms and comparable to the best eunektonic ones; xeronekton are mainly average swimmers, adapted to movement in the regime $Re < 10^7$. The manoeuvrability of xeronektonic animals is on the whole high. Camouflage is of the usual nektonic type, but is manifested mainly in colouring only, cryptic features of body shape usually being cancelled out by the development of adaptations facilitating crawling on land and associated with the flattening of the ventral side of the body (Sauropterygia, Placodontia, Testudinata, Pinnipedia, etc.) (Aleyev 1973a).

Xeronekton are undoubtedly the youngest form of nekton. Xeronektonic amphibians, if they existed at all, which we so far cannot prove, probably appeared in the Devonian; it is from this time that the group Ichthyostegalia is known, which is the most probable source of xeronektonic forms among amphibians. Other xeronektonic animals appeared not earlier than the Lower Permian (Mesosauria), and in the main not earlier than the Triassic (Sauropterygia, Placodontia and Thalattosauria).

Xeronektonic amphibians, if one admits their existence, most probably evolved from benthonektonic forms. Xeronektonic reptiles, bird and mammals originated from terrestrial nektoxeric ancestors, and all represent the first stage of adaptation by secondary aquatic animals to the nektonic mode of life.

The ontogeny of xeronektonic animals includes terrestrial stages. Development usually follows the pattern terrestrial stage – xeronekton, as one can see in the examples of Chelonioidea, Sphenisciformes and Pinnipedia.

3. Origins and history of development of nekton

As one can see from the above, the paths whereby nekton became established are many and varied. In the course of adaptation by animals to life in the pelagic zone there appeared diverse variants of the nektonic type, i.e. different ecomorphological classes of nekton, adapted to existence in diverse conditions.

According to geological age, nekton are much younger than plankton and much younger than all benthic forms. The first nektonic animals whose existence may be considered proved appeared in the Silurian (Anaspida and Ischnacanthida), when varied benthofauna already existed together with no less varied planktofauna. Thus nekton are one of the latest variants in the development of animal life in the pelagic zone. This can be explained by the fact that the nektonic body structure turns out to be suitable only for comparatively large and at the same time fairly highly organized animals with a rather effective locomotor apparatus and well-developed sense organs. At the same time it has not been ruled out that some benthonektonic and planktonektonic invertebrates (in particular from among Chaetognatha) existed already in the Lower Paleozoic (Cambrian-Ordovician; see Aleyev 1973a).

The most ancient form of nekton, according to all indications, is benthonekton, which have existed since Silurian times. Nearly as ancient, probably, are planktonekton, which have existed at least since the Devonian and possibly since the Cambrian. Eunekton and xeronekton are younger than this: eunekton appeared not earlier than the Devonian and xeronekton only during the Permian (though xeronektonic amphibians could have lived in the Devonian).

Two periods stand out in the history of the development of nekton, differing in the genesis of new nektonic groups: 1. the period of primary aquatic nektogenesis and 2. the period of secondary aquatic nektogenesis.

The period of primary aquatic nektogenesis, from the Silurian (Cambrian?) till the Upper Carboniferous period, is characterized by the emergence of nektonic groups exclusively from primary aquatic fauna: primary benthic fauna or primary plankton. In this period there took shape benthonektonic, planktonektonic and eunektonic complexes consisting of Chaetognatha, Cephalopoda, various ichthyofauna and, possibly, some groups of amphibians. The leading nektonic forms in this period, both in the ocean and in inland waters, were fishes. During this period there appeared, according to every indication, representatives of all nine classes of primary aquatic animals ever represented in nekton: Cephalopoda, Sagittoidea, Diplorhina,

Monorhina, Placodermi, Acanthodei, Chondrichthyes, Osteichthyes and Amphibia (if this class has ever in fact contained nekton).

The period of secondary aquatic nektogenesis, from the Upper Carboniferous period down to the present, is characterized by the evolution of new nektonic complex not from primary aquatic groups alone, as in the previous case, but also and even predominantly from terrestrial semi-aquatic fauna, i.e. by the formation of secondary aquatic nekton. In this period there emerged all the secondary aquatic nektonic groups originating from reptiles, birds, and mammals. Initially xeronektonic groups emerged, some of which later on provided the ancestors of eunektonic forms (Ichthyosauroidei, some Hydrophidae, most cetaceans, etc.) and benthonektonic forms (Omphalosauroidei, and of the *Sirenia, Dugong*, etc). The leading oceanic nektonic forms in this period, along with fishes, were cephalopods, reptiles and cetaceans, while in inland waters the leading forms were fishes of the Osteichthyes class.

Four phases of nektogenesis may be distinguished in the most diverse groups of nektonic animals: 1. anektonic, 2. pronektonic, 3. protonektonic and 4. eunektonic. These are shown schematically in Figure 147.

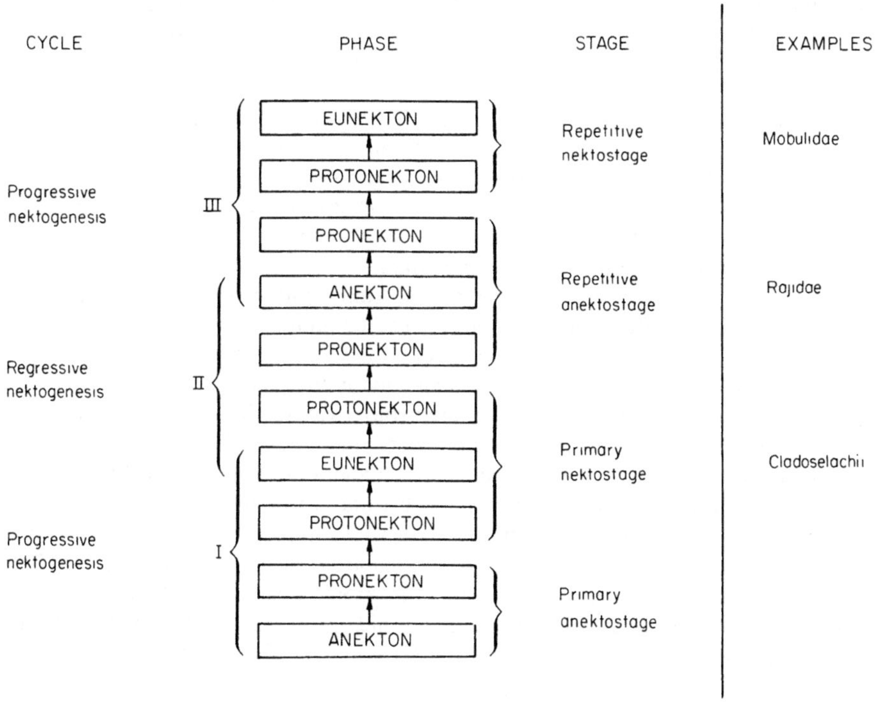

Figure 147. Diagram of nektogenesis. See text.

The anektonic phase of nektogenesis is always the initial phase in the process of adaptation to the nektonic mode of life. We do not yet see at this stage any tendencies towards the development of nekton either on the ecological or on the morphological plane. This is a phase in which we find benthic, planktonic and terrestrial forms which do not reveal any tendencies towards nektonic development; all these are merely forms which may sooner or later become involved in nektogenesis.

The pronektonic phase of nektogenesis is characterized ecologically by the appearance of a tendency towards nektonic development, which may occur in benthic, planktonic or terrestrial semi-aquatic (nektoxeric) forms. Morphologically this phase is characterized by the formation of an elongated body shape, which facilitates swimming under water. Here the conditions $U \leqslant 0.40$ and $S_0 \leqslant 4.50$ may or may not be met, depending on whether or not they are met in the corresponding anektonic form. In general, no specifically nektonic propulsive mechanisms appear in this phase of nektogenesis, and if they do they are rather imperfect; usually neutral buoyancy is absent, and when it does occur (in secondary aquatic forms), special hydrostatic adaptations for its regulation are lacking; as a rule there are no adaptations providing nektonic camouflage. The pronektonic phase of nektogenesis corresponds to nektobenthic, nektoplanktonic and nektoxeric forms; as example, we may point to Cephalaspidida, Antiarchi (*Pterichthys*), *Amblyrhynchus*, Crocodylidae, *Enhydra*, *Lutra*, etc.

The protonektonic phase of nektogenesis is ecologically characterized by transition to the nektonic mode of life, though with the preservation of either ecological connections with immersed solid substrata or land (benthonekton and xeronekton), or planktonic features in the ecology (planktonekton). Morphologically the protonektonic phase is characterized primarily by the formation of a nektonic body shape, expressed by compliance with the conditions $U \leqslant 0.40$, $S_0 \leqslant 4.50$ and $C_{Dp} \leqslant 0.010$ (Aleyev 1972b). Later on, the protonektonic phase is characterized by the formation of specifically nektonic propulsors (of the undulatory, paddling or hydrojet types), the development of neutral buoyancy and special adaptations controlling it, a noticeable improvement in the animal's manoeuvrability, and the development of the nektonic type of camouflage. Found in the protonektonic phase of nektogenesis are all the benthonektonic, planktonektonic and xeronektonic animals.

The eunektonic phase of nektogenesis is ecologically characterized by the complete loss of ecological ties with solid substrata. Morphologically this phase is characterized by the further perfection of the entire complex of nektonic adaptations, primarily the body shape, expressed by compliance with the condition $C_{Dp} \leqslant 0.005$. All eunektonic animals correspond to the eunektonic phase of nektogenesis.

The four phases described above, anektonic, pronektonic, protonektonic and eunektonic, together form one cycle of nektogenesis. Each cycle consists of an anektostage, which in turn includes the anektonic and pronektonic phases of nektogenesis, and a nektostage, which consists of the protonektonic and eunektonic phases of nektogenesis.

Nektogenesis may be either progressive or regressive. Progressive nektogenesis means the progressive development of nektonic adaptations, i.e. the formation of nektonic forms from anektonic ones (e.g. eunektonic cetaceans from anektonic mammals). Regressive nektogenesis means the regression of nektonic adaptations, i.e. the formation of anektonic forms from nektonic ones (e.g. benthic Batoidei from primary nektonic Elasmobranchii).

It should be stressed that regressive nektogenesis always corresponds to the progressive development of some anektonic complex of adaptations, for example planktonic or benthic. Likewise, progressive nektogenesis inevitably corresponds to regression of some anektonic complex of adaptations. For example, instead of using the expression 'regressive nektogenesis' when speaking of the transformation of a nektonic form into a planktonic one, we could use 'progressive planktogenesis'. By using the term 'regressive nektogenesis', we are merely trying to stress the direction of changes in the nektonic adaptations, whose study constitutes the basic task of this investigation.

Like any process, nektogenesis may be complete or incomplete. All eunektonic forms are the result of complete progressive nektogenesis. One may observe the various moments of incomplete nektogenesis, both progressive and regressive, in pronektonic and protonektonic phases. Complete loss of nektonic adaptations, as is seen, for example, in benthic Batoidei and Pleuronectiformes, is the result of complete regressive nektogenesis.

During the complicated historical development of nektonic animals, periods of progressive and regressive nektogenesis sometimes alternated repeatedly, i.e. many nektogenetic cycles occurred. Accordingly, one may distinguish primary nektonic and secondary nektonic form. As has been noted earlier (Aleyev 1963a), primary nektonic forms have no previous nektostages in their history separated from the nektostage under examination by any anektostages, i.e. there is only one nektogenetic cycle in their history. Many primary aquatic groups are also primary nektonic, particularly, in all probability some Pachyosteida (Synaucheniidae, Oxyosteidae, etc.), all Acanthodida, all Cladoselachii and most of the extant nektonic Elasmobranchii and Actinopterygii. Secondary nektonic forms have previous nektostages in their history separated from the nektostage under consideration by anektostages, i.e. there are at least two nektogenetic cycles in their history. As examples of secondary nektonic forms, one may point to eunektonic Mobulidae and some planktonektonic Gobiidae (*Aphya*).

The process of nektogenesis, with all its different cycles, phases, stages, directions and degrees of completeness, is illustrated schematically in Figure 147. When looking at this figure, one should bear in mind that a change in the direction of nektogenesis, i.e. a transition from progressive to regressive genesis or vice versa, does not necessarily occur as shown in the diagram, i.e. in the anektonic or eunektonic phase. In a number of cases progressive nektogenesis is followed by regressive nektogenesis in the protonektonic phase, and so on. Apparently, the most diverse variations are possible in this sense, being due in each instance to the particular situation arising in the phylogeny of group in question. One thing, however, remains unaltered: **the sequence of nektogenetic phases**, which is preserved both in the case of progressive and in the case of regressive development of nektonic adaptations, as, indeed, is shown schematically in Figure 147.

As the above material shows, nekton originate from three sources: benthic, planktonic and terrestrial fauna.

X. NEKTON AND THE BODY OF WATER

In chapters I–IX we examined the specific features of nekton as an ecomorphological type of animal. Now we shall dwell on some general characteristics of the formation of nekton dependent on the properties of the body of water in which they dwell.

As one can see in part from what has already been said in chapters II–IX, the characteristics of the body of water determine entirely both the very possibility of nektogenesis and its direction. From an analysis of extant nekton, it follows that, among the factors influencing nektogenesis and the form of the nekton, one should mention first of all 1. the dimensions of the body of water, 2. its salinity and 3. its temperature.

The size of the body of water is the basic, paramount factor controlling the direction and degree of development of nekton. The greater the size of the body of water, the greater is the overall qualitative diversity of its nekton and the greater the relative importance of eunekton, owing to the progressive weakening of ecological connections between nektonic animals and solid substrata. A large enough pelagic zone is the basic prerequisite for the development of primary aquatic and secondary aquatic eunekton and secondary aquatic xeronekton. Thus, in very small or very shallow bodies of water nekton are generally absent, being replaced by nektobenthic and nektoplanktonic forms. We observe this in such bodies of water as pools, marshes which have no areas of free water more than 1 m deep, very small streamlets, etc. Small nektonic animals appear in such bodies of water for only a limited time, getting there as a result of accidental migration (e.g. during the overflow of a river, etc.). Primary aquatic hydrobionts living in this type of body of water are inevitably closely bound with the bottom. The bottom and underwater vegetation connected with it create an environment with numerous shelters of all kinds, which basically hinders the appearance of the chief prerequisite for transition to the nektonic mode of life since it makes escaping from predators most feasible not by fleeing into open water, but primarily by hiding in shelters. Moreover, abounding at the bottom and among underwater vegetation in such bodies of water are small invertebrates, the food of larger animals that might provide the initial material for the formation of nekton. Therefore in such bodies of water, only nektobenthic, planktobenthic and benthoplanktonic forms appear among primary aquatic hydrobionts. Since plankton are smaller than nektonic animals, it is possible for plankton to become ecologically detached from solid substrata when the absolute dimensions of the pelagic zone are too small for the emergence of autochthontic nektonic forms. Secondary aquatic fauna found in these bodies of water lack sufficient prerequisites for effecting ecological detachment from land and remain, therefore, at the nektoxeric stage, without going over to the xeronektonic phase.

A larger pelagic zone makes it possible for nektonic animals to dwell permanently in the body of water. This takes place in such bodies of water as small lakes, normal and karst, marshes with deeper areas of free water, larger streams and creeks, etc. Here we find benthonekton (*Dallia, Carassius carassius m. humilis, Phoxinus percnurus stagnalis,* Poeciliidae, etc.) and planktonekton (*Leucaspius,* etc.), but eunekton and xeronekton are absent.

In still larger bodies of water, i.e. lakes, rivers, shallow marine lagoons, etc., there appear eunekton, at first represented by smaller forms then, as the pelagic zone further increases in size, by ever larger ones. Xeronekton, as a rule, are still absent in these bodies of water.

Finally, in large lakes (Baikal, the Caspian Sea, etc.), large rivers and the offshore parts of the ocean, there appear xeronekton and large eunektonic forms for which $Re > 10^6$. The biggest and fastest representatives of nekton, for which $Re > 2.0 \times 10^7$, are found only in the open sea.

One of the most important factors determining the size of the pelagic zone is the depth of the water, which affects how much of the bottom is accessible to nektonic animals. All other conditions being unchanged, greater depth always tends to disrupt ecological connections with solid substrata, i.e. accelerates progressive nektogenesis, and vice versa. In other words, increasing depth results in the end in the appearance of an ecologically 'bottomless' body of water, which for many nektonic animals leads to the severance of any ecological connections with the bottom, i.e. to the process known as 'pelagization' (Burdak 1959, 1964, and others). Very interesting in this regard is the ecologically 'bottomless' Black Sea, whose depths are devoid of animal life because of hydrogen sulphide contamination, and where noticeable pelagization of some benthonektonic fishes (*Odontogadus*) takes place as compared with the populations of neighbouring habitats of the same species which are 'normal', i.e. free from hydrosulphuric contamination (Burdak 1959).

The salinity of a body of water, like its size, determines to a considerable extent the composition of its nekton. For a number of primary aquatic sthenohaline groups of animals a marked deviation in salinity from the oceanic norm makes their natural occurrence in the body of water absolutely impossible. It is for this reason that nektonic animals of the groups Chaetognatha (Sagittoidea) and Cephalopoda are absent altogether in inland waters, as we noted in chapter II. Similarly, a salinity above the oceanic level precludes for the same reason the development of nekton in such waters as the upper reaches of the Sivash and the Kara-bogaz-gol Bay of the Caspian.

The temperature of a body of water is of the utmost importance in determining whether or not the formation of nekton in it is possible. Being pelagic animals, always or nearly always remaining in the water, nektonic animals are entirely dependent on the temperature of the water: its magnitude, its vertical distribution, its seasonal variation, the size and duration of any ice cap, etc. In harsh thermal conditions where the water temperature does not rise above +2 °C throughout the greater part of the year and occasionally drops below or is even always below zero (in the sea), just as in waters whose temperature exceeds 50 °C (*Cyprinodon maculatus* dwells in California's hot springs at +52 °C), there exist only comparatively few nekton, consisting of aberrant forms specially adapted to the extreme conditions. Examples are the high-Arctic and Antarctic parts of the ocean, as well as some small inland bodies of water (rivers and lakes) in the polar zone of Northern Europe and North America, whose nekton are very sparse. The overwhelming majority of nektonic animals exist in conditions where the temperature ranges from about +2 to +30 °C in the ocean and from +4 °C to +40 °C in inland waters.

Apart from directly influencing the most vital functions, the temperature of the water affects poikilothermic nektonic animals. In most cases it controls to some extent their growth rate and consequently the linear dimensions of individuals (Aleyev 1956b, 1958d), and hence their Reynolds number, i.e. their regime of locomotion. In this connection one may note that the specific nektofaunal features of individual bodies of water in the Mediterranean geosyncline have been shown (Aleyev 1956b, 1958d) to be determined in part by the different degree of continentality (under annual variation) of the thermal conditions in these waters: for genetically marine nektonic animals the continentality of the temperature pattern shortens the period of intensive feeding by poikilothermic forms and retards their growth, resulting in smaller individuals. Also (Burdak 1966), the growth rates of some benthonektonic Black Sea fishes (Mugilidae) have undergone changes directly connected with historical changes in the thermal conditions in the sea due to the changing climate. Since the sizes of poikilothermic nektonic animals determine to a certain degree their morphology (Aleyev 1958b, 1963a, and others), the water temperature also has some direct influence on the development of a number of nektonic adaptations.

The role of light as a factor restricting or stimulating nektogenesis is much less important than those of the size, salinity and temperature of the body of water. This is proved by the fact that the ocean contains comparatively varied deep-water nekton, existing in conditions of total darkness except for bioluminescence.

The gas content of the water does not as a rule affect nektogenesis.

All other abiotic factors influence nektogenesis still less.

An important role in defining the nektofaunal picture is played, naturally, by the various **biotic relationships,** the most important of which are those of a trophic character. These relations influence most decisively the size of nektonic populations and in some measure the qualitative diversity of nekton.

In summing up it should be stressed that in every case the nektofaunal picture is determined by the entire set of factors affecting nektogenesis, both abiotic and biotic, though in particular cases the various factors may acquire greater or smaller importance. It should be borne in mind, however, that the decisive role in controlling nektogenesis is played by the first of the above-listed factors, i.e. the absolute size of the pelagic zone. If the pelagic zone is not large enough nektogenesis is impossible, even if all the other conditions for the life of hydrobionts are satisfied. In the final analysis it is always the size of the pelagic zone that determines the very possibility and direction of nektogenesis, and in this sense it is the main controlling factor, i.e. the chief prerequisite for the formation of nekton.

CONCLUSION

Life in the pelagic zone has many special features. Here, as nowhere else, one feels the 'three-dimensionality' of the region. There is no firm support for organisms, no shelter and hardly anything visible by which to take one's bearings. Neither, as a rule, are there any sharp fluctuations in the abiotic environmental conditions in space and time. Conforming to these features peculiar to the pelagic zone are the no less special features of pelagic animals, both planktonic and nektonic.

Among the factors determining the formation and development of nekton as a specific ecomorphological type, two are decisive: 1. the fact that the animals are always in the body of the water, without any connection with solid substrata, and 2. the absence in the pelagic zone of any shelters.

Living continuously in the body of the water creates prerequisites for the progressive development in the morphology of nektonic animals of all kinds of adaptations associated with the maintenance of neutral or nearly neutral buoyancy, and facilitates the emergence of specific features associated with camouflage, reception and transmission of information, grasping food, etc.

For animals which are sufficiently large, as are most nektonic forms, the absence of any shelters in the pelagic zone makes it necessary to escape from predators by fleeing and, at the same time, to pursue food objects. Both these facts lead to a build-up of swimming speeds, which in turn prompts the progressive development of the entire complex of adaptations ensuring forward movement.

The three-dimensionality of the environment as such is conducive to greater mobility of pelagic animals. No wonder animals capable of movement without any contact with solid substrata, usually make longer migrations, like those so characteristic of fishes, birds and cetaceans. The ability to negotiate large distances calls for the development of adaptations providing adequate methods of navigation, owing to which the means of receiving, processing and transmitting diverse information attain exceptional perfection. A permanently 'exposed' existence compels improvement of camouflage and ways of escaping from predators by fleeing.

Thus appear the characteristics of nektonic animals that constitute the core of the similarity among animals belonging to different, and at times very remote, systematic groups. This similarity, as we have seen, appears mainly through the development of a complex of adaptations associated with movement. This is the complex of adaptations in which the lines of convergence of nektonic animals from different groups are especially numerous and strong, which corresponds to the particular importance of this group of adaptations in the formation of nekton as a specific type of aquatic animal. Also strong is the convergence of adaptations associated with camouflage, especially cryptic colouring. However convergence in other groups of adaptations is not so marked.

The evolution of nekton exemplifies how deep convergence may be and how important convergent processes are in the evolution of the animal kingdom. While by no means belittling the importance of adaptive divergence in evolution, we have every reason to recognize these two processes equally in their own right in phylogeny: divergence and convergence always coexist as two sides of the complex process of development of organic forms, appearing as two equally rational types of adaptive change in organisms as they interact with the environment. However, whereas the phylogenetic development of individual systematic groups is a predominantly divergent process, owing mainly to a group 'spreading out' into diverse biotopes, the development of ecomorphological types such as nekton or plankton is primarily a convergent process, owing to the adaptation of qualitatively heterogenous initial genetic and morphological material to existence in a specific environment within a definite ecomorphological framework. Thus the formation of nekton as a specific ecomorphological type is a basically convergent process.

Along with this, an essential role in the development of nekton is played by divergent processes, particularly in connection with adaptation to living at different depths, at different temperatures, where the substances dissolved in the water are different, etc. The pelagic zone is by no means uniform: although steep gradients in the individual environmental factors are absent, the range of their overall variation is very great, which is sufficient for strongly divergent evolution of nektonic forms within the framework of the general features characteristic of nekton as a whole.

The development of features common to nektonic animals reveals great similarity between ontogenetic and phylogenetic aspects. This uniformity is explained by the analogy between morphological functional changes in phylogeny and ontogeny respectively, as was noted earlier for fishes (Aleyev 1963a). Morphological changes in nektonic animals with age in the range $L_a < 450$ cm are based, as a rule, on changes associated with adaptation to movement at ever increasing speeds, which is characteristic also of the phylogeny of most nektonic groups.

We must note as a phenomenon common to all nektonic animals the multifunctionality of the majority of structures, which reflects the completeness of the organisms, a result of the interdependence of all their functional and morphological aspects.

Adaptation to the nektonic mode of life, associated with the development of a capacity for active forward movement in the water, was possibly only for comparatively large and highly organized animals, predominantly vertebrates. This explains the comparatively late appearance of nekton, as compared with plankton, in the life of the pelagic zone. Among the invertebrates, nektonic variants were contributed only by Cephalopoda and Sagittoidea, and eunektonic forms by Cephalopoda alone. From the Silurian to the present, nekton at different periods were dominated by representatives of different classes of fishes, reptiles and cephalopods.

All in all, our varied studies of the functional morphology and biohydrodynamics of pelagic animals have led us to the conclusion that differentiation and specialization of forms of life in the pelagic zone proceeded along the lines of adaptation to existence in two different regimes of fluid flow: laminar and turbulent.

The two forms of life in the pelagic zone, plankton and nekton, correspond to the two regimes of fluid flow, laminar and turbulent, respectively. This is the essence of plankton and nekton.

Being the first special investigation into nekton, this book by no means covers all the possible aspects of the uniformity of nektonic animals as a specific ecomorphological type. The author would be deeply gratified were it to help in drawing the attention of researchers to the questions discussed, and hopes that the next few years will see the appearance of new investigations along these lines. The need for them at present is clearly apparent.

LITERATURE

Abel, O. 1907. Der Anpassungtypus von Metriorhynchus, Zbl. Mineral., Geol. und Palaontol.: 225–235.

Abel, O. 1912. Grundzuge der Palaeobiologie der Wirbeltiere: 1–708. Nagele und Sproesser, Stuttgart.

Abel, O. 1916. Palaeobiologie der Cephalopoden aus der Gruppe Dibranchiaten: 1–281. Fischer, Jena.

Abel, O. 1922. Lebensbilder aus der Tierwelt der Vorzeit: 1–643. Fischer, Jena.

Affleck, R. J. 1950. Some points in the function, development and evolution of the tail in fishes. *Proc. Zool. Soc. London, 120,* 2: 349–368.

Agarkov, G. B. & Ferents, Z. I. 1967. To the question of intraorganic innervation of the skin musculature of the dolphin – *Tursiops truncatus ponticus* Barabasch. *Vest. Zool.,* 3: 71–75.

Agarkov, G. B. & Khadzhinski, V. G. 1970. To the question of the structure and innervation of the skin of Black Sea dolphins in relation to its protective function, *Bionika,* 4: 64–69.

Agarkov, G. B. & Lukhanin, V. Ya. 1970. To the question of the propulsive musculature of the caudal sector of the white-sided dolphin. *Bionika,* 4: 61–64.

Ahlborn, F. 1895, Ueber die Bedeutung der Heterocerkie und ahnlicher unsymmetrischer Schwanzformen schwimmender Wirbelthiere fur die Ortsbewegung. *Zschr. Will. Zool., 41,* 1: 15.

Akimushkin, I. I. 1963. Cephalopods of the seas of the USSR: 1–234. USSR Ac. of Sci. Press, M.

Akimushkin, I. I. 1970. Cephalopods and their distribution and trophic relations with the rest of nekton. A programme and methodology for studying the biogeocoenoses of the aquatic environment. Biocenoses of the seas and oceans: 137–149. "Nauka" Publishers, M.

Alexander, R. M. 1959a. The physical properties of the swimbladder in intact Cypriniformes. *J. Exp. Biol., 36,* 2: 315–332.

Alexander, R. M. 1959b. The densities of Cyprinidae. *J. Exp. Biol., 36,* 2: 333–340.

Alexander, R. M. 1959c. The physical properties of the isolated swimmbladder in Cyprinidae. *J. Exp. Biol., 36,* 2: 341–346.

Alexander, R. M. 1959d. The physical properties of the swimmbladder of fish other than Cypriniformes. *J. Exp. Biol., 36,* 2: 347–355.

Alexander, R. M. 1961. The physical properties of the swimbladder of some South American Cypriniformes. *J. Exp. Biol., 38,* 2: 403–410.

Alexander, R. M. 1965. The lift produced by the heterocercal tails of *Selachii. J. Exp. Biol., 43,* 1: 131–138.

Alexander, R. M. 1966a. Physical aspects of swimbladder function. *Biol. Rev., 41,* 1–2: 141–176.

Alexander, R. M. 1966b. Lift produced by the Heterocercal Tail of *Acipenser. Nature (London), 210,* 5040: 1049–1050.

Alexander, R. M. 1967. Functional Design in Fishes: 1–160. Hutchinson, London.

Alexander, R. M. 1968. Animal mechanics: 1–339. Sidgwick and Jackson, London.

Alexander, W. B. 1955. Birds of the Ocean: 1–182. Putnam, London.

Alexeyev, E. V. 1966. Some morphological changes of the White Sea harp seal (*Pagophoca groenlandica* Erxleben) with age. Ecomorphological research into nektonic animals.: 82–88. "Naukova Dumka" Publishers, Kiev.

Aleyev, Yu. G. 1952. On the ecology of the propagation of the Black Sea horsemackerel (*Trachurus* L.). *Papers of the USSR Ac. of Sci., 83,* 5: 753–755.
Aleyev, Yu. G. 1955. On the functional and phylogenetical significance of some morphological peculiarities of fishes of the sub-family Caranginae (Carangidae, Perciformes). *Papers of the USSR Ac. of Sci., 100,* 2: 377–380.
Aleyev, Yu. G. 1956a. On the functional significance of the lateral (horizontal) body position in flatfishes (Pleuronectiformes). *Papers of the USSR Ac. of Sci., 110,* 4: 707–709.
Aleyev, Yu. G. 1956b. On some regularities in the growth of fishes. *Vopr. Ikhtiol.*, 6: 75–95.
Aleyev, Yu. G. 1957a. The characteristics and functional topography of the fins of fishes. *Vopr. Ikhtiol.*, 8: 55–76.
Aleyev, Yu. G. 1957b. Horsemackerels (*Trachurus*) of the seas of the USSR. *Tr. Sevastop. Biol. St.*, 9: 274–281.
Aleyev, Yu. G. 1957c. On the evolution of pelagic Caranginae (Carangidae, Perciformes). *Tr. Sevastop. Biol. St.*, 9: 274–281.
Aleyev, Yu. G. 1958a. On the movements of *Zeus faber* L., *Zool. J. 37,* 3: 463–465.
Aleyev, Yu. G. 1958b. On changes in the relative size of fins in the ontogeny and phylogeny of fishes. *Papers of the USSR Ac. of Sci., 120,* 1: 204–207.
Aleyev, Yu. G. 1958c. Adaptation to movement and turning capacity in fishes. *Papers of the USSR Ac. of Sci., 120,* 3: 510–513.
Aleyev, Yu. G. 1958d. On some morphological peculiarities of Caspian, Azovian and Black Sea fish forms and their causes. *Tr. Sevastop. Biol. St., 10,* 83–89.
Aleyev, Yu. G. 1959a. The turning capacity of fishes. *Tr. Sevastop. Biol. St., 12,* 259–270.
Aleyev, Yu. G. 1959b. The structure and functions of the caudal fin of fishes. *Tr. Sevastop. Biol. St., 12:* 219–258.
Aleyev, Yu. G. 1959c. On the structure and functions of dorsal fins of Squalidae (Squaloidei, Squaliformes). *Tr. Sevastop. Biol. St., 11:* 153–158.
Aleyev, Yu. G. 1959d. On the functional significance of alae and structures homologous to them in fishes. *Tr. Sevastop. Biol. St. 11:* 161–163.
Aleyev, Yu. G. 1960a. On the location of the main lateral channel of the lateral line in fishes. *Tr. Sevastop. Biol. St. 13:* 159–162.
Aleyev, Yu. G. 1960b. On the functional importance of the ventral keel in fishes. *Tr. Sevastop. Biol. St., 13:* 155–158.
Aleyev, Yu. G. 1962. On the location of maximum body height in fishes. *Zool. J., 41,* 9: 1429–1431.
Aleyev, Yu. G. 1963a. The functional principles of external structure of fishes: 1–247. USSR Ac. of Sci. Press, M.
Aleyev, Yu. G. 1963b. On the buoyancy of fishes. *Tr. Sevastop. Biol. St. 16:* 375–382.
Aleyev, Yu. G. 1963c. Location of maximum body height in fishes. *Tr. Sevastop. Biol. St. 16:* 369–374.
Aleyev, Yu. G. 1964a. Methods of reducing drag in the external organisation of fishes. *Tr. Sevastop. Biol. St., 15:* 288–291.
Aleyev, Yu. G. 1964b. Why the humpback salmon needs a hump? *Nauka i Zhizn, 7:* 156–157.
Aleyev, Yu. G. 1965a. The dolphin body as a carrying plane. *Zool. J. 44,* 4: 626–630.
Aleyev, Yu. G. 1965b. Investigations into the functional morphology of fishes. Bionika, 192–198. "Nauka" Publishers, M.
Aleyev, Yu. G., 1965c. On the creation of vertical transverse forces by the body of nektonic animals. Bionics Research: 31–36, "Naukova Dumka" Publishers, Kiev.
Aleyev, Yu. G. 1966a. The buoyancy and hydrodynamic function of the trunk of nektonic animals. *Zool. J., 45,* 4: 575–584.
Aleyev, Yu. G. 1966b. Statodynamic types of nektonic animals. Ecomorphological investigations of nektonic animals: 3–13. "Naukova Dumka" Publishers, Kiev.

Aleyev, Yu. G. 1969a. Topography of locomotory function in nektonic animals. Functional-morphological investigations of nektonic animals: 3–12. "Naukova Dumka" Publishers, Kiev.
Aleyev, Yu. G. 1969b. On the hydrodynamic function of breeding dress in salmon. Functional-morphological research into nektonic animals: 12–21. "Naukova Dumka" Publishers, Kiev.
Aleyev, Yu. G. 1969c. Function and gross morphology in fish: III–IV+1–268. Translated from Russian by Israel program for scientific translations. Jerusalem. Smithsonian Inst. and National Science Foundation, Washington.
Aleyev, Yu. G. 1969d. The results and prospects of nekton studies. In the book: Biooceanographic research on southern seas: 50–59. "Naukova Dumka" Publishers, Kiev.
Aleyev, Yu. G. 1970a. Mobile roughness on body surface of nektonic animals as a means of reducing hydrodynamic resistance. *Zool. J. 49,* 8: 1173–1180.
Aleyev, Yu. G. 1970b. The specific hydrodynamics in fishes of the Xiphioidae group. *Zool. J., 49,* 11: 1676–1684.
Aleyev, Yu. G. 1972a. On the biohydrodynamic differences between plankton and nekton. *Zool. J., 51,* 1: 5–12.
Aleyev, Yu. G. 1972b. Hydrodynamic qualities of body shape of nektonic animals. *Zool. J., 51,* 7: 949–953.
Aleyev, Yu. G. 1973a. The genesis and ecological divergence of nekton. *Zool. J. 52,* 1: 5–14.
Aleyev, Yu. G. 1973b. Movement and propulsive mechanisms of nektonic animals. *Zool. J. 52,* 8: 1132–1141.
Aleyev, Yu. G. 1973c. Morphological principles of the mobile roughness effect in skin covers of nektonic animals. IV. All Union Conference on Bionics. *6.* Biomekhanika: 6–11. M.
Aleyev, Yu. G. 1973d. Dolphins and women. *Khimia i zhizn,* 9: 8–10.
Aleyev, Yu. G. 1974. Hydrodynamic resistance and swimming speeds of nektonic animals. *Zool. J., 53,* 4: 5–11.
Aleyev, Yu. G. & Kurbatov, B. V. 1972. The biohydrodynamic channel at the nekton department, Institute of Biology of the South Seas, Ukrainian SSR., Ac. of Sci., *Hydrobiol. J., 8,* 1: 111–113.
Aleyev, Yu. G. & Kurbatov, B. V. 1973. The biohydrodynamic experimental complex based on automatic systems. IV All-Union Conference on Bionics, *6,* Biomekhanika: 12–17, M.
Aleyev, Yu. G. & Kurbatov, B. V. 1974. Hydrodynamic resistance of living fishes and some other nektonic animals during inertia movement. *Vopr. Ikhtiol., 14,* 1 (84): 173–176.
Aleyev, Yu. G. & Mordvinov, Yu. E. 1971. Studies into the functional morphology of nektonic animals. Problemy morskoi biologii: 240–246. "Naukova Dumka" Publishers, Kiev.
Aleyev, Yu. G. & Ovcharov, O. P. 1969. On the development of vortex formation processes and the nature of the boundary layer during the movement of fishes. *Zool. J., 48,* 6: 781–790.
Aleyev, Yu. G. & Ovcharov, O. P. 1971. On the role of vortex formation in fish locomotion and the effect on the borderline of two media on the flow picture. *Zool. J., 50,* 2: 228–234.
Aleyev, Yu. G. & Ovcharov, O. P. 1973a. The picture of flow over a moving fish. IV All Union Conf. on Bionics. 6. Biomekhanika: 18–20.
Aleyev, Yu. G. & Ovcharov, O. P. 1973b. A three-dimensional picture of flow over fish. *Vopr. Ikhtiol. 13,* 6 (83): 1112–1115.
Aleyev, Yu. G. & Vodyanitsky, V. A. 1966. Ships and fishes. Budushchee nauki: 162–172. "Znanie" Publishers, M.
Allen, J. 1880. History of North American Pinnipeds. *U.S. Geol. and Geogr. Surv. Terr., Misc. Publ.,* 12: 1–785. Washington.
Alvariño, A. 1963. Quetognatos epiplanctonicos del mar de Cortes. Revista Soc. *Mexicana Hist.* Nat., *24:* 97–203.
Alvariño, A. 1966. Chaetognaths. *Oceanogr. mar. Biol. Annu. Rev.*: 115–194.

Amans, P. 1888. Comparaison des organes de la locomotion aquatique. *Ann. Sci. Nat. Paris, 7-e ser. Zool., 6,* 1: 1–164.
Andrews, C. W. 1910. A descriptive catalogue of the marine reptiles of the Oxford Clay. 1: 1–205. London.
Andriashev, A. P. 1944. Determination of the natural specific weight of fishes. *Papers of the USSR Ac. of Sci., 43,* 2: 84–86.
Andriashev, A. P. 1946. On the function of the pectoral fin in fishes. *Priroda,* 1: 80–84.
Andriashev, A. P. 1954. The fishes of the Northern seas of the USSR: 1–564. USSR Ac. of Sci. Press, M.-L.
Andriashev, A. P. 1955. A review of ell-like eelpouts (*Lycenchelys* Gill (Pisces, Zoarcidae) and related forms in the seas of the USSR and adjacent waters. *Tr. Zool. Inst., USSR Ac. of Sci., 43:* 349–384.
Andriashev, A. P. 1964. A review of the fish fauna of the Antarctic. *Marine fauna research, 2,* 10: 335–386. "Nauka" Publishers, M.-L.
Andriashev, A. P. 1968. The problem of the life community associated with the Antarctic fast ice. Scar Symposium on Antarct. Oceanogr., Santiago, Chile: 147–155 Santiago.
Andriashev, A. P. 1970. Cryopelagic fishes of the Arctic and Antarctic, and their significance in the polar ecosystem. In: "Antarct. Ecology", 1: 297–304. Acad. Press (M. W. M. Holdgate ed.), London.
Andriashev, A. P. 1971. The morphological justification of the genetic separation of antarctic sculpin (*Trematomus o borchgrevinki* Boulenger and *T. brachysoma* Pappenheim) and the new status of the genus Pagothenia Nichols et Lamonte (Nototheniidae). *Zool. J. 50,* 7: 1041–1055.
Anonymous 1955. Feeding marlin uses its head. *Internat. Oceanogr. Found. Bull.* 1(3): 38–39.
Anonymous 1966. Porpoise Give Piping Tip. *Chemical Week, 98,* 3: 84–85.
Arambourg, C. & Bertin, L. 1958. Sous-classe des Selaciens (Selachii). In: Grasse P.-P., *Traite de Zoologie, 13,* 3: 2016–2056. Masson, Paris.
Arata, G. F. J. 1954. A note on the flying behaviour of certain squids. *Nautilus, 69:* 1–3.
Arzhanikov, N. S. & Maltsev, V. M. 1952. Aerodynamics: 1–480. "Oborongiz" Publishers, M.
Babenko, V. V., Gnitetsky, N. A. & Kozlov, L. F. 1969. Preliminary results of investigations into the elastic properties of the skin of living dolphins. *Bionika,* 3: 12–19.
Babenko, V. V., Gnitetsky, N. A. & Kozlov, L. F. 1970. Investigation into the temperature distribution on the body surface of marine animals. Biocybernetics – Biosystem Modelling – Bionics. IV Ukrainian Republ. Scient. Conference: 126. Press of the Cybernetics Institute, Ukrainian SSR Ac. of Sci., Kiev.
Babenko, V. V., Kozlov, L. F. & Pershin, S. V. 1972. On the variable damping of dolphin skin at different swimming speeds. *Bionika,* 6: 42–52.
Babenko, V. V. & Morozov, D. A. 1968. Some physical regularities in the diving of dolphins. Mechanisms of locomotion and orientation of animals: 97–57. "Naukova Dumka" Publishers, Kiev.
Babenko, V. V. & Surkina, R. M. 1969. Some hydrodynamic features of dolphin swimming. *Bionika,* 3: 19–26.
Backhouse, K. M. 1960. Locomotion and direction finding in whales and dolphins. *The New Scientist, 7:* 26–28.
Backhouse, K. M. 1961. Locomotion of seals with particular reference to the forelimb. *Symp. Zool. Soc. London,* 5: 59–75.
Bainbridge, R. 1958a. The speed of swimming of fish as related to size and to the frequency and amplitude of tail beat. *J. Exp. Biol., 35,* 1: 109–133.
Bainbridge, R. 1958b. The locomotion of fish. *The New Scientist, 4:* 476–478.
Bainbridge, R. 1960. Speed and Stamina in three fish. *J. Exp. Biol., 37,* 1: 129–153.
Bainbridge, R. 1961. Problems of fish locomotion. *Symp. Zool. Soc. London,* 5: 13–32.
Bainbridge, R. 1962. Training, speed and stamina in trout. *J. Exp. Biol., 39,* 3: 537–555.

Bainbridge, R. 1963. Caudal fin and body movement in the propulsion of some fish. *J. Exp. Biol.*, *40*, 1: 23–56.
Bammes, G. 1964. Die Gestalt des Menschen. Hand- und Lehrbuch der Anatomie fur Kunstler: 1–578. VEB Verlag der Kunst, Dresden.
Bandini, R. 1933. Swordfishing. *Calif. Fish and Game*, *19*, 4: 241–248.
Bannikov, A. G. & Denisova, M. N. 1969. The class of amphibians (Amphibia). In: Bannikov A. G. (Ed.), Animal Life, *4*, Part 2, Amphibians, Reptiles: 7–33. "Prosveshchenie" Publishers, M.
Barabash, I. I. 1937. Sea-otters, fur seals, blue foxes: 1–96. Vsesoyuzn. Koop. Objedin. Publishers, M.-L.
Barabash-Nikiforov, I. I. 1933. The Sea otter (Sea beaver). Biological essay: 1–96. "Sovetskaya Azia" Publishers, M.
Barabash-Nikiforov, I. I. 1947. The Sea otter (*Enhydra lutris* L.). its biology and economy. Sea otter: 3–201. Gl. Upr. po Zapovedn. Publishers, M.
Barnard, K. H. 1926. The eggs of the Pilot-fish (*Naucrates ductor*). *Nature* (London), *118*, (2963): 228.
Baron, J. 1950. Contribution a l'etude de l'equilibre chez les poissons. *Bull. Soc. Zool. France*, *75*, 5–6: 247–252.
Barsukov, V. V. 1959a. The wolffish family (Anarhicha-didae) Fauna of the USSR. *Fishes. 5*, 5: 1–171. USSR Ac. of Sci. Press, M.-L.
Barsukov, V. V. 1959b. On the hydrodynamic properties of the caudal fin of Atlantic wolffish (Anarhichadidae). *Papers of the USSR Ac. of Scie.*, *129*, 3: 695–697.
Barsukov, V. V. 1960a. Swimming speeds of fishes. *Priroda*, 3: 103–104.
Barsukov, V. V. 1960b. On the age of the Ob muksun and some theoretical questions. *Zool. J.*, *39*, 10: 1525–1531.
Barsukov, V. V. 1962. Are fishes capable of active flight? *Priroda*, 6: 101–102.
Barsukov, V. V. 1969. Why a fish needs scales. *Priroda*, 4: 74–78.
Barthelmes, D. 1957. Zur Abgrenzung des Planktons von den Nachbarbiocoenosen. *Zschr. Fischerei*, N. E., *6*, 1–7: 441–452.
Bauer, F. 1898. Die Ichthyosaurier des oberen weissen Jura. *Palaeontographica.*, *44*: 283–328.
Beauchamp, P., de 1960. Classe des Chaetognathes (*Chaetognatha*). In: Grassé, P.-P., Traité de Zoologie, *5*, 2: 1500–1520. Masson, Paris.
Bekker, V. Z. 1971. The order Osteoglossiformes. In: "Animal Life", *4*, Fisches: 237–242. "Prosveshchenie" Publishers, M.
Beklemishev, K. V. 1969. The ecology and biogeography of the pelagic zone: 1–291. "Nauka" Publishers, M.
Belkovich, V. M. 1962. Adaptive features in skin cover structure of aquatic mammals. Author's abstract of dissertation for Cand. of Biol. Sci.: 1–20, Zool. Inst., USSR Ac. of Sci., Moscow.
Belkovich, V. M. 1964. The structure of the skin cover of some pinnipeds. Morphological features of aquatic mammals: 5–47. "Nauka" Publishers, M.
Belkovich, V. M. 1965. Specific features of thermoregulation in water (on the example of mammals). Bionika. 215–219 "Nauka" Publishers, M.
Bellairs, A. 1969. The life of reptiles, 1: 1–282. Weidenfeld and Nicolson, London.
Belyayev, V. V. & Koval, A. P. 1972. To the question of the hydrodynamic function of the mucus of some bony fishes. *Bionika*, 6: 78–83.
Berg, A. I. 1965. Bionics and its importance for technological development. Bionika: 3–10. "Nauka" Publishers, M.
Berg, L. S. 1940. The system of ichthyoids and fishes, extant and fossil. *Tr. of Zool. Inst., USSR Ac. of Sci. 5*, 2: 87–517.
Berg, L. S. 1948. Fishes of the fresh waters of the USSR and adjacent countries, 1: 1–466. USSR Ac. of Sci. Press, M.-L.

Berg, L. S. 1949a. Fishes of the fresh water bodies of the USSR and adjacent countries, 2: 467–925. USSR Ac. of Scie. Press, M.-L.
Berg, L. S. 1949b. Fishes of the fresh water bodies of the USSR and adjacent countries, 3: 927–1382. USSR Ac. of Sci. Press, M.-L.
Berg, L. S. 1955. The system of ichthyoids and fishes, extant and fossil (2nd Ed.) *Tr. of Zool. Inst., USSR Ac. of Sci., 20:* 3–286.
Berg, L. S., Kazantseva, A. A. & Obruchev D. V. 1964. The superorder Palaeonisci. In: Principles of paleontology'; Agnatha, Fishes: 336–370. "Nauka" Publishers, M.
Berg, L. S. & Obruchev, D. V. 1964. The order of Ospiida. In: Principles of Paleontology: Agnatha, Fishes: 379–380. "Nauka" Publishers, M.
Berlioz, J. 1950a. Systématique. In: Grassé, P.-P., Traité de Zoologie, *Oiseaux, 15:* 845–1055. Masson, Paris.
Berlioz, J. 1950b. Distribution géographique. In: Grassé, P.-P., Traité de zoologie, *Oiseaux, 15:* 1056–1073.
Berry, S. S. 1920. Light production in cephalopods. *Biol. Bull., Woods Hole, 38:* 141–195.
Bertin, L. 1948. Considérations biogéographique sur les Poissons d'eau douce de Madagascar. *Mém. Inst. Scient. Madag.*, (A), *1,* 2: 169–176.
Bertin, L. 1951. Ichthyogéographie de l'Afrique du Nord. C. R. Seances Soc. Biogeogr.: 79–82.
Bertin, L. 1958a. Modifications des nageoires. In: Grassé, P.-P., *Traité de Zoologie, 13,* 1: 748–782. Masson, Paris.
Bertin, L. 1958b. Squelette appendiculaire. In: Grassé, P.-P., *Traité de Zoologie, 13,* 1: 711–748. Masson, Paris.
Bertin, L. 1958c. Ecailles et sclerifications dermiques. In: Grassé, P.-P., *Traité de Zoologie, 13,* 1: 482–504. Masson, Paris.
Bertin, L. 1958d. Denticules cutanés et dents. In: Grassé, P.-P., *Traité de Zoologie, 13,* 1: 505–531. Masson, Paris.
Bertin, L. 1958e. Form actuelles. In: Grassé, P.-P., *Traité de Zoologie, 13,* 3: 2165–2172. Masson, Paris.
Bertin, L. 1958f. Ecologie. In: Grassé, P.-P., *Traité de Zoologie, 13,* 1: 1885–1933. Masson, Paris.
Bertin, L. 1958g. Peau et Pigmentation. In: Grassé, P.-P., *Traité de Zoologie, 13,* 1: 433–458. Masson, Paris.
Bertin, L, 1958h. Glandes cutanées et organes lumineux. In: Grassé, P.-P., *Traité de Zoologie, 13,* 1: 459–481. Masson, Paris.
Bertin, L. & Arambourg, C. 1958a. Super-ordre des Téléostéens (Teleostei). In: Grassé, P.-P., *Traité de Zoologie, 13,* 3: 2204–2500. Masson, Paris.
Bertin, L. & Arambourg, C. 1958b. Ichthyogéographie. In: Grassé, P.-P., *Traité de Zoologie, 13,* 3: 1944–1966. Masson Paris.
Berzin, A. A. 1971. The Cashalots: 1–367. "Pishchevaya Promyshlennost" Publishers, M.
Bidder, A. M. 1962. Use of the tentacles swimming and buoyancy control in the pearly nautilus. *Nature* (London), *196,* 4853: 451–454.
Bieri, R. 1957. The chaetognath fauna of Peru in 1941. *Pacific Science, 11,* 3: 255–264.
Bigelow, H. B. & Schroeder, W. C. 1948. Fishes of the Western North Atlantic. I. Lancelets, Cyclostomes, Sharks. *Mem. Sears Foundation Mar. Res.*, 1: 1–576. New Haven.
Bigelow, H. B. & Schroeder, W. C. 1953a. Fishes of the Western North Atlantic. 11. Sawfishes, Guiterfisches, Skates and Rays; Chimeroids. *Mem. Sears Foundation Mar. Res.*, 1: 1–588. New Haven.
Bigelow, H. B. & Schroeder, W. C. 1953b. Fishes of the Gulf of Maine. *Fish. Bull. Fish and Wildl. Serv., 53,* 74: 1–577.
Black, V. S. 1948. Changes in density, weight, chloride and swimbladder gas in the killifish, *Fundulus heteroclitus*, in fresh water and sea water. *Biol. Bull., 95:* 83–93.
Blanc, M. 1954. La répatrition des Poissons d'eau douce africains. Bull. I.F.A.N.: 600–628.

Blaxter, J. H. S. 1967. Swimming speeds of fish. FAO Review: FB/67/R/3, FAO conference on fish behaviour in relation to fishing techniques and tactics, Bergen, Norway, 19-27/10/67: 1–32.
Blaxter, J. H. S. 1969. Swimming speeds of fish. *FAO Fisheries Rep., 2,* 62: 69–100.
Blaxter, J. H. S. & Dickson W. 1959. Observations on the swimming speeds of fish. *J. du Conseil, 24,* 3: 472–479.
Bobrinsky, N. A. 1946. Terrestrial fauna. In: Bobrinsky, N. A., Zenkevich, N. A., Birshtein, Ya. A., Zoogeography; 4: 250–450 "Sovetskaya Nauka" Publishers, M.
Bobrinsky, N. A. 1951. Zoogeography: 1–384 USSR Ministry of Education Press, M.
Boddeke, R., Slijper, E. J. & van der Stelt, A. 1959. Histological characteristics of the body musculature of fishes in connection with their mode of life. *Proc. Konink. Nederl. Akad. Wetensch.,* C, *62:* 576–588.
Bohun, S. & Winn, H. E. 1966. Locomotor activity of the American eel (*Anguilla rostrata*). *Chesapeake., Sci.,* 7: 137–147.
Bois-Reymond, R., du 1891. Über die Bewegung der fliegenden Fische. *Zool. Jb., Abt. Syst., Geogr. und Biol., 5,* 5: 822–823.
Bone, Q. & Roberts, B. Z. 1969. The density of elasmobranchs. *J. Mar. Biol. Ass., 49,* 4: 913–937.
Bonthron, R. J. & Fejer, A. A. 1962. A hydrodynamic study of fish locomotion. Depr from Proc. fourth U.S. Nat. congr. of applied mechanics: 1249–1255. Berkeley, California.
Borkhvardt, V. G. 1965. Features of the axial skeleton of aquatic vertebrates associated with their mobility. In: Physiological foundations of the ecology of aquatic animals. Paper abstracts. Sevastopol: 14–16. "Naukova Dumka" Publishers, Kiev.
Boulenger, G. A. 1905. The distribution of African fresh-water Fishes. *Brit. Ass. Adv. Sc., 72:* 413–421.
Bourdelle, E. & Grassé, P.-P. 1955. Ordre des Cetaces. In: Grassé, P.-P., *Traité de Zoologie. 13,* 1: 341–450. Masson, Paris.
Branca, W. 1908a. Sind alle in Inneren von Ichthyosaurus liegenden Jungen ausnahmslos Embryonen? *Abh. Kgl. Preuss. Akad. Wiss.,* Jahrg. 1907: 1–34.
Branca, W. 1908b. Nachtrag zur Embryonenfrage bei Ichthyosarus. *Sitzungsber. Kgl. preuss. Akad. Wiss., Jahrg.* 1908, *18:* 392–396.
Brauer, A. 1906. Die Tiefsee-Fische. I. Systematischer Teil. Deutsche Tiefsee-Expedition 1898–1899, *15,* 1: 1–432. Fischer, Jena.
Brauer, A. 1908. Die Tiefsee-Fische. II. Anatomischer Teil. Deutsche Tiefsee-Expedition 1898–1899, *15,* 2: 1–266. Fischer, Jena.
Breder, C. M. 1924. Respiration as a factor in locomotion of fishes. *Amer. Nat.,* 58: 145–155.
Breder, C. M. 1926. The locomotion of fishes. *Zoologica* (USA), *4,* 5: 159–297.
Breder, C. M. 1929. Field observations on flying fishes: a suggestion of methods. *Zoologica* (USA), *9:* 295–312.
Breder, C. M. 1930. On the structural specialization of flying fishes from the standpoint of aerodynamics. *Copeia,* 4: 114–121.
Breder, C. M. 1934. The oceanographic vessel "Atlantis" in the West Indies. *Bull. N.Y. Zool. Soc., 37:* 30–39.
Breder, C. M. 1937. The perennial flying fish Controversy. *Science, 86,* 2236: 420–422.
Breder, C. M. 1965. Vortices and fish schools. *Zoologica* (USA), *50,* 2: 97–114.
Brooks, W. S. 1917. Notes on some Falkland Islands birds. *Cambridge Mass. Bull. Mus. Comp. Zool., 61:* 135–160.
Brown, S. G. 1960. Sverdfisk og hval. *Norsk hvalfangst-tid., 49,* 10: 457.
Bruun, A. F. 1935. Flying-fishes (Exocoetidae) of the Atlantic. *Dana-Rep.,* Copenhague. *6:* 1.106
Budker, P. 1938. Les cryptes sensorielles et les denticules cutanés des Plagiostomes. *Ann. Inst. Ocean.,* Paris, *18:* 207–288.

Budylenko, G. A. 1973. "Attacks" by sword-fishes on whales. *Priroda*, 8: 101–102.
Bulatova, N. N. & Korzhuyev, P. A. 1952. Erythrocytes and hemoglobin of the Black Sea bonito. Tr. of Inst. of Animal Morphology, USSR Ac. of Sci., 6, 165–167.
Burdak, V. D. 1957. Peculiarities of the ontogenetic development and phylogenetic relationships of Black Sea mullets (*Mugil saliens* Risso, *Mugil auratus* Risso and *Mugil cephalus* Linne). *Tr. Sevastop. Biol. St., 9:* 243–273.
Burdak, V. D. 1959. On the pelagization of the whiting (*Odontogadus merlangus* L.) in the Black Sea. *Tr. Sevastop. Biol. St., 12:* 338–344.
Burdak, V. D. 1964. The biology of the Black Sea whiting (*Odontogadus merlangus euxinus* (Nordmann). *Tr. Sevastop. Biol. St., 15:* 196–278.
Burdak, V. D. 1966. On changes in the growth rate of the Black Sea mullet in a historical time. Papers of the USSR Ac. of Sci., 167, *5:* 1156–1158.
Burdak, V. D. 1968. On the functional significance of ctenoids on fish scales. *Zool. J., 47:* 5: 732–738.
Burdak, V. D. 1969a. The ontogenetic development of the scale cover of the mullet *Mugil saliens* Risso. *Zool. J., 48,* 2: 242–248.
Burdak, V. D. 1969b. On structural changes in the microrelief of the scaly cover of the surmullet (*Mullus barbatus ponticus* Essipov) with age. Functional-morphological research into nektonic animals: 46–52. "Naukova Dumka" Publishers, Kiev.
Burdak, V. D. 1969c. On the functioning of the ctenoid apparatus of fishes in conditions of turbulent boundary layer. *Zool. J., 48,* 7: 1053–1055.
Burdak, V. D. 1970. On the ratios of the hydrodynamic functions of cycloid and ctenoid scale in fishes. *Zool. J., 49,* 6: 869–871.
Burdak, V. D. 1972. On the hydrodynamic function of cycloid scales in fishes. *Zool. J., 51,* 7: 1086–1089.
Burdak, V. D. 1973a. On the hydrodynamic function of the scale cover of fishes. IV. All-Union Conf. on Bionics, *6,* Biomekhanika: 31–34. M.
Burdak, V. D. 1973b. Scale types as stages in the historical development of the hydrodynamic function of the skin cover in fishes. *Zool. J., 52,* 8: 1208–1213.
Burdak, V. D. 1973c. Simply scales. *Chemistry and Life. 10:* 30–31.
Burdak, V. D. 1974. Flow troughs on fish scale. Vopr. Ikhtiol, 14, 3 (86): 520–521.
Burfield, S. T. 1927. Sagitta. *Liverpool Mar. Biol. Comm. Mem., 28:* 1–104.
Butschli, O. 1910. Vorlesungen über vergleichende Anatomie: 1–644. Engelmann, Leipzig.
Bystrov, A. P. 1949. Phlebolepis elegans Pander. Papers of the USSR Ac. of Sci., *64,* 2: 245–247.
Calderwood, W. L. 1931. Notes on flying fish. *Scottish Nat.*, 190: 121–123.
Carpenter, G. C. 1966. The marine iguana of the Galapagos Islands, its behaviour and ecology. *Proc. Californ. Acad. Sci., 34,* 4: 329–376.
Carter, G. S. 1945. The flight of flying-fishes. *Endeavour.* London, *4,* 16: 136–140.
Carter, J. T. 1919. On the occurence of denticles on the snout of *Xiphias Gladius. Proc. Zool. Soc.*, London: 321–326.
Chabanaud, P. 1943. Le frein de la thoracopterygie et les caracteres adaptifs poissons de l'order des Scombroidea. *Bull. Soc. Zool. France, 68:* 110–113.
Chabanaud, P. 1944. La nageoire caudale du hareng. *C. R. Acad. Sci. Paris*, 218: 523–525.
Chaikovskaya, A. V. & Sedykh, T. G. 1972. On the aminoacid content of the mucous substances of fish skin. Summary of papers, V Ukr. Republ. Conf. on Bionics: 43. A. S. Popov Technico-Scientific Radioengineering and Electronics Society, Kiev.
Chapsky, K. K. 1963. The order Pinnipedia. Mammals of the USSR fauna, 2: 895–964. USSR Ac. of Sci. Press, M.-L.
Chepurnov, A. V. 1968. Body form of some cetaceans related to their swimming speeds. Locomotor and orientation mechanisms in animals: 72–77. "Naukova Dumka" Publishers, Kiev.

Chestnoi, V. N. 1961. Maximum swimming speeds in fishes. *Rybn. Khoz-vo.*, 9: 22–27.
Chevrel, R. 1913. Essai sur la morphologie et la physiologie du muscle lateral chez poissons osseux. *Arch. Zool. exp. et gen.*, *52:* 473–607.
Chudinov, P. K. 1964a. Superfamily Varanoidea. In: Principles of Paleontology; Amphibians, Reptiles and Birds: 473. "Nauka" Publishers, M.
Chudinov, P. K. 1964b. The family Aigialosauridae Kramberger, 1892. In: Principles of Paleont. Amphibians, Reptiles and Birds: 474–475. "Nauka" Publishers, M.
Chudinov, P. K. 1964c. The family Dolichosauridae Gervais, 1852. In: Principles of Paleont. Amphibians, Reptiles and Birds: 481. "Nauka" Publishers, M.
Chudinov, P. K. 1964d. The suborder Cholophidia. In: Principles of Paleont. Amphibians, Reptiles and Birds: 181–482. "Nauka" Publishers, M.
Chun, C. 1900. Aus den Tiefen des Weltmeeres: 1–592. Fischer, Jena.
Chun, C. 1903. Über Leuchtorgane und Augen vor Tiefsee Cephalopoden. *Verhandl. Dtsch. Zool. Ges.*, *13:* 67–91.
Chun, C. 1910. Die Cephalopoden. *Wiss. Ergeb. Deutsch. Tiefsee Exped.* ("*Valdivia*"), *18:* 1–552.
Chun, C. 1913. Cephalopoda from the "Michael Sars" North Atlantic Deep-Sea Expedition, 1910. Rep. Scient. Results Michael Sars N. Atlant. deep Sea Exped., *3:* 1–28.
Clarke, M. R. 1962. Respiratory and swimming movements in the Cephalopod *Cranchia acabra. Nature* (London), *196*, 4852: 351–352.
Clarke, M. R. 1966. A review of the systematics and ecology of oceanic squids. Advances Marine Biol., *4:* 91–300. Academic Press (ed. Russel, F. S.), London and New York.
Conrad, G. M. & La-Monte, F. 1937. Observations on the body form of the blue marlin (*Makaira nigricans ampla* Poey). *Bull. Amer. Mus. Nat. Hist.*, *74*, 4: 207–220.
Cott, Ch. 1950. Adaptive colouring of animals. 1–534. Foreign Literature Publishers, M.
Cutchen, C. W. M. 1970. The trout tail fin: a self-cambering hydrofoil. *J. Biomech.*, *3*, 3: 271–281.
Dabelow, A. 1925. Die Schwimmanpassung der Vogel. Ein Beitrag zur biologischen Anatomie der Fortbewegung. *Gegenbaurs Morphologisches Jahrbuch eine Zschr. für Anat. und Entwicklungsgesch.*, *54:* 288–321.
Daget, J. 1948. Les Synodontis (Siluridae) a polarité pigmentaire inverseé. *Bull. Mus. Hist. Nat. Paris.* (2), *20:* 239–243.
Daget, J. 1952. Revision des affinités phylogénétiques des Polyptéridés. *Mem. Inst. Fr. Afr. Noire*, Dakar, 11: 1–178.
Dahl, F. 1892a. Die Bewegung der fliegenden Fische durch die Luft. *Zool. Ib., Abt. Syst., Georgr. und Biol.*, *5*, 4: 679–688.
Dahl, F. 1892b. Zur Frage der Bewegung fliegenden Fische. *Zool. Anz.*, 386: 106–108.
Dammerman, K. W. 1924. On Globicephala and some other Delphinidae from the Indo-Australian Archipelago. *Treubia*, *5:* 340–352.
Danilchenko, P. G. 1964a. Superorder Holostei. Bony ganois (Protospondyli-Halecostomi, Ppars). In: Principles of paleontology; Agnatha, Fishes: 378, 380–392. "Nauka" Publishers, M.
Danilchenko, P. G. 1964b. Superorder Teleostei. Bony fishes. In: Principles of paleontology, Agnatha, Fishes: 396–484. "Nauka" Publishers, M.
Danilchenko, P. G. & Yakovlev, V. N. 1964. The order Pholidophorida. In: Principles of paleontology; Agnatha, Fishes: 392–395. "Nauka" Publishers, M.
David, P. M. 1955. The distribution of *Sagitta gazellae* Ritter-Zahony. *Discovery Rep.*, *27:* 235–278.
David, P. M. 1958. The distribution of the Chaetognatha of the southern ocean. *Discovery Rep.*, *29:* 199–228.
Davis, C. C. 1955. The marine and fresh-water plankton: 1–562. Michigan State Univ. Press.

Davitashvili, L. Sh. 1949. Course of paleontology. 1–833. Geological Literature Publishers, M.-L.
Day, F. 1885. Relationship of the Indian and African fresh-water fish-faunas. *J. Linn. Soc.* (*Zool.*), London, *18:* 308–317.
Dementyev, G. P. 1964. The class Aves. Birds. In: Principles of Paleontology. Amphibians, Reptiles and Birds: 660–699. "Science" Publishers, M.
Denton, E. 1961. The buoyancy of fish and cephalopods. *Progr. Biophys.*, *2:* 179–236.
Denton, E. J. & Gilpin-Brown, J. B. 1959. Buoyancy of the Cuttlefish. *Nature* (London), *184,* 4695: 1330–1331.
Denton, E. & Gilpin-Brown, J. 1960. Daily changes in the cuttlefish. *Proc. physiol. Soc.*, *151:* 36–37.
Denton, E. J. & Gilpin-Brown, J. B. 1961a. The buoyancy of the cuttlefish, *Sepia officinalis* (L.). *J. Mar. Biol. Ass.* U.K., *41:* 319–342.
Denton, E. J. & Gilpin-Brown, J. B. 1961b. The effect of light on the buoyancy of the cuttlefish. *J. Mar. Biol. Ass. U.K.*, *41:* 343–350.
Denton, E. J. & Gilpin-Brown, J. B. 1961c. The distribution of gas and liquid within the cuttlebone. *J. Mar. Biol. Ass. U.K.*, *41:* 365–381.
Denton, E. J. & Gilpin-Brown, J. B. 1966. On the buoyancy of the pearly Nautilus. *J. Mar. Biol. Ass. U.K.*, *46:* 723–759.
Denton, E. J., Gilpin-Brown, J. B. & Howarth, J. V. 1961. The osmotic mechanism of the cuttlebone. *J. Mar. Biol. Ass. U.K.*, *41:* 351–363.
Denton, E. & Marshall, N. 1958. The buoyancy of bathypelagic fishes without a gas-filled swimbladder. *J. Mar. Biol. Ass. U.K.*, *37,* 3: 753–767.
Denton, E. J., Shaw, T. I. & Gilpin-Brown, J. B. 1958. Bathyscaphoid squid. *Nature* (London), *182,* 4652: 1080–1081.
Desbrosses, P. 1945. Le merlan (*Gadus merlangus* L.) de la côte francaise de l'Atlantique. *Rev. Trav. Peches Maritimes, 13,* 1–4: 177–195.
Devnin, S. I. 1967. Aerodynamic analysis of poorly stream-lined ship structures: 1–223. "Sudostroeniye" Publishers, L.
Dexler, H. & Freund, L. 1906. Zur Biologie und Morphologie von Halicore dugong. Arch. Naturgesch., 72. Jahrg., *1,* 2: 77–106.
Dijkgraaf, S. 1962. The function and significance of the lateral line organs. *Biol. Rev.*, *38:* 51–105.
Disler, N. N. 1960. The sense organs of the lateral line system and their importance in fish behaviour: 1–310. USSR Ac. of Sci. Press, M.
Douxchamps, P. 1969. Les pinnipedes. *Naturalistes belg. 50:* 2–21.
Dragunov, A. M. 1950. The weight and chemical composition of the Black Sea mackerel. *Tr. of AzCherNIRO*, 14: 143–150.
Drozdov, N. N. 1969. The family of sea snakes (Hydrophidae) In: Bannikov, A. G. (Ed.), Life of animals, *4,* Part 2, Amphibians, reptiles: 404–409. "Prosveshchenie" publishers, M.
Dugès, A. 1905. Rôle des nageoires chez les poissons. *Bull. Soc. Zool. France*, *30:* 107–110.
Edgerton, H. E. & Breder, C. M. 1941. High speed photographs of flying fishes in flight Zoologica (USA), *24,* 4, 30: 311–314.
Ege, V. 1918. Stomiatidae (Stomias). *Rep. Dan. oceanogr. exp.* 1908–1910, *2. Biol.*, A, 4: 1–28.
Ege, V. 1931. Sudidae (Paralepis). *Rep. Dan. oceanogr. exp.* 1908–1910, *2, Biol.*, A. 13: 1–193.
Ekman, S. 1953. Zoogeography of the sea: 1–417. Sidgwick and Jackson, London.
Essapian, F. S. 1955. Speed-induced skin folds in the bottle-nosed porpoise, *Tursiops truncatus. Breviora Mus. Comp. Zool.*, *43:* 1–4.
Evans, H. M. & Damant, G. C. C. 1928. Observation on the physiology of the swimbladder in cyprinoid fishes. *J. Exp. Biol.*, *6:* 42–55.
Evans, W. E. & Prescott, J. H. 1962. Observations on the sound production capabilities of the bottlenose porpoise: a study of whistles and clicks. *Zoologica* (USA), *47* 3: 121–128.

Evermann, B. W. & Shaw, T. 1927. Fishes from Eastern China with descriptions of new species. *Proc. Calif. Acad. Sci.* San Francisco, *16:* 97–122.
Fage, L. 1920. Engraulidae, Clupeidae. *Rep. Dan. Oceanogr. exp.* 1908–1910, *2, Biol.*, A. 9: 1–140.
Farrington, S. K. 1937. Atlantic game fishing: 1–298. Kennedy Bras., N. Y.
Farrington, S. K. 1942. Pacific game fishing: 1–290. Coward-McCann, N. Y.
Fawcett, D. W. 1942. A comparative study of blood vascular bundles in the Florida manatee (*Trichechus latirostris*) and certain Cetaceans and Edentates. *J. Morphol.*, *75,* 1: 105–117.
Fay F. H. 1960. Structure and function of the pharyngeal pouches of the walrus (*Odobenus rosmarus* L.). *Mammalia* (Paris), *24,* 3: 361–371.
Fejer, A. A. & Backus, R. H. 1960. Porpoises and the bow-riding of ships under way. *Nature* (London), *188,* 4752: 700–703.
Felts, W. J. L. 1966. Some functional and structural characteristics of cetacean flippers and flukes. In: Norris, K. S. (Ed.). Whales, dolphins and porpoises: 255–276. Univ. of California Press. Berkeley-Los Angeles.
Fierstine, H. L. 1966. Studies in locomotion and anatomy of Scombroid fishes. *Diss. Abstr.*, *26:* 4124.
Fierstine, H. L. & Walters, V. 1968. Studies in locomotion and anatomy of scombroid fishes. Mem. South. Calif. Acad. Sci., *6.*
Fish, M. P. 1954. The character and significance of sound production among fishes of the Western North Atlantic. *Bull. Bingham. Oceanogr. Coll.*, *14:* 1–109.
Fish, M. P. & Mowbray, W. H. 1962. Production of underwater sound by the white whale or beluga, *Delphinapterus leucas* (Pallas). *J. Mar. Res.*, *20:* 149–162.
Fish, M. P. & Mowbray, W. H. 1970. Sounds of western north Atlantic fishes: 1–207. Hopkins Press, Baltimore and London.
Fitch, J. E. 1958. Offshore fishes of California: 1–80. Depart. of fish and game. California.
Focke, H. 1965. Über die Ursachen der hohen Schwimmgeschwindigkeiten der delphine. *Zschr. Flugwiss.*, *13,* 2: 54–61.
Foerste, A. F. 1928. American Arctic and related cephalopods. *J. Sci. Lab. Denison U.*, *23:* 1–110.
Fontaine, M. 1958. Formes actuelles super-ordres des Petromyzonoidea et des Myxinoidea. Muscles. In: Grassé P.-P., *Traité de Zoologie, 13,* 1: 39–43. Masson, Paris.
Ford, E. 1930. Herring investigations at Plymouth. 8. The transition from larva to adolescent. *J. Mar. Biol. Ass.*, *16,* 3: 723–752.
Fowler, H. W. 1928. The fishes of Oceania. *Mem. Bishop Mus.*, 10:
Fowler, H. W. 1936. The Marine Fishes of West Africa. *Bull. Am. Mus. Nat. Hist.*, *70,* 1–2: 1–1496.
Fowler, H. W. 1938. A list of the fishes known from Malaga. *Fisher. Bull.*, Singapore, 1: 1–268.
Fraas, E. 1891. Die Ichthyosaurier der suddeutschen Triasund Juraablagerungen: 1–34. Tubingen.
Fraas, E. 1902. Die Meer-Crocodilier (*Thalattosuchia*) des oberen Jura unter spezieller Berucksichtigung von Dacosaurus und Geosaurus. *Palaeontographica, 49:* 52–54.
Fraas, E. 1910. Plesiosaurier aus dem oberen Lias von Holzmaden. *Palaeontographica, 57:* 105–140.
Fraas, E. 1911. Embryonaler Ichthyosaurus mit Hautbekleidung. Jahreshefts Ver. vaterl. Naturk. Wurttemberg: 480–490.
François, Y. 1955. Croissance larvaire et migration de la nageoire dorsale chez un clupeidé. *Bull. Soc. Zool. France, 80,* 2–3: 105–106.
Fraser, F. C. 1959. Some aquatic adaptations of shales and Dolphins. *Proc. roy. Inst. Gr. Brit.*, *37:* 319–333.
Fraser-Brunner, A. 1950. The Fishes of the Family Scombridae. *Ann. Mag. Nat. Hist.*, ser. 12, *3,* 26: 131–163.

Frechkop, S. 1955. Ordre des Pinnipedes. In: Grassé P.-P., *Traité de Zoologié, 17,* 1: 292–340. Masson, Paris.
Furnestin, M. -L. 1957. Chaetognathes et zooplancton du sector Atlantique Marocain. *Rev. trav. Inst. pêches marit., 21,* 1–2.
Furnestin, M.-L. 1959. Campagne de la "Calypso": Golf de Guinée. 8. Chaetognathes. Result. Sci. Campagn. "Calypso", 4: 219–233.
Furnestin, M.-L. 1962a. Péches planctoniques, superficielles et profondes, en mediterranee occidentale. 8. Chaetognathes. *Rev. Trav. Inst. Péches Marit., 26,* 3: 357–368.
Furnestin, M.-L. 1962b. Chaetognathes des côtes africaines Exped. Oceanogr. Eaux Cotieres Africaines de l'Atlantique Sud. 1948–49, *Res. Sci., 3,* 9: 1–54.
Gadd, G. E. 1963. The hydrodynamics of swimming. *New Scientist,* 19 (355): 483–485.
Gambarjan, P. P. & Karapetjan, W. S. 1961. Besonderheiten im Rau des Seelowen (*Eumetopias californianus*), der Baikal robbe (*Phoca sibirica*) und des Seeotters (*Enhydra lutris*) in Anpassung an die Fortbewegung im Wasser. *Zool. Jb.* (*Anat.*), *79:* 123–148.
Garnett, C. S. 1929. Some notes and observations on the flight of flying-fishes. *Proc. Zool. Soc. London,* 1: 39–41.
Gawn, R. W. L. 1948. Aspects of the Locomotion of Whales. *Nature* (London), *161,* 4080: 44–46.
Gerasimov, V. N. & Droblenkov, V. F. 1962. Submarines of imperialist states: 1–302. USSR Ministry of Defence Publishing House, M.
Gero, D. R. 1952. The hydrodynamic aspects of fish propulsion. *Amer. Mus. Novitat.,* 1601: 1–32.
Ghirardelli, E. 1952. Osservazioni biologiche e sistematiche sui Chetognati del Golfo di Napoli. *Publ. Staz. Zool. Napoli, 23:* 296–312.
Ghirardelli, E. 1968. Some aspects of the biology of the Chaetognaths. *Adv. mar. biol.,* 6: 271–376. Acad. Press., London and N.Y.
Gilchrist, J. D. F. 1918. The eggs and spawning habits of the Pilot fish. *Ann. Mag. Nat. Hist., 2,* 9: 114–118.
Gill, E. L. 1935. Flying Fishes. *Nature* (London), *136,* 3438: 478–479.
Gill, T. 1905. Flying fishes and their habits. *Ann. Rep. Smiths. Inst.*: 495–515.
Glikman, L. S. 1964. The subclass Elasmobranchii. Sharks and rays. In: Principles of paleontology; Agnatha, Fishes. 196–237. "Nauka" Publishers, M.
Gnadeberg, W. 1926. Untersuhungen über den Bau der Placoidschuppen der Selachier. *Jen. Zschr. Med. Naturw.,* Jena, *42:* 473–500.
Goodrich, E. S. 1907. On the scales of Fish, living and extinct, and their importance in classification. *Proc. Zool. Soc. London*: 751–774.
Gordon, M. 1935. Swordfish Lore. *Nat. Hist., 36:*
Gordon, M. S. 1961. Wave-Riding Dolphins. *Science, 133,* 3447: 204–205.
Gosline, W. A. & Brock V. E. 1960. Handbook of Hawaiian fishes: 1–372. Univ. Hawaii Press, Honolulu.
Graham-Smith, W. B. A. 1936. The tail of Fishes. *Proc. Zool. Soc. London,* 3: 595–608.
Grassé, P.-P. 1955. Ordre des fissipedes. In: Grassé P.-P., *Traité de Zoologie, 17:* 194–291. Masson, Paris.
Gray, J. 1930. The mechanism of animal movements, with special reference to fish. *Rep. Brit. Assoc. Adv. Sci.* London, *102:* 334–335.
Gray, J. 1933a. The muscular movements of fish. *Proc. Roy. Inst. Gt. Britain, 27,* 5, 131: 849–874.
Gray, J. 1933b. Muscular movement of fish. *Nature* (London), *131,* 3319: 825–828.
Gray, J. 1933c. Studies in animal locomotion. I. The movement of fish with special reference to the eel. *J. Exp. Biol., 10,* 1: 88–104.
Gray, J. 1933d. Studies in animal locomotion. II. The relationship between waves of muscular contraction and the propulsive mechanism of the eel. *J. Exp. Biol., 10,* 4: 386–390.

Gray, J. 1933e. Studies in animal locomotion. III. The propulsive mechanism of the whiting (*Gadus merlangus*), *J. Exp. Biol.*, *10*, 4: 391–400.
Gray, J. 1933f. Direction control of fish movement. *Proc. Roy. Soc. London*, ser. B, *113:* 115–125.
Gray, J. 1936a. Studies in animal locomotion. IV. The neuromuscular mechanism of swimming in the eel. *J. Exp. Biol.*, 13, 2: 170–180.
Gray, J. 1936b. Studies in animal locomotion. VI. Propulsive powers of dolphin. *J. Exp. Biol.*, *13:* 192–208.
Gray, J. 1937. Pseudo-rheotropism in fishes. *J. Exp. biol.*, *14*, 1: 95–103.
Gray, J. 1949. Aquatic locomotion. *Nature* (London), 164: 1073–1075.
Gray, J. 1953a. How animals move: 1–114. Cambridge Univ. Press. Cambridge.
Gray, J. 1953b. The locomotion of fishes. Essays in marine biology being the Richard Elmhirst. Memorial Lectures: 1–16. Oliver and Boyd, Edinburgh-London.
Gray, J. 1957. How fishes swim. *Scient. American*, *197*, 2: 48–54.
Gray, J. 1968, Animal locomotion: 1–479. Weidenfeld and Nicolson, London.
Gray, J. & Parry, D. A. 1948. Aspects of the locomotion of whales. *Nature* (London), *161*, 4084: 199–200.
Greene, C. W. 1913. An undescribed longitudinal differentiation of the great lateral muscle of the king salmon. *Anat. Rec.* *7:* 99–101.
Greenway, P. 1965. Body Form and behavioural types in fish. *Experientia*, *21*, 9: 489–498.
Greenwood, P. H., Rosen, D. E., Weitzman, S. H. & Myers, G. S. 1966. Phyletic studies of Teleostean fishes, with a provisional classification of living forms. *Bull. Amer. Mus. Nat. Hist.*, *131*, 4: 341–355.
Greenwood, P. H. & Thompson, K. S. 1960. The pectoral anatomy of *Pantodon buchhlolzi* Peters (a freshwater flying fish) and the related osteoglossidae. *Proc. Zool. Soc.* London, *135*, 2: 283–301.
Gregory, W. K. 1928. Studies on the body-forms of fishes. *Zoologica* (USA), *8*, 6: 325–421.
Gregory, W. K. & Conrad, G. M. 1937. The comparative osteology of the swordfish (*Xiphias*) and the sailfish (*Istiophorus*). *Amer. Mus. Novitat.*, 925: 1–25.
Gregory, W. K. & La Monte, F. 1947. The world of fishes: 1–96. Publ. Amer. Mus. Nat. Hist., N.Y.
Grey, Z. 1926. Tales of the Angler's Eldorado – New Zealand: 1–228. Kennedy Bras., N.Y.
Grey, Z. 1928. Big game fishing in New Zealand seas. *Natur. Hist.*, *28*, 1: 46–52.
Griffin, D. R. 1950. Underwater sound and the orientation of marine animals, a preliminary survey. Techn. Rep., 3, Proj. N.R. 162–429, t.o. 9, between O.N.R. and Cornell Univ.: 1–26.
Grinberg, M. M. 1950. On the dependence of scale size in bony fishes on body shape and nature of movement. *Zool. J.*, *29*, 5: 446–459.
Gronningsaeter, A. 1946. Sjoormen-blekkspruten. *Naturen*, *70:* 379–380.
Gross, W. 1947. Die Agnathen und Acanthodier des obersilurischen Beyrichienkalks. *Palaeontographica*, A, *96:* 91–158.
Grove, A. J. & Newell, G. E. 1936. A mechanical investigation into an effectual action of the caudial fin of some aquatic chordates. *Ann. Mag. Nat. Hist.*, *17*, 98: 280–290.
Grove, A. J., & Newell, G. E. 1939. The relation of the tail-form in cyclostomes and fishes to specific gravity. *Ann. Mag. Nat. Hist.*, ser. 11, 4: 401–430.
Grushanskaya, J. Ya. & Korotkin, A. I. 1973. Some questions of the hydrodynamics of the dolphin. IV All-Union Bionics Conf., *6*, Biomekhanika: 37–41. M.
Gudger, E. W. 1940. The alleged pugnacity of the swordfish and the spearfishes as shown by their attacks on vessels. *Mem. Roy. Asiatic Soc.*, Bengal Branch, *12*, 2: 215–315.
Günther, A. 1877. Report on the deep-sea fishes collected by H.M.S. "Challenger" during the years 1873–1876. Zool. "Challenger" Exped. Rep., 22, 57: 1–286. London.
Günther, A. C. L. 1880. An introduction to the study of fishes: 1–720. Adam and Charles Black, Edinburgh.

Günther, A. 1920. Freshwater Fishes from Madagascar, *Ann. Mag. Nat. Hist.*, (9), *5:* 419–424.
Haeckel, E. 1890. Plankton-Studien: 1–105. Fischer, Jena.
Hankin, E. H. 1920. Observations on the flight of flying-fishes. *Proc. Zool. Soc. London*: 467–474.
Hans, A. 1960. Different types of body movement in this hagfish, *Myxine glutinosa* L., *Nature* (London), *188,* 4750: 595–596.
Hardenberg, L. D. F. 1941. Fishes of N. Guinea. *Treubia Buitenzorg,* 18: 217–231.
Harris, J. E. 1934. The swimming movements of fishes. *Ann. Rept. Tortugas Lab., Carnegie Inst.*, 1933–1934: 251–253.
Harris, J. E. 1936. The role of the fins in the equilibrium of the swimming fish. I. Wind-tunnel test on a model of *Mustelus canis* (Mitchell). *J. Exp. Biol., 13,* 4: 476–493.
Harris, J. E. 1937a. The mechanical significance of the position and movements of the paired fins in the Teleostei. *Pap. Tortugas Lab., Carnegie Inst., 31:* 173–189.
Harris, J. E., 1937b. The role of fin movements in the equilibrium of the fish. *Ann. Rep. Tortugas Lab., Carnegie Inst.* 1936–1937: 91–93.
Harris, J. E. 1938. The role of the fins in the equilibrium of the swimming fish. II. The role of the pelvic fish. *J. Exp. Biol., 15,* 1: 32–47.
Harris, J. E. 1947. The function of fins in fishes. *Challenger Soc.*, London, *2,* 19: 16–26.
Harris, J. E. 1953. Fin patterns and mode of life in fishes. Essays in marine biology being the Richard Elmhirst Memorial Lectures: 17–28. Oliver and Boyd, Edinburgh-London.
Harrison, R. J. & King, J. E. 1965. Marine mammals: 1–192. Hutchinson and Co., London.
Hayes, W. D. 1953. Wave riding of Dolphins. *Nature* (London), *172,* 4388: 1060.
Heilner, V. C. 1953. Salt Water Fishing: 1–330. Knopf. N.Y.
Hensen, V. 1887. Ueber die Bestimmung des Planktons oder des im Neere treibenden Materials an Pflanzen und Tieren. 5. Ber. Kommiss. wissensch. Unters. deutsch. Meere, Kiel. 1882–1886, 12–16: 1–108.
Herald, E. S. 1962. Living Fishes of the World: 1–304. Doubleday, N.Y.
Herre, A. W. 1953. Check list of Philippine fishes. *Res. U.S. Fish and Wildl. Serv., 20:* 1–977.
Hertel, H. 1963. Structur, Form, Bewegung (Biologic and Technik): 1–244. Krausskopf-Verlag, Mainz.
Hertel, H. 1966. Structure, Form and Movement: 1-250. Reinhold Publ. Corp., N.Y.
Hertel, H. 1967a. Schwigende Antribe und Flatterschwingungen in Nature und Technik. *VDI-Zschr.*, 109, 24: 1133–1138.
Hertel, H. 1967b. Gekoppelte Biege- und Rehschwingungen als Antrieb. *VDI-Zschr., 109,* 26: 1215–1221.
Hertel, H. 1967c. Biologish-Technische Forschungen uber stromungstechnisch optimale Formen. I. *VDI-Zschr., 109,* 19: 837–840.
Hertel, H. 1967d. Biologish-technische Forschungen uber stromungstechnisch optimale Formen. 2. *VDI-Zschr., 109,* 22: 1051–1055.
Hertwig, O. 1874. Ueber Bau und Entwickelung der Placoid-schuppen und der Zähne der Selachier. *Jen Zschr. Med. Naturw.*, Jena, *8:* 331–404.
Hertwig, O. 1876, 1879, 1882. Ueber das Hautskelet der Fische. *Morph. Jb., 2:* 328–395, *5;* 1–21, *7:* 1–42.
Heumann, G., Jacobsohn, J. & Vilter, V. 1941. Inversion pigmentaire chez *Synodontis batensoda* Rupp., Poisson nageant sur le dos. (Etude microscopique). *Compt. Rend. Soc. Biol. France, 135:* 361–363.
Heyerdahl, Th. 1962. Journey on Kon-Tiki: 1–530. "Molodaya Cvardia" Publishers, M.
Hildebrand, S. F. 1946. A descriptive catalogue of shore fishes of Peru. *Bull. U.S. nat. mus.*, 189: 1–530.
Hill, A. V. 1950. The dimensions of animals and their muscular dynamics. *Sci. Progress* (London), *38,* 50: 209–230.

Hoedeman, J. J. 1952. Surinam Representatives of *Gasterophelecus* and *Carnegiella*. *Beaufotia* (Amsterdam), 20: 1–16.
Hofer, B. 1890. Ueber den Bau und die Entwicklung der Cycloid- und Ctenoidschuppen. *Sitzb. Ges. Morph. Physiol. Munchen, 6:* 103–118.
Holmgren, U. 1958. On the pineal organ of the tuna, *Thunnus thynnus* L., *Breviora,* 100: 1–5.
Hora, S. L. 1935. Ancient Hindu Conception of Correlation between Form and Locomotion of Fishes. *J. Asiat. Soc. Bengal Sci., 1,* 1: 1–7.
Hora, S. L. 1937. Geographical distribution of Indian fresh-water fishes and its bearing on the probable land connections between India and adjacent countries. *Curr. Sc.*: 351–356.
Houghton, G. 1964. Simulation of fluttering lift in a bird, locust, moth, fly and bee. *Nature* (London), *202,* 4938: 1183–1185.
Houssay, F. 1909a. Sur les conditions hydrodynamique de la forme chez les poissons. *Compt. Rend. Acad. Sci. Paris, 138:* 1076–1078.
Houssay, F. 1909b. Nouvelles expériences sur la forme et la stabilité des poissons. *Rev. gen. Sci. pures et appliquées. 20:* 943–948.
Houssay, F. 1912. Forme, Puissance et Stabilité des Poissons. *Collect. morphol. dinamique, 4:* 1–372. Hermann, Paris.
Howell, A. B. 1927. Contribution to the anatomy of the Chinese Finless Porpoise, *Neomeris phocoenoides, Proc. U.S. Nat. Mus., 70,* 13: 1–43.
Howell, A. B. 1929. Contribution to the comparative anatomy of the eared and earless seals (genera Zapophus and Phoca). *Proc. U.S. Nat. Mus., 73,* 15: 1–142.
Howell, A. B. 1930. Aquatic mammals. Their adaptations to life in the water. I–XII: 1–338. Thomas, Springfield-Baltimore.
Hubbs, C. L. 1933. Observations on the flight of fishes. *Pap. Mich. Ac. Sc. Artz Sett., 17:* 575–611.
Hubbs, C. L. 1935. Nature's own seaplanes. *Ann. Rep. Smiths. Inst.*: 333–348.
Hubbs, C. L. 1936. Further observations and statistics on the flight of fishes. *Pap. Mich. Ac. Sc. Arts Lett., 22:* 641–660.
Hubbs, C. L. & Wisner, R. L. 1953. Food of marlin in 1951 off San Diego, California. *Fish and Game, 39,* 1: 127–131.
Huene, F. 1916. Beitrag zur Kentnis der Ichthyosaurier im deutschen Muschelkalk. *Palaeontographica, 62:* 1–68.
Huene, F. 1923. Lines of phyletic and biological development of the Ichthyopterygia. *Bull. Geol. Soc. America, 34:* 463–468.
Huene, F. R. F. 1956. Palaontologie und Phylogenie der Niederen Tetrapoden: 1–176. Fischer, Jena.
Hughes, G. M. 1960a. The mechanism of gill ventilation in the dogfish and skate. *J. Exp. Biol., 37,* 1: 11–27.
Hughes, G. M. 1960b. A comparative study of gill ventilation in marine teleost. *J. Exp. Biol., 37,* 1: 28–45.
Hughes, G. R., Bass, A. J. & Mentis, M. T. 1967. Further studies on marine turtles in Tongland. 1,2. *Lammergeyer,* 7: 7–72.
Hyman, L. H. 1959. Phylum Chaetognatha. In: The Invertebrates, *5:* 1–71. McGraw-Hill, N.Y.
Ihle, J. E. W., Kampen, P. N., van Nierstrasz, H. F. & Versluys, J. 1927. Vergleichende Anatomie der Wirbeltiere: 1–906. Springer, Berlin.
Ilyin, M. M. & Rass, T. S. 1971. Suborder Characinoidei. In: Life of animals, *4,* Fishes: 261–265. "Prosveshchenie" Publishers, M.
Jaeckel, O. 1907. *Placochelys placodonta* aus der Obertrias des Bakony. *Result. wiss. Erforsch. Balatonsees., 1,* 1, Anhang: Palaeontol.: 1–90.
Jaeckel, S. G. A. 1958. Cephalopoden. *Tierwelt des Nordund Ostsee,* Lief. *37:* Teil 9, *3:* 479–723.

Jarvik, E. 1950. On some osteolepiform crossopterygians from the Upper Old Red Sandstone of Scotland. *Kdl. Svenska Vetenskapsakad. Handl.* (4), *2,* 2: 1–35.
Jatta, G. 1893. Sopra L'organo dell'imbuto nei Cephalopodi. *Boll. Soc. Nat. Napoli,* (1), *7:* 45–60.
Johannessen, C. L. & Harder, J. A. 1960. Sustained Swimming Speeds of Dolphins. *Science, 132,* 12: 1550–1551.
Jones, F. R. H. 1949. The Teleostean Swimbladder and Vertical Migration. *Nature* (London), *164,* 4176: 847.
Jones, F. R. H. 1951. The swimbladder and the vertical movement of teleostean fishes. 1. Physical factors. *J. Exp. Biol.,* 28: 553–566.
Jones, F. R. H. 1952. The swimbladder and the vertical movement of teleostean fishes. II. The restriction to rapid and slow movements. *J. Exp. Biol., 29:* 94–109.
Jones, F. R. H. 1957. The swimbladder. In: Brown, M. E., The physiology of fishes, *2,* Chap. 4: 305–322. Academic Press, N.Y.
Jones, F. R. H. & Marshall, N. B. 1953. The structure and functions of the teleost swimbladder. *Biol. Rev., 28:* 16–83.
Jonsgard, A. 1959. Nytt funn av sverd fra sverdfisk (*Xiphias gladius*) i blahval (*Balaenoptera musculus*) i sydishavet. *Norsk hvalfangst-tid., 48,* 7: 352.
Jonsgard, A. 1962. Tre funn av sverd fra sverdfisk (*Xiphias gladius*) i antarktisk finnhval (*Balaenoptera physalus* (L.). *Norsk hvalfangst-tid., 51,* 7: 287–291.
Jordan, D. S. & Evermann, B. W. 1896–1900. The fishes of North and Middle America: a descriptive catalogue of the species of fish-like vertebrates found in the waters of North America, north of the Isthmus of Panama. *Bull. U.S. Nat. Mus., 47,* 1–4: 1–3313.
Jordan, D. S., Evermann, B. W. & Clark, H. W. 1930. Check list of the fishes and fishlike vertebrates of North and Middle America north of the northern boundary of Venezuela and Colombia. *Rep. U.S. Comm. Fisher.,* 1928: 1–670.
Jordan, D. St. Tanaka, S. & Snyder, J. O. 1913. A catalogue of the fishes of Japan. *J. Coll. Sci. Imp. Univ. Tokyo, 33,* 1: 1–497.
Joubin, L. 1893. Recherches sur la coloration du tegument chez les cephalopodes. *Arch. Zool. exp. et gen., 2,* 10: 277–303.
Joysey, K. A. 1961. Life and its environment in ancient seas. *Nature* (London), *192,* 4806: 925–926.
June, F. C. 1951. Note on the feeding habits of the giant white marlin of the Pacific. *Pacif. Sci., 5,* 3: 287.
Kalugin, V. N. & Merkulov, V. I. 1968. Possible mechanism of reducing resistance in fishes. Propulsive and orientation mechanism in animals: 12–21. "Naukova Dumka" Publishers, Kiev.
Kanwisher, J. & Ebeling, A. 1957. Composition of the swimbladder gas in bathypelagic fishes. *Deep-Sea Res., 4,* 3: 211–217.
Kanwisher, J. & Sundnes, G. 1966. Thermal Regulation in Cetaceans. In: Norris, K.S. (Ed.), Whales, dolphins and porpoises: 397–409. Univ. of California Press, Berkeley-Los Angeles.
Karafoli, E. 1956. Aircraft wing aerodynamics. Incompressible fluid: 1–477. Transl. from Rumanian. USSR Ac. of Sci. press, Moscow.
Karandeyeva, O. G., Protasov, V. A. & Semenov, N. P. 1970. To the question of the physiological substantiation of Gray's paradox. *Bionika,* 4: 36–43.
Kazantseva, A. A. 1964. Subclass Actinopterygii. Actinopterygian fishes. General. In: Principles of Paleontology; Agnatha, Fishes: 323–335. "Nauka" Publishers, M.
Kellogg, R. 1928. History of Whales. Their adaptation to life in the Water. *Quart. Rev. Biol., 3:* 29–76, 174–208.
Kellogg, R. 1938. Adaptation of structure to function in Whales. *Carnegie Inst. Washing. Publ.,* 501: 649–682.
Kellogg, W. N. 1953. Ultrasonic hearing in the porpoise. *J. Comp. Physiol. Psych., 46:* 446–450.

Kellogg, W. N. 1958. Echo-ranging in the porpoise. *Science, 128:* 982–988.
Kellogg, W. N. 1959. Size discrimination by reflected sound in a bottlenose porpoise. *J. Comp. Physiol. Psych., 52:* 509–514.
Kellogg, W. N. 1961. Porpoises and sonar: XIV +1–77. Univ. Chicago Press, Chicago.
Kelly, H. R. 1961. Fish Propulsion Hydrodynamics. Developments in Mechanics, Proc. of 7th Midwestern Mechanic. Conf., 1: 442–450. Plenum Press, N.Y.
Kelly, H. R., Rentz, A. W. & Siekmann, J. 1964. Experimental Studies on the Motion of a Flexible Hydrofoil. *J. Fluid Mechanics, 19,* 1: 30–48.
Kenyon, K. W. & Rice, D. W. 1959. Life History of the Kawayian Monk Seal (*Monachus schauinslandi* Matschie, 1905). *Pacific Sci., 13,* 3: 215–252.
Kermack, K. A. 1943. The functional significance of the hypocercal tail in *Pteraspis rostrata. J. Exp. Biol., 20,* 1: 23–27.
Kermack, K. A. 1948. The propulsive powers of Blue and Fin whales. *J. Exp. Biol., 25,* 3: 237–240.
Kerr, T. 1952. The scales of primitive living Actinipterigians. *Proc. Zool. Soc. London, 122:* 55–78.
Khadzhinsky, V. G. 1972. Some morphofunctional peculiarities of the dolphin skin. *Bionika,* 6: 58–66.
Khomenko, B. G. 1970a. Some peculiarities in the histostructure and innervation of the frontal promontori of Black Sea dolphins. *Bionika,* 4: 70–76.
Khomenko, B. G. 1970b. Some morphological aspects of dolphin bionics. Biocybernetics – biosystem modelling – Bionics. IV Ukr. Republ. Scient. Conf.: 129. Inst. of Cybernetics. Ukrainian SSR Ac. of Sci. Press, Kiev.
Khozatsky, A. I. & Yuryev, K. B. 1964. The family Mosasauridae Gervais, 1853. In: Principles of Paleont. Amphibians, Reptiles and Birds. 475–481. "Nauka," M.
Kiaer, J. 1924. The Downtonian fauna of Norway. I. Anaspida, with a geological introduction. *Vidensk. Selsk. Skrift., math.-nat. Kl.,* 6: 1–139.
Kiaer, J. 1932. The Downtonian and Devonian vertebrates of Spitsbergen. IV. Suborder Cyathaspida (A preliminary report edited by A. Heintz). *Skrift. Svalbard Ishavet,* No. 52: 1–26.
King, J. E. 1962. Some of the aquatic modifications of seals. *Norsk Hvalf.-tidende* 1962. 3: 104–120.
King, J. E. 1964. Seals of the world: 1–154. Trustees of the Br. Mus (Nat. Hist.), London.
Kiselev, I. A. 1969. Plankton of the seas and continental water bodies, 1: 1–657, "Nauka" Publishers, L.
Klaatsch, H. 1890. Zur Morphologie der Fischschuppen und der Geschichte der Hartsubstanzgewebe. *Marph. Jb.,* Leipzig, *16:* 97–202, 209–258.
Klausewitz, W. 1959. Fliegende Tiere des Wassers. In: Der Flug der Tière: 145–158. Frankfurt am Maine.
Klausewitz, W. 1962. Wie schwimmen Haifische? *Natur und Museum, 92,* 6: 219–226.
Kleinenberg, S. E. & Kokshaisky, N. V. 1967. Modern problems of biological aero- and hydrodynamics. Voprosy bioniki: 531–539. "Nauka" Publishers, M.
Kobets, G. F. 1968. Effect of surface wetting on resistance of marine animals. Propulsive and orientation mechanisms in animals. 3–11. "Naukova Dumka" Publishers, Kiev.
Kobets, G. F. 1969. The mechanism of the effect of dissolved macromolecules on turbulent friction. *Bionika,* 3: 72–80.
Kobets, G. F. & Komarova, M. L. 1971. The role of the specific features of external structure in the hydrodynamics of fast fishes. *Bionika,* 5: 101–108.
Kobets, G. F., Zavyalova, V. S. & Komarova, M. L. 1969. Effect of fish mucus on turbulent friction. *Bionika,* 3: 80–84.
Kobi, L. & Pristovsek, S. 1959. O mehanizmu plavanja kod hrskavicavih riba *Scyllium stellare* i *Scyllium canicula. Thalassia Jugoslavica, 1,* 6–10: 5–17.

Kochin, N. E., Kibel, I. A. & Rose, N. V. 1963. *Theoretical hydromechanics,* 1: 1–583. Physico-Math. Lit. Publishers, M.
Komarov, V. T. 1970. Experimental methods and equipment for determining swimming speeds of nektonic animals. *Zool. J.,* 49, 6: 923–927.
Komarov, V. T. 1971. Swimming speeds and non-uniformity in nektonic animals: 1–20. Author's abstract of dissertation for Cand. of Biol. Sci, Rostov State University, Rostov-on-Don.
Kondakov, N. N. 1940. The class Cephalopoda. *Zoological manual, 2,* 548–609. USSR Ac. of Sci. Press, M.-L.
Kondakov, N. N. 1941. Cephalopods of the Far Eastern seas of the USSR. Exploring the Far Eastern seas of the USSR, *1:* 216–255. USSR Ac. of Sci. Press, M.-L.
Kondakov, N. N. 1948. Class Cephalopoda. Key to the fauna and flora of the northern seas of the USSR: 447–450. "Soviet-skaya Nauka" Publishers, M.
Kondratieva, N. G. & Obruchev, D. V. 1955. On the age of the Bavlino series of the Saratov right bank area. *Papers of the USSR Ac. of Sci., 105,* 5: 1074–1075.
Konstantinov, K. G. 1965. Changing functions of the swimbladder in the evolution of bony fishes (Osteichthyes). *J. Obshch. Biol., 26,* 5: 538–545.
Konzhukova, E. D. 1964a. Subclass Apsidospondyli. In: Principles of Paleontology; Amphibians, Reptiles and Birds: 60–64. "Nauka" Publishers, M.
Konzhukova, E. D. 1964b. Superorder Labyrinthodontia. In: Principles of Paleontology; Amphibians, Reptiles and Birds. 64–66. "Nauka" Publishers, M.
Konzhukova, E. D. 1964c. Order Temnospondyli. In: Principles of Paleontology; Amphibians, Reptiles and Birds: 66. "Nauka" Publishers, M.
Konzhukova, E. D. 1964d. Suborder Ichthyostegalia. In: Principles of Paleontology; Amphibians, Reptiles and Birds: 66–67. "Nauka" Publishers, M.
Konzhukova, E. D. 1964e. Suborder Rhachitomi. In: Principles of Paleontology; Amphibians, Reptiles and Birds: 67. "Nauka" Publishers, M.
Konzhukova, E. D. 1964f. Subclass Batrachosauria. In: Principles of Paleontology: Amphibians, reptiles and Birds: 133–136. "Nauka" Publishers, M.
Konzhukova, E. D. 1964g. Order Anthracosauria. In: Principles of Paleontology: Amphibians, Reptiles and Birds: 136. "Nauka" Publishers, M.
Konzhukova, E. D. 1964h. Suborder Embolomeri. In: Principles of Paleont.: Amphibians, Reptiles and Birds: 136–138. "Nauka" Publishers, M.
Konzhukova, E. D. 1964i. Superorder Crocodilia. In: Princ. of Paleont.: Amphibians, Reptiles and Birds: 506–523. "Nauka" Publishers, M.
Kooyman, G. L. 1966. Maximum Diving Capacities of the Weddell Seal *Leptonychotes weddelli. Science, 151,* 3717: 1553–1554.
Korotkin, A. M. 1973. On the propulsive mechanism of dolphins riding ship and wind-induced waves. *Bionika,* 7: 27–31.
Korzhuev, P. A. 1965. Gravitation forces and the phylogeny of vertebrates. *Usp. Sovr. Biol., 60,* Issue 2 (5).
Korzhuev, P. A., Balabanova, L. V., Evstropova, S. N. & Moderatova, Z. M. 1965 Hemoglobin and marrow assaying in the common Black Sea dolphin. Marine mammals: 258–260. "Nauka" Publishers, M.
Korzhuev, P. A. & Glazova, T. N. 1965. Water as a hypogravitational habitat of animals. Vopr. Gydrobiol. Papers of All-Union Hydrobiological Congress.
Korzhuev, P. A. & Glazova, T. N. 1969. Morpho-functional features of the Caspian Seal. Marine mammals: 187–191. "Nauka" Publishers, M.
Korzhuev, P. A. & Glazova, T. N. 1971. The biochemical aspect in the adaptation of cetaceans. Morphology and ecology of marine mammals (dolphins): 130–135. "Nauka" Publishers, M.
Kossovoi, L. S. & Obruchev, D. V. 1962. On the lower Devonian of the northern Timan. *Papers of the USSR Ac. of Sci., 147,* 5: 1147–1150.

Koval, A. P. 1972. Roughness and some peculiarities of skin structure of the swordfish. *Bionika*, 6: 73–77.
Koval, A. P. 1973. Some peculiarities in the skin cover structures of tuna fishes and possible mechanisms of reducing hydrodynamic resistance. IV All-Union Bionics Conf., *6.* Biomekhanika: 55–59. M.
Kovalevskaya, L. A. 1953. The energetics of swimming fish. Authors' abstract of dissertation for Candidate of Physics and Mathematics: 1–8. USSR Ac. of Sci. Press, M.
Kozlov, L. F. & Pyatetsky, V. E. 1968. The effect of copolymers and fish mucus on the hydrodynamic resistance of models and fishes. Propulsive and orientation mechanisms of animals: 22–28. "Naukova Dumka" Publishers, Kiev.
Kozyrev, G. S. 1950. Position of center of gravity in fishes. *Uch. Zap.* (Kharkov State University), *33, Tr. of Biol. Research Inst. 14–15:* 251–256.
Kramer, E. 1959. Modellversuche zur Bewegung der Fische. *Zool. Anz.*, Suppl. 22: 327–329.
Kramer, E. 1960. Zur form und Funktion des Lokomotion-sapparates der Fische. *Zschr. Wiss. Zool., 163,* 1–2: 1–36.
Kramer, M. O. 1957. Boundary Layer Stabilization by Distributed Damping. *J. Aeronaut. Sci., 24,* 6: 459–460.
Kramer, M. O. 1960a. The dolphins' secret. *The New Scientist, 7,* 181: 1118–1120.
Kramer, M. O. 1960b. Boundary Layer Stabilization by Distributed Damping, *J. Aero/Space Sci., 27,* 1: 1071–1072.
Kramer, M. O. 1960c. Boundary Layer Stabilization by Distributed Damping. *Naval Engin. J., 72,* 1: 25–3*L.*
Kramer, M. O. 1962. Boundary Layer Stabilization by Distributed Damping. *Naval Engin. J., 74,* 2: 341–348.
Kramer, M. O. 1964. U.S. patent, cl. 244–130, N 313185, claim. 15.06.60, published 15.12.64.
Kramer, M. O. 1965. Hydrodynamics of the dolphin. In: Advances in Hydroscience, *2:* 111–130. Acad. Press. N.Y. and London.
Krettmann, L. 1932. La vitesse de nage des poissons. *Bull. Fr. Pescicult.*, Orleans: 145–151, 186–197.
Kripp, D. 1954. Schwimmtechnische Betrachtungen dei grossen Hochseeschwimmern mit terminalen Antrieb, vorhehmlich bei Ichtyosauriern. *Österr. Zool. Zschr., 4,* 4–5: 460–480.
Kruger, W. 1958. Bewegungstypen. Handbuch der Zoologie, *8,* 15, 6(3): 1–56. De Gruyter, Berlin.
Krymgolts, G. Ya. 1958. Subclass Endocochlia. In: Principles of paleont.: cephalopods. 2: 145–179. "Gosgeoltekhizdat" Publishers, M.
Kryzhanovsky, S. G. 1956. Material on the development of herring fishes. *Tr. Inst. Morph. Zhivotn. USSR Ac. of Sci.*, 17: 1–254.
Kudryashov, A. F. 1969a. On the swimming mechanism of fishes and dolphins. Marine biology, 16, Functional-Morphological Research on Nektonic Animals: 59–69. "Naukova Dumka" Publishers, Kiev.
Kudryashov, A. F. 1969b. On the resistance of water to fish movement. Marine biology, 16; Functional-morphological research on nektonic animals: 21–38. "Naukova Dumka" Publishers, Kiev.
Kudryashov, A. F. & Barsukov, V. V. 1967a. On the hydrodynamic role of the scale cover of fishes as an analogue of surfaces directly formed by vortex flow. Communication 1. Similarity of scaly cover roughnesses with those formed by streams on river beds. *Zool. J.*, 46, 3: 393–403.
Kudryashov, A. F. & Barsukov, V. V. 1967b. On the hydrodynamic role of the scale cover of fishes as an analogue of surfaces directly formed by vortex flow. Communication 2. The hydrodynamic function of the scaly cover. *Zool. J.*, 46, 4: 566–566.
Kuhl, W. 1938. Chaetognatha. *Bronns KL. Ordn. Tierreichs, 4:* 1–126.
Kuhn, O. 1937. Die fossilen Reptilien: 1–121. Gebr. Borntraeger. Berlin.

Kurbatov, B. V. 1973. The hydrodynamic resistance of living nektonic animals. IV All Union Bionics Conference., 6. Biomekhanika: 72–77.
Lane, F. W. 1957. Kingdom of the octopus: 1–287. Jarrolds, London.
Lang, T. G. 1963. Porpoise, whales and fishes. Comparison of predicted and observed speeds. *Naval Engin. J.*, *75*, 2: 437–441.
Lang, T. G. 1966a. Hydrodynamic analysis of cetacean performance. In: Norris, K. S. (Ed.), Whales, dolphins and porpoises: 410–432. Univ. Calif. Press., Berkeley-Los Angeles.
Lang, T. G. 1966b. Hydrodynamic analysis of dolphin fir profiles. *Nature* (London), *209:* 1110–1111.
Lang, T. G. & Norris, K. S. 1966. Swimming Speed of a Pacific Bottlenose Porpoise. *Science, 151,* 3710: 588–590.
Lang, T. G. and Pryor, K. 1966. Hydrodynamic Performance of Porpoises (*Stanella attenuata*). *Science, 152,* 3721: 531–533.
Langton, J. L. 1949. Aquatic locomotion. *Nature* (London). *164.* 4182: 1073–1076.
Lawrence, B. & Schevill, W. E. 1956. The functional anatomy of the Delphinid nose. *Bull. Mus. Compar. Zool.*, *114*, 4: 103–151.
Lebour, M. V. 1921. The larval and postlarval stages of the pilchard, sprat and herring from Plymouth district. *J. Mar. Biol. Ass. 12,* 3: 427–457.
Le Danois, Ed. 1939–1943. Remarques ichthyologiques, *Rev. Trav. Pech. Marit.*, *13,* 1–4, 49–52: 55–175.
Le Danois, Y. 1958. Systeme musculaire. In: Grassé, P.-P., *Traité de Zoologie, 13,* 1: 783–817. Masson, Paris.
Legendre, R. 1934. La faune pelagique de l'Atlantique, au large du Golfe de Gascogne, recueillié dans des estomacs de Germons. Première partie: Poissons. *An. Inst. Oceanogr.*, Ser. 2, *14,* 6: 247–418.
Le Mare, D. W. 1936. Reflex and rhythmical movement in the dogfish. *J. Exp. Biol.*, *13,* 4: 429–442.
Lendenfeldt, R., von 1887. Report on the structure of the phosphorescent Organs of Fishes. *Challenger Report, Zool.*, *22:* 277–329.
Lendenfeldt, R., von 1905. The Radiating organs of the deep sea Fishes. *Mem. Mus. Comp. Zool.*, Cambridge, *30:* 169–213.
Liebe, W. 1963. Der Schwanzschlag der Fische. *VDI-Zschr.*, *105,* 28: 1298–1302.
Lighthill, M. J. 1960. Note on the swimming of slender fish. *J. Fluid Mech. 9,* part: 305–317.
Lighthill, M. J. 1969. Hydromechanics of aquatic animal propulsion. Ann. Fluid. Mech., *1,* Palo Alto Calif. Ann. Rev.: 413–446.
Lighthill, M. J. 1971. Large-amplitude elongated-body theory of fish locomotion. Proc. Roy. Soc. Lond., Ser. B, *179,* 1055.
Lindberg, G. U. & Legeza, M. I. 1959. Fishes of the Sea of Japan and adjacent parts of the Olhotsk and Yellow Seas 1: 1–208. USSR Ac. of Sci. Press, M.-L.
Lindberg, G. U. & Legeza M. I. 1965. Fishes of the Sea of Japan and adjacent parts of the Okhotsk and Yellow Seas. 2: 1–391. USSR Ac. of Sci. Press, M-L.
Lindberg, G. U. & Krasyukova, Z. V. 1969. Fishes of the Sea of Japan and adjacent parts of the Okhotsk and Yellow Seas. 3: 1–379. Nauka Publishers, L.
Lissmann, H. W. 1961. Zoology, Locomotory adaptations and the problem of electric fish. The cell and the organism: 301–317.
Lockley, R. M. 1968. Animal navigation: 1–207. Baker, London.
Logvinovich, G. V. 1970. The hydrodynamics of a fin flexible body (evaluation of fish hydrodynamics). *Bionika*, 4: 5–11.
Logvinovich, G. V. 1973. Hydrodynamics of fish swimming. *Bionika*, 7: 3–8.
Lowndes, A. G. 1942. The displacement method of weighing living aquatic organisms. *J. Mar. Biol. Ass.*, 25: 555–574.
Lowndes, A. G. 1955. Density of fishes. *Ann. Mag. Nat. Hist.*, Ser. 12, *8,* 88: 241–256.

Lowndes, A. G. 1956. Swimming of fishes. *Nature* (London), *177,* 4500: 194.
Lozano Cabo, F. 1953. Monografia de los centracantidos mediterraneos, con un estudio especial de la biometria, biologia y anatomia de *Spicara smaris* (L.). Mem. Real Acad. Cienc. exact., fisic. y Nat., Madrid, ser. de cienc. Nat., *17,* 2, *Bol. Inst. esp. Oceanogr.*, 59: 1-126.
Lund, R. 1967. An analysis of the propulsive mechanisms of fishes with reference to some fossil actinopterygians. *Ann. Carneg. Mus., 39:* 195–218.
Luppov, N. P., Kiparisova, L. D. & Krymgolts, G. Ya. 1958. Superorder Ammonoidea (Mezozoic Ammonoidea). In: Principles of Paleont. Cephalopods. 2: 15–144. Gosgeoltechizdat Publishers, M.
Lyapin, M. A. 1971. On some peculiarities in the hydrodynamics of nektonic animals. (fishes and dolphins). *Zool. J., 50,* 11: 1686–1694.
Magnan, A. 1929. Les caractéristiques géométriques et physiques des poissons, avec contribution a l'étude de leur équilibre statique et dynamique. I. *Ann. Sci. Nat., Zool.*, ser. 10, *12:* 5–135.
Magnan, A. 1930. Les caractéristiques géométriques et physiques des poissons, avec contribution a l'étude de leur équilibre statique et dynamique. 2 *Ann. Sci. Nat., Zool.*, Sér. 10, *13:* 355–490.
Magnan, A. & Lariboisiére, J., de 1912. La densité de quelques espose de Poisson. Mém. prés. Congr. Soc. Sav. Paris: 202–204.
Magnan, A. & Lariboisiére, J., de 1914. Le maitre-couple des poissons. Mém. Prés. Congr. Soc. Sav. Paris:
Magnan, A. & Sainte-Laguë, A. 1928a. Sur la determination expérimentale de la résistance à l'avancement des poissons. *Compt.-Rend. Acad. Sci. Paris, 187,* 24: 1163–1165.
Magnan, A. & Sainte-Laguë, A. 1928b. Sur une méthode de morphométrie de poissons. *Comp.-Rend. Acad. Sci. Paris, 187,* 25: 1316–1318.
Magnan, A. & Sainte-Laguë, A. 1929a. Essai de théorie du poissons. *Bull. Serv. techniq. aéronaut.*, 58: 1–180.
Magnan, A. & Sainte-Laguë, A. 1929b. Nouvelles expériences sur la résistance à l'avancement des poissons dans l'eau. *Compt.-Rend. Acad. Sci. Paris, 189,* 20: 798–799.
Magnan, A. & Sainte-Laguë, A. 1930. Résistance à l'avancement et puissance des poissons. *Bull. Serv. techniq. aéronaut.*, 71: 1–107.
Magnusson, J. J. 1963. Tuna behaviour and physiology, a review. *FAO Fisheries Reports,* 6, *3:* 1057–1066.
Magnusson, J. J. 1966. Continuous locomotion in scombroid fishes. *Amer. Zool., 6:* 503–504.
Magnusson, J. J. & Prescott, J. H. 1966. Courtship, locomotion, feeding and miscellaneous behaviour of pacific bonito (*Sarda orientalis*). *Anim. Behav., 14:* 54–67.
Maleyev, E. A. 1964a. Suborder Choristodera. In: Principles of Paleont. Amphibians, Reptiles and Birds: 453–455. "Nauka" Publishers, M.
Maleyev, E. A. 1964b. Family Tangassauridae Haughton, 1924. In: Principles of Paleonthology. Amphibians, Reptiles and Birds: 459. "Nauka" Publishers, M.
Maleyev, E. A. 1964c. Family Thalattosauridae Merriam, 1905. In: Principles of Paleont. Amphibians, Reptiles and Birds: 460. "Nauka" Publishers, M.
Maleyev, E. A. 1964d. Superorder Thecodontia. In: Principles of Paleonthology. Amphibians, Reptiles and Birds: 497. "Nauka" Publishers, M.
Maleyev, E. A. 1964e. Order Phytosauria (Parasuchia). In: Principles of Paleonthology. Amphibians, Reptiles and Birds: 503–506. "Nauka" Publishers, M.
Mann, F. G. 1946. Ojo y vision de las Ballenas. *Biol. Trab. Inst. biol. Santiago, 4:* 23–71.
Mansfield, A. W. 1963. Seals of Arctic and Eastern Canada. *Bull. Fish. Res. Board Canada, 137:* 1–30.
Marakov, S. V. 1964. Fur seals, *Priroda,* 9: 57–64.
Marey, E. J. 1894. Le movement: 1–323. Masson, Paris.
Marshall, N. B. 1958. Aspects of deep sea biology: 1–380. Hutchinson, London.

Marshall, N. B. 1960. Swimbladder structure of deep-sea fishes in relation to their systematics and biology. *Discovery rep. 31:* 1–122.
Marshall, T. C. 1965. Fishes of the Great Barrier Reef and coastal waters of Queensland: 1–566. Livingston, Sydney.
Martynov, A. K. 1958. Experimental erodynamics: 1–348. "Oborongiz" Publishers, M.
Maslov, N. K. 1970. On the manoeuvrability and contral-lability of dolphins. *Bionika,* 4: 46–50.
Matthews, L. H. 1948. The Swimming of Dolphins. *Nature* (London), *161,* 4097, 731.
Matthews, L. H. 1970. The life of mammals, *1:* 1–340. Universe books, N.Y.
Matthiessen, L. 1886. Über den physikalischoptischen Bau des Auges der Cetacean und der Fische. *Pflugers Arch., 38:* 521–528.
Matveyeva, A. A. 1958. Paleoniscides of the Issyk-Chul horizon of the Minusinsk basin. *Vopr. Ichtiol.,* 11: 154–161.
Matyukhin, V. A. 1973. The bioenergetics and physiology of swimming in fishes: 1–254. "Nauka" Publishers, M.
Matyukhin, V. A., Alikin, Yu. S., Stolbov, A. Ya. & Turetsky, V. I. 1973. Experimental studies into the coefficient of performance of fishes. *Bionika,* 7: 31–37.
Matyukhin, V. A. & Turetsky, V. I. 1972. Research into the resistance affecting the body of a swimming fish. *Bionika,* 6: 3–6.
McAlister, D. E. 1960. List of the marine fishes of Canada. *Bull. Nat. Mus. Canada,* 168: I–IV 1–76.
McCulloch, A. R. 1929. A checklist of the fishes recorded from Australia. I. *Austr. Mus.,* Sydney, *Me.m* 5: 1–144.
Merkulov, V. I. & Khotinskaya, V. D. 1969. The mechanism of reducing hydrodynamic resistance in some fish species. *Bionika,* 3: 96–101. "Naukova Dumka" Publishers, Kiev.
Merriam, J. C. 1902. Triassic Ichthyopterygia from California and Nevada, *Univ. Calif. Publ., Bull. Dept. Geol.,* 3: 63–108.
Millot, J. & Anthony, J. 1958. Crossopterygeins actuels. *Latimeria Chalumnae* dernier des crossopterygiens. In: Grassé, P.-P., *Traité de Zoologie, 13,* 3: 2552–2597. Masson, Paris.
Möbius, K. 1878. Die Bewegungen der fliegenden Fische durch die Luft. *Zschr. Wiss. Zool., 30,* 2: 343–382.
Mohr, E. 1954. Fliegende Fische: 1–55. Ziemsen, Wittenberg Lutherstadt.
Monod, T. 1950. Notes d'ichthyologie africaine. *Bull. Inst. Franc. d'Afrique Noire, 12:* 1–71.
Monod, T. 1959. Les nageories ventrales du Balistes forcipatus Gmelin 1789. *Bull. Inst. Franc. d'Afrique Noire,* ser. A, *Sci. nat.,* 21, 2: 695–709.
Moore, H. L. 1950. The occurrence of a black marline *Tetrapterus mazara* (Jordan and Snyder) without a spear. *Pacific Sci.,* 4, 2: 164.
Mordvinov, Yu. E. 1968. Observations of the locomotion of some pinnipeds. *Zool. J.,* 47, 9: 114–122.
Mordvinov, Yu. E. 1969. The trunk as carrying plane and the buoyancy of some pinnipeds. *Vestn. Zool.,* 2: 10–14.
Mordvinov, Yu. E. 1972. The manoeuvrability of pinnipeds. *Vestn. Zool.,* 1: 22–26.
Mordvinov, Yu. E. 1973. On the hydrodynamic resistance of some secondary aquatic nektonic animals. IV All-Union Bionics Conf., 6, Biomekhanika: 78–82. M.
Mordvinov, Yu. E. & Kurbatov, B. V. 1972. The effect of the hair cover of some true seal species (Phocidae) on the total hydro dynamic resistance. *Zool. J.,* 51, 2: 242–247.
Moreau, A. 1876. Recherches expérimentales sur les functions de la vessie natatoire. *Ann. Sci. Nat., Zool.,* Ser. 6, *4,* 9: 1–85.
Mori, M. 1958. The skeleton and musculature of Zalophus. *Okajimas Fokia anat. jap., 31:* 203–284.
Morozov, D. A. & Tomilin, A. G. 1970. Elements of hydrostatics in dolphins. *Bionika,* 4: 50–54.

Morrow, J. E. 1951. A striped marlin (*Makaira mitsukurii*) without a spear. *Copeia,* 4: 303–304.

Morrow, J. E. 1952. Food of the striped marlin, *Makaira mitsukurii,* from New Zealand. *Copeia,* 3: 143–145.

Morrow, J. E. & Mauro, A. 1950. Body Temperature of some Marine Fishes. *Copeia,* 2: 108–116.

Müller, A. H. 1968. Lehrbuch der Palaozoologie, B. 111. Vertebraten, Teil 2. Reptilien und Vogel: 1–657. Fischer, Jena.

Murie, J. 1871. Researches upon the anatomy of the Pinnipedia. Part I. On the walrus (*Trichechus rosmarus,* Linn.). *Trans. Zool. Soc. Lond., 7,* 6: 411–464. Mattews, London.

Murie, J. 1872 Researches upon the anatomy of the Pinnipedia. Part. 11. Descriptive anatomy of the Sea lion (*Otaria jubata*). *Trans. Zool. Soc. Lond. 7,* 8: 527–596.

Murie, J. 1874. Researches upon the anatomy of the Pinnipedia. Part III. Descriptive anatomy of the sea lion (*Otaria jubata*). *Trans. Zool. Soc. Lond., 8,* 9: 501–582.

Murie, J. 1880. Further observations on the Manatee. *Trans. Zool. Soc. Lond., 11;* 48–50.

Murphy, R. C. 1915. The Penguins of South Georgia. *Mus. Brooklyn Inst. Arts and Sci., Sci. Bull., 2,* 5: 103–133.

Murphy, R. C. 1936. Oceanic birds of South America, 1–2: 1–450. Publ. Amer. Mus. Nat. Hist., N.Y.

Murray, J. & Hjort, J. 1912. The depths of the ocean: 1–821. MacMillan, London.

Murray-Levick, G. 1914. Antarctica Penguins. A Study of their sociable Habits: 1–140. McBride, Nast et Co., N.Y.

Murray-Levick, G. 1915. Natural History of the Adelie Penguin. Brit. Antarctic ("Terra Nova") Exped. 1910. *Nat. Hist. Rep. Zool.,* 1: 55–84.

Myers, G. S. 1937. Freshwater Fishes and West Indian Zoogeography. *Smiths. Inst. Rep.*: 339–364.

Myers, G. S. 1950. Flying Characid fishes. *Standard Ichthyol. Bull., 3,* 4: 182–183.

Naaktgeboren, C. 1960. Die Entwicklungsgeshichte der Haut des Finwals, *Balaenoptera physalus* (L.). Zool. Anz., *165,* 5–6, 159–167.

Nachtigall, W. 1961. Funktionelle Morphologie, Kinematik und Hydromechanik des Ruderapparates von Cyprinus. *Zschr. vergl. Physiol., 45,* 2: 193–266.

Naef, A. 1922. Die fossilen Tintenfische. Eine palaozoologische Monographie: 1–322. Fischer, Jena.

Naef, A. 1923. Die Cephalopoden. Fauna und Flora des Golfs von Neapel. *Monogr. 35:* 1–863.

Narkhov, A. S. 1937. The morphology of the musculature of the caudal region in *Delphinus delphis* and *Tursiops tursio. Zool. J., 16,* 4:

Narkhov, A. S. 1939. On the swimming movements of *Delphinus Delphis* and *Tursiops tursio. Zool. J.,* 18, 2: 326–330.

Nemoto, T. 1959. Food of baleen whales with reference to whale movements. *Sci. Rep. Whales Res. Inst.,* 14: 149–290.

Nesis K. N. 1973. Taxonomy, phylogeny and evolution of squids of the family Gonatidae (Cephalopoda). *Zool. J., 52,* 11: 1626–1638.

Nesis K. N. 1974. The system of recent Cephalopoda. *Bull. Mosc. Obshch. Ispyt. Prir., Otdel. Biol., 76,* 5: 81–93.

Nestrukh, M. F. 1970. The origin of man. 2nd Ed.: 1–439. "Nauka" Publishers, M.

Neu, W. 1931. Die Schwimmbewegungen der Tauchvogel (Blasshuhn und Pinguıne). *Zschr. Vergl. Physiol., 14,* 4: 682–708.

Newmann, B. & Tarlo, B. 1967. A giant marine reptile from Bedfordshire. *Animals, 10:* 61–63.

Nichols, J. T. 1943. The fresh-water fishes of China. *Am. Mus. Nat. Hist.,* N.Y. 4: 1–322.

Nichols, J. T. & Griscom, L. 1917. Fresh-water fishes of the Congo Basin. *Bull. Am. Mus. Nat. Hist.,* 37: 653–756.

Nichols, J. T. & Breder, C. M. 1928. About flying fishes. *Nat. Hist., 28:* 64–77.
Niiler, P. P. & White, H. G. 1969. Note on the swimming deceleration of the dolphins. *J. Fluid Mech., 38,* 3: 613–617.
Nikolsky, G. V. 1940. Fishes of the Aralian Sea: 1–216. Mosk. Obshch. Isp. Prir. Publishers, M.
Nikolsky, G. V. 1954. Particular Ichthyology. 2nd Ed.: 1–458. "Sovetskaya Nauka" Publishers, M.
Nikolsky, G. V. 1963. Fish ecology: 1–368. "Vysshaya Shkola" Publishers, Moscow.
Nikulin, P. G. 1937. Observations of the pinnipeds of the Okhotsk and Japanese Seas. *Izv. Tikhook. Inst. Rybn. Khoz. and Oceanogr., 10:* 49–59.
Nishimura, S. 1968. A preliminary list of the pelagic Cephalopoda from the Japan. *Sea. Publ. Seto Mar. Biol. Lab., 16:* 1: 71–83.
Norman, J. R. 1935. Coast fishes. I. The South Atlantic. *Discovery Rep., 12:* 1–58.
Norman, J. R. 1937. Coast fishes II. The Patagonian region. *Discovery Rep., 16:* 1–150.
Norris, K. S. 1961. An experimental demonstration of echolocation behaviour in the porpoise. *Biol. Bull. 120:* 163.
Norris, K. S. 1964. Some problems of echolocation in cetaceans. In: Tavolga, W. (Ed.), Marine Bio-Acoustics: 317–336. Perg. Press: N.Y., London, Oxford, Paris.
Norris, K. S. 1965. Trained Porpoise Released in the Open Sea. *Science, 147,* 3661: 1048–1050.
Norris, K. S. (Ed.) 1966. Whales, Dolphins and Porpoises: 1–789. Univ. Calif. Press, Berkeley–Los Angeles.
Norris, K. S. 1967. Some observations on the migration and orientation of marine mammals. In: Storm, R. M., Animal orientation and navigation: 101–125. Oregon State Univ. Press.
Norris, K. S. 1968. The evolution of acoustic mechanisms in odontocete cetaceans. In: Evolution and environment: 297–324. New Haven–London.
Norris, K. S. 1969. The echolocation of marine mammals. In: Andersen, H. T. (Ed.), The biology of marine mammals: 391–424. Acad. Press, N.Y.–London.
Norris, K. S. & Prescott, J. H. 1961. Observations on pacific cetaceans of Californian and Mexican waters. *Univ. Calif. Publ., Zool., 63:* 291–402.
Norris, K. S., Prescott, J. H., Asa-Dorian, P. V. & Perkins, P. 1961. An experimental demonstration of echolocation behaviour in the porpoise, *Tursiops truncatus* (Montagu). *Biol. Bull., 120:* 163–176.
Novitskaya, L. I. & Obruchev, D. V. 1964. Class Acanthodei. In: Principles of Paleonthology, Agnatha, Fishes: 175–194. "Nauka" Publishers, M.
Novozhilov, N. 1964a. Order Sauropterygia. In: Principles of Paleontology, Amphibians, Reptiles and Birds: 309–310. "Nauka" Publishers, M.
Novozhilov, N. 1964b. Superfamily Pistosauroidea. In: Principles of Paleontology, Amphibians, Reptiles and Birds: 317–318. "Nauka" Publishers, M.
Novozhilov, N. 1964c. Superfamily Plesiosauroidea. In: Principles of Paleontology. Amphibians, Reptiles and Birds: 318–327. "Nauka" Publishers, M.
Novozhilov, N. 1964d. Superfamily Pliosauroidea. In: Principles of Paleontology. Amphibians, Reptiles and Birds: 327–332. "Nauka" Publishers, M.
Nursall, J. R. 1956. The lateral musculature and the swimming of fish. *Proc. Zool. Soc. London, 126,* 1: 127–143.
Nursall, J. R. 1958a. The caudal fin as a hydrofoil. *Evolution,* 12, 1: 116–120.
Nursall, J. R. 1958b. A method of analysis of the swimming of fish. *Copeia,* 2: 136–141.
Nursall, J. R. 1962. Swimming and the origin of paired appendages. *Am. Zoologist,* 2: 127–141.
Nursall, J. R. 1963a. The caudal musculature of *Hoplopagrus guntheri* Gill (Parciformes: Lutjanidae). *Canad. J. Zool., 41:* 865–880.
Nursall, J. R. 1963b. The hypurapophysis, an important element of the caudal skeleton. *Copeia,* 2: 458–459.

Nybelin, O. 1947. Antarctic fishes. *Sci. Res. Norweg. Antarctic Exped. 1927–1928, 26:* 1–76.
Nybelin, O. 1951. Subantarctic and Antarctic fishes. *Sci. Res. Bratteg Exped.* 1947–48, 1–32.
Obruchev, D. V. 1947. The type Chordata. Atlas-Manual of Fossil Fauna of the USSR, 3, Devonian system: 190–206. "Gosgeoltechizdat" Publishers, M.
Obruchev, D. V. 1956. Cephalaspids from the Lower Devonian of Tuva. *Papers of the USSR Ac. of Sci., 106,* 5: 917–919.
Obruchev, D. V. 1964a. The branch Agnatha. In: Principles of Paleontology, Agnatha, Fishes: 34–116. "Nauka" Publishers, M.
Obruchev, D. V. 1964b. Class Placodermi. In: Principles of Paleontology, Agnatha, Fishes: 118–172. "Nauka" Publishers, M.
Obruchev, D. V. 1964c. Subclass Holocephali (or Chimeras), In: Principles of Paleontology. Agnatha, Fishes: 288–266. "Nauka" Publishers, M.
Obruchev, D. V. & Kazantseva, A. A. 1964a. Superorder Chondrostei. In: Principles of Paleontology, Agnatha, Fishes: 371–375. "Nauka" Publishers, M.
Obruchev, D. V. & Kazantseva, A. A. 1964b. Superorder Polypteri (Cladistia, Brachiopterygii). In: Principles of Paleontology, Agnatha, Fishes: 376–377. "Nauka" Publishers, M.
Obrucheva, O. P. 1955. Upper Devonian fishes of Central Kazakhstan. *Sov. Geol.* 45: 84–99.
Obrucheva, O. P. 1959. Two species of Plourdosteus (*Arthrodira*) from the Upper Devonian deposits of the USSR. *Paleont. J.*, 3: 78–94.
Oehmichen, E. 1938. Essai sur la dinamique des Ichthyosauriens longipinnati et particulièrment *d'Ichthyosaurus burgundiae* (Gaud.). *Ann. Paleontol., 28:* 91–114.
Oehmichen, E. 1950a. Le vol des oiseaux. In: Grassé, P.-P., *Traite de zoologie, 15:* 131–170. Masson, Paris.
Oehmichen, E. 1950b. Locomotion terrestre, natationplongee. In: Grassé, P.-P., *Traité de zoologie, 15:* 171–184. Masson, Paris.
Oehmichen, E. 1958. Locomotion des Poissons. In: Grassé, P.-P., *Traité de Zoologie, 13,* 1: 818–853. Masson, Paris.
Ognev, S. I. 1935. The beasts of the USSR and adjacent countries. 3: 316–600 "Biomedgiz" Publishers, M-L.
Ognev, S. I. 1941. The zoology of vertebrates: 1–665. "Sovetskaya Nauka" Publishers, M.
Ohlmer, W. 1964. Untersuchungen ueber die Beziehungen zwischen Korpevform und Bewegungsmedium bei Fischen aus stehenden. *Binnengewassern. Zool. Jb., Anat., 81:* 2: 151–240.
Ohlmer, W. & Schwartzkopff, J. 1959. Schwimmgeschwindigkeiten von Fischen aus stehenden Binnengewassern. *Naturwissenschaften, 46,* 10: 362–363.
Okada, Y. 1955. Fishes of Japan: 1–434. Tokyo.
Okada, Y. & Matsubara, K. 1938. Keys to the fishes and fish-like animals of Japan including Kuril Islands, Southern Sakhalin, Bonin Islands, Ryukyu Islands, Korea and Formosa: 1–584. Tokyo and Osaka.
Orvig, T. 1957. Remarks on the vertebrate fauna of the Lower Upper Devonian of Escuminas Bay, P. Q., Canada, with special reference to the prorolepiform crossopterygians, *Arkiv. zool.* (2), *10,* 6: 367–426.
Osborn, M. F. 1961. The hydrodynamical performance of migratory salmon. *J. Exp. Biol., 38,* 2: 365–390.
Osburn, R. C. 1906. The Functions of the Fins of Fishes. *Science, 23:* 585–587.
Ostheimer, M. A. 1966. Why ... is a hatchetfish? *The Aquarium, 35,* 3: 5–12.
Ovcharov, O. P. 1971. On vortex formation in the hydrodynamic wake of moving fishes. *Zool. J.*, 50, 12: 1766–1769.
Ovcharov, O. P. 1974. On the manoeuvrability of fishes. Vopr. Ikhtiol; 14, 4: 679–686.
Ovchinnikov, V. V. 1966a. Boundary layer turbulization as a means of reducing drag during movement of some fishes. *Biofizika,* 11, 1: 186–188.

Ovchinnikov, V. V. 1966b. A morphological and functional characteristic of the rostrum of Xiphioidae (Perciformes, Fisces). Ecomorphological research into nektonic animals: 42–52. "Naukova Dumka" Publishers, Kiev.
Ovchinnikov, V. V. 1966c. The functional significance of the fins of Xiphioidae. Ecomorphological research into nektonic animals: 53–62. "Naukova Dumka" Publishers, Kiev.
Ovchinnikov, V. V. 1967a. On the hydrodynamic characteristics of the swordfish (*Xiphias gladius L.*) Sci. papers of the higher school, *Biol. Sci.*, 1: 22–25.
Ovchinnikov, V. V. 1967b. On the functional significance of the dorsal fin of the sailfish – *Histiophorus americanus* Cuv. et Val., *Vopr. Ikhtiol.*, *7*, 1: 194–196.
Ovchinnikov, V. V. 1968. Locomotor adaptations of the swordfish and related species. Propulsive mechanism and orientation of animals: 65–71. "Naukova Dumka" Publishers, Kiev.
Ovchinnikov, V. V. 1969. Studies into morphological features, connected with ensuring the functions of sense organ receptors, in members of the superfamily Xiphioidae and some other scombroid fishes. Marine biology, 16; Functional-morphological research into nektonic animals: 70–81. "Naukova Dumka" Publishers, Kiev.
Ovchinnikov, V. V. 1970. The swordfish and the sail-fishes. (The Atlantic. Econogy and Functional Morphology): 1–106. "AtlantNIRO" Publishers, Kaliningrad.
Palmer, E. & Weddel, G. 1964. The relationship between structure, innervation and function of the skin of the bottlenose dolphin (*Tursiops truncatus*). *Proc. Zool. Soc. London*, 143, 4: 553–568.
Pao, S. & Siekmann, J. 1964. Note on the Smith-Stone theory of Fish Propulsion, *Proc. Roy. Soc. London, A280,* 1382: 398–408.
Parin, N. V. 1961a. To studies into the fauna of flying fishes (family Exocoetidae) of the Pacific and Indian Oceans. *Trans. of the Inst. of Oceanology, USSR Ac. of Sci.*, 43: 40–91.
Parin, N. V. 1961b. The principles of the system of flying fishes (families Oxyporhamphidae and Exocoetidae). *Trans. of the Inst. of Oceanol., USSR Ac. of Sci., 43:* 92–183.
Parin, N. V. 1968. The ichthyofauna of the oceanic epipelagic zone: 1–185. "Nauka" Publishers, M.
Parin, N. V. 1971a. The order Beloniformes. In: Life of animals, 4, Fishes: 359–366. "Prosveshchenie" Publishers, M.
Parin, N. V. 1971b. The suborder Dactylopteridae. In: Life of animals, 4, Fishes: 581–582. "Prosveshchenie" Publishers, M.
Parin, N. V. 1971c. The suborder Scombroidei. In: Life of animals, 4, Fishes: 535–542. "Prosveshchenie" Publishers, M.
Parker, G. H. 1930. Chromatophores. *Biol. Rev.*, *5:* 59–90.
Parry, D. A. 1949a. The anatomical basis of swimming in whales. *Proc. Zool. Soc. London, 119:* 49–60.
Parry, D. A. 1949b. The swimming of whales and discussion of Gray's Paradox. *J. Exp. Biol.*, *26*, 1: 24–34.
Parry, D. A. 1949c. The structure of whale blubber and discussion of its thermal properties. *Quart. J. Micr. Soc.*, 90, 1: 13–26.
Parsons, H. B. 1888. The displacements and the area curves of fish. *Trans. Am. Soc. Mechan. Engin.*, 9, 1: 1–17.
Patrashev, A. N. 1953. Hydroaeromechanics: 1–719. USSR Ministry for the Navy Publishing House, M.
Pavlenko, G. E. 1953. The resistance of water to ship movement. 1–507. Vodn. Transp. Water Transport Publishing House, M.
Pellegrin, J. 1911. La distribution géographique de Poissons d'eau douce en Afrique. *Comp.-Rend. Ac. Sci. Paris, 153:* 297–299.
Pellegrin, J. 1933. Le Poissons des eaux douces de Madagascar et des iles voisines. *Mem. Ac. Malgache, 14:* 1–222.

Perry, B., Acosta, A. J. & Kiceniuk, T. 1961. Simulated wave-riding Dolphins. *Nature* (London), 192, 4798: 148–150.
Pershin, S. V. 1965a. On the hydrodynamic characteristics of the movement of some aquatic animals. Bionics research: 5–15. "Naukova Dumka" Publishers, Kiev.
Pershin, S. V. 1965b. The hydrodynamic aspects of studying the movements of aquatic animals. Bionika: 205–215. "Nauka" Publishers, M.
Pershin, S. V. 1967. Biohydrodynamic regularities in the swimming of aquatic animals as principles of optimisation of immersed bodies' movement in nature. Voprosy bioniki: 555–560. "Nauka" Publishers, M.
Pershin, S. V. 1969a. The hydrodynamic characteristic of cetaceans and swimming speeds of dolphins recorded in the natural environment and in captivity. *Bionika,* 3: 5–12.
Pershin, S. V. 1969b. Optimisation of the rear fin propulsor in nature on examples of cetaceans. *Bionika,* 3: 26–36.
Pershin, S. V. 1970. On the frequency characteristics of hydrobionts. *Bionika,* 4: 27–31.
Pershin, S. V., Sokolov, A. S. & Tomilin, A. G. 1970. On dolphin flipper elasticity being regulated by special vascular organs. *Papers of the USSR Ac. of Sci., 190,* 3: 709–712.
Petit, G. 1955. Ordre des sireniens. In: Grassé, P.-P., *Traite de Zoologie, 13,* 1: 918–993, Masson, Paris.
Petrova, I. M. (Author's review) 1970. Hydrobionics in shipbuilding (review of foreign investigations in the period from 1963 to early 1969): 1–271. "TSNIITEIS" Publishing House, Moscow.
Pettigrew, J. B. 1873. Animal locomotion: 1–264. King, London.
Pierce, E. L. 1951. The Chactognatha of the West coast of Florida. Biol., *100,* 3: 206–228.
Pierce, E. L. 1953. The Chaetognatha over the continental shelf of North Carolina with attention to their relation to the hydrography of the area. Sears Found. *J. Mar. Res., 12,* 1: 75–92.
Pierce, E. L. 1958. The Chaetognatha of the inshore waters of North Carolina, *Limnot. Oceanogr.,* 3: 166–170.
Pierce, E. L. 1962. Chaeotognatha from the Texas coast. *Publ. Inst. Mar. Sci. Univ. Texas, 8:* 147–152.
Pikharev, G. A. 1940. Far Eastern marine mammals. *Izv. Tikhook. nauchn.-issl. inst. morsk. rybn. khoz. i okeanogr., 20:* 61–99.
Pilleri, G. 1964. Zur Morphologie des Auges vom Wesswal, *Delphinapterus leucas* Pallas (1776). *Hvalradets akr.,* 47: 1–17.
Pilleri, G. 1972. Investigations on Cetacea, *4:* 1–299. Berne (Switzerland).
Piveteau, J. 1950. Origine et évolution des oiseaux. In: Grassé, P.-P., *Traité de Zoologie, 15:* 792–844. Masson, Paris.
Plattner, W. 1941. Étude sur la fonction hydrostatique de la vessie natatoire des poissons. *Rev. suisse Zool., 48:* 201–338.
Poll, M. 1950. Histoire du peuplement et origine des especes de la faune ichthyologique du lac Tanganika. *Ann. Soc. Zool. Belg., 81:* 111–140.
Popta, C. M. 1910. Étude sur la vessie aérienne des poissons. *Ann. Sci. Nat. Zool.,* ser. 9, *12:* 1–160.
Portier, P. 1903. Sur la températur du *Thynnus alalonga. Bull. Soc. Zool. France, 28:* 79–81.
Poulter, T. C. 1963. Sonar signals of the sea lion. *Science, 139,* 3556: 753–754.
Poulter, T. C. 1965. Location of the point of origin of the vocalization of the California sea lion, *Zalophus californianus.* In: Rice, C. E. (Ed.), Proc. of the second animal conf. on biol. sonar and living mammals: 41–48. Stanford Res. Inst., Menlo Park, California.
Poulter, T. C. 1966. The use of active sonar by the California sea lion. *J. Auditory Res., 6:* 165–173.
Poulter, T. C. 1967. Systems of Echolocation. Animal sonar systems. Biology and Bionics, 1. Ed. R. G. Busnel, INRA-CNRZ: 157–186. Jouy-Josas.

Prandtl, L. 1951. Hydroaeromechanics: 1–572. 2nd ed. New York, Dover.
Priol, E.-P. 1939–1943. Observation sur les germons et les thons rouges captures par les pecheurs Bretons. *Rev. Trav. Pech. Mar.*, *13*, 1–4, 49–52, 387–440.
Protasov, V. R. 1965. Bioacoustics in fishes: 1–207. "Nauka" Publishers, M.
Pumphrey, R. J. & Young, J. Z. 1938. The rates of conduction of nerve fibres of various diameters in cephalopods. *J. Exp. Biol.*, *15:* 453–466.
Purves, P. E. 1963. Locomotion in whales. *Nature* (London), *197*, 4865: 334–337.
Pütter, A. 1903. Die Augen der Wassersaugetiere. *Zool. Jb.*, *Anat.*, *17:* 99–402.
Pyatetsky, V. E. 1970. Hydrodynamic swimming characteristics of some fast marine fishes. *Bionika*, 4: 20–27.
Pyatetsky, V. E. & Kayan, V. P. 1971. Swimming kinematics and hydrodynamics of Black Sea garfishes. *Bionika*, 5: 5–11.
Pyatetsky, V. E. & Kayan, V. P. 1972a. On the swimming characteristics of the lake trout. *Bionika*, 6: 13–18.
Pyatetsky, V. E. & Kayan, V. P. 1972b. Some kinematic swimming characteristics of the Azovka dolphin. *Bionika* 6: 18–21.
Pyatetsky, V. E. & Savchenko, Yu. N. 1969. The influence of mucus on hydrodynamic resistance of fishes. *Bionika*, 3: 90–91.
Radakov, D. V. & Protasov, V. R. 1964. Swimming speeds and some peculiarities of fish vision: 1–48. "Nauka" Publishers, M.
Rass, T. S. 1971. The suborder Galaxioidei. In: Life of Animals, 4, Fishes: 206–208. "Prosveshchenie" Publishers, M.
Ray, C. 1963. Locomotion in Pinnipeds. Swimming methods related to food habits. *Nat. Hist.*, *72*, 3: 10–21.
Reece, J. W., Uldrick, J. P. & Siekmann, J. 1965. Some recent developments in theoretical and applied mechanics, proceed. 2nd Southeastern Conference on Theoretical and Applied Mechanics, Atlanta, Georgia, 1964, *2:* 337–349. Pergamon Press; Oxford, London, N.Y.
Rees, W. J. 1949. Note on the hooked squid *Onychoteuthis banksii. Proc. Malacol. Soc. London*, *28:* 43–45.
Reeve, M. R. 1966. Observations on the biology of chaetognath. In: Barnes, H., Contemporary Studies in Marine Science: 613–630. Allen, London.
Regan, C. T. 1910a. The Caudal Fin of the Elopidae and of Some Other Teleostean Fishes. *Ann. Mag. Nat. Hist.*, ser. 8, *5*, 28: 354–358.
Regan, C. T. 1910b. On the caudal fin of the Clupeidae and on the teleostean urostyle. *Ann. Mag. Nat. Hist.*, ser. 8, *5*, 30: 531–533.
Regan, C. T. 1925. Dwarfed males parasitic on the females in oceanic angler-fishes. *Proc. Roy. Soc. London*, ser. B, *97:* 386–400.
Regan, C. T. 1926. The Pediculate fishes of the suborder Ceratioidea. Danish "Dana" Exp. 1920–1922. *Oceanogr. Rep.*, 2: 1–45.
Regan, C. T. & Trewavas, E. 1932. Deep-sea Angler-fishes (Ceratioidea). Carlsberg found. *Oceanogr. Exp. round the world 1928–1930.* 2: 1–113. Copenhagen.
Reysenbach de Haan, F. W. 1957. Hearing in whales. *Acta otolaryngol.*, 134, Suppl.: 1–114.
Richardson, E. G. 1936. The physical aspect of fish locomotion. *J. Exp. Biol.*, *13*, 2: 63–74.
Ridewood, W. G. 1913. Notes on the South-American freshwater flyingfish, Gasteropelecus, and the common flying-fish, Exocoetus. *Ann. Mag. Nat. Hist.*, (8), *12:* 544–548.
Rivas, L. R. 1953. The pineal apparatus of tunas and related scombroid fishes as a possible light receptor controlling phototactic movements. *Bull. Mar. Sci. Gulf. Caribb.*, *3*, 3: 168–180.
Rochon-Duvigneaud, A. J. 1940. L'oeil des cetaces. *Arch. Mus. hist. nat. Paris*, *16:* 59–90.
Rochon-Duvigneaud, A. 1943. Les yeux et la vision des Vértebrés: 1–716. Masson, Paris.
Rochon-Duvigneaud, A. 1950. Les yeux et la vision. In: Grassé, P.-P., *Traité de Zoologie*, *15:* 221–242. Masson, Paris.

Rochon-Duvigneaud, A. 1954. L'oeil des vertebres. In: Grassé, P.-P., *Traité de Zoologie, 12:* 333–452.
Rochon-Duvigneaud, A. 1958. L'oeil et la vision. In: Grassé, P.-P., *Traité de Zoologie,* 13, 2: 1099–1142, Masson, Paris.
Rockwell, H., Evans, F. G. & Pheasant, H. C. 1938. The comparative morphology of the vertebrate spinal column. Its form as related to function. *J. Morph., 63:* 87–117.
Romanenko, E. V., Tomilin, A. G. & Artemenko, B. A. 1965. On sound formation and sound orientation in dolphins. Bionika: 269–273. "Nauka" Publishers, M.
Romanenko, E. V. & Yanov, V. G. 1973. Experimental results of studying the hydrodynamics of dolphins. *Bionika,* 7: 52–56.
Romer, A. Sh. 1939. Paleozoology of vertebrates: 1–414. transl. from Engl. "Gos. Nauchn.-Techn. Izd-vo Neftyanoi i Gornotoplivnoi Lit.", M-L.
Romer, A. Sh. R. 1956. Osteology of the Reptiles: 1–772. The Univ. of Chicago Press, Chicago.
Rose, M. 1957. Chétognathes. In: Trégouboff, G. & Rose, M., Manuel de planctonologie Méditerraneenne, *1:* 477–484. Ed. Centre Nat. de la Rech. Sci., Paris.
Rosen, M. W. 1959. Water flow about a swimming fish. Stat. Tech. Pull. U.S. Naval Ordn. Test Station, California, NOTS TP2298, I–V, 1–94.
Rosen, M. W. 1961. Experiments with swimming fish and dolphins. Paper. Amer. Soc. Mech. Engrs., NWA-203, 1–11.
Rosen, M. W. & Cornford, N. E. 1971. Fluid friction of fish slimes. *Nature* (London), *234,* 5323: 45–47.
Rosen, N. 1913. Studies on the Plectognaths. III. The integument. *Ark. Zool.,* Stockholm, *8,* 10: 1–29.
Rosen, N. 1914. Wie wachsen die Ktenoidschuppen. *Ark. Zool.* Stockholm, *9,* 20: 1–6.
Rotta, I. K. 1967. Turbulent boundary layer in incompressible fluid: 1–232. "Sudostroenie" Publishing House, L.
Roule, L. 1924. Étude sur l'ontogenèse et la croissance avec hypermétamorphose de *Luvarus imperialis. Ann. Inst. Ocean.,* Paris, *1:* 121–157.
Royce, W. F. 1957. Observation on the spearfishes of the Central Pacific. *U.S. Fish and Wildl. Serv., Fish. Bull., 57:* 497–544.
Rozhdestvensky, A. K. 1964a. The class Reptilia. Reptiles or reptilians. In: Principles of Paleont. Amphibians, Reptiles and Birds: 190–213. "Nauka" Publishers, M.
Rozhdestvensky, A. K. 1964b. The suborder Archosauria. In: Principles of Paleontology; Amphibians, Reptiles and Birds: 493–496. "Nauka" Publishers, M.
Rush, W. 1892. Letter. *Nautilus, 6:* 81–82.
Ryder, J. A. 1892. On the mechanical genesis of the scales of Fishes. *Proc. Ac. Sci.,* Philadelphia: 219–224.
Saburenkov, E. N. 1966. Some specific features in fish swimming speeds. Summary of papers, All-Union Conference on Ecological Physiology of Fishes: 145–146. "Nauka" Publishers, M.
Sachs, C. 1881. Untersuchungen am Zitteraal Gymnotus electricus: 1–446. Von Veit, Leipzig.
Satchell, G. H. 1966. Sink or Swim: some consequences of buoyancy in fish. *Aust. Nat. Hist., 15:* 201–206.
Schaeffer, B. 1952. Rates of evolution in the coelacanth and dipnoan fishes. *Evolution, 6,* 1: 101–111.
Scheffer, V. B. 1950. Probing the life secrets of the Alaska fur seal. *Pacific Discovery, 3:* 22–30.
Scheffer, V. B. 1958. Seals, sea lions and walruses. A review of the Pinnipedia: 1–179. Stanford University Press. London.
Scheuring, L. 1929–1930. Die Wanderungen der Fische. *Ergeben. Biol.,* 5 (1929): 405–591, 6 (1930): 4–304.
Schevill, W. E., Watkins, W. A. & Ray, C. 1963. Underwater sounds of Pinnipeds. *Science, 141,* 3575: 50–53.

Schevill, W. E., Watkins, W. A. & Ray, C. 1966. Analysis of underwater odobenus calls with remarks on the development and function of the pharyngeal Pouches. *Zoologica* (USA), *51*, 3: 103–106.
Schlesinger, G. 1910a. Die Gymnoten. *Zool. Jb., Syst., Georg. und Biol., 29,* 6: 613–640.
Schlesinger, G. 1910b. Die Locomotion der Notopteriden. *Zool. Jb., Syst., Georg. und Biol., 29,* 6: 681–687.
Schlesinger, G. 1911a. Die Locomotion der täniformes fische. *Zool. Jb., Syst., Georg. und Biol., 31:* 469–490.
Schlesinger, G. 1911b. Über undulatorische Bewegung bei Fischen. *Verhandl. Zool.-bot. Gesellsch. Wien, 61:* 301–322.
Schlichting, H. 1956. Boundary layer theory: 1–528. Pergamon, Oxford.
Schmalhausen, I. I. 1913. Unpaired fins and their phylogenetic development. *Zap. Kievsk. Obshch. Yestestvoisp.* 23: 1–256.
Schmalhausen, I. I. 1916. On the functional significance of fins in fishes. *Russk. Zool. J., 1,* 6–7: 185–214.
Schmassmann, W. 1928. Ueber den Formwiderstand des Fischkorpers bei verschidenen Wassergeschwindigkeiten. *Verhandl. Schweiz. Naturf. Gesselsch., 109:* 196–197.
Schmidt, J. 1912. Contributions to the biology of some North Atlantic species of Eels. *Vid. Med. Naturh. Foren., 64:* 39–51.
Schmidt, J. 1918. Argentinidae, Microstomidae, Opisthoproctidae, Mediterranean Odontostomidae. *Rep. Dan. oceanogr. exp. 1908, 1910, 2,* Biol., A. 5: 10.
Schmidt, J. 1925. The breeding places of the eel. *Smithson. Rep.,* 1924: 279–316.
Schmidt, P. Yu. 1950. Fishes of the Sea of Okhotsk. *Tr. Tikhookeansk. Komiteta,* 6: 1–370.
Scholander, P. F. 1940. Experimental investigation on the respiratory function in diving mammals and birds. *Hvalradets skr.,* 22: 1–131.
Scholander, P. F. 1959. Wave-Riding Dolphins: How do they do it? *Science, 129,* 3356: 1085–1087.
Scholander, P. F. & Schevill, W. E. 1955. Counter-current vascular heat exchange in the fins of whales. *J. Appl. Physiol., 8,* 3: 279–282.
Schubert-Soldern, R. 1966. Hydrostatische und hydrodynamische Bewegungsformen bei Schildkroten. *Zool. Anz., 29:* 540–541.
Schulze, F. E. 1894. Über die Abwartsbiegung des Scwanztheiles der Wirbelsaule bei Ichthyosauren. *Sitzungsber. Acad. Wiss. Berlin, 43–44:* 513–514.
Seitz, A. 1891. Das Fliegen der Fische. *Zool. Jb., Syst., Georg. und Biol., 5,* 2: 361–372.
Semenov, B. N. 1969. On the existence of the hydrodynamic phenomenon of the bottlenose dolphin. *Bionika,* 3: 54–61.
Sewertzov, A. N. 1932. Die Entwicklung der Knochenschuppen von Polypterus. *Zschr. Naturw.,* Leipzig, 67: 387–418.
Shakalo, V. M. & Buryanova, L. D. 1973. Instrumental studies of velocity pulsations in the boundary layer of a dolphin relative to character of swimming. IV Vsesojuzn. Conf. po Bionike, *6.* Biomekhanika: 112–115. M.
Shebalov, A. M. 1969. Some questions of the influence of non-stationarity on the "Mechanism" of resistance formation. *Bionika,* 3: 61–66.
Sheer, D. 1963. Neues vom Schuertfisch. *Aquar. und Terrar., 8,* 12: 370–372.
Sheer, D. 1964. Die Flossel der Makrelen. *Wiss. Zschr. Humboldt-Univ. Berlin, math.-nat., 13,* 1: 77–84.
Shimansky, V. N. 1962. The class Cephalopoda. In: Principles of Paleont.; Cephalopods, 1: 15–18. "Gosgeoltechizdat" Publishers, M.
Shimkevich, V. 1912. Course in comparative anatomy of vertebrate animals: 1–634. Publisher Wolf. St. Petersburg-Moscow.
Shishkin, M. A. 1964. The suborder Stereospondyli. In: Principles of Paleont.; Amphibians, Rept., and Birds. 83–122. "Nauka" Publishers, M.

Shuleikin, V. V. 1928. The aerodynamics of the flying fish. Izv. AN SSSR, Ser. 7, Otd. Mat. i Estest. Nauk, 6–7: 573–582.
Shuleikin, V. V. 1929. Aerodynamics of the flying fish. *Int. Res. Ges. Hydrob. Hydrogr. Leipzig, 22:* 102–110.
Shuleikin, V. V. 1933. Materials on the optics of highly diffuse medium with reference to sea water, fogs and clouds. *J. Geofiziki*, 3, 3–5.
Shuleikin, V. V. 1934. The internal and external dynamics of the fish. *Izv. AN SSSR, Ser. 7, Otd. Mat i Estest. Nauk*, 8: 1151–1186.
Shuleikin, V. V. 1935. Dolphin kinematics. *Izv. AN SSSR, OMEN*, 3: 651.
Shuleikin, V. V. 1949. Essays on marine physics: 1–334. USSR Ac. of Sci. Press, M-L.
Shuleikin, V. V. 1953. Marine Physics: 1–989. 3rd ed. USSR Ac. of Sci. Press, M.
Shuleikin, V. V. 1958a. How the pilot fish moves with the speed of the shark. *Papers of the USSR Ac. of Sci., 119*, 5: 929–932.
Shuleikin, V. V. 1958b. Some investigations carried out on the oceanographic vessel "Sedov." *Priroda*, 10: 59–67.
Shuleikin, V. V. 1965. The energetics of marine animals. *Papers of the USSR Ac. of Sci., 163*, 3: 754–757.
Shuleikin, V. V. 1966. The energetics and migration speeds of fishes, dolphins, whales. *Tr. Vsesoyuzn. Nauchn. Issl. Inst. Morsk. Rybn. Khoz. i Okeanogr., 60:* 27–39.
Shuleikin, V. V. 1968. Marine physics: 1–1083. 4th Ed. "Nauka" Publishers, M.
Shuleikin, V. V., Lukjanova, V. S. & Stas, I. I. 1937. The hydrodynamic properties of fishes and the dolphin. *Izv. AN. SSSR, Ser. 7, Otd. Mat. i Estest. Nauk.*, 3: 581–592.
Shuleikin, V. V., Lukjanova, V. S. & Stas, I. I. 1939. Comparative dynamics of marine animals. *Papers of the USSR Ac. of Sci., 22*, 7: 424–429.
Siekmann, J. 1962a. Untersuchungen uber die Bewegung Schwimmender Tiere. *Verein Deutsche Ingen. Zschr., 104:* 433–468.
Siekmann, J. 1962b. Theoretical studies of sea animal locomotion. 1. *Engen. Arch., 31*, 3: 214–228.
Siekmann, J. 1963a. Theoretical studies of sea animal locomotion. 2. *Engen, Arch., 32*, 1: 40–50.
Siekmann, J. 1963b. On a pulsating jet from the end of a tube with application to the propulsion of certain aquatic animals. *J. Fluid Mech., 15*, 3: 399–418.
Siekmann, J. 1965. Zur Theorie der Bewegung Schwimmender Tiere. *Forsch. Ingen., 31*, 6: 192–197.
Siekmann, J. 1966. Zur Theorie der Bewegung schwimmender Tiere. *Forsch. Ingen., 32*, 1: 8–10.
Simpson, G. G. 1946. Penguins. *Bull. Amer. Mus. Nat. Hist., 87*, 1: 1–99.
Sleptsov, M. M. 1940. On the adaptation of pinnipeds to swimming. *Zool. J., 19*, 3: 379–386.
Sleptsov, M. M., 1952. The pinnipeds of Far Eastern Seas. *Izv. Tikhook. Nauchno-Issl. Inst. Rybn. Khoz. i Okeanogr. 38:* 1–166.
Slijper, E. J. 1936. Die Cetaceen, vergleichend-anatomisch und systematisch. *Capita zoologica*, La Haye, *7:* 1–590.
Slijper, E. J. 1958. Das Verhalten der Wale (Cetacea). Handbuch der Zoologie, *8*, 15, 10(14): 1–32. De Gruyter, Berlin.
Slijper, E. J. 1961. Locomotion and locomotory organs in whales and dolphins (Cetacea). In: Vertebrate locomotion, *Symp. Zool. Soc. London, 5:* 77–94. London.
Slijper, E. J. 1962. The whales: 1–474. Basic books, N.Y.
Smirnov, N. A. 1914. On the distribution of Pinnipedia in the northern hemisphere. *Zap. Novoross. Obshch. Estestvoisp., 39:* 305–323.
Smirnov, N. A. 1929. Key to pinnipeds of Europe and Northern Asia. *Izv. Otd. Prikladn. Ikht. i Nauchno. Prom. Issled., 9*, 3: 231–268.
Smith, E. H. & Stone, D. E. 1961. Perfect fluid forces in fish propulsion. *Proc. Roy. Soc. London, 261*, 1306: 316–328.

Smith, J. L. B. 1956. Pugnacity of marlins and swordfish. *Nature* (London), *178*, 4541: 1065.
Smith, J. L. B. 1961. The sea fishes of Southern Africa: 1–580. Central News Agency, LTD, Cape Town.
Smith, J. L. B. & Smith, M. M. 1963. The fishes of Seychelles: 1–215. Publ. Der. Ichthyol. Rhodes Univ., Grahamstown.
Sokolov, A. S. 1955. Some functional-morphological and age peculiarities of the Ladoga seal, connected with the aquatic mode of life: 1–20. Author's abstract of Dissert. for Cand. of Biol. Sci., Leningrad State Pedagogical Inst., L.
Sokolov, A. S. 1966. Specific features of the musculature of the propulsive organs in pinnipeds. Paper summaries. Third All-Union Conference on Marine Mammals: 42–43. "Nauka" Publishers, M.
Sokolov, A. S. 1969a. The weight of the muscles of the propulsive organs in pinnipeds. Marine mammals: 47–55. "Nauka" Publishers, M.
Sokolov, A. S. 1969b. Some results and prospects of ecomorphological studies into pacific pinnipeds. Marine mammals: 237–244. "Nauka" Publishers, M.
Sokolov, A. S. 1970. Some peculiarities of the propulsive organs of the river otter and the sea otter due to mode of existence. *Bull. Mosc. Obshch. Ispyt. Prir., Otd. Biol.*, *75*, 5: 5–17.
Sokolov, A. S., Kosygin, G. N. & Shustov, A. P. 1968. The structure of the lungs and trachea in Bering Sea pinnipeds. Pinnipeds of the Northern Pacific: 252–263. "Pishchevaya Promyshlennost" Publishers, M.
Sokolov, A. S., Kosygin, G. N. & Tikhomirov, E. A. 1966. Some data on the weight of the internal organs of Bering Sea pinnipeds. *Izv. Tikhook. Inst. Rybn. Khoz. in Oceanogr.*, *58*, 129–235.
Sokolov, V. E. 1953. The structure of the skin cover of mammals. 1–20. Author's abstract of Dissert. for Cand. of Biol. Sci., Moscow State University Press, M.
Sokolov, V. E. 1955. The structure of the skin cover of some cetaceans. *Bull. Mosc. Obshch. Ispyt. Prir., Otdel. Biol.*, *60*, 6: 45–60.
Sokolov, V. E. 1960a. Structural peculiarities of the skin covers of baleen whales, increasing skin elasticity. *Zool. J.*, *39*, 2: 307–308.
Sokolov, V. 1960b. Some similarities and dissimilarities in the structure of the skin among the members of the Suborders Odontoceti and Mystacocoeti (Cetacea). *Nature*, Lond., *185*, 4715: 745–747.
Sokolov, V. E., 1962a. The structure of the skin cover of some cetaceans. Communication 2. *Nauchn. Dokl. Vyssh. Shk., Biol. Nauki*, 3: 45–55.
Sokolov, V. 1962b. Adaptations of the mammalian skin to the aquatic mode of life. *Nature*, Lond., *195:* 464–466.
Sokolov, V. E. 1965a. Adaptive peculiarities of the skin of aquatic mammals. Marine mammals: 266–272. "Nauka" Publishers, M.
Sokolov, V. E. 1965b. The structure of the epidermis in three dolphin species (*Delphinus delphis, Tursiops truncatus* and *Lagenorhynchus obliquidens*). Marine mammals: 273–274. "Nauka" Publishers, M.
Sokolov, V. E. 1971. The skin structure of some cetaceans. Communication 3. Morphology and ecology of marine mammals (dolphins): 47–63. "Nauka" Publishers, M.
Sokolov, V., Bulina, I. & Rodionov, V. 1969. Interaction of Dolphin Epidermis with Flow Boundary Layer. *Nature* (London) *222*. 5190: 267–268.
Sokolov, V. E., Kalashnikova, M. M. & Rodionov, V. A. 1971. The micro- and ultrastructure of the skin of the harbour porpoise (*Phacaena phacaena relicta* Abel). Morphology and ecology of marine mammals (dolphins): 26–46. "Nauka" Publishers, M.
Sokolov, V. E. & Kuznetsov, G. V. 1966. Direction of dermal tori in dolphin skin in connection with water flow over body. Paper summaries. Third All-Union Conference on Marine Mammals: 45–46. "Nauka" Publishers, M.

Sokolov, V. E., Kuznetsov, G. V. & Rodionov, V. A. 1968. Direction of dermal tori in dolphin skin in connection with water flow over body. *Bull. Mosk. Obshch. Ispyt. Prir., Otd. Biol.*, 3: 123–126.

Spanovskaya, V. D. 1971. The suborder of the pikes and allies (Esocoidei). In: Life of Animals, 4, Fishes: 201–206. "Prosveshchenie" Publishers, M.

Stas, I. I. 1939a. Recording dolphin movements at sea. *Papers of the USSR Ac. of Sci., 24,* 6: 534–537.

Stas, I. I. 1939b. More on recording dolphin movements at sea. *Papers of the USSR Ac. of Sci., 25,* 8: 669–672.

Stas, I. I. 1941. Some physical characteristics of Black Sea fishes. *Papers of the USSR Ac. of Sci.*, 30, *8:* 710–712.

Stensiö, E. A. 1932. The Cephalaspids of Great Britain: 1–220. Brit. Mus. (Nat. Hist.), London.

Stephens, W. M. 1964. Flying fishes. *Sea Frontiers, 10:* 66–72.

Steven, G. A. 1950. Swimming of dolphins. *Science. Progr., 38:* 524–525.

Stewart, K. W. 1962. Observations on the Morphology and Optical Properties of the Adipose eyelid of fishes. *J. Fish. Res. Bd Canada, 19,* 6: 1161–1162.

Strasser, H. 1882. Zur Lehre von der Ortsbewegung der Fisch durch Biegungen der Leibes und der unpaaren Flossen: 1–124. Enke, Stuttgart.

Suares Caabro, J. A. 1955. Quetognatos de los mares Cubanos. *Mem. Soc. Cubana Hist., Nat., 22,* 2: 125–180.

Suares Caabro, J. A. & Madruga, J. E. 1960. The Chaetognatha of the north-eastern coast of Honduras, Central America. *Bull. Mar. Sci. Gulf Caribb., 10,* 4: 421–429.

Sukhanov, V. B. 1964. The subclass Testudinata. In: Principles of Paleontology; Amphibians, Reptiles and Birds: 354–438. "Nauka" Publishers, M.

Surkina, R. M. 1968. The structure of the connective-tissue support of the dolphin skin. Propulsive and orientation mechanisms in animals: 78–86. "Naukova Dumka" Publishers, Kiev.

Surkina, R. M. 1970. Some data on the functional morphology of the skin musculature of dolphins. Biocybernetics - Biosystems Modelling - Bionics. IV Ukr. Respubl. Nauchn. Konfer.: 131–132. Cybernetics Inst., Ukrainian SSR Ac. of Sci. Press, Kiev.

Surkina, R. M. 1971a. On the structure and functions of the skin musculature in dolphins. *Bionika,* 5: 81–87.

Surkina, R. M. 1971b. The location of dermal teri on the body of the white-sided dolphin. *Bionika,* 5: 88–94.

Surkina, R. M., Uskova, E. T. & Momot, L. M. 1972. On some features of the epidermis and the composition of the surface secretions of dolphin skin. *Bionika,* 6: 52–57.

Suvorov, E. K. 1948. Principles of Ichthyology: 1–580. "Sovetskaya Nauka" Publishers, L.

Svetovidov, A. N. 1948. Fishes of the order Gadiformes. *Fauna of the USSR, Fishes, 7,* 4: 1–221, USSR Ac. of Sci. Press, M-L.

Svetovidov, A. N. 1949. On the specific features of some bipolar habitats of marine fishes and their causes. *Izv. Vsesojuzn. Geogr. Obshch., 81,* 1: 44–52.

Svetovidov, A. N. 1952. Fishes of the Herring Family (Clupeidae). *Fauna of the USSR, Fishes,* 2, 1: 1–331. USSR Ac. of Sci. Press, M-L.

Talbot, F. H. & Penrith, J. J. 1962. Spearing behaviour in feeding in the black marlin *Istiompax marlina. Copeia,* 2: 468.

Tåning, A. V. 1918. Mediterranean Scopelidae. *Rep. Dan. oceanogr. exp. 1908–1910. 2, Biol.,* A, 7: 1–154.

Tarlo, L. B. 1962. The classification and evolution of the Heterostraci. *Acta palaentol. Polonica, 7:* 249–290.

Tatarinov, L. P. 1964a. The class Amphibia. In: Principles of Paleontology; Amphibians, Reptiles and Birds: 25–59. "Nauka" Publishers, M.

Tatarinov, L. P. 1964b. The superfamily Colosteoidea. In: Principles of Paleonthology; Amphibians, Reptiles and Birds: 67–69. "Nauka" Publishers, M.
Tatarinov, L. P. 1964c. The order Plesiopoda. In: Principles of Paleontology; Amphibians, Reptiles and Birds: 123–124. "Nauka," M.
Tatarinov, L. P. 1964d. The subclass Leptospondyli. In: Principles of Paleontology; Amphibians, Reptiles and Birds: 144–164. "Nauka" Publishers, M.
Tatarinov, L. P. 1964e. The order Microsauria. In: Principles of Paleontology; Amphibians, Reptiles and Birds. 164–171. "Nauka" Publishers, M.
Tatarinov, L. P. 1964f. The subclass Synaptosauria. In: Principles of Paleontology; Amphibians, Reptiles and Birds: 299–303, "Nauka" Publishers, M.
Tetarinov, L. P. 1964g. The order Araeoscelidia. In: Principles of Paleontology; Amphibians, Reptiles and Birds: 303–309. "Nauka" Publishers, M.
Tatarinov, L. P. 1964h. The suborder Nothosauria. In: Principles of Paleontology; Amphibians, Reptiles and Birds: 310–316. "Nauka" Publishers, M.
Tatarinov, L. P. 1964i. The suborder Plesiosauria. In: Principles of Paleontology; Amphibians, Reptiles and Birds. 316–317. "Nauka" Publishers, M.
Tatarinov, L. P. 1964j. The Order Placodontia. In: Principles of Paleontology; Amphibians, Reptiles and Birds: 322–338. "Nauka" Publishers, M.
Tatarinov, L. P. 1964k. The suborder Ichthyopterygia. In: Principles of Paleontology; Amphibians, Reptiles and Birds: 338–354, "Nauka" Publishers, M.
Tatarinov, L. P. 1964l. The suborder Lepidosauria. In: Principles of Paleontology; Amphibians, Reptiles and Birds: 439–444. "Nauka" Publishers, M.
Tatarinov, L. P. 1964m. The order Eosuchia. In: Principles of Paleontology; Amph., Rept., and Birds: 446. "Nauka" Publishers, M.
Tatarinov, L. P. 1964n. The suborder Rhynchocephalia. Principles of Paleontology; Amphibians, Reptiles and Birds: 447–451. "Nauka" Publishers, M.
Tatarinov, L. P. 1964o. The order Lacertilia. Lizards. In: Principles of Paleontology; Amphibians, Reptiles and Birds: 455–456. "Nauka" Publishers, M.
Tatarinov, L. P. 1964p. The suborder Prolacertilia. In: Principles of Paleontology; Amphibians, Reptiles and Birds: 455. "Nauka" Publishers, M.
Tatarinov, L. P. 1964q. The suborder Talattosauria. In: Principles of Paleontology, Amphibians, Reptiles and Birds: 459. "Nauka" Publishers, M.
Tatarinov, L. P. 1964r. Family: Askeptosauridae Khn-Schnyder, 1952. In: Principles of Paleontology; Amphibians, Reptiles and Birds: 459–460. "Nauka" Publishers, M.
Tatarinov, L. P. 1964s. The suborder Anguinomorpha (Diploglossa). In: Principles of Paleontology; Amphibians, Reptiles and Birds: 469. "Nauka" Publishers, M.
Tatarinov, L. P. 1964t. The order Ophidia. Snakes. In: Principles of Paleontology; Amphibians, Reptiles and Birds: 484–493. "Nauka" Publishers, M.
Tavolga, W. N. 1960. Sound production and underwater communication in fishes. In: Lanyon, W. E. and Tavolga, W. N. (Eds.), Animal Sounds and Communication; Publ. Amer. Inst. Biol. Sci., 7: 93–136.
Tavolga, W. N. (ed.) 1964. Marine bio-acoustics: 1–432. Pergamon Press; Oxford, London, N.Y., Paris.
Tavolga, W. N. (ed). 1967. Marine bio-acoustics, II: 1–Pergamon Press, Oxford.
Tavolga, W. N. (ed.) 1969. Marine bio-acoustics: 1–424. Transl. from Engl. "Sudostroenie" Publishing House, L.
Tawara, T. 1950. On the respiratory pigments of whale. (Studies on whale blood. II). *Sci. Rep. Whales Res. Inst.*, 3: 96–101.
Taylor, H. F. 1921. Airbladder and specific gravity. *Bull. U.S. Bureau Fish.*, *38:* 121–126.
Terentyev, P. V. 1961. Herpetology: 1–336. "Vysshaya Shkola" Publishers, Moscow.
Tester, A. 1940. A Specific Gravity Method for determining Fatness (Condition) in Herring. (Clupea pallasi). *J. Fish. Res. Bd Canada*, *4*, 5: 461–471.

Tewes, H. 1936. Zur Gliedmassenfunktion der Scleroparei mit Ausblicken auf das Flugproblem von Dactylopterus. *Zschr. Naturwiss.*, *70*, 3: 429–483.
Thayer, A. H. 1909. An Arrangement of the Theories of Mimicry and Warning Colours. *Pop. Sci. Mon. N.Y.*, *75:* 550–570.
Thiel, M. E. 1938. Die Chaetognathen—Bevolkerung des Sudatlantischen Ozeans. *Wiss. Ergeb. Dtsch. Atlant. Exp. "Meteor" 1925, 1927, 13,* 1: 1–40.
Thomson, J. M. 1947. The Chaetognatha of southeastern Australia. *Council Sci. Ind. Res. Australia, Bull,* 222: 1–43.
Tikhomirov, E. A. 1964. Incidence of fur seals in the sea of Okhotsk. *Izv. Tikhook. Inst. Rybn. Khoz. i Okeanogr.*, 54, 65–66.
Tikhomirov, E. A. 1966. Determining Far Eastern Seal Species from an Aircraft. *Izv. Tikhook. Inst. Rybn. Khoz. i Okeanogr.*, *58,* 163–172.
Timofeyeva, V. A. 1951. Distribution of brightness in highly diffusing media. *Papers of the USSR Ac. of Sci.*, *76,* 5: 677–680.
Tokarev, A. K. 1949. Biological groups of Caspian "common kilka" (sprat) and a method of lasting observation: 1–19. Author's abstract of Dissert. for Cand. of Biol. Sci., Dalnevost. n-p. baza, USSR Ac. of Sci., Vladivostok.
Tokioka, T. 1952. Chaetognaths of the Indo-Pacific. *Annot. Zool. Japan.*, *25,* 1–2: 307–316.
Tokioka, T. 1956. On chaetognaths and appendicularians collected in the central part of the Indian Ocean. *Publ. Seto Marine Biol. Lab.*, *5,* 2: 65–70.
Tokioka, T. 1959. Observations on the taxonomy and distribution of chaetognaths of the North Pacific. *Publ. Seto Mar. Biol. Lab.*, *7,* 3: 349–456.
Tokioka, T. 1965a. The taxonomical outline of Chaetognatha. *Publ. Seto Mar. Biol. Lab.*, *12,* 5: 335–357.
Tokioka, T. 1965b. Supplementary notes on the Systematics of Chaetognatha. *Publ. Seto Mar. Biol. Lab.*, *13,* 3: 231–242.
Tolgay, Z. 1957. Investigations into the Chemical composition of the Pelamid. *Debats et docum. techn. Conseil gen. peches Medit.*, *4:* 16–18.
Tomilin, A. G. 1940. Some questions pertaining to the ecology of cetaceans (adaptations to ambient temperature). *Bull. Mosk. Obshch. Ispyt. Prir., Otd. Biol.*, *49,* 5–6: 112–124.
Tomilin, A. G. 1950. Flippers – thermoregulatory organs of cetaceans. *Rybn. Khoz-vo.* 12: 50.
Tomilin, A. G. 1951. On thermoregulation in cetaceans. *Priroda,* 6: 55–58.
Tomilin, A. G. 1957. Cetaceans. Beasts of the USSR and adjacent countries, 9: 1–756. USSR Ac. of Sci. Press, M.
Tomilin, A. G. 1962a. Cetaceans in the marine fauna of the USSR: 1–211, USSR Ac. of Sci. Press, M.
Tomilin, A. G. 1962b. Bionics and the cetaceans. *Priroda,* 10, 101–103.
Tomilin, A. G. 1962c. On the adaptations of cetaceans to rapid swimming and on possibilities of using these adaptations in shipbuilding. *Bull. Mosk. Obshch. Ispyt. Prir., Otd. Biol.*, *67,* 5.
Tomilin, A. G. 1964. The ultrasonic projector and extra-long-distance signals of whales. *Tr. Vses. S. Kh. Inst. Zaochn. Obr.*, 17, 2: 102–108.
Tomilin, A. G. 1965. The story of a blind cachalot: 1–19. "Nauka" Publishers, M.
Tomilin, A. G. 1968. Why dolphins have an enormous brain? *Okhota i okhotn. khoz.* 9: 38–39.
Tomilin, A. G. 1970. Marine mammals and their biogeocenotic connections with other marine faunal groups. Programme and methods for studying aquatic biogeocenosis. Biocenosis of the seas and oceans: 169–193. "Nauka" Publishers, M.
Tomilin, A. G. 1971. The order Sirenians (Sirenia). In: Life of animals, 6, Mammals and beasts: 406–410. "Prosveshchenie" Publishers, M.
Tompsett, D. H. 1939. Sepia. *Liverpool Mar. Biol. Comm. Memoirs, 32:* 1–184.
Tortonese, E. 1950. Alcune osservazioni sul volo dei Cipseluri (*Pesci volanti*). *Natura* (Milan), *41,* (1/2): 34–36.

Tortonese, E. 1956. Elenco dei pesci teleostei viventi nel Mediterraneo. *Fauna e flora Golfo Napoli, monogr. 38:*

Traquair, R. H. 1899a. *On Thelodus pagei* (Powrie) from the Old Red Sandstone of Forfarshire. *Trans. Roy. Soc. Edinburgh, 39,* 3, 21: 591–602.

Traquair, R. H. 1899b. Report on fossil fishes collected by the Geological Survey of Scotland in the Silurian rocks of the South of Scotland. *Trans. Roy. Soc. Edinburgh, 39,* 3, 32: 827–864.

Tweedie, M. 1969. Adaptations. Part 4. The swimming animal. *Animals, 11:* 475–476.

Ueyanagi, S. & Watanabe, H. 1965. On certain specific differences in the vertebrae of the Istiophorids. *Repr. Rep. Nankai Region. Fish. Res. Lab.,* 22: 1–8.

Utrecht, W. L., van 1958. Temperaturregulierende Gefässysteme in der Haut und anderen epidermalen Strukturen bei Cetacean. *Zool. Anz., 161,* 3–4: 77–82.

Van Beneden, P. J. & Gervais, P. 1868–1880. Osteographie des Cetaces vivants et fossiles: 1–634. Paris.

Vannucci, M. & Hosoe, K. 1952. Resultados cientificos do cruzeiro do "Baependi" e do "Vega" a I. da Trinidade. Chaetognatha. *Bolet. Inst. Oceanogr. Sao Paulo, 3,* 1–2: 5–30.

Van Oosten, J. 1957. The skin and scales. In: Brown, M. E., The physiology of fishes, *1,* Chap. 5: 207–244. Acad. press, N.Y.

Varich, Yu. N. 1971. An analysis of the flow over living spiny dogfish (*Squalus acanthias*) by means of an information system. *Zool. J., 50,* 1: 126–129.

Vasnetsov, V. V. 1941. Fin function in bony fishes. *Papers of the USSR Ac. of Sci., 31,* 5: 503–506.

Vasnetsov, V. V. 1948. Specific features of fin movement and action in the bream, roach and carp in connection with feeding. The morphological peculiarities determining the feeding of the bream, roach and carp at every developmental stage: 7–53. The USSR Ac. of Sci. Press, M-L.

Verrill, A. E. 1879–1882. The Cephalopoda of the north-eastern coast of America. *Trans. Connecticut Acad. Art and Sci., 5:* 177–446.

Voitkunsky, Ya. I. 1964. Resistance of water to movement of ships: 1–412. "Sudostroyenie" Publishers, L.

Vorobyova, E. I. 1963. The genus Porolepis from the Devonian of the USSR. *Paleont. J.,* 2: 83–92.

Vorobyova, E. I. & Obruchev, D. V. 1964. Subclass Sarcopterygii. In: Principles of paleontology, Agnatha, Fishes: 268–322. "Nauka" Publishers, M.

Voss, G. 1956. Solving life secrets of the sailfish. *Nat. Georg. Mag., 109,* 6: 859–872.

Vyushkov, B. P. 1964. The subclass Proganosauria. Proganosaurs. In: Principles of paleont. Amphibians, Reptiles and Birds: 298–299. "Nauka" Publishers, M.

Wahlert, G., von 1964. Passiv Atmung bei Haien. *Naturwissenschaften, 51,* 12: 297–298.

Wahlert, G., von & Wahlert, H., von 1964. Eine bemerkenswerte Anordnung von Schuppen fortsätzen beim Knochenfisch *Balistes bursa* und ihre hydrodynamische Erklarung. *Naturwissenschaften, 51,* 2: 45.

Walford, L. A. 1937. Marine game fishes of the Pacific Coast from Alaska to the Equator: 1–208. Univ. California Press, Berkeley.

Walls, G. L. 1942. The vertebrate eye and its adaptive radiation. *Bull. Cranbrook Inst. Sci., 19,* 1–785.

Walters, V. 1962. Body form and swimming performance of the scombroid fishes. *Amer. Zoologist,* 2: 143–149.

Walters, V. 1963. The Trachypterid integument and an hypothesis on its hydrodynamic function. *Copeia,* 2: 260–270.

Walters, V. 1966. The "problematic" hydrodynamic performance of Gero's great Barracuda. *Nature* (London), *212,* 5058: 216–217.

Walters, V. & Fierstine, H. L. 1964. Measurements of the swimming speeds of yellowfin tuna and wahoo. *Nature* (London), *202,* 4928: 208–209.

Watson, D. M. S. 1937. The Acanthodian fishes. *Philos. Trans. Roy. Soc. London (B), 228,* 549: 49–146.

Weaver, C. R. 1962. Influence of water velocity upon orientation and performance of adult migrating salmonis. *Fishery Bull., 63,* 1: 97–121.

Weber, M. & Beaufort, L. F. 1911. The fishes of the Indo-Australian archipelago. I. Index of the Ichthyological papers of P. Bleeker: I–XI + 1–410. Leyden.

Weber, M. & Beaufort, L. F. 1913. The fishes of the Indo-Australian archipelago. II. Malacopterygii, Myctophoidae, Ostariophysi: Siluroidea: I–XX + 1–404. Leyden.

Weber, M. & Beaufort, L. F. 1916. The fishes of the Indo-Australian archipelago. III. Ostarioohysi: Cyprinoidea, Apodes, Synbranchi: I–XV + 1–455. Leyden.

Weber, M. & Beaufort, L. F. 1922. The fishes of the Indo-Australian archipelago. IV. Heteromi, Solenichthyes, Synentognathi, Percesoces, Labyrinthici, Microcyprini: I–XIII – 1–410. Leyden.

Weber, M. & Beaufort, L. F. 1929. The fishes of the Indo-Australian archipelago. V. Acanthinim, Allotriognathi, Hetereosomata, Berycomorphi, Percomorphi: I–XIV + 1–458. Leyden.

Weber, M. & Beaufort, L. F. 1931. The fishes of the Indo-Australian archipelago. VI. Perciformes (continued): I–XII + 1–448. Leyden.

Weber, M. & Beaufort, L. F. 1936. The fishes of the Indo-Australian archipelago. VII. Perciformes (continued): I–XVI + 1–607. Leyden.

Weber, M. & Beaufort, L. F. (Beaufort) 1940. The fishes of the Indo-Australian archipelago. VIII. Percomorphi (continued); I–XV + 1–508. Leyden.

Weber, M. & Beaufort, L. F. (Beaufort and Chapman). 1951. The fishes of the Indo-Australian archipelago. IX. Percomorphi (continued): Blennioidea: I–XL + 1–484. Leyden.

Weitzman, S. H. 1954. Osteology and the Relationships of the South American Characid fishes of the subfamily Gasteropelecinae. *Stanford Ichthyol. Bull., 4,* 4: 213–263.

Weitzman, S. H. 1960. Relationships and Classification of Characid Fishes of the Subfamily Gasteropelecinae. *Stanford Ichthyol. Bull., 10,* 217–239.

Westoll, T. S. 1944. The Haplolepidae, a new family of late Carboniferous bony fishes. A study in taxonomy and evolution. *Bull. Amer. Mus. Nat. Hist., 83,* 1: 1–122.

White, E. I. 1935. The Ostracoderm *Pteraspis Kner* and the relationships of the Agnathous Vertebrates. *Phil. Trans. Roy. Soc. London,* ser. B, *225:* 381–457.

Whitear, M. 1970. The skin surface of bony fishes. *J. Zool.,* 4: 437–454.

Whitehouse, R. H. 1910a. The caudal fin of the Teleostomi. *Proc. Zool. Soc. London,* 2: 590–626.

Whitehouse, R. H. 1910b. Some remarks on the Teleostean Caudal Fin. *Ann. Mag. Nat. Hist.,* ser. 8, *5,* 29: 426–428.

Whitehouse, R. H. 1918. The evolution of the caudal fins of fishes. *Res. Indian Mus., 15:* 135–142.

Whitley, G. P. 1940. The fishes of Australia, I. The sharks, rays, devil fish and other primitive fishes of Australia and New Zealand. *Roy. Zool. Soc. N. S. Wales-Austr.,* Sydney: 1–280.

Whitley, G. P. 1956. Namelist of New Zealand Fishes. In: D. H. Graham, A treasury of New Zealand fishes: 397–414. Wellington, N. Zeal.

Wickler, W. 1960. Die Stammesgeschichte typischer Bewegungsformen der Fisch-Brustflosse. *Zschr. Tierpsychol., 17,* 1: 31–66.

Wiedersheim, R. 1906. Vergleichende Anatomie der Wirbeltiere: 1–799. Fischer, Jena.

Wiman, C. 1922. Some Reptiles from the Niobara Group in Kansas. *Bull. Geol. Inst. Uppsala, 18:* 9–18.

Wisner, R. L. 1958. Is the spear of istiophorid fishes used in feeding? *Pacific sci., 12,* 1: 60–70.

W. M. 1953. Fische als Schwimmer. *Schweiz. Fischerei-Zeit.,* 6: 150.

Wood, F. G. 1954. Underwater Sound Production and Concurrent Behaviour of Captive Porpoises, *Tursiops truncatus* and *Stenella plagiodon. Bull. Mar. Sci. Gulf and Caribbean,* 5(2): 120, 133.

Woodcock, A. H. 1948. The Swimming of Dolphins, *Nature* (London), *161*, 4094: 602.
Woodcock, A. H. & McBride, A. F. 1951. Wave-riding Dolphins. *J. Exp. Biol.*, *28*, 2: 215–217.
Woodward, A. S. 1901. Catalogue of the fossil fishes in the British Museum (Natural History), *4:* 1–636.
Worthington, L. V. & Schevill, W. E. 1957. Underwater sound heard from sperm whales. *Nature* (London), *180:* 291.
Yablokov, A. V. 1961. Functional morphology of respiratory organs. in dentate cetaceans. Tr. Soveshch. po ekologii i promyslu mor. mlekop.: 79–86. USSR Ac. of Sci. press, M.
Yablokov, A. V. 1965. Some aspects of the problem of deep autonomous diving by man in the light of research into the biology of diving mammals. *Bionika:* 220–227. "Nauka" Publishers, M.
Yablokov, A. V., Belkovich, V. M. & Borisov, V. I. 1972. Whales and dolphins: 1–472. "Nauka" Publishers, M.
Yakovlev, V. N. 1966. The functional evolution of the fish skeleton. *Paleont. J.*, 3: 3–13.
Young, J. Z. 1938. The functioning of the gland nerve fibres of the squid. *J. Exp. Biol.*, 15: 170–185.
Young, J. Z. 1944. Giant nerve-fibres. *Endeavour*, 3: 108–113.
Yuen, H. S. 1961. Bow Wave Riding of Dolphins. *Science*, *134*, 3484: 1011–1012.
Zayets, V. A. 1972. To the question of variable roughness of shark skins. *Bionika*, 6: 67–73.
Zayets, V. A. 1973a. The distribution of placoid scale on the body of sharks. *Bionika*, 7: 83–87.
Zayets, V. A. 1973b. The peculiarities in the skin covers structures of fast swimming sharks. IV All-Union Bionics Conf., *6*. Biomekhanika: 45–50. M.
Zenkevich, L. A. 1944. Essays on the evolution of the propulsive mechanism of animals. *J. Obshch. Biol.*, *5*, 3.
Zenkevich, L. A. 1951. The fauna and biological productivity of the sea. 1. World Ocean: 1–506, "Sovetskaya Nauka" Publishers, L.
Zernov, S. A. 1949. General hydrobiology. 2nd Ed.: 1–587. USSR Ac. of Sci. Press, M-L.
Zuyev, G. V. 1963. On the specific weight of the squid *Ommastrephes sagittatus* Lamarck. *Tr. of Sevastop. Biol. St.*, *16:* 383–386.
Zuyev, G. V. 1964a. On the capacity of cephalopod larvae to active movement. *Zool. J.*, *43*, 10: 1440–1445.
Zuyev, G. V. 1964b. On adaptation of cephalopods to movement. *Zool. J.*, *43*, 9: 1304–1308.
Zuyev, G. V. 1964c. Swimming speeds of squids. *Priroda*, 6: 96–97.
Zuyev, G. V. 1964d. On body shape in cephalopods. Tr. of Sevastop. *Biol. St.*, *17:* 379–387.
Zuyev, G. V. 1965. On the soaring mechanism of pelagic squids. *Biofizika*, *10*, 1: 190–192.
Zuyev, G. V. 1966. The functional principles of external structure in cephalopods: 1–139. "Naukova Dumka" Publishers, Kiev.
Zuyev, G. V. 1969. The specific features in the structure of the mantle in cephalopods (Mantle of *Symplectoteuthis oualaniensis* (Lesson). Marine biology, 16: Functional-morphological research into nektonic animals: 102–110. "Naukova Dumka" Publishers, Kiev.
Zuyev, G. V. & Kudriashov, A. F., 1968. On the manoeuvrability of aquatic animals. *Vopr. Ikhtiol.*, *8*, 6(53): 1057–1062.

NAME INDEX

ANIMAL LATIN NAMES INDEX

SUBJECT INDEX

GPSR Compliance

The European Union's (EU) General Product Safety Regulation (GPSR) is a set of rules that requires consumer products to be safe and our obligations to ensure this.

If you have any concerns about our products, you can contact us on ProductSafety@springernature.com

In case Publisher is established outside the EU, the EU authorized representative is:

Springer Nature Customer Service Center GmbH
Europaplatz 3
69115 Heidelberg, Germany

Batch number: 10607212

Printed by Printforce, the Netherlands